A DIY Smart Home Guide

A DIY Smart Home Guide

Tools for Automating Your Home Monitoring and Security Using Arduino, ESP8266, and Android

Robert Chin

New York Chicago San Francisco Athens London Madrid
Mexico City Milan New Delhi Singapore Sydney Toronto

Library of Congress Cataloging-in-Publication Data

Names: Chin, Robert, author.
Title: A DIY Smart Home Guide: Tools for Automating Your Home Monitoring and Security Using Arduino, ESP8266, and Android / Robert Chin.
Description: New York : McGraw-Hill Education, [2020] | Includes index. | Summary: "Setting up a "smart home" can be costly, intimidating, and invasive. This hands-on guide presents readers with an accessible and cheap way to do it themselves using free software that will enable their homes and their mobile devices to communicate. Tools for Automating Your Home Monitoring and Security Using Arduino, ESP8266, and Android contains step-by-step plans for easy-to-build projects that work through your phone to control your home environment remotely"–Provided by publisher.
Identifiers: LCCN 2019042170 | ISBN 9781260456134 (paperback ; acid-free paper) | ISBN 9781260456141 (ebook)
Subjects: LCSH: Home automation–Amateurs' manuals.
Classification: LCC TK7881.25 .C56 2020 | DDC 006.2/5–dc23
LC record available at https://lccn.loc.gov/2019042170

McGraw-Hill Education books are available at special quantity discounts to use as premiums and sales promotions or for use in corporate training programs. To contact a representative, please visit the Contact Us page at www.mhprofessional.com.

A DIY Smart Home Guide: Tools for Automating Your Home Monitoring and Security Using Arduino, ESP8266, and Android

1 2 3 4 5 6 7 8 9 LOV 23 22 21 20 19

ISBN 978-1-260-45613-4
MHID 1-260-45613-7

This book is printed on acid-free paper.

Sponsoring Editor
Lara Zoble

Editorial Supervisor
Donna M. Martone

Production Supervisor
Lynn M. Messina

Acquisitions Coordinator
Elizabeth Houde

Project Manager
Ishan Chaudhary, MPS Limited

Copy Editor
Kirti Dogra, MPS Limited

Proofreader
Vandana Gupta, MPS Limited

Art Director, Cover
Jeff Weeks

Composition
MPS Limited

About the Author

Robert Chin has a Bachelor of Science degree in computer engineering and is experienced in developing projects on the ESP8266, TI CC3200 SimpleLink, Android, Arduino, Raspberry Pi, and PC Windows platforms using C/C++, Java, Python, Unreal Script, DirectX, OpenGL, and OpenGL ES 2.0. He is the author of *Arduino and Raspberry Pi Sensor Projects for the Evil Genius* published by McGraw-Hill. He is also the author of *Beginning Android 3D Game Development* and *Beginning iOS 3D Unreal Games Development*, both published by Apress and distributed by Springer Nature. *Beginning Android 3D Game Development* was licensed to Tsinghua University through Tsinghua University Press. He was the technical reviewer for *UDK Game Development,* published by Course Technology Cengage Learning. His home security related books include *Home Security System DIY PRO Using Android and TI CC3200 SimpleLink, Home Security Systems DIY Using Android and Arduino*, and *Beginning Arduino ov7670 Camera Development*.

About the Source Code and Project files

The source code and Android APK files for this book are located at:

https://www.mhprofessional.com/Chin_SmartHome

For your convenience, all the Android APK installation files have been placed in the Chapter 9 download section.

Contents

A DIY Smart Home Guide

CHAPTER 1

Introduction to the Arduino

IN THIS CHAPTER I INTRODUCE YOU to the Arduino. I first give a brief explanation of what the Arduino is. I then specifically concentrate on the Arduino Mega 2560. I discuss the general features of the Arduino Mega 2560 including the capabilities and key functional components of the device. Next, I give a quick summary of languages that the Arduino uses which are C and C++. Next, I discuss the Arduino Integrated Development Environment (IDE) software that is needed to develop programs for the Arduino. I cover each key function of the Arduino IDE and then conclude with a hands-on example where I give detailed step-by-step instructions on how to set up the Arduino for development and how to run an example program using the Arduino IDE.

What Is an Arduino?

The Arduino is an open-source microcontroller that uses the C and C++ languages to control digital and analog outputs to devices and electronics components and to read in digital and analog inputs from other devices and electronics components for processing. For example, the Arduino can read in information from a sensor to a home security system that would detect the heat that a human being emits and sends a signal to the Arduino to indicate that a human is in front of the sensor. After receiving this information the Arduino can send commands to a camera to start taking pictures of the intruder or intruders and save these images to an SD card for later viewing. It can also transmit a message over Wi-Fi to an Android cell phone so that a text alarm notification message is sent to the homeowner. The official web site of the Arduino is http://www.arduino.cc.

Why the Arduino Mega 2560?

There are many different Arduino models out there. However, in order to perform the examples in this book you will need an Arduino Mega 2560. The reason for this is that the Arduino Mega 2560 has many hardware serial ports that can communicate at high speed with the ESP-01 ESP8266 module that will be responsible for Wi-Fi communication. The Arduino UNO only has one hardware serial port that is already used for debugging with the Arduino IDE's serial monitor. A separate software serial port can be created on the Arduino UNO but it is unreliable at high speeds with many people reporting that any speed above 9600 baud is unreliable. Most of the recent ESP-01 modules being produced now are set to the default speed of 115,200 baud. Another consideration is the large amount of current that the ESP-01 module can draw. The ESP-01 module can draw up to 170 mA when operating under normal circumstances. If you add in the current drawn by other components attached to the Arduino, such as sensors, then the total might quickly add up to the maximum current allowed for the Arduino UNO which is around 400 mA. However, for the Arduino Mega 2560 the maximum allowed current is 800 mA which is considerably more. Also, please note that these maximum currents should not be sustained for long periods of time otherwise there could be damage to the Arduino board. This information comes from the official Arduino web site at https://playground.arduino.cc/Main/ArduinoPinCurrentLimitations.

Figure 1-1 Contents of the official Arduino Mega 2560 that I purchased.

The Arduino Mega 2560 Specifications

- Microcontroller: ATmega2560
- Operating voltage: 5 V
- Input voltage (recommended): 7 to 12 V
- Input voltage (limit): 6 to 20 V
- Digital I/O pins: 54 (of which 15 provide PWM output)
- Logic level: 5 V is true or high
- Analog input pins: 16
- DC current per I/O pin: 20 mA
- DC current for 3.3 V pin: 50 mA
- Total maximum current: 800 mA
- Total maximum current (USB powered): 500 mA (>500 mA will trip the Arduino's fuse)
- Flash memory: 256 kB of which 8 kB used by bootloader
- SRAM: 8 kB
- EEPROM: 4 kB
- Clock speed: 16 MHz
- LED_BUILTIN: 13
- Length: 101.52 mm
- Width: 53.3 mm
- Weight: 37 g

The Official Arduino Mega 2560

I purchased an official Arduino Mega 2560 from the official Arduino web site that was listed earlier for about $40.00 and the unit arrived in early October 2018. Included in the package was a new Arduino 2560 microcontroller, a thank you note, a sheet of stickers, and a plastic stand that was attached to the bottom of the Arduino. See Figure 1-1.

The backside of the Arduino 2560 is shown in Figure 1-2.

Since the Arduino Mega 2560 is an open-source hardware, anyone can make and sell their own Arduino Mega 2560 board legally as long as they don't use certain logos that are trademarked. Thus, there are many unofficial Arduino Mega 2560 boards available on sites such as Amazon that are a fraction of the cost of an official Arduino. For example, on Amazon I just searched for an Arduino Mega 2560 and the search results returned many unofficial boards which cost around $15.00 USD each. This is less than half the price of an official

Figure 1-2 Backside of the Arduino Mega 2560.

Arduino Mega 2560 bought from the official Arduino store.

Arduino Mega 2560 Components

This section covers the functional components of the Arduino Mega 2560.

USB Connection Port

The Arduino Mega 2560 has a USB connector that is used to connect the Arduino to the main computer development system via standard USB A male to B male cable so it can be programmed and debugged. See Figure 1-3.

9-V Battery Connector

The Arduino Mega 2560 has a 9-V battery connector where you can attach a 9-V battery to power the Arduino. See Figure 1-4.

Reset Button

There is a Reset button on the Arduino Mega 2560 where you can press the button down to reset the board. This restarts the program contained in the Arduino's memory. See Figure 1-5.

Digital Pulse Width Modulation (PWM)

The Arduino Mega has many digital pins capable of simulating analog output through the process of PWM. For example, a light-emitting diode (LED) light generally has only two modes which are on (full brightness) and off (no light emitted). However, with digital PWM the LED light can appear to have a brightness in between on and off. For instance, with PWM an LED can start from an off state and slowly brighten until it is at its highest brightness level and then slowly dim until back to the off state. The digital pins on the Arduino Mega 2560 that support PWM are pins 2 through 13. These PWM capable digital pins are circled in Figure 1-6.

Communication

The communication section of the Arduino Mega 2560 contains pins for serial communication between the Arduino and another device such as a Wi-Fi adapter like the ESP-01 module or your personal computer. The Tx0 and the Rx0 pins are connected to the USB port and serve as communication from your Arduino to your computer through your USB cable. The Serial Monitor that can be used for sending data to the Arduino and reading data from the Arduino uses the Tx0 and Rx0 pins. Thus, you should not connect anything to these pins if you want to use the Serial Monitor to debug your Arduino programs or to receive user input. I will talk more about the Serial Monitor later in this book. In addition, the Arduino Mega 2560 has three more sets of serial communication pins that are labeled Tx1/Rx1, Tx2/Rx2, and Tx3/Rx3. See Figure 1-7.

Figure 1-3 The USB port on the Arduino Mega 2560.

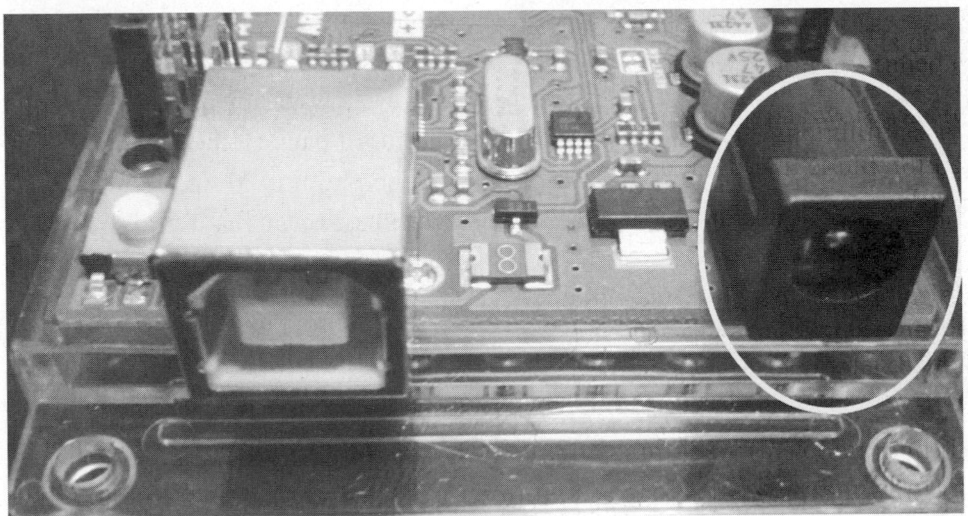

Figure 1-4 The 9-V battery connector for the Arduino Mega 2560.

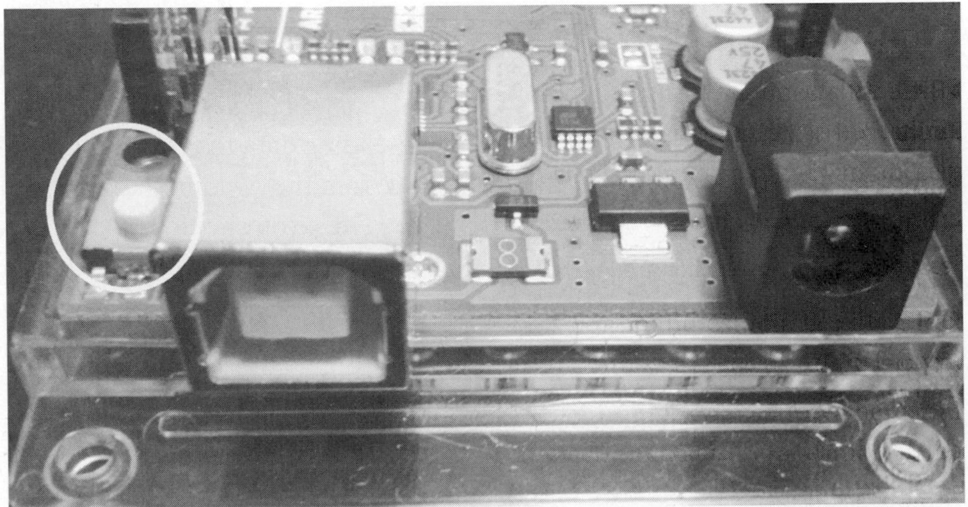

Figure 1-5 The Reset button on the Arduino Mega 2560.

Figure 1-6 Pins capable of pulse width modulation.

Figure 1-7 Hardware serial communication pins.

The I2C Interface

The I2C interface is a communications interface that consists of an SDA pin which is pin 20 and is used for data and an SCL pin which is pin 21 and is used for clocking or driving the device or devices attached to the I2C interface. The SDA and SCL pins are circled in Figure 1-8.

Digital Output/Input

The Arduino Mega 2560 has many more digital output/input pins than the Arduino Uno which is a popular Arduino model for beginners with a limited number of digital and analog output/input pins. Pins 22 through 53 are digital pins on the Arduino Mega 2560. Pins discussed earlier that are capable of PWM are also capable of normal digital output/input. See Figure 1-9.

Analog Input

The Arduino Mega 2560 has 16 analog input pins that can read in a range of values instead of just digital values of 0 or 1. The analog input pin uses a 10 bit analog to digital converter to transform voltage input in the range of 0 to 5 V into a number in the range between 0 and 1023. See Figure 1-10.

Figure 1-8 The I2C interface.

Figure 1-10 Analog input pins.

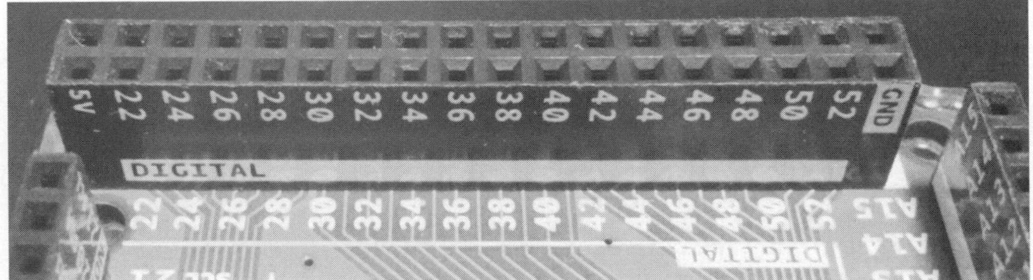

Figure 1-9 The Arduino Mega 2560 has many digital pins.

Power

The Arduino Mega 2560 has outputs for 3.3 and 5 V. One section that provides power is located on the side of the Arduino. Another section is on the bottom of the Arduino which has two 5-V power pins available for use. There is a total of one 3.3-V power pin and three 5-V power pins on the Arduino Mega 2560 that you can use to power your electronics circuit. You can also

Figure 1-11 Power pins on the Arduino Mega 2560.

provide your own power source by connecting the positive terminal of the power source to the Vin pin and the ground of the power source to the Arduino's ground. Make sure the voltage being supplied is within the Arduino board's voltage range. See Figure 1-11.

Ground Pins

There are 5 ground pins on the Arduino Mega 2560 and are circled in Figure 1-12.

The Custom Fuse

The most unique component of the Arduino Mega 2560 is the golden resettable fuse with

Figure 1-12 The ground pins on the Arduino Mega 2560.

the Arduino infinity logo on it. It is located between the USB port and the 9-V battery power input jack. This fuse is tripped if the Arduino uses more than 500 mA of current when being powered through the USB port. See Figure 1-13.

Overview of the C/C++ Language for the Arduino

The Arduino uses C and C++ in its programs, which are called sketches. This section briefly summarizes key language elements. This is

Figure 1-13 The custom Arduino fuse.

not meant as a reference guide, and ideally, you should have some experience with a programming language similar to C and/or C++.

Comments

//. This signifies a single-line comment that is used by the programmer to document the code. These comments are not executed by the Arduino.

/* */. These enclose a multiline comment that is used by the programmer to document the code. These comments are not executed by the Arduino.

Data Types

- **`void`.** This type being used with a function indicates that the function will not return any value to the function caller. For example, the `setup()` function that is part of the standard Arduino code framework has a return type of void.

```
void setup()
{
 // Initialize the Arduino, sensors,
 // and Wifi adapter here
}
```

- **`boolean`.** A `boolean` variable can hold either the value of true or false and is 1 byte in length. For example, in the following code, the variable `result` is declared of type `boolean` and is initialized to false:

```
boolean result = false;
```

- **`char`.** The `char` variable type can store character values and is 1 byte in length. The following code declares that `tempchar` is of type char and is an array with 50 elements:

```
char tempchar[50];
```

- **`unsigned char`.** The unsigned char data type holds 1 byte of information in the range of 0 through 255.
- **`byte`.** The `byte data` type is the same as the `unsigned char` data type. The following code declares a variable called `data` of type byte that is initialized to 0:

```
byte data = 0;
```

- **`int`.** The `int` data type holds a 2-byte number in the range of −32,768 to 32,767.
- **`unsigned int`.** This data type is 2 bytes in length and holds a value from 0 to 65,535.
- **`word`.** This data type is the same as the `unsigned int` type.

- `long`. This data type is 4 bytes in length and holds a value from −2,147,483,648 to 2,147,483,647.
- `unsigned long`. This data type is 4 bytes in length and holds a value between 0 and 4,294,967,295.
- `float`. This is a floating-point number that is 4 bytes in length and holds a value between −3.4028235E+38 and 3.4028235E+38.
- `double`. On the current Arduino implementation, `double` is the same as float with no gain in precision.
- `String`. This is a class object that allows the user to easily manipulate groups of characters. In the following code, a new variable called `ssid`, which represents the name of an access point, is given the name of ESP-01.

```
String ssid = "ESP-01";
```

- `array`. An array is a continuous collection of data that can be accessed by an index number. Arrays are 0 based, so the first element in the array has an index of 0. Common types of arrays are character arrays and integer arrays. The following code creates an array of `ClientInfo` class objects called `Clients`. The number of elements in the array is held in the constant integer variable `MAX_CLIENTS` which is set to 4.

```
const int MAX_CLIENTS = 4;
ClientInfo Clients[MAX_CLIENTS];
```

Constants

- `INPUT`. This is an Arduino pin configuration that sets the pin as an input pin that allows you to easily read the voltage value at that pin with respect to ground on the Arduino. For example, the following code sets the pin defined as VSYNC on the Arduino as an INPUT pin, which allows you to read the voltage value of the pin. The function `pinMode()` is an Arduino function included in the built-in library.

```
pinMode(VSYNC, INPUT);
```

- `OUTPUT`. This is an Arduino pin configuration that sets the pin as an output pin that allows you to drive other electronics components such as an LED or to provide digital input to other devices in terms of HIGH or LOW voltages. In the following code, the pin that is defined as WEN is set to `OUTPUT` using the built-in `pinMode()` function:

```
pinMode(WEN , OUTPUT);
```

- `HIGH (pin declared as INPUT)`. If a pin on the Arduino is declared as an `INPUT`, then when the `digitalRead()` function is called to read the value at that pin, a HIGH value would indicate a value of 3 V or more at that pin.
- `HIGH (pin declared as OUTPUT)`. If a pin on the Arduino is declared as an `OUTPUT`, then when the pin is set to HIGH with the `digitalWrite()` function, the pin's value is 5 V.
- `LOW (pin declared as INPUT)`. If a pin on the Arduino is declared as an `INPUT`, then when the `digitalRead()` function is called to read the value at that pin, a LOW value would indicate a value of 2 V or less.
- `LOW (pin declared as OUTPUT)`. If a pin on the Arduino is declared as an `OUTPUT`, then when the `digitalWrite()` function is called to set the pin to LOW, the voltage value at that pin would be set to 0 V.
- `true`. True is defined as any nonzero number such as 1, −1, 200, 5, etc.
- `false`. False is defined as 0.

The define Statement

The `define` statement assigns a name to a constant value. During the compilation process, the compiler will replace the constant name with the constant value:

```
#define constantName value
```

The following code defines the software serial data receive pin as pin 6 on the Arduino and the software serial data transmit pin as pin 7 on the Arduino:

```
#define RxD 6
#define TxD 7
```

These definitions are used in defining which pins are to receive and transmit data via the software serial method, which is initialized as follows and can be used to communicate with a Wi-Fi adapter:

```
SoftwareSerial WifiAdapter(RxD,TxD);
```

Note: For this book I recommend using hardware serial ports rather than software serial ports because software serial ports are unreliable at high speeds with some users reporting that any speed over 9600 baud generates many errors in transmission. I also personally tried to use software serial at a high baud rate to communicate with the ESP-01 Wi-Fi module (115,200 baud) and the results were unreliable and unusable.

The include Statement

The `#include` statement brings in code from outside files and "includes" it in your Arduino sketch. Generally, a header or `.h` file is included that allows access to the functions and classes inside that file. For example, we can include in a program a `Wire.h` file, which lets us use the `Wire` library. The `Wire` library has functions to initialize, to read data from, and to write data to a device connected to the I2C interface. We need the `Wire` library to use a device that uses the I2C bus.

```
#include <Wire.h>
```

The Semicolon

Each statement in C/C++ needs to end with a semicolon. For example, when declaring and initializing a variable, you will need a semicolon:

```
const int chipSelect = 48;
```

When you use a library that you included with the `#include` statement, you will need a semicolon at the end when you call a function:

```
Wire.begin();
```

Curly Braces

Curly braces such as { and } specify blocks of code and must occur in pairs. That is, for every opening brace, there must be a closing brace to match. A function requires curly braces to denote the beginning and end of the function:

```
void Function1()
{
  // Body of Function
}
```

Program loops such as the `for` statement may also need curly braces:

```
for (int i = 0; i < 9; i++)
{
  // Body of loop
}
```

It is also good practice to use braces in control structures such as the `if` statement:

```
if (i< 0)
{
  // Body of If statement
}
```

Arithmetic Operators

- **=.** The equals sign is the assignment operator used to set a variable to a value. For example, the following code sets the values of the access point name and the access point password. The variable `ssid` is set to ESP-01 and the `pwd` variable is set to esp8266esp01.

```
String ssid = "ESP-01";
String pwd = "esp8266esp01";
```

- **+.** The plus sign performs addition between numbers. It can also be used in other contexts depending on how the addition operator is defined. For example, the following code builds a command string used to execute a command on the ESP-01. The plus sign is used to concatenate different String objects to form a final string. Here the ssid and the password for an access point are used to build the final string held in the variable `APConfigCommand`.

```
String APConfigCommand = "AT+CWSAP_CUR=";
APConfigCommand += "\"" + ssid + "\"" +
 "," + "\"" + pwd + "\"" + ",";
```

- **-.** The minus sign performs subtraction. For example, the following code calculates the time that has passed since the last ping was sent to the server. The time that has passed is held in the variable `TimeDelayed`. If the time that has elapsed is greater than the value in the `PingInterval` variable then a ping is sent to the server.

```
TimeDelayed = millis() - TimeStarted;
if (TimeDelayed > PingInterval)
{
  // Send Ping to server
}
```

- ***.** The asterisk sign performs multiplication. For example, the following code calculates the ping interval in milliseconds which totals 30 seconds or 30,000 milliscconds.

```
unsigned int PingInterval = 1000 * 30;
```

- **/.** The back slash sign performs division. For example, the speed in miles per hour of an object is calculated by dividing the number of miles the object has traveled by the number of hours that it took to travel that distance:

```
float Speed = NumberMiles / NumberHours;
```

- **%.** The percent sign is the modulo operator that returns the remainder from a division between two integers. For example:

```
int remainder = dividend % divisor;
```

Comparison Operators

- **==.** The double equals sign is a comparison operator to test whether the argument on the left side of the double equals sign is equal to the argument on the right side. If the arguments are equal, then it evaluates to true. Otherwise, it evaluates to false. For example, the following code goes through a list of clients that may be connected to a server and looks for a client with a specific connection ID. If this connection ID is found then the client is processed.

```
for (int i = 0; (i < MAX_CLIENTS) &&
 !done; i++)
{
  if (Clients[i].ConnectionID ==
  ConnectionID)
  {
    // Client found
    // Process Client
  }
}
```

- **!=.** The exclamation point followed by an equals sign is the not-equal-to operator that evaluates to true if the argument on the left is not equal to the argument on the right side. Otherwise, it evaluates to false. For example, the following code searches through the array that keeps track of the clients that are connected to an access point. If a valid client is found then it is processed. Specifically if the client's connection ID is not equal to the null string which means that the client is connected to the access point then process this valid client.

```
// Check for active clients
for (int i = 0; i < MAX_CLIENTS; i++)
{
  if (Clients[i].ConnectionID != "")
  {
    // A valid client is found
  }
}
```

- **<.** The less than operator evaluates to true if the argument on the left is less than the argument on the right. For example, in the code below, the `for` loop is executed as long as i is less than `MAX_CLIENTS`.

```
for (int i = 0; i < MAX_CLIENTS; i++)
{
  // Loop is executed as long as i is
    less than MAX_CLIENTS
}
```

- **>.** The greater than operator evaluates to true if the argument on the left side is greater than the argument on the right side. For example, in the following code, if the `available()` function returns a result that is greater than 0 then there are characters to read from the serial port, so the code block executes. That is, the number of available characters to read must be greater than 0.

```
if (Serial3.available() > 0)
{
  // Process available characters
  // from Serial port
}
```

- **<=.** The less than sign followed by an equals sign returns true if the argument on the left side is less than or equal to the argument on the right side. It returns false otherwise.
- **>=.** The greater than sign followed by an equals sign returns true if the argument on the left side is greater than or equal to the argument on the right side. It returns false otherwise.

Boolean Operators

- **&&.** This is the AND `boolean` operator that only returns true if both the arguments on the left and right sides evaluate to true. It returns false otherwise. For example, in the following code, the `for` loop is executed as long as the variable `i` is less than the maximum number of clients and the processing is not done yet.

```
for (int i = 0; (i < MAX_CLIENTS) &&
  !Done; i++)
{
  // Process Clients
}
```

- **||.** This is the OR operator and returns true if either the left-side argument or the right-side argument evaluates to true. Otherwise, it returns false. For example, in the following code, if the incoming data from a client is equal to `ledon` or `ledoff` then the code block is executed and the command from the client is processed.

```
if ((Data == "ledon") || (Data ==
 "ledoff"))
{
  String response =
  ProcessBuiltInLED(Data);
  if (!SendDataToClient(ClientID,
  response))
  {
   Serial.println(F("***************
    ERROR ... Client Not Found"));
  }
}
```

- **!.** The `NOT` operator returns the opposite `boolean` value. The not value of true is false, which is 0, and the not value of false is true, which is nonzero. In the following code, a file is opened on the SD card, and a pointer to the file is returned. If the pointer to the file is `NULL`, which has a 0 value, then not `NULL` would be 1, which is true. The `if` statement is executed when the argument is evaluated to true, which means that the file pointer is `NULL`. This means that the Open operation has failed, and an error message needs to be displayed.

```
// Open File
InfoFile = SD.open(Filename.c_str(),
FILE_WRITE);
// Test if file actually open
if (!InfoFile)
{
 Serial.println(F("\nCritical ERROR
  ...Can not open Photo Info File for
  output ... "));
 return;
}
```

Bitwise Operators

- `&`. This is the bitwise AND operator between two numbers, where each bit of each number has the AND operation performed on it to produce the result in the final number. The resulting bit is 1 only if both bits in each number are 1. Otherwise, the resulting bit is 0.
- `|`. This is the bitwise OR operator between two numbers, where each bit of each number has the OR operation performed on it to produce the result in the final number. The resulting bit is 1 if the bit in either number is 1. Otherwise, the resulting bit is 0.
- `^`. This is the bitwise XOR operator between two numbers, where each bit of each number has the exclusive OR operation performed on it to produce the result in the final number. The resulting bit is 1 if the bits in each number are different and 0 otherwise.
- `~`. This is the bitwise NOT operator, where each bit in the number following the NOT symbol is inverted. The resulting bit is 1 if the initial bit was 0 and is 0 if the initial bit was 1.
- `<<`. This is the "bitshift left" operator, where each bit in the left operand is shifted to the left by the number of positions indicated by the right operand. For example, in the following code, a 1 is shifted left `PinPosition` times, and the final value is assigned to the variable `ByteValue`:

```
ByteValue = 1 << PinPosition;
```

- `>>`. This is the "bitshift right" operator, where each bit in the left operand is shifted to the right by the number of positions indicated by the right operand. For example, in the following code, bits in the number 255 are shifted to the right `PinPosition` times, and the final value is assigned to the variable `ByteValue`:

```
ByteValue = 255 >> PinPosition;
```

Compound Operators

- `++`. This is the increment operator. The exact behavior of this operator also depends on whether it is placed before or after the variable being incremented. In the following code, the variable `PhotoTakenCount` is incremented by 1:

```
PhotoTakenCount++;
```

- If the increment operator is placed after the variable being incremented, then the variable is used first in the expression it is in before being incremented. For example, in the following code, the `height` variable is used first in the `for` loop expression before it is incremented. So the first iteration of the `for` loop below would use height = 0. So the first iteration of the `for` loop below would assign `h` = 0. The `h2` variable would be incremented after being used in the expression and assigned to `h`.

```
int h2 = 0;
for (int height = 0; height < PHOTO_
 HEIGHT; height++)
{
  // Process row of image
  h = h2++;
}
```

- If the increment operator is placed before the variable being incremented, then the variable is incremented first, and then it is used in the expression that it is in. For example, in the following code, the `h2` variable is incremented first before it is used in the `for` loop. This means that in the first iteration of the loop, the `h` variable is 1.

```
int h = 0;
int h2 = 0;
for (int height = 0; height < PHOTO_
 HEIGHT; height++)
{
  // Process row of image
  h = ++h2;
}
```

- **--.** The decrement operator decrements a variable by 1, and its exact behavior depends on the placement of the operator either before or after the variable being decremented. If the operator is placed before the variable, then the variable is decremented before being used in an expression. If the operator is placed after the variable, then the variable is used in an expression before it is decremented. This follows the same pattern as the increment operator discussed previously.
- **+=.** The compound addition operator adds the right operand to the left operand. This is actually a shorthand version of `operand1 = operand1 + operand2`, which is the same as the version that uses the compound addition operator:

  ```
  operand1 += operand2;
  ```
- **-=.** The compound subtraction operator subtracts the operand on the right from the operand on the left. For example, the code for a compound subtraction would be

  ```
  operand1 -= operand2;
  ```

  This is the same as

  ```
  operand1 = operand1 - operand2;
  ```
- ***=.** The compound multiplication operator multiplies the operand on the right by the operand on the left. The code for this is

  ```
  operand1 *= operand2;
  ```

  This is also equivalent to

  ```
  operand1 = operand1 * operand2;
  ```
- **/=.** The compound division operator divides the operand on the left by the operand on the right. For example,

  ```
  operand1 /= operand2;
  ```

  This is equivalent to

  ```
  operand1 = operand1 / operand2;
  ```
- **&=.** The compound bitwise AND operator is equivalent to

  ```
  x = x & y;
  ```
- **!=.** This compound bitwise OR operator is equivalent to

  ```
  x = x | y;
  ```

Pointer Access Operators

- ***.** The dereference operator allows you to access the contents to which a pointer points. For example, the following code declares a variable `pdata` as a pointer to a byte and creates storage for the data using the new command. The pointer variable `pdata` is then dereferenced to allow the actual data to which the pointer points to be set to 1:

  ```
  byte *pdata = new byte;
  *pdata = 1;
  ```
- **&.** The address operator creates a pointer to a variable. For example, the following code declares a variable data of type byte and assigns the value of 1 to it. A function called `FunctionPointer()` is defined that accepts as a parameter a pointer to a byte. In order to use this function with the variable `data`, we need to call that function with a pointer to the variable data:

  ```
  byte data = 1;
  void FunctionPointer(byte *data)
  {
    // body of function
  }
  FunctionPointer(&data);
  ```

Variable Scope

- **Global variables.** In the Arduino programming environment, global variables are variables that are declared outside any function and before they are used. The following variables are global variables that represent the information needed in order to configure an access point on an ESP-01 module that uses the ESP8266 Wi-Fi communications chip.

```
// AP configuration parameters
String ssid = "ESP-01";
String pwd = "esp8266esp01";
int chl = 1;
int ecn = 2;
int maxconn = 4;
int ssidhidden = 0;
```

- **Local variables.** Local variables are declared inside functions or code blocks and are only valid inside that function or code block. For example, in the following function, the variable `localnumber` is only visible inside the `Function1()` function:

```
void Function1()
{
 int localnumber = 0;
}
```

Conversion

- **`char(x)`.** This function converts a value x into a char data type and then returns it.
- **`byte(x)`.** This function converts a value x into a byte data type and then returns it.
- **`int(x)`.** This function converts a value x into an integer data type and then returns it.
- **`word(x)`.** This function converts a value x into a word data type and then returns it.
- **`word(highbyte,lowbyte)`.** This function combines two bytes, the high-order byte and the low-order byte, into a single word and then returns it.
- **`long(x)`.** This function converts a value x into a long and then returns it.
- **`float(x)`.** This function converts a value x into a float and then returns it.

Control Structures

- **`if` (comparison operator).** The `if` statement is a control statement that tests whether the result of the comparison operator or argument is true. If it is true, then execute the code block. For example, in the following code, the `if` statement tests to see whether more data from a Wi-Fi connection needs to be read. If so, then read the data in, and assign it to the Out character variable. There is more data available if the return value from `Serial3.available()` is greater than 0.

```
if (Serial3.available()>0)
{
 Out = (char)Serial3.read();
 ESP01ResponseString += Out;
}
```

- **`if` (comparison operator) `else`.** The `if else` control statement is similar to the `if` statement except with the addition of the else section, which is executed if the previous if statement evaluates to false and is not executed. For example, in the following code, if the command to the ESP-01 module is successfully performed then print out a message to the Arduino's Serial Monitor indicating that the module's state was changed to allow multiple client connections. Otherwise, print out a message indicating that the command failed.

```
// Enable Multiple Connections
if (SendATCommand(F("AT+CIPMUX=1\r\n")))
{
 Serial.println(F("ESP-01 mode set to
  allow multiple connections."));
}
else
{
 Serial.println(F("ERROR - ESP-
  01 FAILED to allow for multiple
  connections."));
}
```

- **`for` (initialization; condition; increment).** The `for` statement is used to execute a code block usually initializing a counter and then performing actions on a group of objects indexed by that incremented value. For example:

```
for (int i = 0; i < MAX_CLIENTS; i++)
{
 // Loop as long as i is less than
  MAX_CLIENTS
}
```

- **`while` (expression).** The `while` statement executes a code block repeatedly until the expression evaluates to false. In the following code, the `while` code block is executed as long as an "OK" is not received from the ESP-01 module and the operation has not timed out due to exceeding the maximum allowable time.

```
while (!OKReceived && !TimedOut)
{
  // Perform this loop as long as an
   "OK" is not received
  // from the ESP-01 and the operation
   has not exceeded the maximum
  // allowable time.
}
```

- **`break`.** A `break` statement is used to exit from a loop such as a `while` or `for loop`. In the following code, the `while loop` causes the code block to be executed forever. If data are available from the Serial Monitor, then they are processed, and then the `while loop` is exited:

```
while (1)
{
  if (Serial3.available() > 0)
  {
   // Process the data
   break;
  }
}
```

- **`return` (value).** The `return` statement exits a function. It also may return a value to the calling function:

```
return;
return false;
```

Object-Oriented Programming

The Arduino programming environment also supports the object-oriented programming aspects of C++. An example of Arduino code that uses object-oriented programming is the WiFiClient class that supports Wi-Fi communication on the Arduino. A C++ class is composed of data, functions that use that data, and a constructor that is used to create an object of the class. An example of class data is

```
uint16_t _socket;
```

An example of a class constructor is

```
WiFiClient(uint8_t sock);
```

An example of a class function is

```
virtual void stop();
```

The following code is the `WiFiClient` class from the standard built-in Arduino Wi-Fi library.

```
class WiFiClient : public Client {

public:
  WiFiClient();
  WiFiClient(uint8_t sock);

  uint8_t status();
  virtual int connect(IPAddress ip,
    uint16_t port);
  virtual int connect(const char *host,
    uint16_t port);
  virtual size_t write(uint8_t);
  virtual size_t write(const uint8_t
  *buf, size_t size);
  virtual int available();
  virtual int read();
  virtual int read(uint8_t *buf, size_t
    size);
  virtual int peek();
  virtual void flush();
  virtual void stop();
  virtual uint8_t connected();
  virtual operator bool();

  friend class WiFiServer;

  using Print::write;

private:
  static uint16_t _srcport;
  uint8_t _sock; //not used
  uint16_t _socket;

  uint8_t getFirstSocket();
};
```

Arduino Development System Requirements

Developing projects for the Arduino can be done on the Windows, Mac, and Linux operating systems. The software needed to develop programs that run on the Arduino can be downloaded from the main web site at http://www.arduino.cc/en/Main/Software.

The following is a summary of the different types of Arduino IDE distributions that are available for download. You will only need to download and install one of these files. The file you choose will depend on the operating system your computer is using.

Windows

- Windows Installer – This is a .exe file that must be run to install the Arduino IDE.
- Windows ZIP file for non admin install – This is a zip file that must be uncompressed in order to install the Arduino IDE. 7-zip is a free file compression and uncompression program is available at http://www.7-zip.org.

Mac

- Mac OS X 10.7 Lion or newer – This is a zip file that must be uncompressed and installed for users of the Mac operating system.

Linux

- Linux 32 bits – Installation file for the Linux 32 bit operating system.
- Linux 64 bits – Installation file for the Linux 64 bit operating system.

The Arduino Web Editor

You can now create Arduino sketches online through your web browser and save your program to cloud storage. The only downside is that you don't have control over your programs or the Arduino IDE so that if the web editor is down or the cloud storage system is down you will either lose your data or not have access to your data until the site is back up.

Arduino Software IDE

This is the program that is used to develop the program code that runs on and controls the Arduino. For example, in order for you to have the Arduino control the lighting state of an LED you will need to write a computer program in C/C++ using the Arduino IDE. Then, you will need to compile this program into a form that the Arduino is able to execute and then transfer the final compiled program using the Arduino IDE. From there the program automatically executes and controls the LED that is connected to the Arduino. New versions of the IDE are compiled daily or hourly and are available for download. Older versions of the IDE are also available for downloading at http://www.arduino.cc/en/Main/OldSoftwareReleases.

In this section we will go over the key features of the Arduino Software IDE. The IDE you are using may be slightly different than the version discussed in this section but the general functions we cover here should still be the same. We won't go in depth into every detail of the IDE since this book is meant as a quick start guide and not a reference manual. We will cover the critical features of the Arduino IDE that you will need to get started on the projects in this book. See Figure 1-14.

The "Verify" button checks to see if the program you have entered into the Arduino IDE is valid and without errors. These programs are called "sketches." See Figure 1-15.

The "Upload" button first verifies that the program in the IDE is a valid C/C++ program with no errors, compiles the program into a

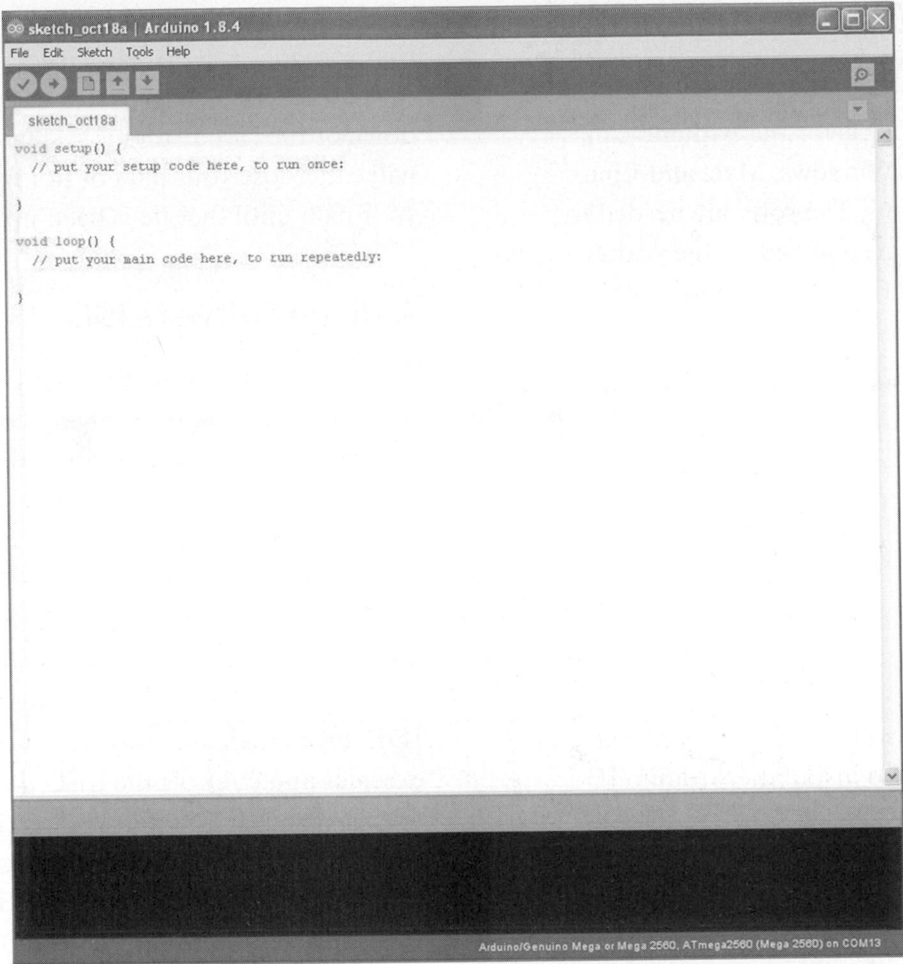

Figure 1-14 The Arduino IDE.

Figure 1-15 The Verify button.

form the Arduino can execute, and then finally transfers the program via the USB cable that is connected from your computer to your Arduino board. See Figure 1-16.

The "New" button creates a new blank file or sketch inside the Arduino IDE where the user can create his own C/C++ program for verification, compilation, and transferring to the Arduino. See Figure 1-17.

The "Open" File button is used to open and load in the Arduino C/C++ program source code from a file or load in various sample source codes from example Arduino projects that are included with the IDE. See Figure 1-18.

The "Save" button saves the sketch you are currently working on to disk. A file save dialog is brought up first and then you will be able to

Figure 1-16 The Upload button.

Figure 1-17 The New File button.

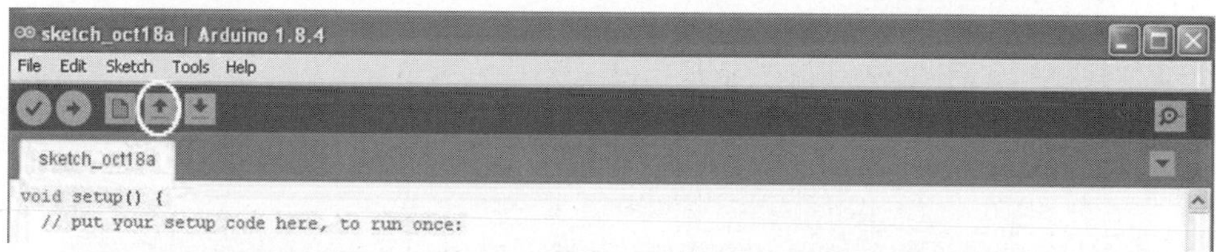

Figure 1-18 The Open File button.

Figure 1-19 The Save File button.

Figure 1-20 The Serial Monitor button.

save the file on your computer's hard drive. See Figure 1-19.

The "Serial Monitor" button brings up the Serial Monitor debug program where the user can examine the output of debug statements from the Arduino program. The Serial Monitor can also accept user input that can be processed by the Arduino program. See Figure 1-20.

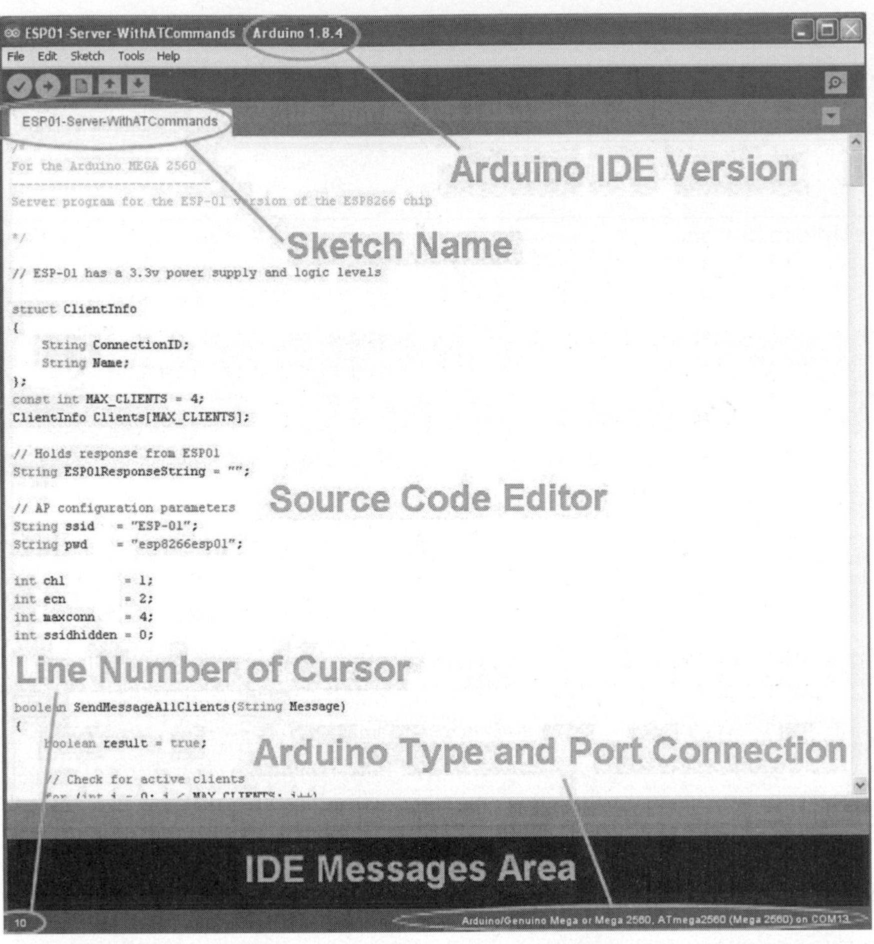

Figure 1-21 The Arduino IDE.

There are also other important features of the main window of the Arduino IDE. The title bar of the IDE window contains the Arduino IDE version number. In Figure 1-21 the Arduino version number is 1.8.4. The sketch name is displayed in the source code tab and is "ESP01-Server-WithATCommands." The source code area, which is the large white area with scrollbars on the right side and bottom, is where you enter your C/C++ source code that will control the behavior of the Arduino. The bottom black area in the IDE is where warning and errors are displayed from the code verification process. At the bottom left-hand corner of the IDE is a number that represents the line number in the source code where the user's cursor is currently located. In the lower right-hand corner of the IDE is the currently selected Arduino model and COM port that the Arduino is attached to. See Figure 1-21.

Hands-on Example: A Simple Arduino "Hello World" Program with an LED

In this hands-on example I show you how to set up the Arduino development system on your Windows-based PC or Mac. I first discuss where you can get an Arduino board and USB cable. Then I discuss the installation of the Arduino IDE and Arduino hardware device drivers. I then discuss how to load in the "Blink" sketch example program. Next, I tell you how to verify that the program is without syntax errors, how to upload it onto the Arduino, and how to tell if the program is working.

Get an Arduino Board and USB Cable

You can purchase an official Arduino Mega 2560 board from http://www.arduino.cc. A second option is to buy an unofficial Arduino Mega 2560. These boards are generally a lot cheaper than an "official" Arduino board. However, the quality may vary widely between manufacturers or even between production runs between the same manufacturer. In terms of the USB cable that is used to connect the Arduino to your development computer the official Arduino board generally does not come with a cable but many unofficial boards come with short USB cables. Arduino compatible USB cables of a longer length such as 6 foot or 10 foot can be bought on Amazon.com or ebay.com. Make sure you get the right kind of USB cable with the right connectors on either end. The rectangular end of the USB cable is connected to your computer and the square end is connected to your Arduino. See Figure 1-22.

Install the Arduino IDE

The Arduino IDE has versions that can run on the Windows, Mac, and Linux operating systems. The Arduino IDE can be downloaded from http://www.arduino.cc/en/Main/Software.

Figure 1-22 Arduino USB cable.

I recommend installing the Windows executable version if you have a Windows-based computer. Follow the directions in the pop-up windows.

Note: **The Arduino web site also contains links to instructions for installing the Arduino IDE for Windows, Mac, and Linux on http://www.arduino.cc/en/Guide/HomePage. The installation for Linux depends on the exact version of Linux being used.**

Install the Arduino Drivers

The next step is to connect your Arduino to your computer using the USB cable. If you are using Windows it will try to automatically install your new Arduino hardware. Follow the directions in the pop-up windows to install the drivers. Decline to connect to Windows Update to search for the driver. Select "Install the software automatically" as recommended. If you are using XP ignore the pop-up window warning about the driver not passing windows logo testing to verify its compatibility with XP. If this does not work then instead of selecting "Install the software automatically" specify a specific driver location which is the "drivers/FTDI" directory under your main Arduino installation directory. For some operating systems like Windows 10 the drivers may already be installed by default. I personally noticed that my Android and Arduino devices were automatically recognized by the Windows 10 operating system.

Loading in the Blink Arduino Sketch Example

Next, we need to load the Blink sketch example into the Arduino IDE. Click the "Open" button to bring up the menu. Under "01. Basics" select the "Blink" example to load in. See Figure 1-23

The code that is loaded into the Arduino IDE should look like the code in Listing 1-1.

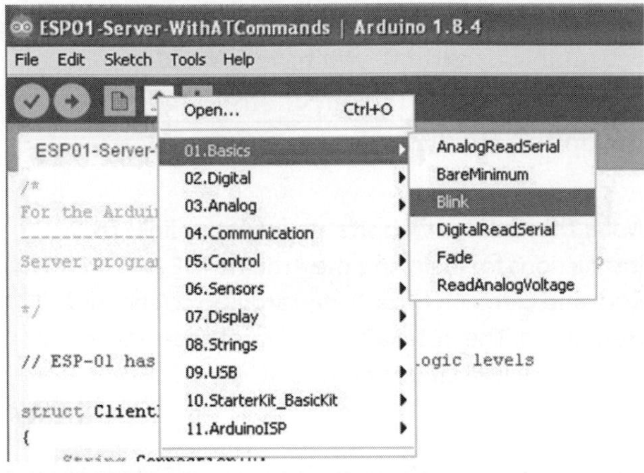

Figure 1-23 Loading in the blink example.

Verifying the Blink Arduino Sketch Example

Click the "Verify" button to verify the program is valid C/C++ code and is error-free. See Figure 1-24.

Uploading the Blink Arduino Sketch Example

Before uploading the sketch to your Arduino make sure the type of Arduino under the "Tools->Board" menu item is correct.

Listing 1-1 Blink Sketch

```
/*
  Blink

  Turns an LED on for one second, then off for one second, repeatedly.

  Most Arduinos have an on-board LED you can control. On the UNO, MEGA and ZERO
  it is attached to digital pin 13, on MKR1000 on pin 6. LED_BUILTIN is set to
  the correct LED pin independent of which board is used.
  If you want to know what pin the on-board LED is connected to on your Arduino
  model, check the Technical Specs of your board at:
  https://www.arduino.cc/en/Main/Products

  modified 8 May 2014
  by Scott Fitzgerald
  modified 2 Sep 2016
  by Arturo Guadalupi
  modified 8 Sep 2016
  by Colby Newman

  This example code is in the public domain.

  http://www.arduino.cc/en/Tutorial/Blink
*/

// the setup function runs once when you press reset or power the board
void setup() {
  // initialize digital pin LED_BUILTIN as an output.
  pinMode(LED_BUILTIN, OUTPUT);
}

// the loop function runs over and over again forever
void loop() {
   digitalWrite(LED_BUILTIN, HIGH); // turn the LED on (HIGH is the voltage level)
   delay(1000);                     // wait for a second
   digitalWrite(LED_BUILTIN, LOW);  // turn the LED off by making the voltage LOW
   delay(1000);                     // wait for a second
}
```

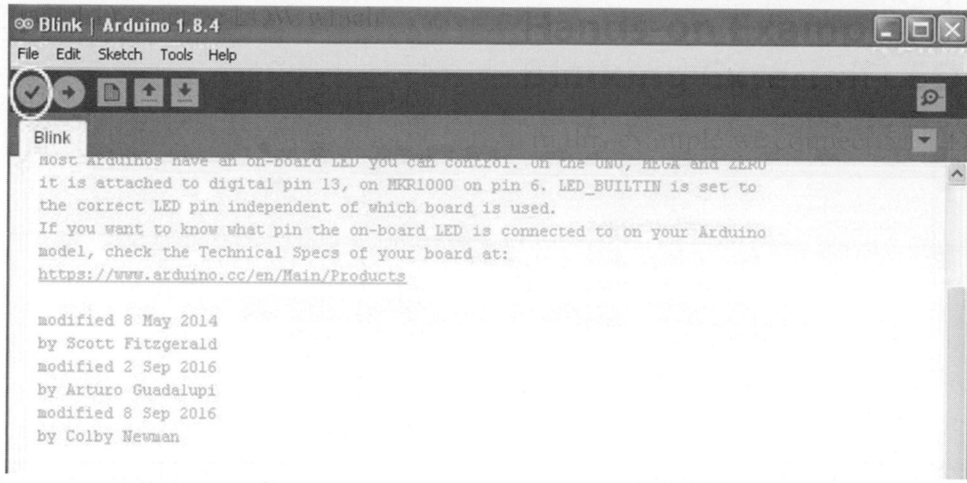

Figure 1-24 Verifying the Blink sketch.

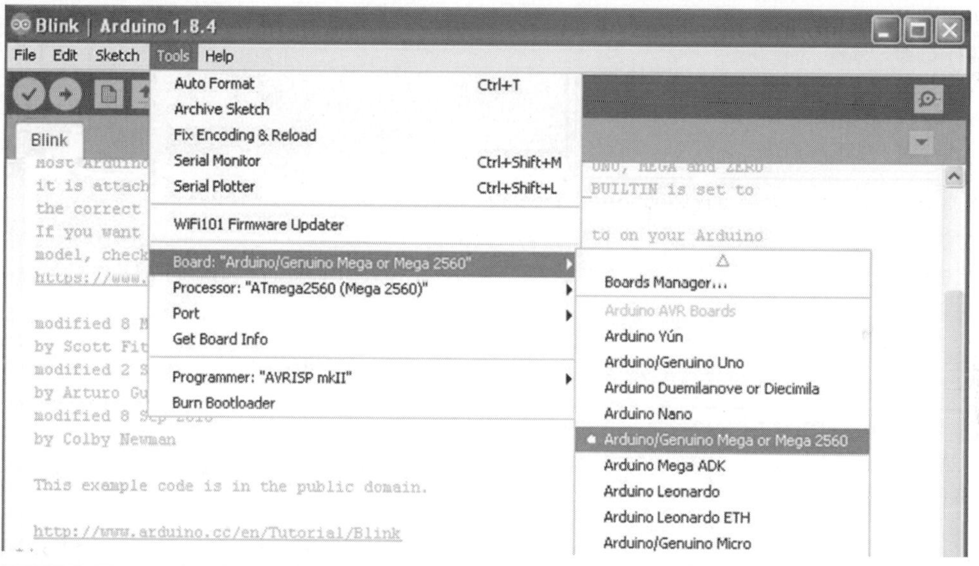

Figure 1-25 Setting the Arduino model.

In our case the board type should be set to be "Arduino/Genuino Mega or Mega 2560." See Figure 1-25.

Next, make sure the serial port is set correctly to the one that is being used by your Arduino. Generally Com1 and Com2 are reserved and the serial port that the Arduino will be connected to is Com3 or higher. See Figure 1-26.

If you are using a Mac then the Serial Port selection should be something like "/dev/tty.usbmodem" instead of a COMXX value.

Next, with the Arduino connected press the "Upload" button to verify, compile, and then transfer the Blink example program to the Arduino. After the program has finished uploading you should see a message that the upload has been completed in the warnings/error window at the bottom of the IDE inside the black window. See Figure 1-27.

Note: **The Upload button does the job of the "Verify" button but also uploads the final compiled program to the Arduino.**

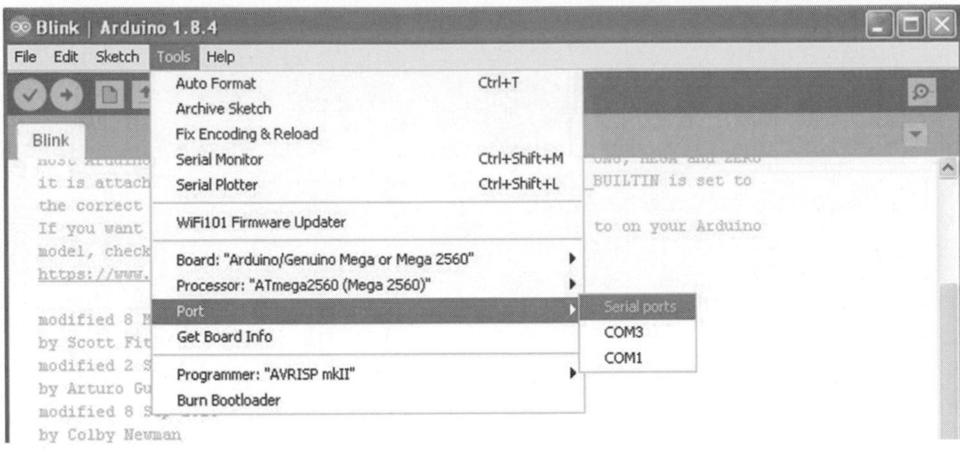

Figure 1-26 Set the com port if needed.

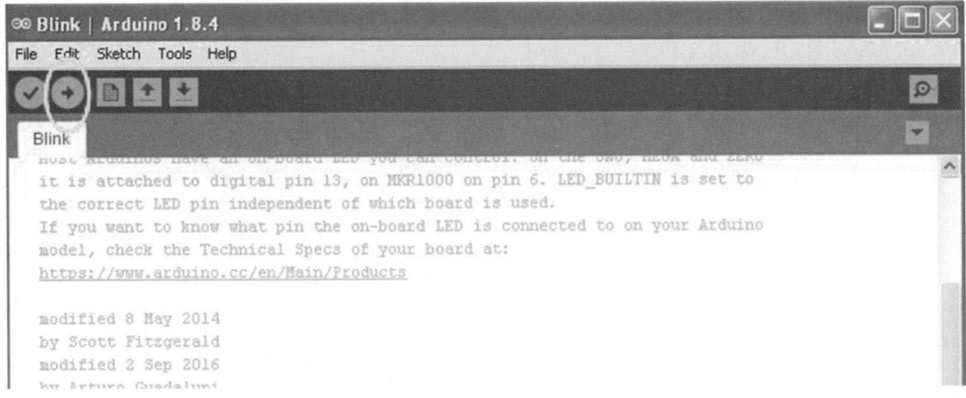

Figure 1-27 Upload to the Arduino.

Figure 1-28 Blinking LED light.

The Final Result

The final result will be a blinking light on the Arduino board near digital pin 13. By design the Arduino board has a built-in LED connected to pin 13. So this example did not require you to connect an actual separate LED to the Arduino board. See Figure 1-28.

Summary

In this chapter I introduced the Arduino to the reader. I concentrated my coverage on the Arduino Mega 2560. I discussed the Mega 2560's basic features and then covered key functional components. I then gave a brief summary of the programming languages you will need to create programs with the Arduino. Next, information about the software needed to develop programs for the Arduino, called the Arduino IDE, was presented, including the different versions of the IDE available for different platforms. Information about key features of the IDE was discussed. Finally, a hands-on example was presented where I took the reader through a step-by-step guide to setting up the Arduino with a development computer system. In this example I also discussed loading in an example program, uploading this example program to the Arduino, and checking for the correct operation of the program.

CHAPTER 2

Introduction to the ESP8266

In this chapter I will cover the ESP-01 module and the NodeMCU which are based on the ESP-12E module. Both the ESP-01 module and the NodeMCU use the ESP8266EX chip. I start by discussing the specifications of the ESP8266EX chip. Next, I discuss the ESP-01 module. I discuss the functions of each of the pins on the ESP-01. I then discuss why a logic level converter and a step voltage regulator are needed when you use the ESP-01 module with an Arduino. Next, I cover the essential AT commands that are needed in order to operate the ESP-01 module with the Arduino over a serial communications port. A hands-on example is presented where I show you how to use an ESP-01 module with an Arduino Mega 2560. Then, I cover the NodeMCU. I first discuss the NodeMCU's specifications. Next, I cover in detail the functions of the NodeMCU's pins including what pins are suitable for input and output operations and under what conditions. I then show you how to update the Arduino IDE in order to develop programs for the NodeMCU. Finally, I present some hands-on examples on how to develop programs on the NodeMCU using the Arduino IDE.

What Is the ESP8266

The ESP8266, also known as the ESP8266EX, is a Wi-Fi chip made by Espressif Systems based in Shanghai, China. It is a system on a chip (SoC) which means that it contains the basic elements of a complete computer system such as processor, memory, storage, and input output ports on a single chip. This design greatly reduces the power needed to operate the chip and is well suited for portable devices. The Espressif Systems official web site is located at https://www.espressif.com/.

Technical Specification of the ESP8266

Wi-Fi Specifications

- Standards: FCC/CE/TELEC/SRRC
- Protocols: 802.11b/g/n/e/i
- Frequency range: 2.4 to 2.5 G (2400–2483.5 M)
- Tx power:

  802.11b: +20 dBm

  802.11g: +17 dBm

  802.11n: +14 dBm
- Rx sensitivity:

  802.11b: −91 dBm (11 Mbps)

  802.11g: −75 dBm (54 Mbps)

  802.11n: −72 dBm (MCS7)
- Antenna: PCB Trace, External, IPEX Connector, Ceramic Chip

Hardware

- CPU: Tensilica L106 32-bit micro controller
- Peripheral interface:

  UART/SDIO/SPI/I2C/I2S/IR Remote Control

  GPIO/ADC/PWM/LED Light and Button

- Operating voltage: 2.5 – 3.6 V
- Operating current average value: 80 mA
- Operating temperature range: −40°C – 125°C
- Storage temperature range: −40°C – 125°C

Software

- Wi-Fi mode: Station/SoftAP/SoftAP+ Station
- Security: WPA/WPA2
- Encryption: WEP/TKIP/AES
- Firmware upgrade: UART Download/OTA (via network)
- Software development: Supports Cloud Server Development/Firmware and SDK for fast on-chip programming
- Network protocols: IPv4, TCP/UDP/HTTP/FTP
- User configuration: AT Instruction Set, Cloud Server, Android/iOS App

Power Consumption Examples

Tx 802.11b, CCK 11 Mbps, P OUT=+17 dBm	−170 mA
Tx 802.11g, OFDM 54 Mbps, P OUT =+15 dBm	−140 mA
Tx 802.11n, MCS7, P OUT =+13 dBm	−120 mA
Rx 802.11b, 1024 bytes packet length, −80 dBm	−50 mA
Rx 802.11g, 1024 bytes packet length, −70 dBm	−56 mA
Rx 802.11n, 1024 bytes packet length, −65 dBm	−56 mA
Modem-sleep	−15 mA
Light-sleep	−0.9 mA
Deep-sleep	−20 μA
Power off	−0.5 μA

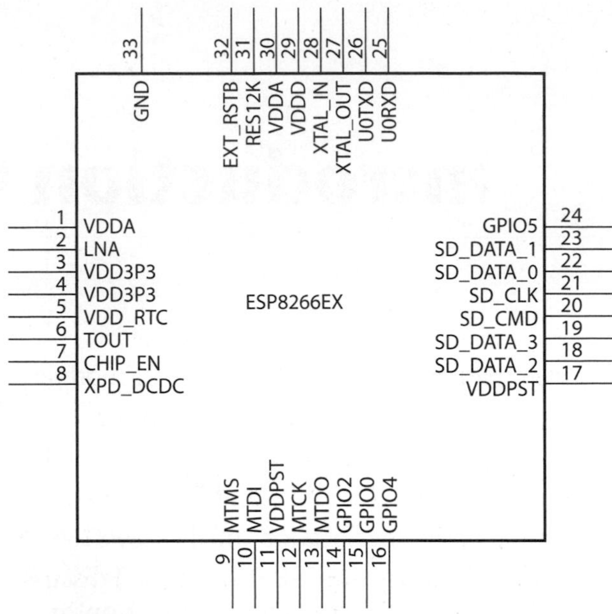

Figure 2-1 The ESP8266EX SoC (system on a chip).

The ESP8266EX SoC

The actual ESP8266EX chip pinout diagram is shown in Figure 2-1.

Pin Definitions for the ESP8266EX

The following are the pin definitions for the ESP8266EX chip.

Pin 1: VDDA (Power)

- Analog power 2.5 – 3.6 V

Pin 2: LNA (Input/Output)

- RF antenna interface
- Chip output impedance = 50 Ohms
- No matching required. It is suggested to retain the Pi-type matching network to match the antenna.

Pin 3: VDD3P3 (Power)

- Amplifier power 2.5 – 3.6 V

Pin 4: VDD3P3 (Power)

- Amplifier power 2.5 – 3.6 V

Pin 5: VDD_RTC (Power)

- NC (1.1 V)

Pin 6: TOUT (Input)

- ADC pin. It can be used to test the power-supply voltage of VDD3P3 (Pin3 and Pin4) and the input power voltage of TOUT (Pin 6). However, these two functions cannot be used simultaneously.

Pin 7: CHIP_PU (Input)

- Chip enable
- High: On, chip works properly
- Low: Off, small current consumed

Pin 8: XPD_DCDC (Input/Output)

- Deep-sleep wakeup (need to be connected to EXT_RSTB);
- GPIO16

Pin 9: MTMS (Input/Output)

- GPIO 14; HSPI_CLK

Pin 10: MTDI (Input/Output)

- GPIO 12; HSPI_MISO

Pin 11: VDDPST (Power)

- Digital/IO power supply (1.8 – 3.3 V)

Pin 12: MTCK (Input/Output)

- GPIO 13; HSPI_MOSI; UART0_CTS

Pin 13: MTDO (Input/Output)

- GPIO 15; HSPI_CS; UART0_RTS

Pin 14: GPIO2 (Input/Output)

- UART Tx during flash programming; GPIO2

Pin 15: GPIO0 (Input/Output)

- GPIO0; SPI_CS2

Pin 16: GPIO4 (Input/Output)

- GPIO 4

Pin 17: VDDPST (Power)

- Digital/IO power supply (1.8 – 3.3 V)

Pin 18: SDIO_DATA_2 (Input/Output)

- Connect to SD_D2 (Series R: 200 Ohms); SPIHD; HSPIHD; GPIO 9

Pin 19: SDIO_DATA_3 (Input/Output)

- Connect to SD_D3 (Series R: 200 Ohms); SPIWP; HSPIWP; GPIO 10

Pin 20: SDIO_CMD (Input/Output)

- Connect to SD_CMD (Series R: 200 Ohms); SPI_CS0; GPIO 11

Pin 21: SDIO_CLK (Input/Output)

- Connect to SD_CLK (Series R: 200 Ohms); SPI_CLK; GPIO 6

Pin 22: SDIO_DATA_0 (Input/Output)

- Connect to SD_D0 (Series R: 200 Ohms); SPI_MSIO; GPIO 7

Pin 23: SDIO_DATA_1 (Input/Output)

- Connect to SD_D1 (Series R: 200 Ohms); SPI_MOSI; GPIO 8

Pin 24: GPIO5 (Input/Output)

- GPIO 5

Pin 25: U0RXD (Input/Output)

- UART Rx during flash programming; GPIO 3

Pin 26: U0TXD (Input/Output)

- UART Tx during flash programming; GPIO 1; SPI_CS1

Pin 27: XTAL_OUT (Input/Output)

- Connect to crystal oscillator output, can be used to provide BT clock input

Pin 28: XTAL_IN (Input/Output)

- Connect to crystal oscillator input

Pin 29: VDDD (Power)

- Analog power 2.5 – 3.6 V

Pin 30: VDDA (Power)

- Analog power 2.5 – 3.6 V

Pin 31: RES12K (Input)

- Serial connection with a 12-k Ohms resistor and connect to the ground

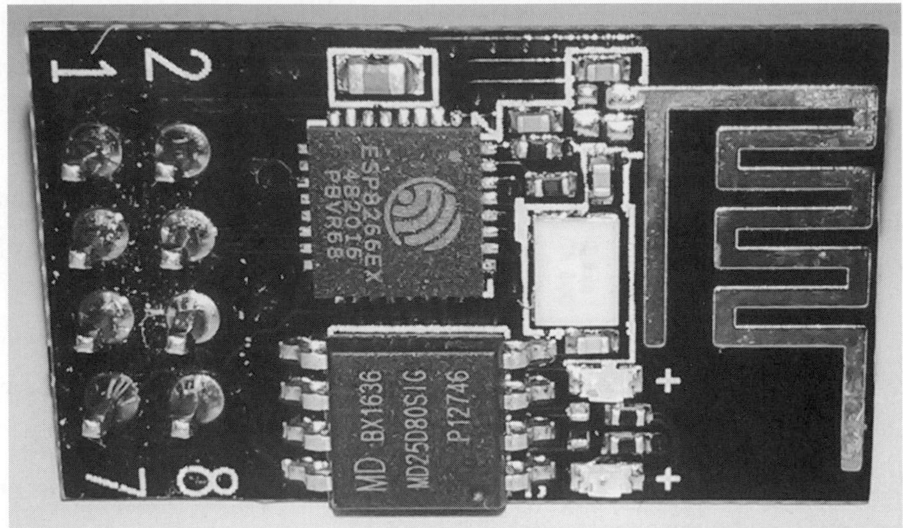

Figure 2-2 Front view of the ESP-01.

Pin 32: EXT_RSTB (Input)

- External reset signal (Low voltage level: Active)

Pin 33: GND

Note: **The GPIO2, GPIO0, and MTDO (GPIO 15) pins are configurable on the PCB as the 3-bit strapping register that determines the booting mode and the SDIO timing mode.**

The ESP-01 Module

General Summary and Specifications

The ESP8266EX chip is used in the ESP-01 module which is made by a company called AI-Thinker. This module is ideally designed to work with another microcontroller such as the Arduino Mega 2560. The module is connected through the Arduino's hardware serial port and is controlled by the use of "`AT`" commands sent by the Arduino through the serial port to the ESP-01 module. See Figure 2-2 for front view of the ESP-01 module.

The back view of the ESP-01 shows a diagram of how the pins on the module are configured. See Figure 2-3.

The following are the pin definitions for the ESP-01 module:

- **3v3:** The 3.3-V power supply input pin.
- **IO16:** Also called the RST pin. This is connected to the ESP8266EX GPIO 16 pin which is the deep sleep wakeup pin and is also connected to the EXT_RSTB pin. The EXT_RSTB pin can be connected to an external reset signal. The EXT_RSTB pin is active low. Thus a low-voltage input signal will reset the ESP-01 module.
- **EN:** Also called the CH_PD pin. This pin is the chip enable pin which is active high. A high voltage on this pin means that the ESP-01 is enabled.
- **Tx:** The data transmit pin.
- **Rx:** The data receive pin.
- **IO0:** The pin connected to ESP8266EX pin GPIO 0. Default set to high voltage through an internal pull-up resistor.
- **IO2:** The pin connected to ESP8266EX pin GPIO 2. Default set to high voltage through an internal pull-up resistor.
- **GND:** The ground pin.

Figure 2-3 Back view of the ESP-01.

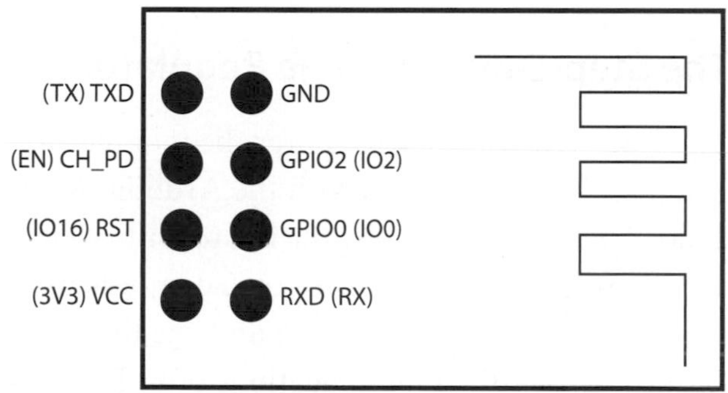

Figure 2-4 Top view of ESP-01 module.

The top view of the ESP-01 with the pins labeled including alternate names for the pins enclosed in parenthesis is shown in Figure 2-4.

Important Note on the GPIO2 and GPIO0 Pins and Normal Boot Mode

Normal boot mode requires that GPIO15 be low, and GPIO02 and GPIO0 are set to high. On my ESP-01 both the GPIO2 and GPIO0 pins can be left unconnected. I was able to use the ESP-01 without connecting these pins because the default voltages on these pins are high as the result of an internal pull-up resistor that makes the default value high. On the ESP-01 the GPIO15 pin is set by default to low voltage. Thus, by default you can boot the ESP-01 normally without connecting a high voltage to the GPIO0 and GPIO2 pins. **Note that if the GPIO0 is low on boot up, and GPIO2 is high, and GPIO15 is low then the ESP-01 would boot up in Flash Mode or UART programming mode.**

The Logic Level Shifter

The Arduino microcontroller has a logic level of 5 V for a true value. The ESP-01 has a 3.3-V level for true. Recall from Chapter 1 that the Arduino reads a pin a high or true value when it is 3 V or greater. Thus, a true value of 3.3 V from the ESP-01 would also read as true for an Arduino input pin and the voltage would be within the pin's voltage tolerance level. However, although a 5-V true value from the Arduino to the ESP-01 would also register as a true value, the 5-V value would be outside the ESP-01's input pin's voltage tolerance.

The official Espressif FAQ for 2016 on the ESP8266 states that the GPIO pins on the ESP8266 are not 5-V compatible and that using the pins with 5 V input may damage the chip in the long run. The FAQ recommends using a

Figure 2-5 The logic level converter or shifter.

logic level converter chip to convert 5-V levels to 3.3-V levels. The FAQ states that "while many applications may get away by using a resistor voltage divider or series resistor, we highly recommend using a proper logic level converter chip to interface with 5V logic." It further states that "not doing so may lead to damage to the ESP8266 in the long run." A dedicated logic level converter would be something like that is shown in Figure 2-5.

The logic level converter shown in Figure 2-5 provides for 8 channel bidirectional voltage translations between two different voltage levels automatically. The voltage levels can be between 1.8 and 6 V. For example, the A side of the converter is connected to the device with 3.3 V as a logic level. The VCCA pin would be connected to 3.3 V. The B side of the converter is connected to the device that has 5 V as a logic level. The VCCB pin would be connected to 5 V. The two GND pins on each side of the converter are connected to a common ground. The pins on the A side which are A0 through A7 would be connected to pins on a device that uses 3.3-V logic. The pins on the B side which are B0 through B7 would be connected to pins on a device that uses 5-V logic. The transmit or Tx pin on the Arduino which uses 5-V logic could be connected to B0 and this signal would be translated to 3.3-V logic and output on A0 which would be connected to the receive or Rx pin on the ESP-01 that uses 3.3-V logic.

The Step-Down Voltage Regulator

The ESP-01 uses 3.3 V to power the module. We will be using this module with the Arduino Mega 2560. The Arduino will be the microcontroller and we will attach other electronic components such as sensors to the Arduino and use the ESP-01 for Wi-Fi communications only. Recall that in Chapter 1 we gave the specifications for the 3.3-V power pin on the Arduino Mega 2560 as providing up to 50 mA of current. However, the ESP-01 normally draws more than 50 mA of current in receiving and sending data over Wi-Fi. One of the typical cases presented in the documentation when sending data over Wi-Fi uses 170 mA of current. This amount is way more current than the 3.3-V power pin on the Arduino can provide. The solution to this problem is to draw current from the 5-V power pin. However, we will need to use a step-down voltage regulator to reduce the voltage of the 5-V power pin to 3.3 V before being applied to the ESP-01 module. When drawing power from the USB port of a device the Arduino can draw up to 500 mA of power before the built-in fuse shuts off the current. The maximum total current

Figure 2-6 The step-down voltage regulator.

that can be drawn by the Arduino Mega 2560 is 800 mA which should be more than enough for the ESP-01 and any electronics component that we will need to attach to the microcontroller. In fact 500 mA should be more than enough current for our needs. The step-down voltage regulator is shown in Figure 2-6.

The 5-V power pin on the Arduino is connected to Vin on the voltage regulator. The Out pin provides the 3.3-V power source with up to 800 mA of current. The GND pin is connected to the common ground of the circuit. The step-down voltage regulator that I used is based on the AMS1117 chip made by Advanced Monolithic Systems. See Figure 2-7.

I purchased this module as well as the logic level converter on Amazon. Ebay is also a good source of these components as well.

Figure 2-7 The front of the step-down voltage regulator.

The ESP-01 `AT` Command Set

The Arduino will communicate with the ESP-01 module using one of the Arduino Mega 2560's hardware serial ports. In order to actually set up and configure the ESP-01 module you will be sending it `AT` commands. An `AT` command is of the general form "`AT+<rest of command>`." Each command begins with an `AT` and then a + or plus sign. The company that makes the ESP-01 appears to regularly update the `AT` command set. Espressif, which is the company that makes the ESP8266, has a pdf file that describes the `AT` instruction set for the ESP8266 entitled "ESP8266 AT Instruction Set" available for download on their web site.

Important Note: **Sometimes `AT` commands are removed and others are added from version to version of the firmware. If you are having trouble getting an `AT` command to work check the version of the SDK firmware on the ESP-01 module to make sure that it supports the command and the exact options you are trying to use.**

Overview of Essential AT Commands

This sections give a brief summary of the some of the essential AT commands that a beginner to the ESP-01 should familiarize himself with.

- **AT** – Tests AT startup.
- **AT+RST** – Restarts module.
- **AT+GMR** – Views version info.
- **AT+CWMODE_CUR** – Sets the Wi-Fi mode (Station/AP/Station+AP) and the configuration is not saved in flash memory.
- **AT+CWSAP_CUR** – Sets the current configuration of the ESP8266 SoftAP and the configuration is not saved in flash memory.
- **AT+CWJAP_CUR** – Connects to an AP and the configuration is not saved in flash memory.
- **AT+CWQAP** – Disconnects from AP.
- **AT+CIPSTART** – Establishes a TCP connection from a client to a server.
- **AT+CIPSTO** – Sets the timeout value when the ESP8266 runs as a TCP server.
- **AT+CWLAP** – Lists available APs or access points.
- **AT+CIPSTATUS** – Gets the connection status.
- **AT+CIFSR** – Gets the local IP address.
- **AT+CIPMUX** – Configures the multiple connections mode.
- **AT+CIPSERVER** – Deletes or creates a TCP server.
- **AT+CIPSEND** – Sends data over a TCP connection.
- **AT+CIPCLOSE** – Closes the TCP connection.

General AT Command Formats

Many AT commands can have different forms with each form serving a different purpose. These forms are called test, query, set, and execute.

The Test AT Command Format

The test command form queries the command or internal parameters and returns the types and ranges of its values.

The general format is:

AT+<x>=?

A more specific example would be AT+CWMODE_CUR=? which retrieves the valid range of input for changing the Wi-Fi mode of the ESP-01 Wi-Fi chip.

The Query AT Command Format

The query command form returns the current value of the command.

The general format is:

AT+<x>?

An example would be AT+CWMODE_CUR? which requests the current Wi-Fi mode that the ESP-01 module is in.

The Set AT Command Format

The set AT command format sets the value of the user-defined parameters associated with a particular AT command.

The general format is:

AT+<x>=<...>

An example would be AT+CWMODE_CUR=2 which sets the Wi-Fi mode of the ESP-01 to an access point mode.

The Execute AT Command Format

The execute AT command format executes a command with no user-defined parameters.

The general format is:

AT+<x>

An example would be AT+GMR which gets the version information for the ESP-01 module.

Other AT Command Format Requirements

The AT command is case sensitive and the "AT" portion must be uppercase. The AT command string must also end in a "\r\n" which is a line wrap followed by a new line special control character.

Default Baud Rate

The default baud rate for serial communication with the ESP-01 is currently set by Espressif to 115,200 baud. However, you should check your specific ESP-01 module specifications to determine the default baud rate.

Quick Start Guide to the AT Command Set

This section will give more details on how to use some of the key AT commands with the ESP-01.

AT

This is a test function that returns an "OK" string value if the ESP-01 module successfully recognizes the "AT" string as the beginning of a valid command sequence.

AT+RST

This command resets the module and returns an "OK" string value if successful.

AT+GMR

This command gets the version information for the firmware on the ESP-01 module. The response from the module when I send my ESP-01 module this command is:

```
AT version:1.3.0.0(Jul 14 2016 18:54:01)
SDK version:2.0.0(656edbf)
compile time:Jul 19 2016 18:44:44
OK
```

AT+CWMODE_CUR

This command sets the Wi-Fi mode (Station/AP/Station+AP) and the configuration is not saved in the ESP-01's flash memory.

Use the query form of the command to find the current value:

AT+CWMODE_CUR?

The response should be of the form:

```
+CWMODE_CUR:<mode>
OK
```

Use the set form of the command in order to set the value for the current Wi-Fi mode:

AT+CWMODE_CUR=<mode>

The successful completion of the command should return an "OK" string.

The <mode> values in both the query and set forms of the AT command should be an integer number which corresponds to the following:

1. station mode
2. softAP mode
3. softAP + station mode

An example of setting the Wi-Fi mode to access point mode or softAP mode would be:

```
AT+CWMODE_CUR=2
```

AT+CWSAP_CUR

This command sets the current configuration of the ESP8266 SoftAP and the configuration is not saved in the flash memory of the ESP-01 module.

Get the current configuration by using the query form of the command:

AT+CWSAP_CUR?

The response from the ESP-01 should be of the form:

```
+CWSAP_CUR:<ssid>, <pwd>, <chl>, <ecn>,
<max conn>, <ssid hidden>
```

Set the configuration of the ESP8266's access point by using the set form of the command:

AT+CWSAP_CUR=<ssid>, <pwd>, <chl>, <ecn>[, <max conn>] [, <ssidhidden>]

The response should be the "OK" string if the parameters were changed or the "ERROR" string if one or more parameters were invalid.

The parameters for this `AT` command are:

- `<ssid>` The ESP8266 softAP's SSID of type string.
- `<pwd>` The ESP8266 softAP's password of type string. The range is between 8 and 64 characters or bytes ASCII.
- `<chl>` The channel ID.
- `<ecn>` Type of password protection if any.

  0: OPEN

  2: WPA_PSK

  3: WPA2_PSK

  4: WPA_WPA2_PSK
- `[<max conn>]` Optional parameter which is the maximum number of stations allowed to connect to this access point. The default is 4 and the range is from 1 through 4.

  `[<ssid hidden>]` Optional parameter which determines if the SSID will be broadcast. The default is set to broadcast the SSID.

  0: Broadcast the SSID of the ESP8266 soft-AP.

  1: Do not broadcast the SSID of the ESP8266 soft-AP.

An example of configuring the access point on the ESP-01 module is the following:

```
AT+CWSAP_CUR="ESP-01","esp8266esp01",1,2,4,0
```

AT+CWJAP_CUR

This command connects an ESP-01 module to an access point or AP. This default configuration is not saved in the ESP-01's flash memory.

Join an access point by using the following command:

```
AT+CWJAP_CUR=<ssid>, <pwd>[, <bssid>]
```

The response will be "OK" if successful or if there is an error then the following will be returned:

```
+CWJAP:<error code>
FAIL
```

The parameters of the command are:

- `<ssid>` The AP's SSID in string format.
- `<pwd>` The AP's password in string format with a maximum of 64 bytes of ASCII characters.
- `[<bssid>]` The AP's MAC address in string format. Several APs may have the same SSID.
- `<error code>` The error code which may not be reliable.
  1. The connection has timed out.
  2. The wrong password was received.
  3. The target AP could not be found.
  4. The connection failed.

Note: **This command requires station mode to be active. Escape character syntax is needed if "SSID" or "password" contains any special characters such as ',' or '"' or '\'.**

An example of this command that would join the access point `ESP8266-12E-AP` using the password `robtestESP12E` is the following:

```
AT+CWJAP_CUR="ESP8266-12E-AP","robtestESP12E"
```

AT+CWQAP

This command disconnects the ESP-01 module from an access point. A string value of "OK" is returned if the command was successful.

AT+CIPSTART

This command establishes a TCP connection, UDP transmission, or an SSL connection.

For a single connection where multiple connections to an ESP-01 module have been disabled by the `AT` command `AT+CIPMUX=0`, use the following AT command format to start a TCP connection:

```
AT+CIPSTART=<type>, <remote IP>, <remote port>[, <TCP keep alive>]
```

For an ESP-01 module that has multiple TCP connections available or active where AT+CIPMUX=1 then use the following AT command format to start a TCP connection:

```
AT+CIPSTART=<link ID>,<type>, <remote IP>, <remote port>[, <TCP keep alive>]
```

The response to this command will be a return string with "OK" if the command was successful or "ERROR" if the command has failed.

If TCP is already connected the module returns

```
"ALREADY CONNECT"
```

The parameters for this command are:

- <link ID> The ID of the network connection (0 through 4) that is used for multi-connections.
- <type> A string that is either "TCP" or "UDP" depending if the connection is a TCP connection or a UDP connection.
- <remote IP> A string that contains the remote IP address to connect to.
- <remote port> A string that contains the remote port number to connect to.
- [<TCP keep alive>] An optional parameter that contains the detection time interval when TCP is kept alive, this function is disabled by default.

  0: Disable TCP keep-alive.

  1 through 7200: The detection time interval in seconds.

An example of using the command to create a TCP connection with IP address 192.168.4.1 on port 80 is the following:

```
AT+CIPSTART="TCP","192.168.4.1",80
```

AT+CIPSTO

This command sets the timeout when the ESP8266 runs as a TCP server.

To get the current timeout value in seconds of the ESP8266 use the following command:

```
AT+CIPSTO?
```

The response to this command if successful will be the following:

```
+ CIPSTO:<time>
OK
```

In order to set the timeout value use the following command:

```
AT+CIPSTO=<time>
```

The response to this command is the "OK" string if successful.

The parameters for this command are as follows:

- <time> The TCP server timeout which ranges from 0 to 7200 seconds.

Note that if the ESP8266 is configured as TCP server then it will disconnect a connected TCP client if that TCP client has not communicated with the server for <time> seconds.

If the <time> parameter is set to 0 then the connection will never time out.

AT+CWLAP

The following command lists available access points:

```
AT+CWLAP
```

The response to this command if it succeeds is:

```
+CWLAP:<ecn>, <ssid>, <rssi>, <mac>,
<ch>, <freq offset>, <freq calibration>
OK
```

The parameters for this command are:

- <ecn> 0: Access point is open with no security.

  1: Access point has WEP security.

  2: Access point has WPA_PSK security.

3: Access point has WPA2_PSK security.

4: Access point has WPA_WPA2_PSK security.

5: Access point has WPA2_Enterprise security. (The ESP8266 can NOT connect to a WPA2_Enterprise AP.)

- `<ssid>` This is the SSID of the access point in string format.
- `<rssi>` The Wi-Fi signal strength of the access point.
- `<mac>` The MAC address of the access point in string format.
- `<freq offset>` The frequency offset of the access point in kHz. The ppm value is equal to the <freq offset> / 2.4.
- `<freq calibration>` This is the calibration for the frequency offset.

AT+CIPSTATUS

The following command gets the network connection status of the ESP-01 module:

AT+CIPSTATUS

The return values should be of the following format:

```
STATUS:<stat>
+CIPSTATUS:<link ID>, <type>, <remote_
IP>, <remote_port>, <local_port>,
<tetype>
```

The return parameters are as follows:

- `<stat>` The status of ESP8266 station interface.

2: The ESP8266 station is connected to an AP and has obtained an IP address.

3: The ESP8266 station has created a TCP or UDP transmission connection.

4: The TCP or UDP transmission of the ESP8266 station has been disconnected.

5: The ESP8266 station did NOT connect to an AP.

- `<link ID>` The ID of the connection (from 0 through 4) for a multi-connect configuration.
- `<type>` The string value “TCP” or “UDP.”
- `<remote_IP>` The remote IP address in string format.
- `<remote_port>` The remote port number.
- `<local_port>` The ESP8266 local port number.
- `<tetype>`

0: The ESP8266 runs as a client.

1: The ESP8266 runs as a server.

AT+CIFSR

The following command gets the local IP address:

AT+CIFSR

The response to this command is of the following format:

```
+ CIFSR:<IP address>
OK
```

The return parameters of this command are:

- `<IP address>` The IP address of the ESP8266’s softAP or

  The IP address of the ESP8266 station depending on the mode.

You should note that the ESP8266 station needs to be connected to an AP before the IP address can be retrieved.

AT+CIPMUX

This command configures the multiple connections mode of the ESP-01 module.

The following query command retrieves the current active mode:

AT+CIPMUX?

The response is of the form:

```
+CIPMUX:<mode>
OK
```

The return parameters are:

- `<mode>`

  0: The current connection is a single connection.

  1: The current connection is a multiple connection.

The command to set the multiple connection status is the following:

AT+CIPMUX=<mode>

The response to this command is either "OK" if the command succeeds or if the ESP-01 is already connected then it returns a "Link is builded" string message.

The return parameters are the same as for the query version of the command. The command "`AT+CIPMUX=1`" can only be set when transparent transmission is disabled which would mean that "`AT+CIPMODE=0`." This mode can only be changed after all the connections to the ESP-01 module are disconnected. If a TCP server is running on the ESP-01 module then it must be deleted before a single connection mode can be activated.

To set the ESP-01 module for single connection mode execute the following command which turns off the multi-connection mode:

```
AT+CIPMUX=0
```

AT+CIPSERVER

This command deletes or creates a TCP server and its format is as follows:

AT+CIPSERVER=<mode>[, <port>]

The response to this command if successful is the "OK" string.

The parameters for this command are:

- `<mode>`

  0: Deletes a server that currently exists.

  1: Creates a new server.

- `<port>` The port number which defaults to 333.

A server can only be created when multiple connections on the ESP-01 module are allowed. This means that `AT+CIPMUX=1`.

AT+CIPSEND

This command sends data over a TCP connection.

For a single connection which means that multiple connections are disabled `(+CIPMUX=0)` you would use the following command to send data over a TCP connection:

AT+CIPSEND=<length>

For multiple connections which means that multiple connections have previously been turned on `(+CIPMUX=1)` you would use the following command:

AT+CIPSEND=<link ID>, <length>

The response to this command if successful would first be a "`>`" character which would indicate that the user should start entering the data. Once the data length has been met then the ESP-01 module starts to send the data over the TCP connection. If the data fails to transfer, such as if the connection cannot be established or gets disconnected during data transfer, the module returns an "ERROR" string message. If the data is transmitted successfully the module returns a "SEND OK" string.

The parameters for the command are:

- `<link ID>` The ID of the connection from 0 through 4 if multiple connections are enabled.
- `<length>` The data length which is a maximum of 2048 bytes.

An example of sending data of length 7 over a single TCP connection is the following:

```
AT+CIPSEND=7
```

AT+CIPCLOSE

This command closes the TCP connection.

For multiple connections use the command format:

AT+CIPCLOSE=<link ID>

The response will be an "OK" string if the command succeeds or if there is no such connection an "ERROR" string will be returned.

The parameters are as follows:

- `<link ID>` The ID number of the connection to close. When ID=5 all connections will be closed. However, ID=5 has no effect when the ESP-01 module is in server mode.

For a single connection use the following format:

AT+CIPCLOSE

The response if the command is successful is the "OK" string or if no such connection is active will return an "ERROR" string.

Hands-on Example: Using an ESP-01 with an Arduino Mega 2560

In this hands-on example I will show you how to use an ESP-01 with an Arduino Mega 2560 and how to communicate in a client/server setup with another Arduino Mega 2560 equipped with an ESP-01 module. You will enter commands directly in the Arduino's Serial Monitor to control the ESP-01 module.

Parts List

- 2 Arduino Mega 2560 microcontrollers
- 2 5- to 3.3-V step-down voltage regulators
- 2 Logic level converters from 5 to 3.3 V
- 2 ESP-01 Wi-Fi modules
- 2 Arduino development stations such as a desktop and a notebook
- 2 Breadboards
- Wires both male-to-male and female-to-male to connect the components

Setting Up the Hardware

Step-Down Voltage Regulator Connections

1. Connect the Vin pin on the voltage regulator to the 5-V power pin on the Arduino Mega 2560.
2. Connect the Vout pin on the voltage regulator to a node on your breadboard represented by a horizontal row of empty slots. This is the 3.3-V power node that will supply the 3.3-V power to the ESP-01 module as well as provide the reference voltage for the logic level shifter.
3. Connect the GND pin on the voltage regulator to the GND pin on the Arduino or form a GND node similar to the 3.3-V power node in the previous step.

ESP-01 Wi-Fi Module Connections

1. Connect the Tx or TxD pin to the Rx3 pin on the Arduino Mega 2560. This is the receive pin for the serial hardware port 3.
2. Connect the EN or CH_PD pin to the 3.3-V power node from the step-down voltage regulator.
3. Connect the IO16 or RST pin to the 3.3-V power node from the step-down voltage regulator.
4. Connect the 3V3 or VCC pin to the 3.3-V power node from the step-down voltage regulator.

5. Connect the GND pin to the ground pin or node.
6. Connect the Rx or RxD pin to pin B0 on the 3.3-V side of the logic level shifter. The voltage from this pin will be shifted from 5 V from the Arduino to 3.3 V for the ESP-01 module.

Logic Level Shifter

1. Connect the 3.3-V pin to the 3.3-V power node from the step-down voltage regulator.
2. Connect the GND pin on the 3.3-V side to the ground node for the Arduino.
3. Connect the B0 pin on the 3.3-V side to Rx or RxD pin on the ESP-01. This corresponds to the A0 pin on the 5-V side.
4. Connect the 5-V pin to the 5-V power pin on the Arduino.
5. Connect the GND pin on the 5-V side to the ground node for the Arduino.
6. Connect the A0 pin on the 5-V side to Tx3 pin on the Arduino. This pin corresponds to the B0 pin on the 3.3-V side.

Now all the required components for this hands-on example should be connected. See Figure 2-8. Repeat the previous connection steps for another Arduino. One Arduino will be the client and the other will be the server in this example.

Setting Up the Software

This section discusses the software that you will need to load onto the Arduino for this hands-on

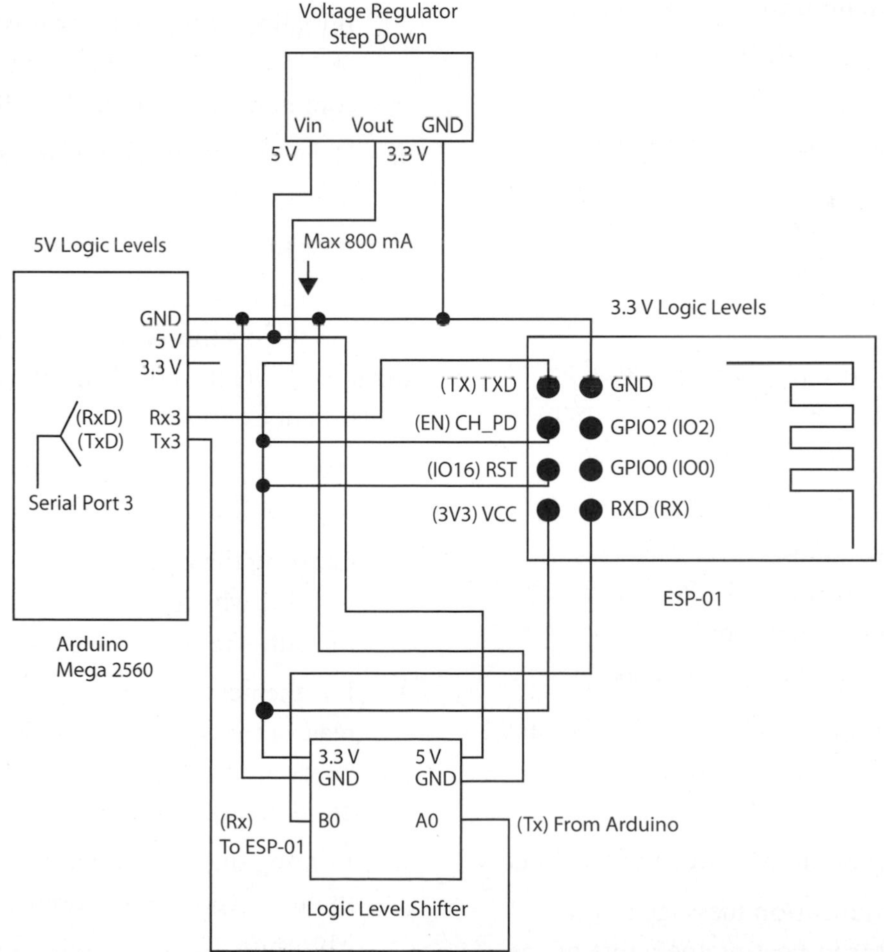

Figure 2-8 Using an ESP-01, step-down voltage regulator, and logic level shifter with an Arduino Mega 2560.

example. You can also download this program from my publisher's web site which currently is McGraw-Hill at https://www.mhprofessional.com/.

You will need to search for the book title and navigate to the downloadable content area.

This program accepts `AT` commands directly as user input through the Arduino's Serial Monitor. These commands are sent through the Arduino's hardware serial port directly to the ESP-01 module. The module will then process the commands and return values as appropriate.

The program first declares the global program variables:

The `BUFFERLENGTH` variable is declared as a constant integer and is set to 255 which indicates the length of the temporary character buffer that is used to read in the user input.

```
const int BUFFERLENGTH = 255;
```

The IncomingByte variable is a character array of length `BUFFERLENGTH` that is the buffer that temporarily holds incoming characters send by the user to the Arduino.

```
char IncomingByte[BUFFERLENGTH];
```

The `RawCommandLine String` variable holds the final command that is input by the user and that will be sent to the ESP-01.

```
String RawCommandLine = "";
```

The `ESP01ResponseString` variable holds the response from the ESP-01 to the `AT` commands entered by the user.

```
String ESP01ResponseString = "";
```

The `setup()` function initializes the Serial Monitor and ESP-01 module by doing the following:

1. Initializes the Serial Monitor to 9600 baud.
2. Prints an initialization message to the Serial Monitor indicating the name of the program.
3. Initializes the Serial 3 port on the Arduino to 115,200. This is the port that will communicate with the ESP-01 module.
4. Delays program execution for 1000 milliseconds or 1 second.

See Listing 2-1.

Listing 2-1 The `setup()` Function

```
void setup()
{
  // Initialize Serial
  Serial.begin(9600);
  Serial.println(F("****** ESP-01
   Arduino Mega 2560 Serial Monitor AT
   Command Program ********"));
  Serial.println();
  Serial.println();
  Serial.println();
  Serial.println();

  // Initialize ESP-01 on Serial Port
   3 on Arduino Mega 2560
  Serial3.begin(115200);
  delay(1000);
}
```

The `PrintESP01Response()` function reads in the responses to the `AT` commands that the user has sent to the ESP-01 module. The function does this by:

1. Assigning the read interval value to 4000 milliseconds or 4 seconds.
2. Clearing the global `ESP01ResponseString` variable which holds the incoming data to the null string.
3. For the next `DelayTime` milliseconds read in the data from the Serial 3 hardware port and add the results to `ESP01ResponseString`.
4. Printing out the text value held in `ESP01ResponseString` to the Serial Monitor.

See Listing 2-2.

Listing 2-2 The `PrintESP01Response()` Function

```
void PrintESP01Response()
{
 char Out;
 unsigned long DelayTimeCounter = 0;
 unsigned long DelayTimeStart = 0;
 int DelayTime = 4000; // in
  milliseconds

 ESP01ResponseString = "";

 DelayTimeStart = millis();
 while(DelayTimeCounter < DelayTime)
 {
  if (Serial3.available()>0)
  {
   Out = (char)Serial3.read();
   ESP01ResponseString += Out;
  }
  DelayTimeCounter = millis() -
   DelayTimeStart;
 }
 Serial.println(ESP01ResponseString);
}
```

The `PrintESP01Notifications()` function prints to the Serial Monitor any data received from the ESP-01. This function is useful in receiving any incoming notifications or error messages that are unexpected. See Listing 2-3.

The `loop()` function holds the main code and is executed repeatedly in an infinite loop. It does the following:

1. A command prompt is displayed when the program is ready to accept new user input.
2. Does the following until a break command is encountered:
    1. If there is data to be read in from the user then do the following:
        1. Read in the data until a newline character is found or a timeout occurs.
        2. Set the `RawCommandLine String` variable to the user input.
        3. Execute a break command to exit the infinite while loop.
    2. Calls the `PrintESP01Notifications()` function to print to the Serial Monitor any notifications that have come in from the ESP-01 module.
3. Displays the user's `AT` command by printing out the text value in the `RawCommandLine String` object to the Serial Monitor.
4. Adds a return character and a newline character to the user's `AT` command and then sends it to the ESP-01 module.
5. Prints out the ESP-01 module's response to the user's command by calling the `PrintESP01Response()` function.
6. Resets the `RawCommandLine` variable by setting it to the null string which is `""`.

Listing 2-3 The `PrintESP01Notifications()` Function

```
void PrintESP01Notifications()
{
 char Out;
 String Notification = "";

 if (Serial3.available()>0)
 {
  Out = (char)Serial3.read();
  Notification += Out;
 }

 Serial.print(Notification);
}
```

See Listing 2-4.

Running the Program

You will need to set up two systems as shown in Figure 2-8 and upload the program discussed

Listing 2-4 The `loop()` Function

```
void loop()
{
 // Wait for Command from Serial Monitor
 Serial.println(F("Ready to Accept new Command => "));
 while (1)
 {
  if (Serial.available() > 0)
  {
   int NumberCharsRead = Serial.readBytesUntil('\n', IncomingByte, BUFFERLENGTH);
   for (int i = 0; i < NumberCharsRead; i++)
   {
         RawCommandLine += IncomingByte[i];
   }
   break;
  }

  // Check to see if the ESP01 has sent any notifications such as when
  // a TCP client has connected to a server that is being run on the ESP01.
  PrintESP01Notifications();
 }

 // Print out the command to the Serial Monitor
 Serial.println();
 Serial.print(F("Raw Command from Serial Monitor: "));
 Serial.println(RawCommandLine);

 // Send command or data to ESP-01
 Serial3.print(RawCommandLine + "\r\n");

 // Read in response from ESP-01
 Serial.print("ESP-01 Response: ");
 PrintESP01Response();

 // Reset Command Line
 RawCommandLine = "";

 Serial.println();
 Serial.println();
}
```

in the previous section to both Arduinos. If you downloaded the program from the McGraw-Hill's web site then the name of the program to use is called "ESP8266MegaCommandLineTest.ino." One of these Arduino's will serve as a server while the other will serve as a client.

Set Up the Server

On the Arduino you want to use as the server type `AT` in the Serial Monitor and hit the return key or press the send key. Make sure the Serial Monitor's baud rate is 9600 and the "No Line

Ending" option is selected. You should see the following response:

```
Raw Command from Serial Monitor: AT
ESP-01 Response: AT

OK
```

Next, get the version information for your ESP-01 module by typing in `AT+GMR` and hitting the return key. You should see the following response:

```
Raw Command from Serial Monitor:
AT+GMR
ESP-01 Response: AT+GMR

AT version:1.3.0.0(Jul 14 2016
18:54:01)
SDK version:2.0.0(656edbf)
compile time:Jul 19 2016 18:44:44
OK
```

Note that your `SDK` and `AT` versions may be different from the above.

Next, get the current Wi-Fi mode by typing in `AT+CWMODE_CUR?` and pressing send or hitting the return key. You should see a response somewhat similar to the following depending on the exact state of your module:

```
Raw Command from Serial Monitor:
AT+CWMODE_CUR?
ESP-01 Response: AT+CWMODE_CUR?

+CWMODE_CUR:1

OK
```

My module is currently in the station mode so this needs to be changed to the access point mode which is 2. So type in `AT+CWMODE_CUR=2` and press the return key. You should see the following response:

```
Raw Command from Serial Monitor:
AT+CWMODE_CUR=2
ESP-01 Response: AT+CWMODE_CUR=2

OK
```

In order to run a server you will need to enable multiple connections. Let's check to see the current status of this by typing in `AT+CIPMUX?` into the Serial Monitor and pressing return. You should receive something like:

```
Raw Command from Serial Monitor:
AT+CIPMUX?
ESP-01 Response: AT+CIPMUX?

+CIPMUX:0

OK
```

On my module multiple connections are disabled. Thus, I will need to turn it on by typing `AT+CIPMUX=1` and pressing return. The response should be:

```
Raw Command from Serial Monitor:
AT+CIPMUX=1

ESP-01 Response: AT+CIPMUX=1

OK
```

Now, let's find out the current access point information such as SSID and password by typing in `AT+CWSAP_CUR?` and hitting the send key. You should see something like the following response:

```
Raw Command from Serial Monitor:
AT+CWSAP_CUR?
ESP-01 Response: AT+CWSAP_CUR?

+CWSAP_CUR:"ESP_A56A0F","",1,0,4,0


OK
```

The SSID of this access point is ***ESP_A56A0F*** and the password is the empty string because the access point is open. Now, let's get the IP address of the access point by typing in `AT+CIFSR` and pressing send. You should see the following response:

```
Raw Command from Serial Monitor:
AT+CIFSR
```

```
ESP-01 Response: AT+CIFSR

+CIFSR:APIP,"192.168.4.1"
+CIFSR:APMAC,"5e:cf:7f:a5:6a:0f"

OK
```

Note that the default IP address of the access point for an ESP8266 is "192.168.4.1" as stated in the documentation. I will assume this default address for the code in this book. If this default address was to change you might have to modify my code accordingly.

Next, create a server on port 80 by typing in the following `AT+CIPSERVER=1,80` and pressing return. The response should be:

```
Raw Command from Serial Monitor:
AT+CIPSERVER=1,80
ESP-01 Response: AT+CIPSERVER=1,80

OK
```

A server should now be running on the ESP-01 module.

Set Up the Client

On the other Arduino you will now set it up as a Client. First check to see if the ESP-01 is working correctly by sending it an `AT` command. The response should be:

```
Raw Command from Serial Monitor: AT
ESP-01 Response: AT

OK
```

Next, determine the version of the ESP-01 module you are using by executing the `AT+GMR` command. The result should be something similar to the following:

```
Raw Command from Serial Monitor:
AT+GMR
ESP-01 Response: AT+GMR

AT version:1.3.0.0(Jul 14 2016
18:54:01)
SDK version:2.0.0(656edbf)
compile time:Jul 19 2016 18:44:44
OK
```

Now, let's demonstrate how to get the valid parameters for the Wi-Fi mode by executing `AT+CWMODE_CUR=?` in the Serial Monitor. You should see the following response:

```
Raw Command from Serial Monitor:
AT+CWMODE_CUR=?
ESP-01 Response: AT+CWMODE_CUR=?

+CWMODE_CUR:(1-3)

OK
```

The valid values for the Wi-Fi mode are 1 through 3. Next, let's find the actual value of the Wi-Fi mode by executing `AT+CWMODE_CUR?` in the Serial Monitor. The response should be something like:

```
Raw Command from Serial Monitor:
AT+CWMODE_CUR?
ESP-01 Response: AT+CWMODE_CUR?

+CWMODE_CUR:1

OK
```

My ESP-01 module is set to 1 which is station mode which can be used for a client. This means that I don't need to change Wi-Fi modes. Next, let's check to see if the module has an IP address by executing the `AT+CIFSR` command. The response should be:

```
Raw Command from Serial Monitor:
AT+CIFSR
ESP-01 Response: AT+CIFSR

+CIFSR:STAIP,"0.0.0.0"
+CIFSR:STAMAC,"5c:cf:7f:a5:6a:0f"

OK
```

Since this station is not connected to an access point the IP address should not be assigned yet or in this case is shown as all 0's. Next, we need to find a list of access points to connect to by executing the `AT+CWLAP` command. The results should be something like the following:

```
Raw Command from Serial Monitor:
AT+CWLAP
ESP-01 Response: AT+CWLAP

+CWLAP:(0,"ESP_A562E0",-56,
"5e:cf:7f:a5:62:e0",1,-29,0)
+CWLAP:(4,"E5CD27",-75,
"3c:04:61:e5:cd:27",1,-34,0)
+CWLAP:(3,"BE0D95",-84,
"b0:93:5b:be:0d:95",1,-37,0)
+CWLAP:(3,"07FX11035899",-36,
"00:12:0e:8a:d0:a3",6,-46,0)
+CWLAP:(4,"This House",-89,
"b0:93:5b:11:8f:fd",6,-34,0)
+CWLAP:(4,"037807",-89,
"c0:c5:22:03:78:07",6,-41,0)
+CWLAP:(4,"11CF72",-92,
"10:56:11:11:cf:72",1,-37,0)
+CWLAP:(3,"82ECF6",-73,"2c:30:33:82:
ec:f6",11,-47,0)

OK
```

Notice that the first entry is our access point that we created on the other Arduino which is "`ESP_A562E0`." Now, join this access point by executing the `AT+CWJAP_CUR="ESP_A562E0", ""` command. The following shows the result:

```
Raw Command from Serial Monitor:
AT+CWJAP_CUR="ESP_A562E0",""
ESP-01 Response: AT+CWJAP_CUR="ESP_
A562E0",""

WIFI CONNECTED
WIFI GOT IP

OK
```

Now, the station is connected to the Wi-Fi access point. Next, let's try to find out the station's new IP address by executing the `AT+CIFSR` command. The result is:

```
Raw Command from Serial Monitor:
AT+CIFSR
ESP-01 Response: AT+CIFSR
+CIFSR:STAIP,"192.168.4.2"
+CIFSR:STAMAC,"5c:cf:7f:a5:6a:0f"

OK
```

The station's IP address is now 192.168.4.2. Now let's connect to the TCP server by executing the command `AT+CIPSTART="TCP"," 192.168.4.1",80`. This command connects to the TCP server on port 80 running on the ESP-01 module that is running the access point.

```
Raw Command from Serial Monitor:
AT+CIPSTART="TCP","192.168.4.1",80
ESP-01 Response: AT+CIPSTART="T
CP","192.168.4.1",80

CONNECT

OK
```

The client/station is now connected to the TCP server running on the other ESP-01 module. On the server side there should be a notification that indicates that the station has just been connected such as:

```
0,CONNECT
```

Now, let's try to send some data to the server by executing the command `AT+CIPSEND=7`. The following should appear:

```
Raw Command from Serial Monitor:
AT+CIPSEND=7
ESP-01 Response: AT+CIPSEND=7

OK

>
```

The > symbol indicates that you should enter the text you want to send to the server. Since the program automatically adds in a return and a newline character the text you enter must be 5 characters long. Enter the word "hello" and press the Send button. The following should appear as a response:

```
Raw Command from Serial Monitor:
hello
```

```
ESP-01 Response:
Recv 7 bytes

SEND OK
```

Now, look on the Serial Monitor of the server and you should see:

```
+IPD,0,7:hello
```

This means that the server on connection 0 has received data of length 7 with text "hello\r\n." Remember that the program automatically adds in the "\r\n" for everything sent to the ESP-01 module. Since this is data and not a command we didn't need to add in the extra control characters but this proves that control characters can be easily sent using the ESP-01.

Next, on the server side we are going to send a message to the client by executing the command AT+CIPSEND=0,9.

```
Raw Command from Serial Monitor:
AT+CIPSEND=0,9
ESP-01 Response: AT+CIPSEND=0,9

OK
>
```

Enter the text "goodbye" and press the send key. The following should appear as a response:

```
Raw Command from Serial Monitor:
goodbye
ESP-01 Response:
Recv 9 bytes

SEND OK
```

This means that the client has received the message. On the Serial Monitor of the client Arduino you should see something like:

```
+IPD,9:goodbye
```

Thus, we are able to communicate from the client to the server as well as from the server to the client.

On the client side close the TCP connection by executing the command AT+CIPCLOSE.

```
Raw Command from Serial Monitor:
AT+CIPCLOSE
ESP-01 Response: AT+CIPCLOSE

CLOSED


OK
```

On the server side you should see a message that the TCP connection is now closed such as:

```
0,CLOSED
```

Now we need to close the Wi-Fi connection to the access point by executing the AT+CWQAP command.

```
Raw Command from Serial Monitor:
AT+CWQAP
ESP-01 Response: AT+CWQAP

OK
WIFI DISCONNECT
```

The client is now disconnected from the access point.

The NodeMCU

The NodeMCU is an open-source Internet of Things (IOT) device that uses the ESP-12E module based on the ESP8266 chip that we have covered earlier in this chapter. The official site is located at http://nodemcu.com/index_en.html.

This device is popular and can be found on online web sites like Amazon or ebay for under $10.00. The front view of the NodeMCU is shown in Figure 2-9.

The back view of the NodeMCU is shown in Figure 2-10.

Figure 2-9 The front view of the NodeMCU.

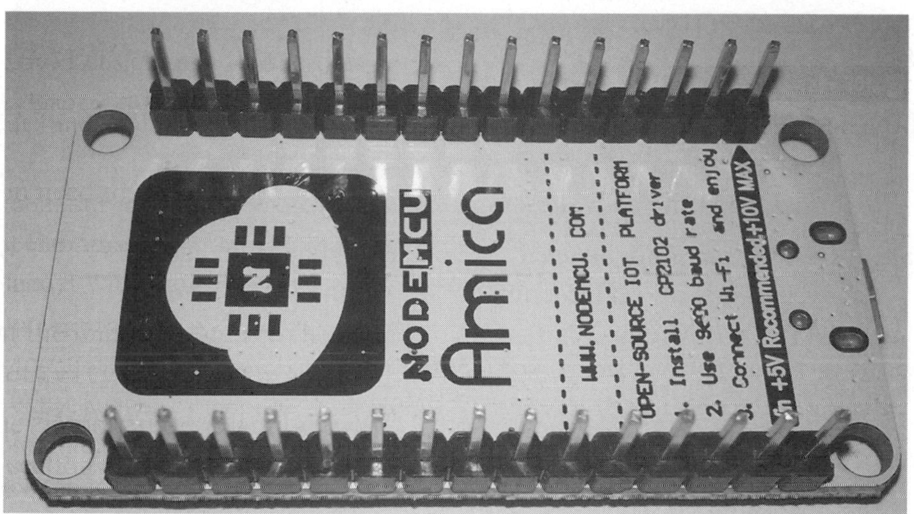

Figure 2-10 The back view of the NodeMCU.

General Summary and Specifications

The version of the NodeMCU shown in Figures 2-9 and 2-10 is the latest version which is 1.0 of this development kit. The associated web site for this version of the development kit is https://github.com/nodemcu/nodemcu-devkit-v1.0.

The NodeMCU development kit is a standalone device that does not need another microcontroller to operate. It has exposed pins for general input and output operations. It has a USB port which can be used to power the NodeMCU and debug the program running on the device. It also has a Vin pin where you can input voltage from a power source, such as a battery, in order to power the device.

Power Specifications

- Working output: 3.3 V, 800 mA
- Working current limit: 1000 mA

- Max current: 1000 mA
- Max supply voltage: 20 V
- Voltage dropout:1.2 V@800 mA

NodeMCU Boot Up Pin Information

On every boot or reset or wakeup:

- If GPIO15 is LOW and GPIO2 is HIGH andGPIO0 is HIGH then the NodeMCU enters normal operations where the user's program will be executed.
- If GPIO15 is LOW and GPIO2 is HIGH and GPIO0 is LOW then the NodeMCU enters FLASH MODE.

Information on Pin D0 (GPIO16)

- For sleep mode the D0 or GPIO16 pin and the RST pin should be connected together. The GPIO16 pin will output a LOW voltage to reset the system at the time of wakeup.
- The D0 or GPIO16 pin can only be used as a GPIO read/write pin with no interrupt supported and no PWM/I2C/One Wire supported.

Recommended Useable Pins for Output and Input on the NodeMCU

- Pins D1(GPIO5) and D2(GPIO4) are always available for use and are not used in other functions.
- Pins D5(GPIO14) and D6(GPIO12) are available for use if the HSPI interface is not being used.
- Pin D7(GPIO13) is available for use if the HSPI interface is not being used and the pin is not being used to receive data from a serial device.
- Pin A0 contains an analog to digital converter which converts a specific voltage to a number. You can use this to measure a range of voltages instead of just a HIGH or LOW voltage value.

Pins You Should NOT Use

Pins D9 and D10 should not be used because they are used when flashing a program to the NodeMCU's internal memory and are used by the Arduino's Serial Monitor program which is needed for debugging.

The Vin Pin

The Vin pin provides a 5-V output when connected to a USB port. Otherwise, this pin accepts a voltage input which will be converted to 3.3 V to power the NodeMCU.

A diagram of the pin specifications for the NodeMCU is shown in Figure 2-11.

Setting Up the Development System for the NodeMCU

To develop programs for the NodeMCU we will be using the Arduino IDE along with additional components called the ESP8266 Core for the Arduino IDE which is available for download at https://github.com/esp8266/Arduino.

The ESP8266 Arduino Core's documentation is located at https://arduino-esp8266.readthedocs.io/en/latest/.

The first thing you will need to do is to download and install the ESP8266 Arduino Core. Generally, the easiest way to install the ESP8266 Arduino Core is by using the Arduino IDE's Board Manager. However, if this fails you will have to install it manually.

ESP8266 Arduino Core Board Manager Installation

1. First download and install version 1.8 of the Arduino IDE or later.
2. Select File->Preferences in the menu to bring up the Preferences window. See Figure 2-12. In the "Additional Boards Manager URLs" section add in the

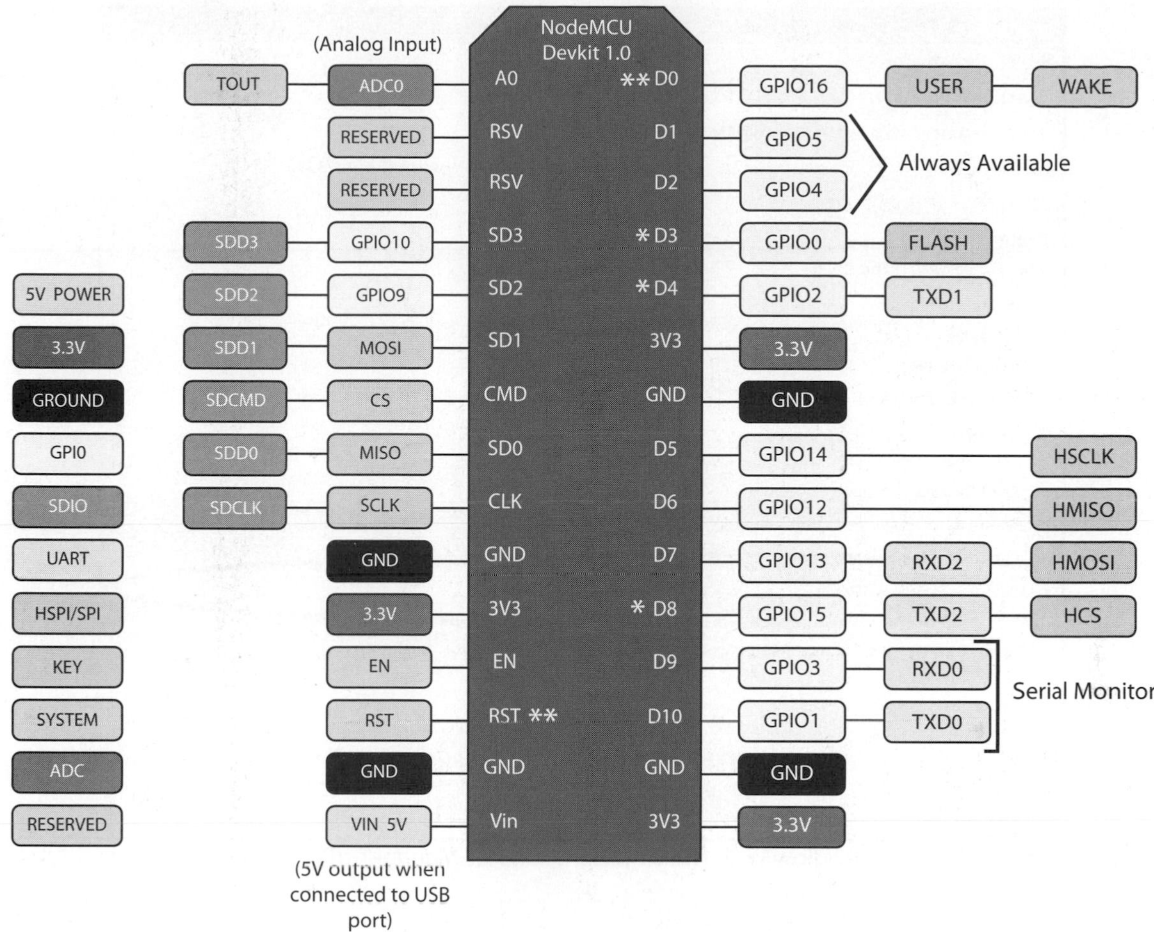

* On every boot/reset/wakeup,
GPIO15 = LOW, GPIO2 = HIGH, GPIO0 = HIGH --> RUN MODE
GPIO15 = LOW, GPIO2 = HIGH, GPIO0 = LOW --> FLASH MODE

** For Sleep Mode : GPIO16 and RST should be connected
GPIO16 will output LOW to reset the system at the time of wakeup.

DO(GPIO16) can only be used as GPIO read/write with no interrupt supported and no PWM / I2C / OW supported.

Figure 2-11 The NodeMCU version 1.0 pinout diagram.

following URL: http://arduino.esp8266.com/stable/package_esp8266com_index.json.

3. Under the Tools->Board menu item select the "Boards Manager" to bring up the Boards Manager.
4. Find the ESP8266 platform and install it.
5. Under the Tools->Board menu item select the "NodeMCU 1.0 (ESP-12E Module)" to select the NodeMCU as the target platform for development. See Figure 2-13.

ESP8266 Arduino Core Manual Installation

If the Boards Manager installation does not work correctly you may have to install the ESP8266 Arduino Core manually. On my Windows XP system the Board Manager installation procedure failed so I had to do a manual install.

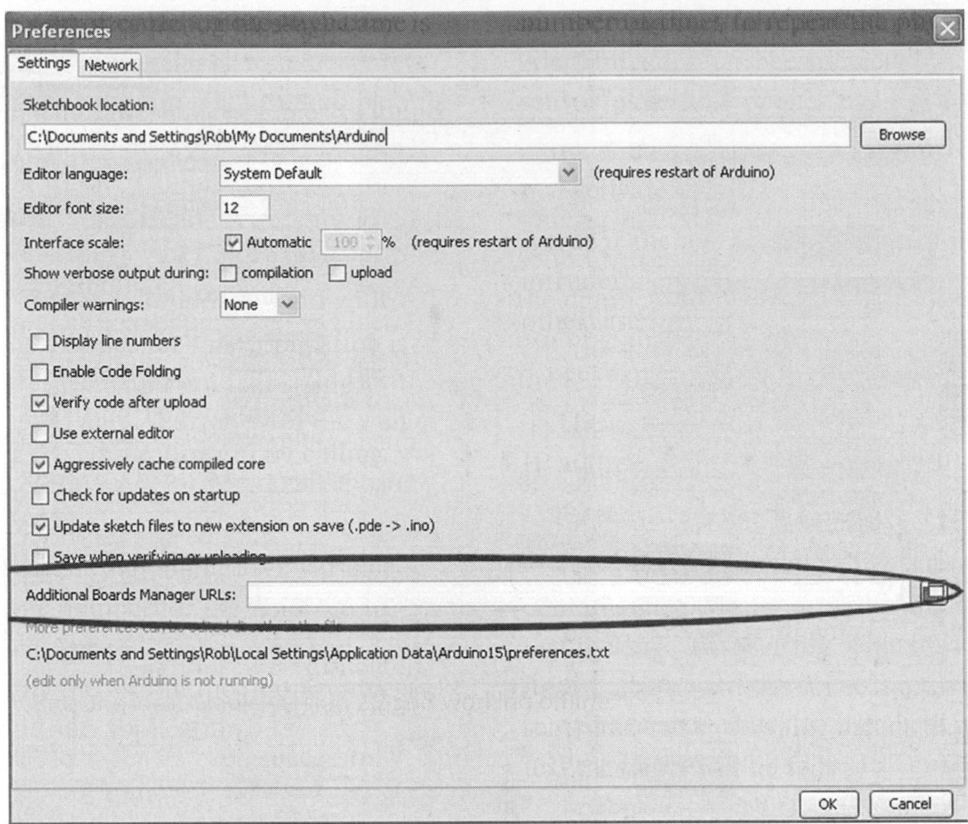

Figure 2-12 The Preferences window with the "Additional Boards Manager URLs" circled.

1. Install version 1.8 or later of the Arduino IDE.
2. Download the ESP8266 Arduino Core file from the github site.
3. Unzip the downloaded file to a temporary directory.
4. Go to the Arduino directory on your computer.
5. Create a hardware/esp8266com/esp8266 directory in the Arduino directory. The hardware directory should already exist. You will need to make the esp8266com and esp8266 directories.
6. Copy the contents from the ESP8266 Arduino Core file you downloaded to the esp8266 directory. See Figure 2-14.
7. Next, you will need to download the binary tools by changing your working directory to the esp8266/tools directory and running python (version 2.7 or later) on the "get.py" script file such as running "python get.py" from the MS-DOS command prompt. This will download the required binaries to your system. See Figure 2-15.
8. Close then restart your Arduino IDE program.
9. Under the Tools->Board menu item select the "NodeMCU 1.0 (ESP-12E Module)" to select the NodeMCU as the target platform for development. See Figure 2-13.

Hands-on Example: The Blink Sketch

In this hands-on example we load in an example sketch that comes with the Arduino IDE and walk you through the code. Load in

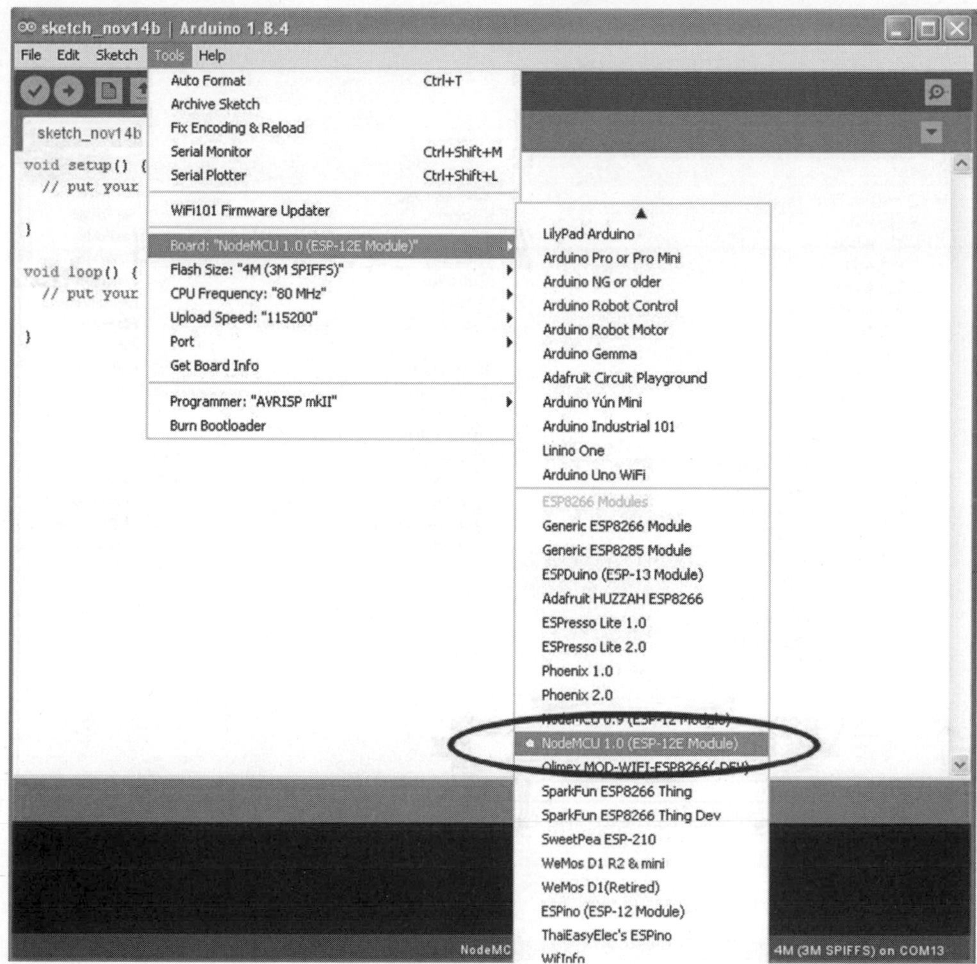

Figure 2-13 Selecting the NodeMCU for development.

the sketch by selecting the File->Examples->ESP8266->Blink menu item to bring up the Blink sketch in a new window. The Blink sketch blinks the LED that is on the NodeMCU.

The `setup()` function declares the pin connected to the built-in LED to be an OUTPUT pin so that the pin can be used to supply current to the electronic component that is attached to the pin. See Listing 2-5.

Listing 2-5 The Blink Example Sketch

```
void setup() {
 pinMode(LED_BUILTIN, OUTPUT);
 // Initialize the LED_BUILTIN pin
 as an output
}
```

The `loop()` function runs repeatedly in an infinite loop and turns on and turns off the LED on the NodeMCU by doing the following:

1. Turns on the LED by writing a LOW voltage level to the digital output pin that is attached to the built-in LED on the NodeMCU.
2. Halts execution of the program for 1000 milliseconds or 1 second.
3. Turns off the LED by writing a HIGH voltage level to the digital output pin that is attached to the built-in LED on the NodeMCU.
4. Halts the execution of the program for 2000 milliseconds or 2 seconds.

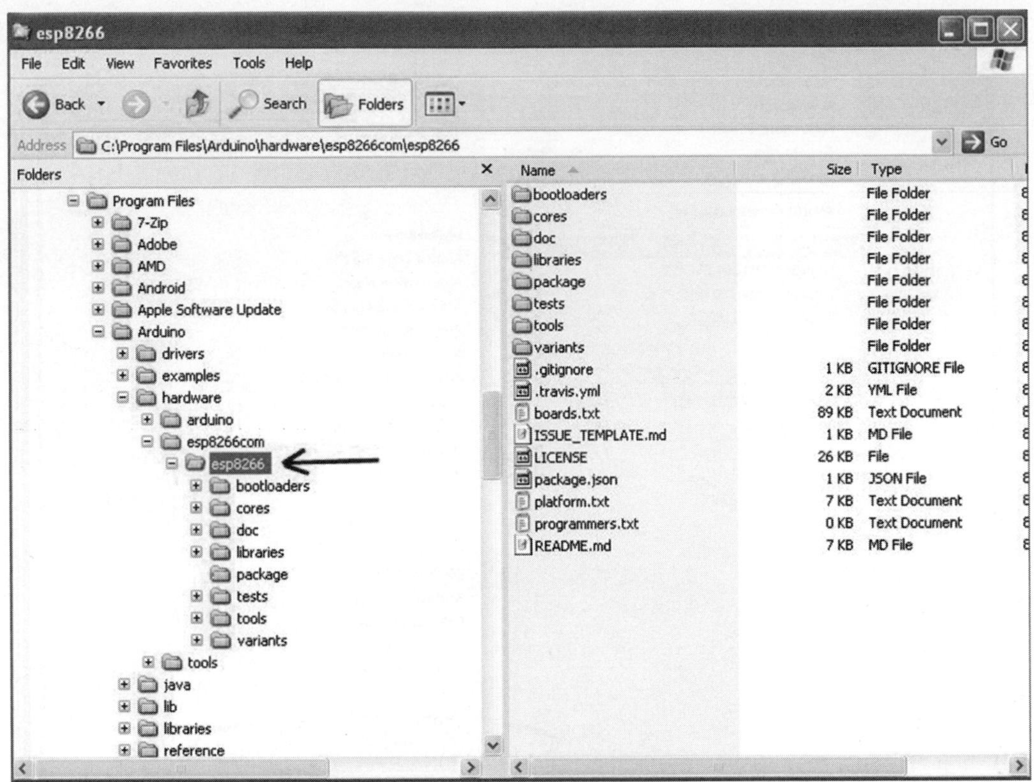

Figure 2-14 Copying the ESP8266 Core file to the esp8266 directory.

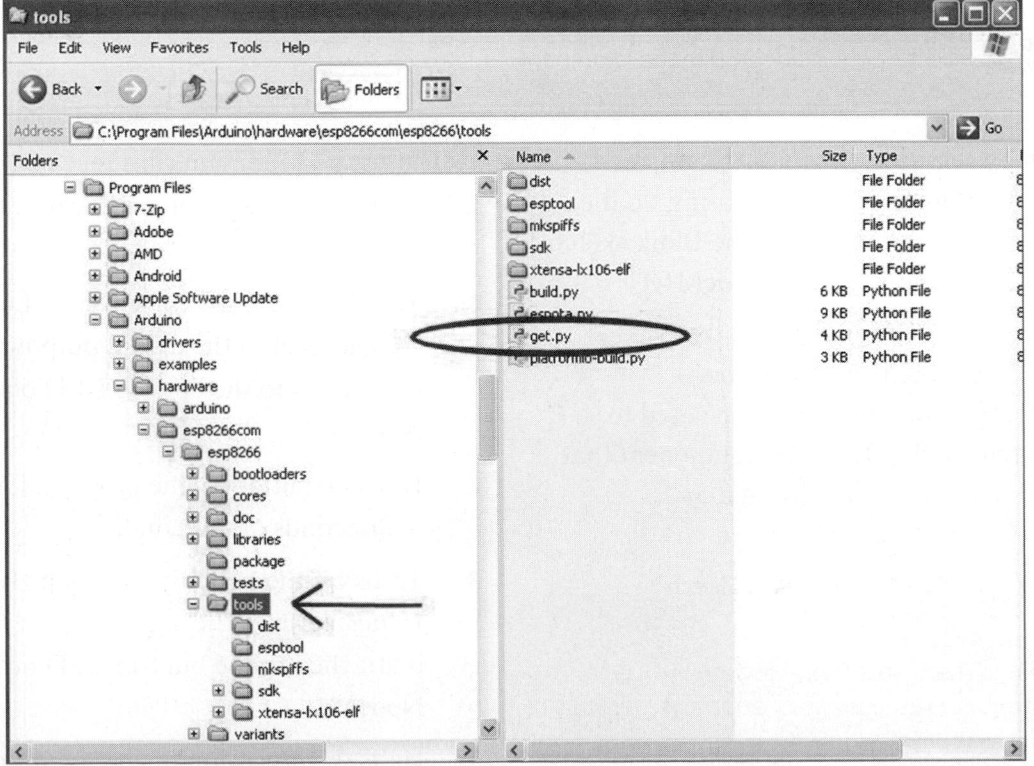

Figure 2-15 Downloading the binary tools to your system.

Note that the LED is active LOW which means that a LOW voltage levels turns the LED on and a HIGH voltage level turns the LED off. See Listing 2-6.

Connect your NodeMCU to your development system, make sure that the NodeMCU 1.0 board is selected as the target for compilation under the "Tools" menu and upload the program to your device by hitting the Upload button on the Arduino IDE. Once the program has been uploaded to the NodeMCU then an LED on the device should start to blink such as in Figure 2-16.

Hands-on Example: Multiple Blinking External LEDs

In this example we connect 5 LEDs to the NodeMCU and blink them in sequence from LED 1 to LED 5. The LED on the NodeMCU itself is also turned on and left on. This example could be adapted to a NodeMCU that is being used as an access point. LEDs 1 through 4 could be used to indicate the status of a Wi-Fi connection between the access point and a station with a lit LED indicating that there is a connection. The fifth LED could

Listing 2-6 The `loop()` Function

```
// the loop function runs over and over again forever
void loop() {
  digitalWrite(LED_BUILTIN, LOW); // Turn the LED on (Note that LOW is the voltage
  level
                    // but actually the LED is on; this is because
                    // it is active low on the ESP 01)
  delay(1000);                     // Wait for a second
  digitalWrite(LED_BUILTIN, HIGH);   // Turn the LED off by making the voltage HIGH
  delay(2000);                       // Wait for two seconds (to demonstrate the active
  low LED)
}
```

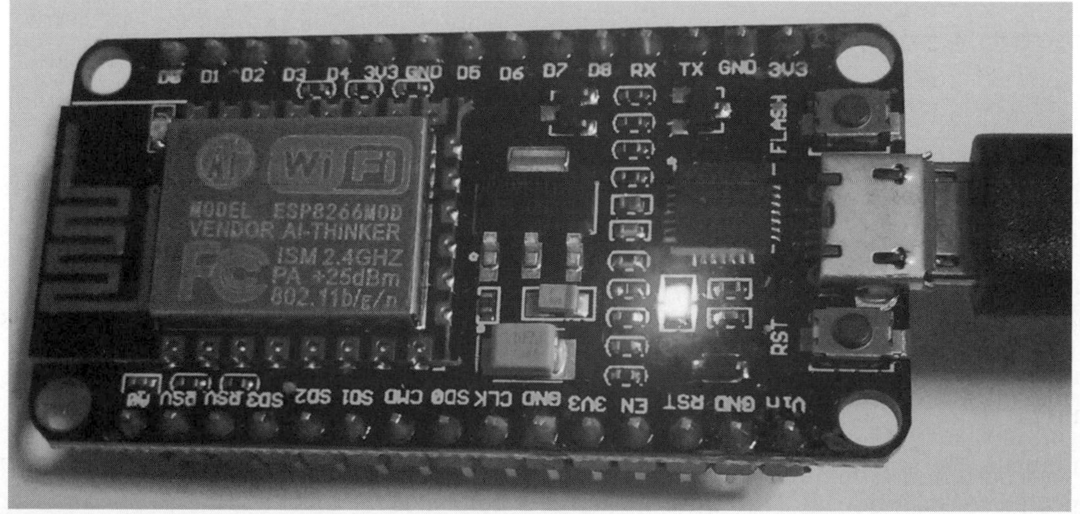

Figure 2-16 The Blink program being executed on a NodeMCU.

indicate if any errors have occurred. The fact that the built-in LED has been turned on could indicate that the access point is now fully started and ready to accept incoming connections from stations. This example also illustrates which GPIO pins are good for normal input/output operations.

Parts List

- 5 LEDs
- 1 NodeMCU
- Wires to connect the project together

Setting Up the Hardware

1. Connect the positive terminal of LED 1 to D1 and the negative terminal to ground.
2. Connect the positive terminal of LED 2 to D2 and the negative terminal to ground.
3. Connect the positive terminal of LED 3 to D5 and the negative terminal to ground.
4. Connect the positive terminal of LED 4 to D6 and the negative terminal to ground.
5. Connect the positive terminal of LED 5 to D7 and the negative terminal to ground.

See Figure 2-17.

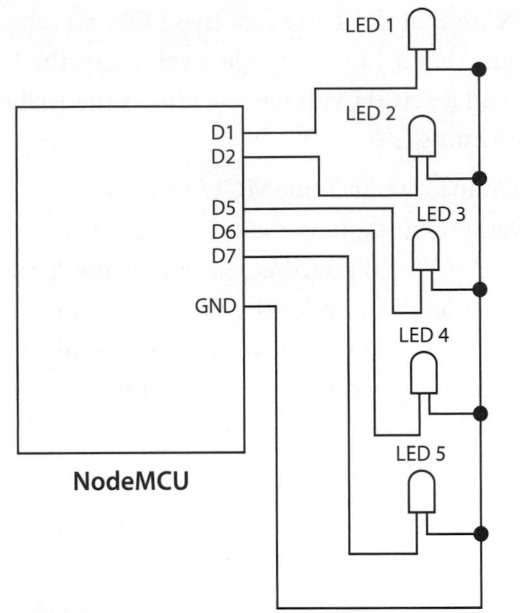

Figure 2-17 Multiple Blinking LEDs.

Setting Up the Software

The `setup()` function initializes the project by:

1. Initializing the serial monitor and setting the communication speed to 9600 baud.
2. Printing out a start-up message to the screen.
3. Configuring the built-in LED and then turning it on.
4. Configuring the external LEDs that are connected to the NodeMCU to outputs that will receive a voltage.

See Listing 2-7.

Listing 2-7 The `setup()` Function

```
void setup()
{
  Serial.begin(9600);
  Serial.println();
  Serial.println(F("**************
   Multiple External LED Blink Test
   *************************"));

  // Initialize LEDS
  pinMode(LED_BUILTIN, OUTPUT);      //
   Initialize the LED_BUILTIN pin as
   an output
  digitalWrite(LED_BUILTIN, LOW);    //
   turn on
    pinMode(D1, OUTPUT);
    pinMode(D2, OUTPUT);
    pinMode(D5, OUTPUT);
    pinMode(D6, OUTPUT);
    pinMode(D7, OUTPUT);
}
```

In the `loop()` function each of the 5 LEDs is turned on for 1 second then turned off in sequence for LED 1 through LED 5. The `loop()` function repeats indefinitely. See Listing 2-8.

Listing 2-8 The `loop()` Function

```
void loop()
{
  // LED 1
  digitalWrite(D1, HIGH);
  delay(1000);
  digitalWrite(D1, LOW);

  // LED 2
  digitalWrite(D2, HIGH);
  delay(1000);
  digitalWrite(D2, LOW);

  // LED 3
  digitalWrite(D5, HIGH);
  delay(1000);
  digitalWrite(D5, LOW);

  // LED 4
  digitalWrite(D6, HIGH);
  delay(1000);
  digitalWrite(D6, LOW);

  // LED 5
  digitalWrite(D7, HIGH);
  delay(1000);
  digitalWrite(D7, LOW);
}
```

Running the Program

Connect your NodeMCU to your development system and then press the "Upload" button on the Arduino IDE to transfer the program to the device. I actually found it easier to upload the program to the NodeMCU first, build the LEDs on a separate breadboard, and then connect the two together. After connecting the LEDs to the NodeMCU connect the NodeMCU to the USB port of your development system. You should see each light turn on for a second then turn off in sequence from LED 1 to LED 5 and then repeat continuously. See Figure 2-18.

Figure 2-18 The Blinking external LEDs.

Summary

In this chapter I covered the ESP8266 chip, the ESP-01, and the NodeMCU. The ESP-01 and NodeMCU both use the ESP8266 chip. I started with detailing the specification of the ESP8266 chip. Next, I covered the specifications of the ESP-01 module including the functions of each of the pins on the ESP-01. I then cover why you will need to use a logic level shifter and voltage regulator in projects that use the ESP-01 and Arduino Mega 2560. Next, I discuss the essential `AT` commands you will need to communicate with the ESP-01 over a serial communications port. A hands-on example follows which shows the reader how to use an ESP-01 with an Arduino 2560. The NodeMCU is discussed next. The specifications including pin functions are covered. I then tell the reader how to update their Arduino IDE so that they will be able to use it to develop programs for the NodeMCU. Finally, I cover some hands-on examples that use the updated Arduino IDE and the NodeMCU.

CHAPTER 3

Introduction to the Android

THIS CHAPTER COVERS THE ANDROID MOBILE operating system and related devices. I first start off discussing what Android actually is and I cover some basic concepts concerning Android cell phones. Next, I discuss the system requirements of the main program that is currently used to develop Android applications which is called Android Studio. I then give a quick start guide tour of the key features of Android Studio you will need to use in order to create your own programs. Next, I give the reader a basic overview of the Java language that is used to create Android applications. I cover key Java concepts such as data types, program flow control statements, and classes. Then, I cover the `Activity` class and the `Activity` class life cycle that are key to understanding the basic Android application. Next, I present a hands-on example which is an exercise on how to create a program using the Android Studio, modify this program, and run it on an Android device. Finally, I discuss some information on how to import an Android Studio project from an older version of Android Studio to a newer version.

What Is Android?

Android is a mobile operating system that is popular on cell phones and tablets. Android is based on the free operating system Linux and other open-source components. There are also other variations of this operating system such as Android TV for televisions, Android Auto for cars, and Wear OS for wrist watches. Other versions of the Android operating system have been developed for game consoles, digital cameras, PCs, and other electronic devices.

In terms of Android cell phone service there are two general types to choose from which are prepaid and postpaid. Prepaid service involves paying before using any cell phone minutes, text messages, or Internet data. In a prepaid service once all the credit allocated for minutes, text messages, and Internet data have been used up then the cell phone service either stops or is reduced. For example, if you run out of credit on a prepaid plan then phone and text message service may stop but Internet data may continue at a reduced speed depending on the policy of the specific carrier. For a prepaid plan there is no credit check or transmission of personal data such as your identity, address, social security number and no long-term or short-term contracts. Thus, it is a more secure and private way of obtaining cell phone service on demand only when you need it. Also, most prepaid plans offer prepaid cards which can be bought for cash at a local store so that even people without access to credit cards can get prepaid phone service. For a postpaid account, you pay each month for the amount of calling minutes, text messages, and Internet data you have used. This type of account generally requires a credit check, and perhaps requires a valid credit card to be used to fund the account.

A mobile network operator (MNO), which is also called a wireless service provider or wireless carrier, owns the required radio frequencies and cellular infrastructure in order to provide cell phone services to users. Some examples of these MNOs are Verizon, AT and T Mobility, T-Mobile, Sprint Corporation, and U.S. Cellular. A mobile virtual network operator (MVNO) buys cell phone services from an MNO that actually owns the wireless cell phone infrastructure in bulk and resells these cell phone services to consumers in their own custom plans and custom prices. Some examples of MVNOs are Boost Mobile, Cricket Wireless, Liberty Wireless, Mint Mobile, Patriot Mobile, Simple Mobile, Tello, and Ting. The biggest advantage of using these virtual network operators is that the cost of cell phone service can be dramatically lower and also much more flexible. For example, some MVNOs like Tello offer a no contract, prepaid, pay as you go plan where you can put money into your account and are only charged for the calling minutes, text messages, and data you actually use. This credit has no expiration date but you must use some credit after a certain period of time in order to keep your account open and active. For example, you can put $25 into your Tello pay as you go account and link your Android cell phone to it. You can use this Android cell phone in combination with an Arduino, ESP-01, and a motion sensor to provide an affordable burglar alarm system with no monthly fees. The only fees would be for such things like text messages that would notify you that the burglar alarm has been tripped. This would be charged on a per text message basis. Thus, this book can help you build a Do it Yourself (DIY) alternative to the high priced home security systems offered by commercial alarm companies.

The official android web site is located at https://www.android.com/.

Developing Programs for the Android Mobile Operating System

Programs for the Android platform are written in Java. The most current development tool for the Android is called Android Studio. The Android Studio program is available for the Windows, Mac, and Linux platforms.

The official android web site for developers is located at https://developer.android.com/.

The web page for downloading Android Studio is located at https://developer.android.com/studio/.

Android Studio Windows 7, 8, and 10 System Requirements

The current version of Android Studio runs on Windows 7, 8, and 10. The following are the system requirements:

- Microsoft Windows 7/8/10 (32- or 64-bit)
- 3 GB RAM minimum, 8 GB RAM recommended; plus 1 GB for the Android Emulator
- 2 GB of available disk space minimum, 4 GB recommended (500 MB for IDE + 1.5 GB for Android SDK and emulator system image)
- 1280 × 800 minimum screen resolution
- Java Development Kit (JDK) 8, however, the use of the bundled OpenJDK (version 2.2 and later) is recommended

Android Studio Windows XP System Requirements (Version 1.5 of Android Studio)

The current version of Android Studio does not support the Windows XP operating system. However, a previous version of Android Studio which is 1.5 supports Windows XP and is still available for download from Google. The following are the system requirements for version 1.5 of Android Studio:

- 3 GB RAM minimum, 4 GB RAM recommended
- 500 MB disk space, at least 1 GB for Android SDK, emulator system images, and caches

- JDK 7 or higher
- 1280 × 800 minimum screen resolution

The Win XP compatible Android Studio version 1.5 download link is http://dl.google.com/dl/android/studio/install/2.1.2.0/android-studio-bundle-143.2915827-windows.exe.

Android Studio Mac System Requirements

- Mac OS X 10.10 (Yosemite) or higher, up to 10.13 (macOS High Sierra)
- 3 GB RAM minimum, 8 GB RAM recommended; plus 1 GB for the Android Emulator
- 2 GB of available disk space minimum, 4 GB recommended (500 MB for IDE + 1.5 GB for Android SDK and emulator system image)
- 1280 × 800 minimum screen resolution

Android Studio Linux System Requirements

- GNOME or KDE desktop
- Tested on Ubuntu® 14.04 LTS, Trusty Tahr (64-bit distribution capable of running 32-bit applications)
- 64-bit distribution capable of running 32-bit applications
- GNU C Library (glibc) 2.19 or later
- 3 GB RAM minimum, 8 GB RAM recommended; plus 1 GB for the Android Emulator
- 2 GB of available disk space minimum, 4 GB recommended (500 MB for IDE + 1.5 GB for Android SDK and emulator system image)
- 1280 × 800 minimum screen resolution

Android Studio Overview

This section will discuss the Android Studio Integrated Development Environment (IDE). Since the Android Studio program is constantly being updated and new versions are being released, the screen captures in this section may not be exactly like the version of Android Studio you are using. So far in 2018 there have been at least six different versions of Android Studio that have been released. In previous years there have been a similar if not greater number of releases. The general appearance of Android Studio is shown in Figure 3-1.

The Project Tab

The Project tab in the Android Studio IDE displays the current Android project. The exact format that the project is displayed in depends on the drop-down selection box located to the right of the tab. The default format is "Android" and shows the current project categorized under two main headings which are "app" and "Gradle Scripts." Under the "app" heading you have the "manifests," "java," and "res" directories. The manifests directory contains the Android program's manifest which specifies such things as permissions that are granted to the program and the minimum Android operating system that is required to run the program. The Java directory contains the program's source code which is written in the Java programming language. The res directory holds things like the layout of the graphical user interface of the program and the layout of the menus in the program. The "Gradle Scripts" directory holds files that are used to build the final Android program. See Figure 3-2.

Source Code Area

The Arduino Studio has an area where source code can be displayed and edited. In Figure 3-3 the source code area is highlighted by a rectangle and the corresponding source code in the Project tab is circled. You can easily bring up a specific source code file by double clicking on it in the Projects tab. You can select from multiple source

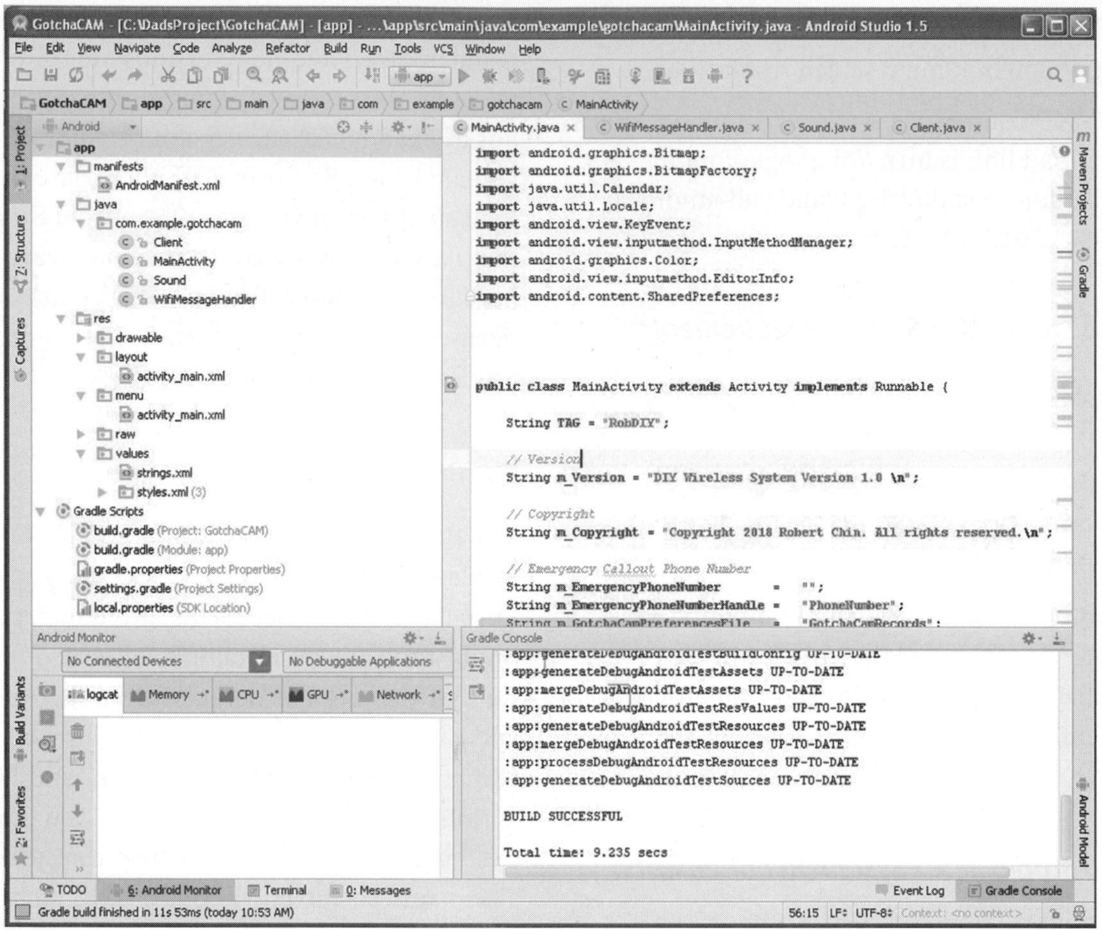

Figure 3-1 The Android Studio Integrated Development Environment (IDE).

code files in the source code by clicking on the corresponding tab.

The Structure Tab

The Structure tab is below the Project tab and displays the functions and variables in the source code file that are currently selected in the source code area. When you click on a function or variable in the Structure tab window the corresponding function or variable within the source code file is displayed. See Figure 3-4.

The Android Monitor

The Android Monitor tool can be used to monitor the information messages, error messages, warning messages, and debug messages being output from the Android device using the logcat tab. Under the Log Level drop-down list box you can select the type of output you want such as verbose, debug, info, warn, error, or assert. See Figure 3-5.

You can also capture a screen shot of an Android device, or record a video from an Android device, using Android Monitor. Just press the screen capture icon which is a picture of a camera on the left-hand side of the Android Monitor window to capture a screen shot of the Android device. See Figure 3-6.

If you want to record the Android screen in the form of a movie then press the screen record icon that is located just below the screen capture icon on the left-hand side of the Android Monitor.

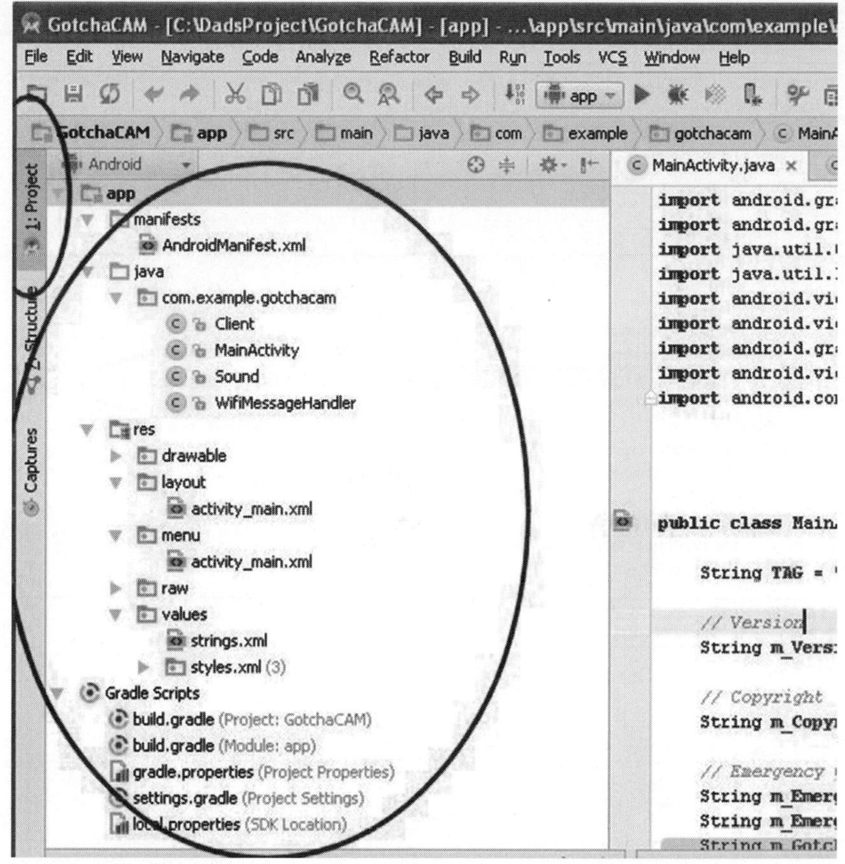

Figure 3-2 The Project tab.

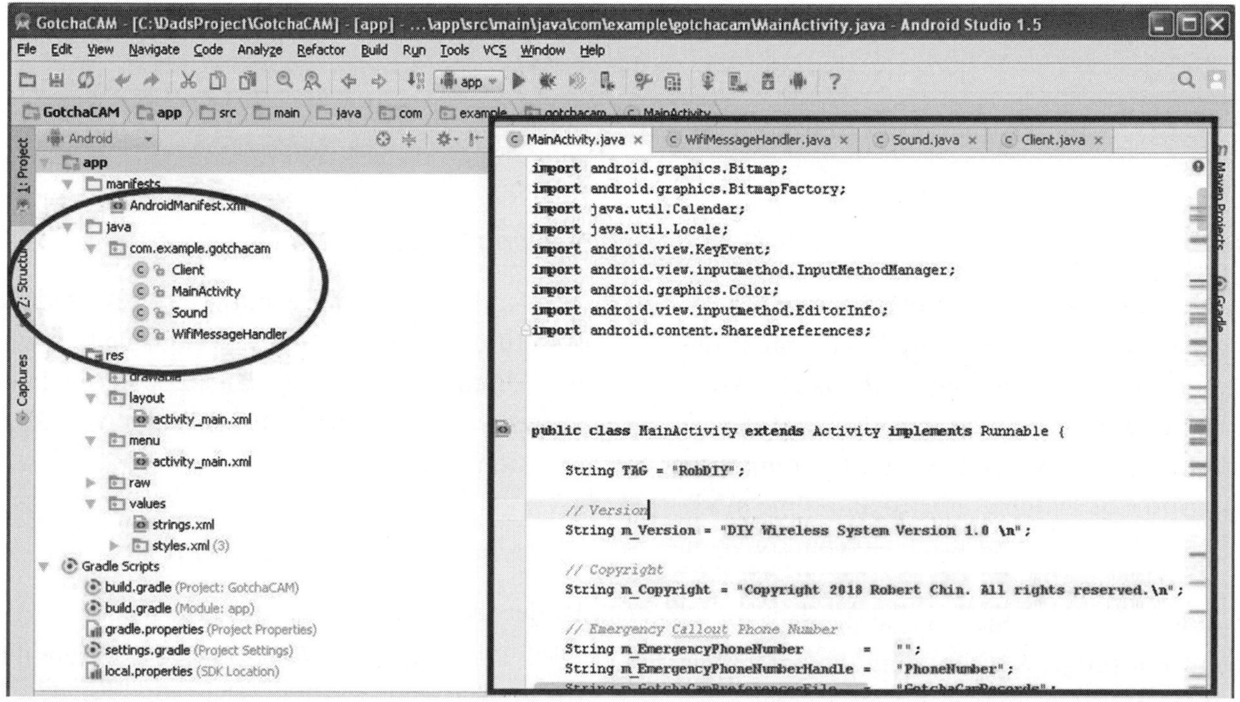

Figure 3-3 The source code area of the Android Studio.

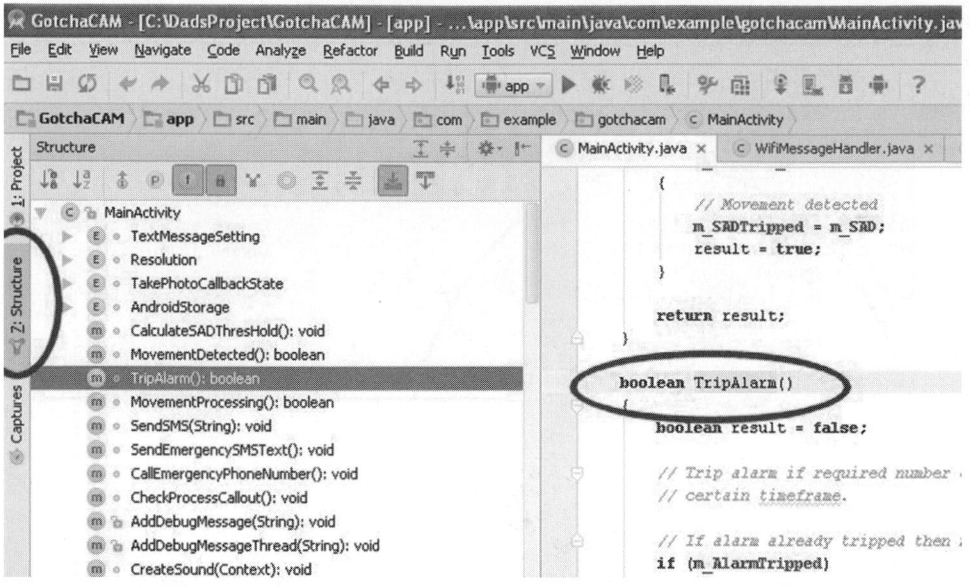

Figure 3-4 The Structure tab.

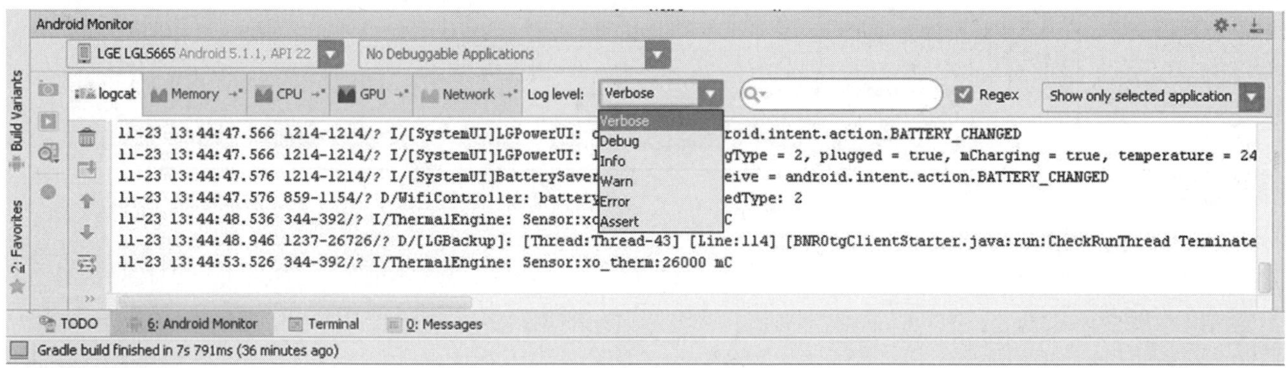

Figure 3-5 The Android Monitor.

The Gradle Console

Gradle is an open-source build tool that Android Studio uses to compile Java programs to a form the Android operating system can understand. Initially, you will need an Internet connection to download key gradle components. Later you can switch gradle to an offline mode and use a local copy of gradle for building your Android application. The Gradle Console is a window in Android Studio that displays the current progress of building the final Android program from Java source code. See Figure 3-7.

Viewing the Tools

The Android Monitor window, project window, structure window, and Gradle Console that we have discussed previously can all be selected and viewed under the View->Tools Windows menu selection. See Figure 3-8.

Cleaning and Rebuilding the Project

Under the build menu we have the menu selections for cleaning the project and rebuilding the project. The "Clean Project" and "Rebuild Project" menu items are highlighted in Figure 3-9.

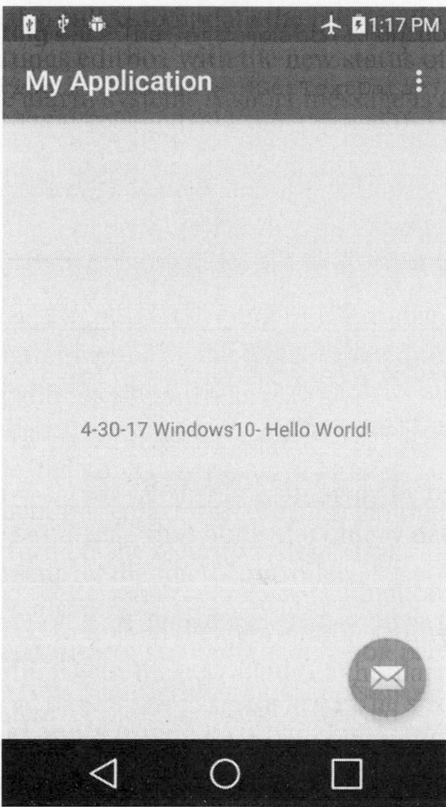

Figure 3-6 Screen shot of Android device using the screen capture feature of the Android Monitor.

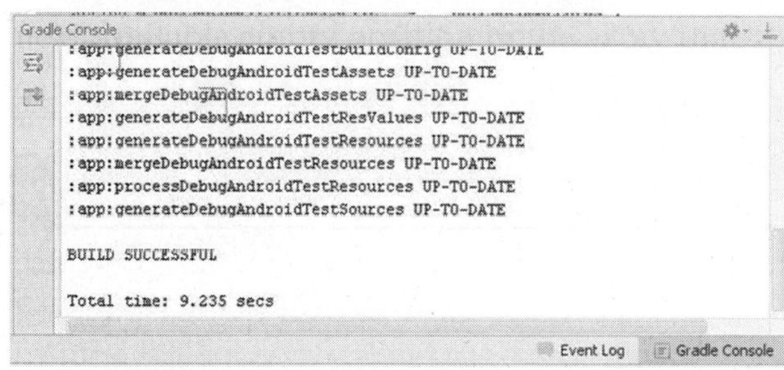

Figure 3-7 The Gradle Console.

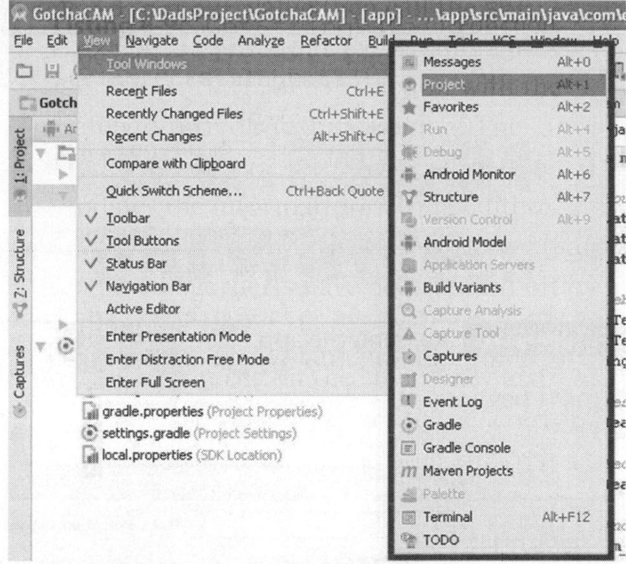

Figure 3-8 Viewing the tools.

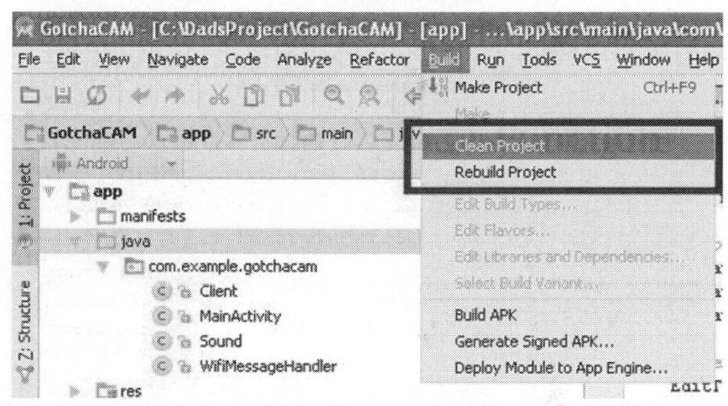

Figure 3-9 Compiling the Android program.

Running the App

To actually transfer the final compiled program to your Android device and then run it you will need to select the Run->Run 'app' menu item. See Figure 3-10.

Setting Up Gradle to Work Offline

By default gradle must use the Internet frequently when you build a project. However, you can select an offline mode to force gradle to use a local gradle distribution on your computer instead of the Internet. Select the File->Settings menu item. This brings up the Settings window. Select "Use local gradle distribution" by clicking on the button next to the item. Select "Offline work" by checking the checkbox next to this item. See Figure 3-11.

Android Device Monitor

The Android device monitor window can be brought up by selecting Tools->Android-> Android Device Monitor from the main menu. Using the Android device monitor you can view the files and the directories that are on the Android device itself as well as upload

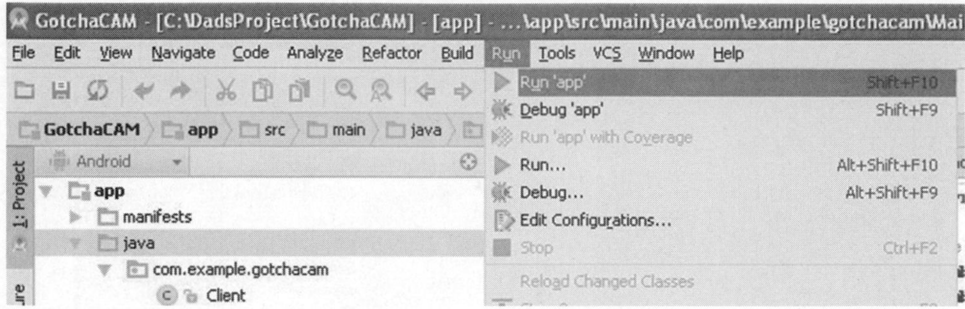

Figure 3-10 Running the app.

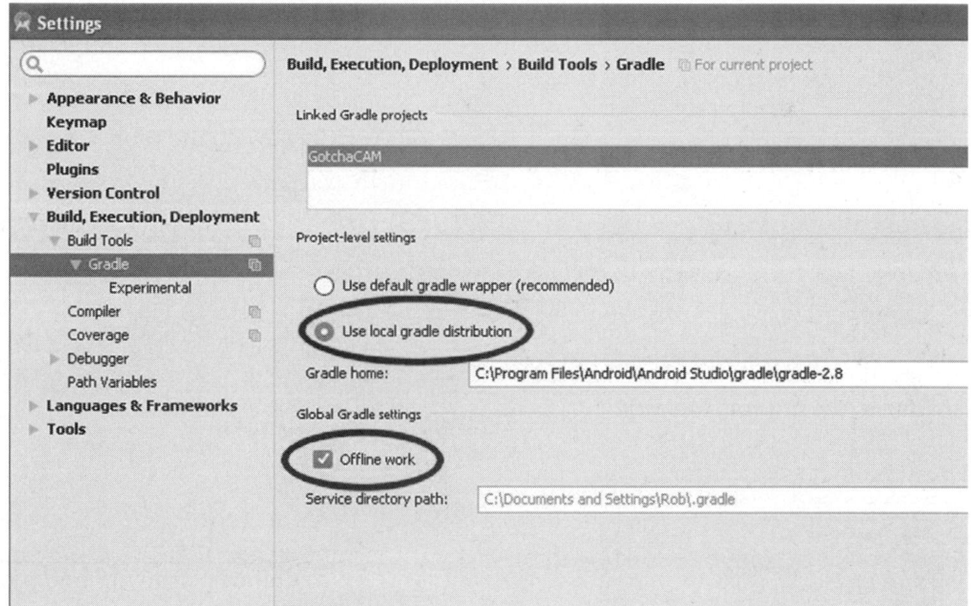

Figure 3-11 Setting up gradle for offline use.

and download files to and from the device. For example, you can upload an Android APK file to your Android device using the Android device monitor function and then install that APK by double clicking on it from the Android device like you would any other APK file that was downloaded from a web site. The Android device if it is connected to your development system is shown under the "Devices" sub window of the Android device monitor. The files and directories that are on the Android device are displayed under the "File Explorer" tab. See Figure 3-12.

Java Language Overview

Java is the programming language that is used to create programs that run on Android-based cell phones and tablets. This section is meant to be a quick overview of the Java programming language and not a reference guide. It would be helpful if you have had experience with other programming languages like C and C++ also.

Java Comments

Java comments are exactly the same as C and C++ comments. Java comments can be in the

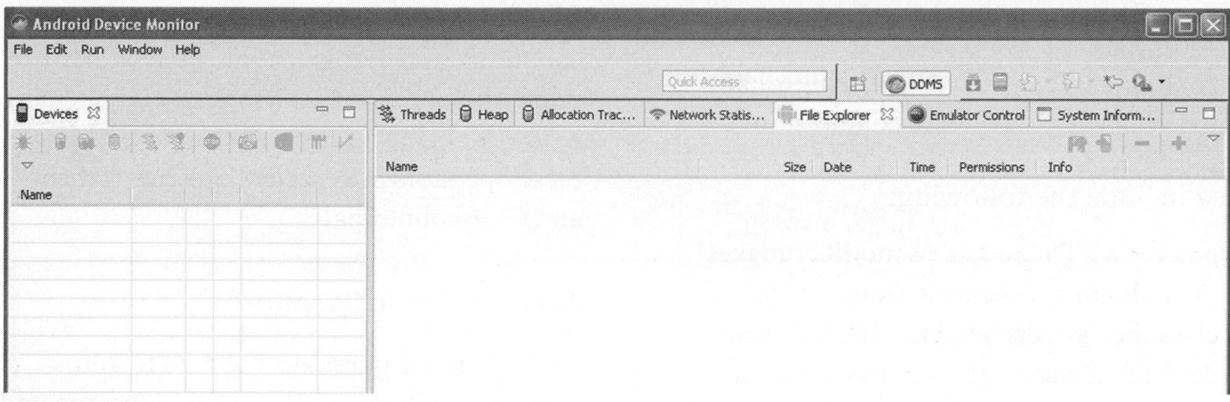

Figure 3-12 The Android device monitor.

form of single-line comments or multi-line comments.

- Single-line comments begin with two slash characters "//".

  // This is a single-line C/C++ and Java comment.

- Multi-line comments begin with a slash followed by an asterisk "/*" and terminate with an asterisk followed by a slash "*/".

```
/*
    This is a multi-line C/C++ or Java
  comment.
*/
```

Java Basic Data Types

This section covers the different data types available in the Java programming language for Android.

- **`byte`.** A `byte` is an 8-bit number with values between −128 and 127, inclusive.
- **`short`.** A `short` is a 16-bit number with values between −32,768 and 32,767, inclusive.
- **`int`.** An integer is a 32-bit number with values between −2,147,483,648 and 2,147,483,647, inclusive.
- **`long`.** A `long` is a 64-bit number with values between −9,223,372,036,854,775,808 and 9,223,372,036,854,775,807, inclusive.
- **`float`.** A `float` is a single-precision 32-bit IEEE 754 floating-point number.
- **`double`.** A `double` is a double-precision 64-bit IEEE 754 floating-point number.
- **`char`.** A **`char`** is a single 16-bit Unicode character that has a range of '\u0000' (or 0) to '\ uffff' (or 65,535, inclusive).
- **`boolean`.** A `boolean` is a data type that has a value of either true or false.

Java Arrays

In Java, just like in C and C++, you can create arrays of items of the basic data types listed in the previous section. However, the format of an array in Java is very different than in C and C++.
The general format of an array declaration is as follows:

```
DataType[] VariableName = new
 DataType[NumberElements];
```

An example of an array of floating point numbers that is 16 elements long is:

```
float[] FloatArray = new float[16];
```

An example of an array of bytes that is declared to be empty is:

```
byte[]  m_PhotoData = null;
```

An example of an array of `String` objects that is `m_MaxFileNames` long is:

```
String[] m_AndroidFileNames = new
 String[m_MaxFileNames];
```

Java Data Modifiers

Java data modifiers allow you to control how variables in a class are accessed and stored.

They include the following:

- **private.** The **private** modifier makes variables only accessible from within the class they are declared in. The following declares `FloatingPointNumberArray` to be private array of eight elements and only accessible from within its own class:

  ```
  private float[] FloatPointNumberArray
   = new float[8];
  ```

- **public.** The `public` modifier allows variables to be accessed from any class. The following variable called `PublicArray` is an array of 23 floating point numbers:

  ```
  public float[] PublicArray = new
   float[23];
  ```

- **static.** The `static` modifier ensures that the variable has only one copy associated with the class they are declared in. The following variable called `StaticData` is declared as a static floating point array with five elements.

  ```
  static float StaticData[] =
  {
   1,2,3,4,5        // Static Data
          };
  ```

- **final.** The `final` modifier indicates that the variable will not change. This is similar to the const keyword in C/C++. For example, the following declares the variable `AppName` is of type `String`. It is initialized to "`WifiServerApp1`" and declared final so that it cannot be changed by other parts of the program.

  ```
  final String AppName = "WifiServerApp1";
  ```

Java Operators

In this section, we cover the basic Java operators that are needed for a useful Java program. These include operators that perform arithmetic, unary operators that only have one argument, conditional operators that are used with program flow control statements, and operators that work on a variable at the bit level. Most of these operators are exactly the same as their C and C++ counterparts.

Java Arithmetic Operators

The arithmetic operators are used to add, subtract, multiply, divide, and find the remainder of two numbers or objects and are exactly the same as in C or C++.

- **+.** The plus sign adds together two numbers and is also used by the `String` class to concatenate two different `String` objects.
- **-.** The minus sign subtracts one number from another.
- ***.** The asterisk sign multiplies two numbers together.
- **/.** The slash sign divides one number by another.
- **%.** The percent sign finds the remainder of a division operation.

Java Unary Operators

Java unary operators are exactly the same as in C and C++.

- **-.** The minus sign in front of a variable or expression negates that variable or expression.
- **++.** Two plus signs next to a variable increment the value of that variable by 1.
- **--.** Two minus signs next to a variable decrement the value of that variable by 1.
- **!.** An exclamation point inverts the value of a `boolean` variable.

Java Assignment and Logical Operators

The assignment and logical operators for Java are exactly the same as in C and C++. The operators in this section have two operands, one on the left side of the operator and another on the right side of the operator.

- `&&`. An expression with two ampersands indicates a logical "AND" and evaluates to true if both operands are true and evaluates to false if either operand is false.
- `||`. An expression with two pipes indicates a logical "OR" and evaluates to true if either operand is true and evaluates to false if both operands are false.
- `=`. The equal sign is the assignment operator that assigns the value on the right of the equal sign to the variable to the left of the equal sign.
- `==`. An expression with two equal signs is a logical equality test and returns true if the two operands are equal and false otherwise.
- `!=`. An expression with an exclamation point followed by an equals sign evaluates to true if the operand on the left is not equal to the operand on the right.
- `>`. An expression with a greater than sign evaluates to true if the operand on the left is greater than the operand on the right and false otherwise.
- `>=`. An expression with a greater than sign followed by an equal sign evaluates to true if the operand on the left is greater than or equal to the operand on the right.
- `<`. An expression with a less than sign evaluates to true if the operand on the left is less than the operand on the right.
- `<=`. An expression with a less than sign followed by an equal sign evaluates to true if the operand on the left is less than or equal to the operand on the right.

Java Bitwise and Bit Shift Operators

- `~`. This is the unary bitwise complement operator.
- `<<`. This is the signed left shift operator.
- `>>`. This is the signed right shift operator.
- `>>>`. This is the unsigned right shift operator.
- `&`. This is the bitwise AND operator.
- `^`. This is the bitwise exclusive OR operator.
- `|`. This is the bitwise inclusive OR operator.

Java Flow Control Statements

The flow control statements in Java are exactly the same as in C and C++.

- **`if then` statement.** The `if` keyword is followed by an expression and if this expression evaluates to true then the rest of the `if` statement is executed or the code block is executed such as:

```
if (expression)
{
  // This block gets executed if the
    expression evaluates to true.
}
```

  A more concrete example is the following where the code block is executed and prints out an error message to logcat if the `m_SoundPool` variable is equal to null which means that the sound pool resource failed to be created.

```
if (m_SoundPool == null)
{
  // This code is executed if the m_SoundPool
    variable is null and the resource fails to be
    created.
  Log.e("Main Activity " , "m_SoundPool
    creation failure!");
}
```

- **if then else statement.** The if keyword is followed by an expression. If this expression evaluates to true then the code block that follows is executed. However, if the expression evaluates to false then the code block that follows the else keyword is executed such as in:

```
if (expression)
{
 // The code in this block is executed
  if the expression evaluates to true
     }
     else
     {
 // The code in this block is executed
  if the expression evaluates to false
}
```

A more concrete example of an "`if then else` statement" is where the `boolean` variable `m_AlarmSet` is tested and if it is true then the `AlarmStatus` variable is set to the string "`AlarmON`." If the `m_AlarmSet` variable is false then the `AlarmStatus` variable is set to "`AlarmOFF`."

```
if (m_AlarmSet)
{
 AlarmStatus = "AlarmON";
}
else
{
 AlarmStatus = "AlarmOFF";
}
```

- **`switch` statement.** The `switch` statement evaluates an expression and tries to match the resulting value with a case block according to the case label such as:

```
switch(expression)
{
 case label1:
  // This code block is executed if
   the expression evaluates to
  // label1
 break;
 case label2:
  // This code block is executed if
   the expression evaluates to
  // label2
 break;

 default:
  // This code block is executed if no
   other labels match.
 break;
}
```

The following is code from an actual Android program. The code is modeled after the standard Android menu callback system. The variable item in the switch statement is a `MenuItem` object and the case labels represent the menu items that are available to be selected by the user. The exact menu item that has been selected is retrieved by calling `item.getItemId()`. If the menu item to activate the alarm has been selected then the case block with the R.id.alarm.activate value is executed. If the menu item to deactivate the alarm is selected then the case block with the R.id.alarm_deactivate label is executed. The code in the default code block is executed if none of the other labels matches.

```
switch (item.getItemId()) {
 ////////////////////////// Alarm
  Activation
 case R.id.alarm_activate:
  SetAlarmStatus(true);
  UpdateCommandTextView();
  Toast.makeText(this, "Alarm
   Activated !!!", Toast.LENGTH_
   LONG).show();
  return true;

 case R.id.alarm_deactivate:
  SetAlarmStatus(false);
  UpdateCommandTextView();
  Toast.makeText(this, "Alarm
   Deactivated !!!", Toast.LENGTH_
   LONG).show();
  return true;

        default:
           return super.
onOptionsItemSelected(item);
}
```

- **`while` statement.** The `while` statement block is executed as long as the expression evaluates to true such as in the following:

```
while (expression)
{
```

```
    // This code block executes as long
     as the expression evaluates to true
}
```

A good example of where a `while` loop is useful is when reading in data over a Wi-Fi connection using the TCP protocol. In the following code while `m_IsConnected` is true then execute the `while` loop code block. In the while code block data is read in from a TCP connection and processed by the `m_WifiMessageHandler.ReceiveMessage()` function.

```
while (m_IsConnected)
{
  try
  {
    // Read from the InputStream
     bytes = m_SocketInputStream.
     read(buffer);

    // Send the obtained bytes to the
     UI activity
     m_WifiMessageHandler.
      ReceiveMessage(bytes, buffer);
                 }
      catch (IOException e)
       {
       Log.e(TAG, "Error Reading
        Wifi DATA ... ERROR = " +
        e.toString());
        break;
       }
}
```

- **`for` statement.** The `for` statement executes a code block as long as the expression is true. Usually the `for` loop initializes a variable and then increments the variable at the end of each loop. The value of this variable is used in the `for` loop block for some processing over a set of objects. The general format for the `for` loop is:

```
for (DataType variable = startvalue;
 expression;
 variable = variable + increment/
  decrement)
 {
    // This code in this block executes
     as long as expression is true
 }
```

A good example of using the `for` loop would be to erase old data from an array. In the following code all the elements in the array `m_DataByte` are set to 0 starting with the value in array index 0.

```
for (int i = 0 ; i < m_DataSize; i++)
{
 m_DataByte[i] = 0;
}
```

- **`try catch`.** The `try catch` block is used for exception handling. The code that is monitored for an exception is located in the try code block. The code to handle any exceptions generated by the code in the try block is located in the catch code block. See the following:

```
try
{
 // Code in this block may generate
  an exception
}
catch (ExceptionType name)
{
 // Code in this block handles
  the exception generated in the try
  block.
}
```

The following is a concrete example of how to use a try catch block in actual Android Java code. The code that is to be monitored for an exception is located in the try code block and creates a new OutputStream file, writes data to it, and then closes it. If there are any problems an exception is thrown and the code in the catch code block is executed. This code write prints the error message so you can view it within the logcat window and displays the error in the program's graphical user interface.

```
try
{
```

```
    OutputStream os = new
     FileOutputStream(file);
    os.write(m_PhotoData, 0, m_PhotoSize);
    os.close();
  }
  catch (IOException e)
  {
    // Unable to create file, likely
     because external storage is
    // not currently mounted.
    Log.e("ExternalStorage", "Error
     writing " + file, e);
    AddDebugMessage(TAG + ": Error
     writing " + file);
  }
```

Java Classes

Java is an object-oriented language like C++ and allows for classes to be derived from other existing classes. The new derived class will inherit the functions and data members of the base class. In this way you can make custom versions of existing classes where you can implement your own functions to customize the behavior of the new derived class. You can tell the compiler to override existing functions in the base class by using the @Override compiler directive. The keyword "extends" is used to indicate that the class that follows is the base class to be extended from. For example, the general format for a Java class `DerivedClass` that is derived from `BaseClass` and overrides `function1()` from the `BaseClass` is as follows:

```
class DerivedClass extends BaseClass
{
  // New code for Derived Class
  @Override
  void function1()
  {
    // This function overrides an
     existing function in the BaseClass
  }
}
```

A good example of a class in Java that inherits from a base class and overrides existing base class functions is the `MainActivity` class shown in the code below that inherits from the `Activity` class. In fact the `Activity` class is the essential class that you will need to use to develop Android programs. The `onCreate()` function is overridden and will contain custom code for your specific Android application. The `onOptionsItemSelected()` function is also overridden and will provide custom code that will process menu items that the user has selected.

public class `MainActivity` extends `Activity`

```
{
  // Code for derived class is put in
   here.
  @Override
  protected void onCreate(Bundle
   savedInstanceState)
  {
    // Custom code block that overrides
     the default function
    // inherited from the Activity class
    // Put custom code here that
     initializes your Android program
  }
  @Override
  public boolean
   onOptionsItemSelected(MenuItem item)
  {
    // This function is called when the
     user selects a menu item.
    // Put your custom code for your own
     specific menu items here.
  }
}
```

Java Packages

Java Packages are a way to group source code and other files that are part of the same application or perform related functions. The package that a source code file belongs to is usually declared at the start of the file with the keyword "package" such as:

```
package com.example.gotchacam;
```

Packages can be imported into a source code file such as:

```
import java.net.Socket;
```

The java.net.Socket package that is imported is a standard Android Java package that comes with the Android library and is needed for TCP communication over Wi-Fi. After you import the package you can easily use the classes within the package by just specifying the class name instead of the package name and then the class name. You can also view the packages in your current Android Studio workspace by selecting the Project tab and then select the "Packages" selection from the drop-down menu.

Java Interfaces

A Java interface consists of function declarations but no actual code. The code must be defined by the class that implements the interface. For example, Runnable is a Java interface that is part of the built-in Android library.

```
public interface Runnable
```

The only function in the Runnable interface is call `run()` and is declared as a public abstract function that does not accept any parameters and does not return any value.

```
public abstract void run ()
```

A class can implement an interface by using the "implements" keyword in the class definition like in the following:

```
public class MainActivity extends
 Activity implements Runnable
{
 public void run()
 {
  // Code that defines the Runnable
   interface function run() is put in
  // this code block.
 }
}
```

Here the `MainActivity` class both inherits from the `Activity` class and implements the Runnable interface. The `run()` function is defined later within the `MainActivity` class.

Java Methods

Java methods are like C and C++ functions. They both refer to the same thing. A Java method is defined by a name and an optional list of parameters, an optional return type, and optional data modifiers such as public, private, etc. If a parameter is an object then it is automatically passed by reference. A basic outline of a method's structure is as follows:

```
DataModifiers ReturnValue
 MethodName(ParameterType1 Parameter1,
 ...)
{
 // Java method's code body
}
```

The following method is called `SetCommand` and takes a `String` object as an input parameter and does not return any value.

```
void SetCommand(String Command)
{
 m_Command = Command;
}
```

The following method is named `GetStringData` and does not take any input parameters and returns a `String` object.

```
String GetStringData()
{
 String temp = m_Data;
 m_Data = "";

 return temp;
}
```

The following method is named `GetBinaryData` that does not take input parameters and returns a pointer to an array of type `byte`.

```
byte[] GetBinaryData()
{
 return m_DataByte;
}
```

Accessing Methods, and Class members in Java

In Java you can access methods, and class members just like in C++ with a dot or "." operator. For example, the m_ClientConnectThread variable is declared to be a class of type Client and initialized to null.

```
Client m_ClientConnectThread=null;
```

A new Client object is created and m_ClientConnectThread is set to refer to this new Client object.

```
m_ClientConnectThread = new
 Client(ServerStaticIP, PortNumber,
 m_WifiMessageHandler, this);
```

If the thread is not null and thus was created then start the thread by calling the m_ClientConnectThread.start() function.

```
if (m_ClientConnectThread != null)
{
 AddDebugMessage(TAG + ": Starting
  ClientConnectThread ...\n");
  m_ClientConnectThread.start();
}
```

Calling the Parent Method in Java

A Java method that overrides a method that is declared in a base class can call the base class's method by using the super.BaseMethod() calling format. For example, the class MainActivity is derived from the base class of Activity which has the following method:

```
protected void onCreate(Bundle
 savedInstanceState)
```

The MainActivity class overrides the onCreate method and is required to call the base class's implementation of this function. In fact the official Android documentation states that for this function:

> "Derived classes must call through to the super class's implementation of this method. If they do not, an exception will be thrown. If you override this method you must call through to the superclass implementation."

The call to the superclass implementation of the onCreate function is the following:

```
super.onCreate(savedInstanceState);
```

The following code shows everything in context:

```
public class MainActivity extends
 Activity implements Runnable
{
 @Override
 protected void onCreate(Bundle
  savedInstanceState)
 {
  super.onCreate(saved InstanceState);
  setContentView(R.layout.activity_
   main);
  // Rest of code for onCreate() function
 }
}
```

The Android `Activity` Class

The one class that is essential for an Android application and that will run on all Android versions is the Activity class which represents the actual Android program or activity. The Activity class is a public class that derives from the ContextThemeWrapper class and implements the interfaces of LayoutInflater.Factory2, Window.Callback, KeyEvent.Callback, View.OnCreateContextMenuListener, and ComponentCallbacks2. See the following:

```
public class Activity extends
 ContextThemeWrapper implements
 LayoutInflater.Factory2, Window.
 Callback, KeyEvent.Callback, View.
 OnCreateContextMenuListener,
 ComponentCallbacks2
```

Generally, the way you will use the Activity class is to define your own custom class that derives from the Activity class such as:

```
public class MainActivity extends
 Activity
{
 @Override
 protected void onCreate(Bundle
  savedInstanceState)
 {
 super.onCreate(savedInstanceState);
```

```
    // Put code to initialize your
     application, graphical user
     interface, storage,
    // etc. in this code block.
  }
}
```

The `MainActivity` class will hold your custom code that will define the custom behavior for your application. Custom behavior can involve the graphical user interface, storage-related operations, as well as other custom initialization operations.

The following statement imports the `Activity` class into your source code file so that it can be easily used:

```
import android.app.Activity;
```

The official documentation on the `Activity` class is located at https://developer.android.com/reference/android/app/Activity.

The Android `Activity` Life Cycle

The Android `Activity` class calls certain functions during its life cycle from the application being created to the application being put into the background and then when the application is terminated.

The key functions that are called during various stages of the `Activity` class's life cycle are the following:

- `onCreate()`
- `onStart()`
- `onRestart()`
- `onStop()`
- `onResume()`
- `onPause()`
- `onDestroy()`

The `MainActivity` class uses these functions and is shown in Listing 3-1.

A flowchart of how and under what conditions these key life-cycle functions are called under is shown in Figure 3-13.

Listing 3-1 The `MainActivity` Class

```
public class MainActivity extends
 Activity
{
 @Override
 protected void onCreate(Bundle
  savedInstanceState)
 {
  super.onCreate(savedInstance
   State);
  // Put code to initialize
   application, graphical user
   interface, storage, etc.
  // in this code block.
 }

 @Override
 protected void onStart()
 {
  super.onStart();
 }

 @Override
 protected void onRestart()
 {
  super.onRestart();
 }

 @Override
 protected void onStop()
 {
  super.onStop();
 }

 @Override
 protected void onResume()
 {
  super.onResume();
 }

 @Override
 protected void onPause()
 {
  super.onPause();
 }

 @Override
 protected void onDestroy()
 {
  super.onDestroy();
 }
}
```

Android Phone Orientation Change Issue

An important thing to note for developers new to Android development is that by default an orientation change in the graphical user interface causes the application to restart and lose any current data. The way to get around this is to the `Activity` class's `setRequestedOrientation()` function to fix the orientation of the application's graphical user interface so that it does not change when the phone's orientation changes. This function is defined as follows:

```
public void setRequestedOrientation
  (int requestedOrientation)
```

The following are all the screen orientation options:

- `SCREEN_ORIENTATION_UNSPECIFIED`
- `SCREEN_ORIENTATION_LANDSCAPE`

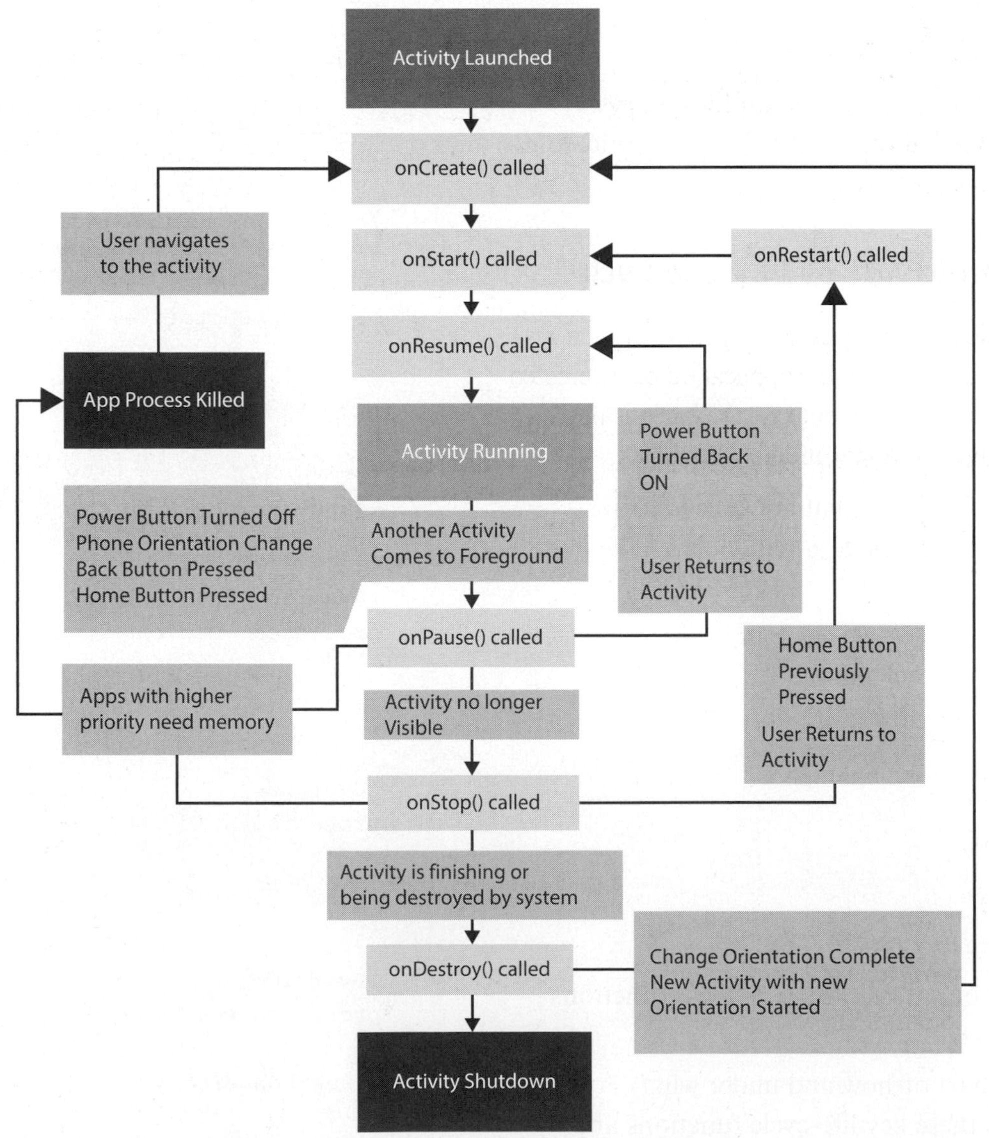

Figure 3-13 The Android Activity life cycle.

- SCREEN_ORIENTATION_PORTRAIT
- SCREEN_ORIENTATION_USER
- SCREEN_ORIENTATION_BEHIND SCREEN_ORIENTATION_SENSOR
- SCREEN_ORIENTATION_NOSENSOR
- SCREEN_ORIENTATION_SENSOR_LANDSCAPE
- SCREEN_ORIENTATION_SENSOR_PORTRAIT
- SCREEN_ORIENTATION_REVERSE_LANDSCAPE
- SCREEN_ORIENTATION_REVERSE_PORTRAIT
- SCREEN_ORIENTATION_FULL_SENSOR
- SCREEN_ORIENTATION_USER_LANDSCAPE
- SCREEN_ORIENTATION_USER_PORTRAIT
- SCREEN_ORIENTATION_FULL_USER
- SCREEN_ORIENTATION_LOCKED

For example, put the following line of code near the beginning of the `onCreate()` function to set the screen orientation to the portrait mode.

```
this.setRequestedOrientation(Activity
 Info.SCREEN_ORIENTATION_PORTRAIT);
```

Hands-on Example: The Hello World Example

In this hands-on example we will be creating a new Android project from scratch, modifying the code and the graphical user interface, and then running it on an actual Android device. First you will need to create the new Android program by selecting the File->New->New Project menu item within Android Studio. This will bring up the new project dialog window. See Figure 3-14.

There should be options where you can enter an application name, a company domain, a package name, and project location. You can click the "Next" button to move on to the next step. The next screen should be for selecting which platforms your Android program will run on. The "Phone and Tablet" option should already be selected by default. Click on the "Next" button. See Figure 3-15.

The next screen should be where you can choose a specific format for your graphical user interface. Select the empty activity graphical user

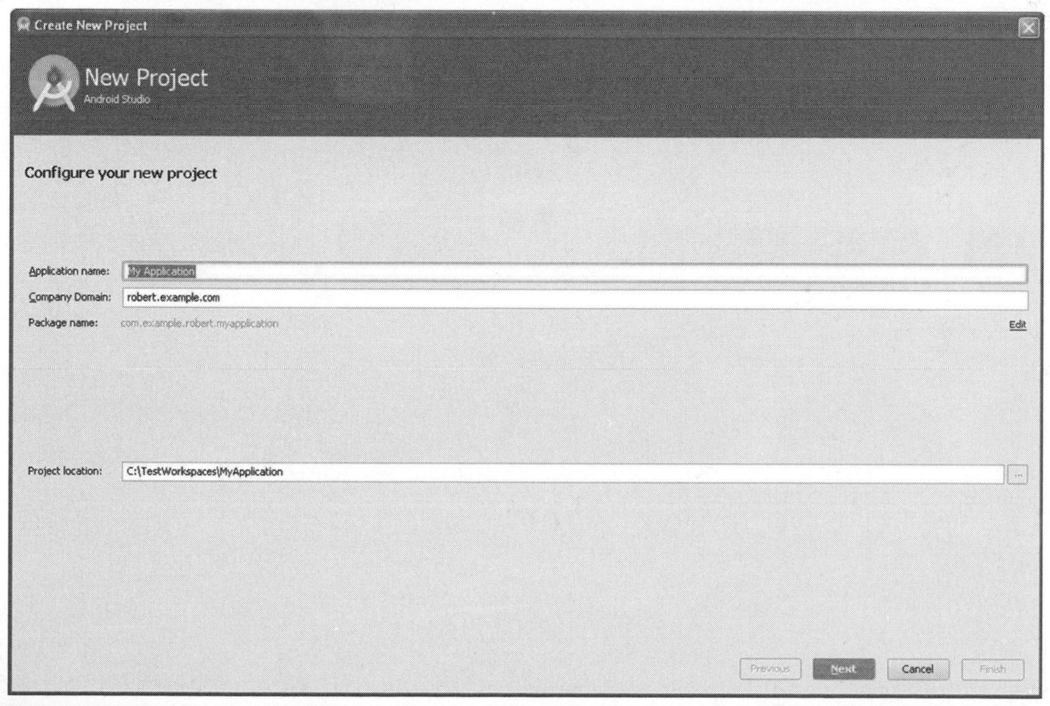

Figure 3-14 New project dialog window.

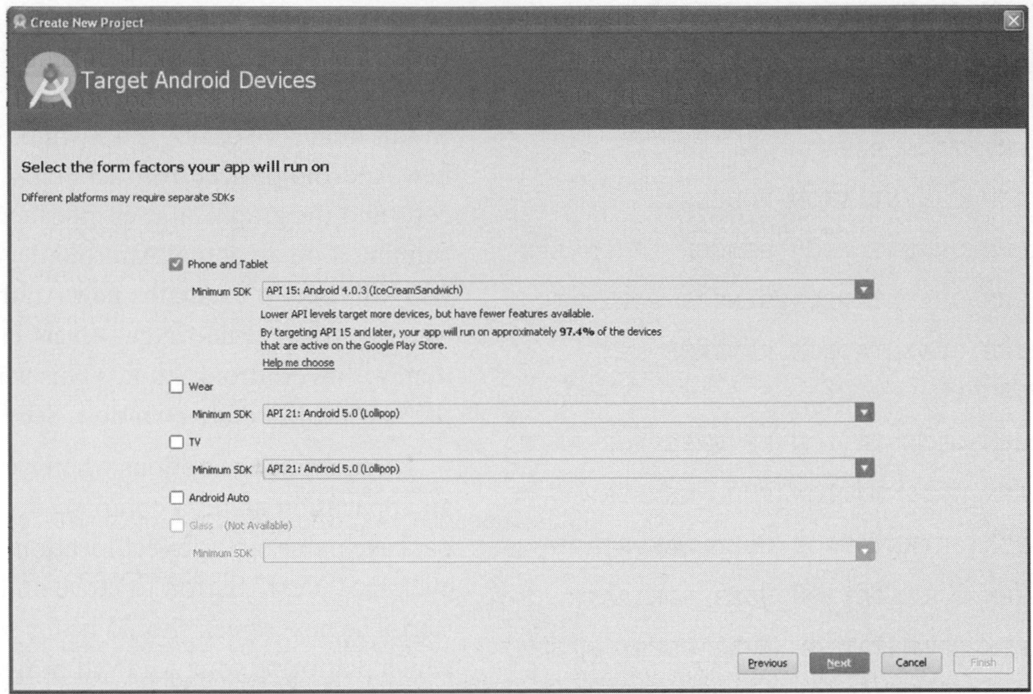

Figure 3-15 The target Android devices window.

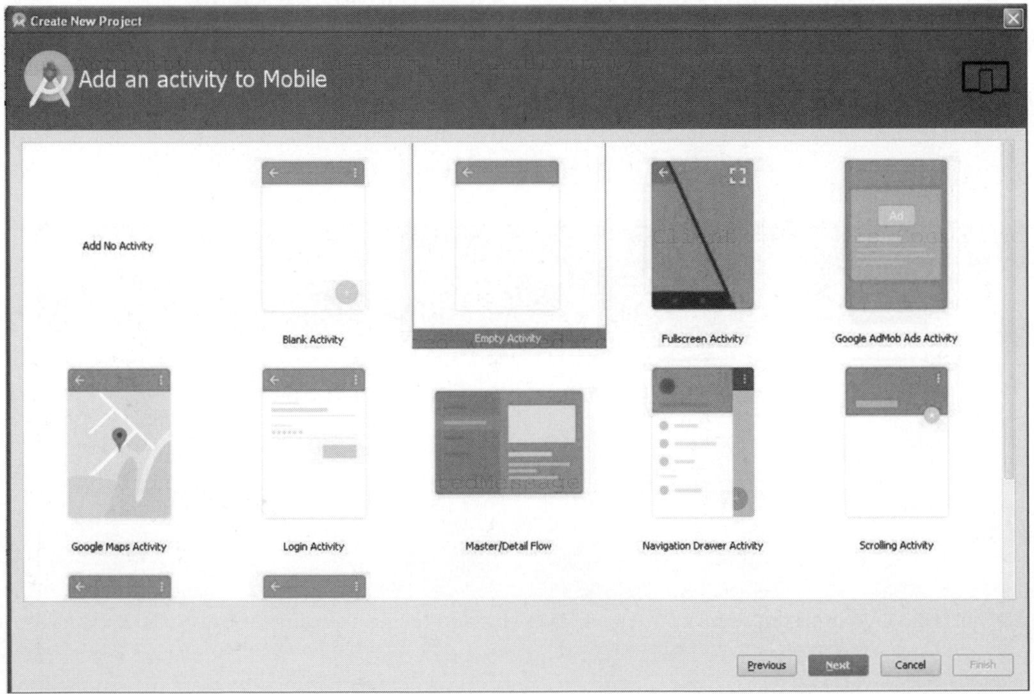

Figure 3-16 Selecting the Android layout type.

interface and click on the "Next" button. See Figure 3-16.

The next screen that is brought up is the "Customize the Activity" screen. Here you can give the activity a unique name. Enter "`MainActivityEmpty`" in the "Activity Name" field and click on the "Finish" button. See Figure 3-17.

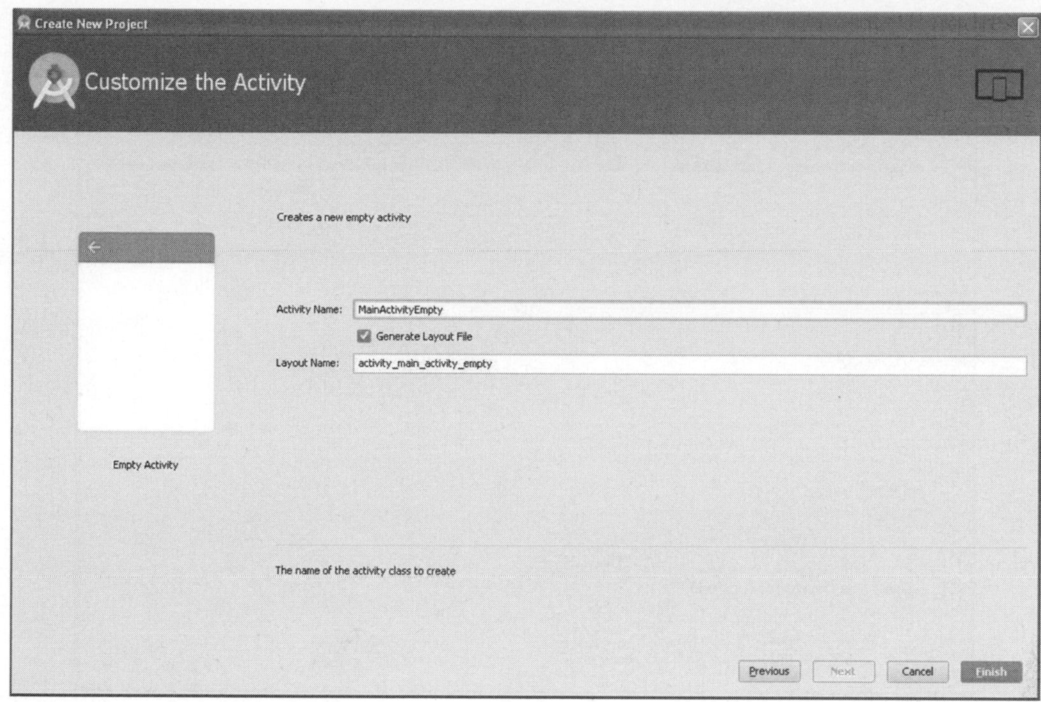

Figure 3-17 The customize activity screen.

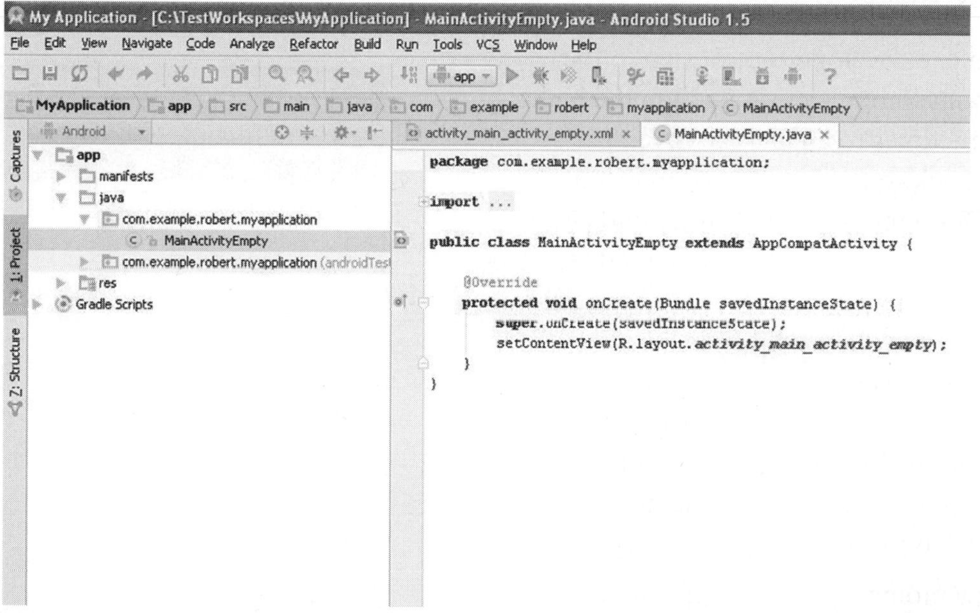

Figure 3-18 The new `MainActivityEmpty` project.

The next screen that should pop up will be the project window for the `MainActivityEmpty` application that you just created. See Figure 3-18.

Next, you should clean the project by selecting Build->Clean Project from the main menu. Then you should rebuild the project by selecting Build->Rebuild Project. After you are able to rebuild the project without getting any compile errors, attach your Android device to your computer using a USB cable. Make sure you have the developer options enabled on your Android

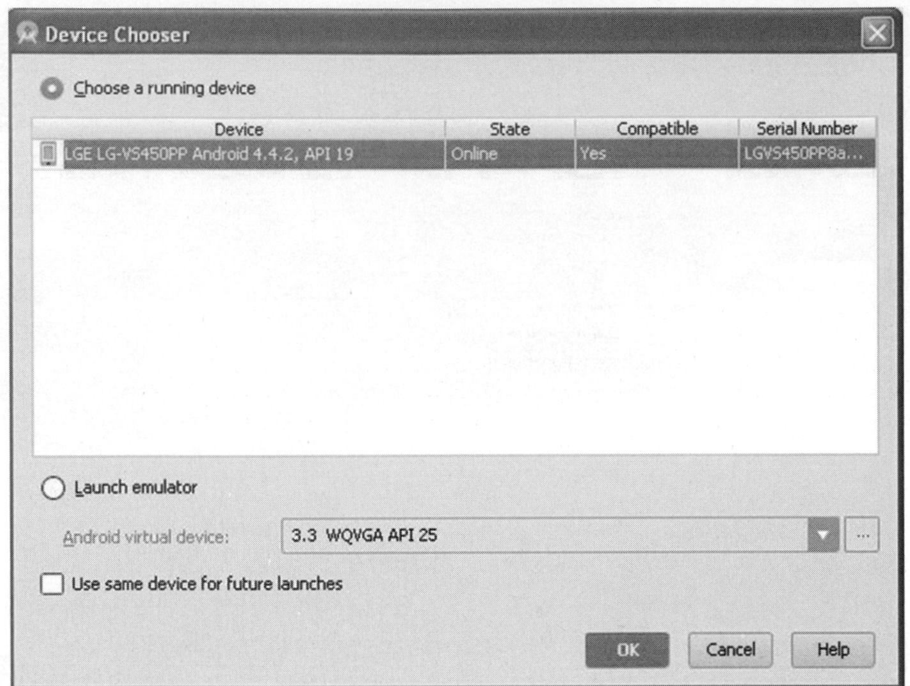

Figure 3-19 Choosing an Android device to run the application on.

phone. To enable developer options go to the Setting->About Phone section and tap the build number item 9 times to enable the developer options. Go back to the Settings main menu under the "Developer Options" menu and enable USB debugging which sets the debug mode to active whenever the Android device is connected via USB. Select Run->Run 'app' to install and run the Android program on your Android device. After you select the Run->Run 'app' menu selection the "Device Chooser" window pops up. Select your device from the list and click the "OK" button. See Figure 3-19.

After a few moments a new application should start running on the Android device that is attached to your development system via USB cable. The program should have the words "Hello World!" displayed within the graphical user interface. See Figure 3-20.

One thing you should notice is that the main activity derives from `AppCompatActivity`

Figure 3-20 The default Hello World application.

```
MainActivityEmpty.java ×
package com.example.robert.myapplication;

//import android.support.v7.app.AppCompatActivity;
import android.os.Bundle;
import android.app.Activity;


public class MainActivityEmpty extends Activity { //AppCompatActivity {

    @Override
    protected void onCreate(Bundle savedInstanceState) {
        super.onCreate(savedInstanceState);
        setContentView(R.layout.activity_main_activity_empty);
    }
}
```

Figure 3-21 Changing from the `AppCompatActivity` class to the `Activity` class.

instead of `Activity`. Let's change this so that the `MainActivityEmpty` class will derive from `Activity`.

The changes you need to make are to:

1. Comment out the `AppCompatActivity` class and replace it with the `Activity` class.
2. Comment out the import related to the `AppCompatActivity` class.
3. Add in the line "import android.app.Activity" in the import section of the code file.

See Figure 3-21.

Run the application with these changes by selecting the Run->Run 'app' menu selection. You should see something like the image in Figure 3-22.

Next, we are going to change the text in the application. In order to do this we need to bring up the layout of the graphical user interface that is located under the res\layout directory. The file is an xml file named "activity_main_activity_empty.xml." Bring this file up by double clicking on it. It should appear in the source code area. See Figure 3-23.

Click on the "Hello World" text in the layout. This should highlight the corresponding text input field in the "Properties" area of the layout where you can enter addition text or change existing text. Enter some new text such as "NEW Hello World with Activity Class!" in the text input field. See Figure 3-24.

Figure 3-22 The Android application derived from the `Activity` class.

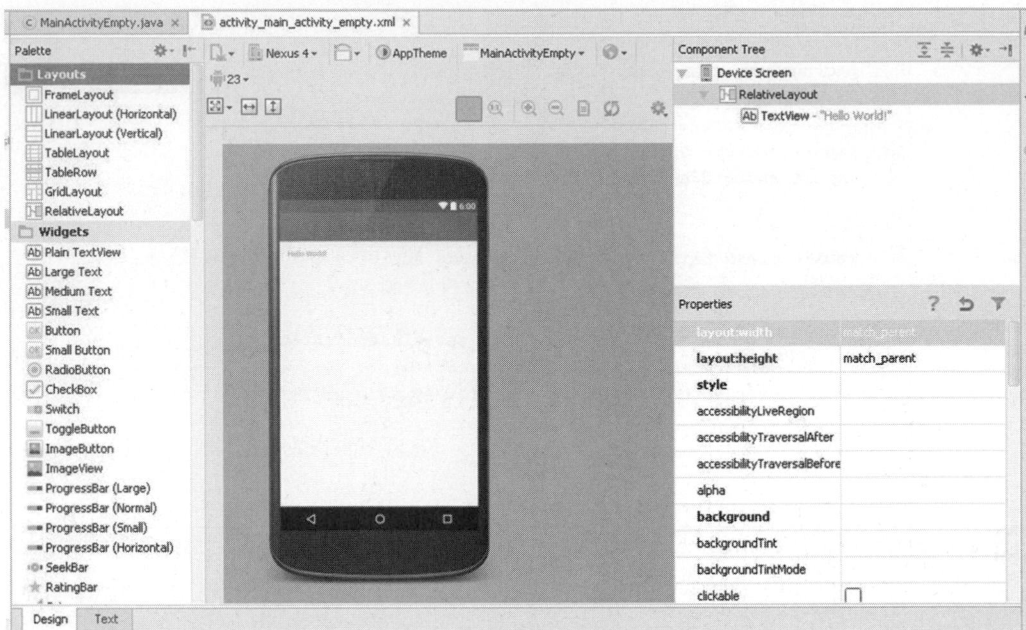

Figure 3-23 The application's graphical user interface.

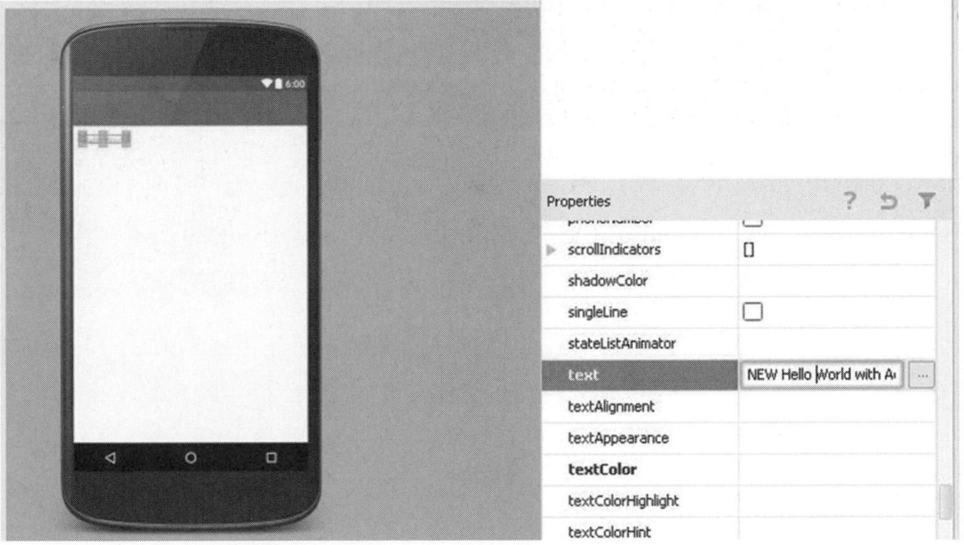

Figure 3-24 Changing the default text in the layout.

After you make the text changes run the application again. The new text changes should show up. See Figure 3-25.

Importing Projects Between Android Studio Versions

New versions of Android Studio are released frequently. Android Studio version 1.5 works on the Windows XP and the most current version of Android Studio is over 3.0. This section will give you some tips on how to convert projects from one version of Android Studio to a more recent version. I tested importing a project from Android Studio version 1.5 to Android Studio version 2.3.1. In order to do this I had to switch the JDK compiler from internal to an external version JDK 8. The reason was that the internal

Figure 3-25 The Android application with the text changes.

JDK compiler could not be located by Android Studio. See Figure 3-26.

Another issue you may encounter is that the project may require a certain version of gradle. My Android Studio 1.5 project required version 2.2 of gradle which I downloaded from gradle's official web site at https://gradle.org/.

All versions of gradle are available for download and are located at https://gradle.org/releases/.

After I unzipped the gradle file I set the Android Studio to use the local gradle distribution and checked the option for gradle to work offline. See Figure 3-27.

Finally, I was able to successfully clean, and rebuild this project for Android Studio 2.3.1. See Figure 3-28.

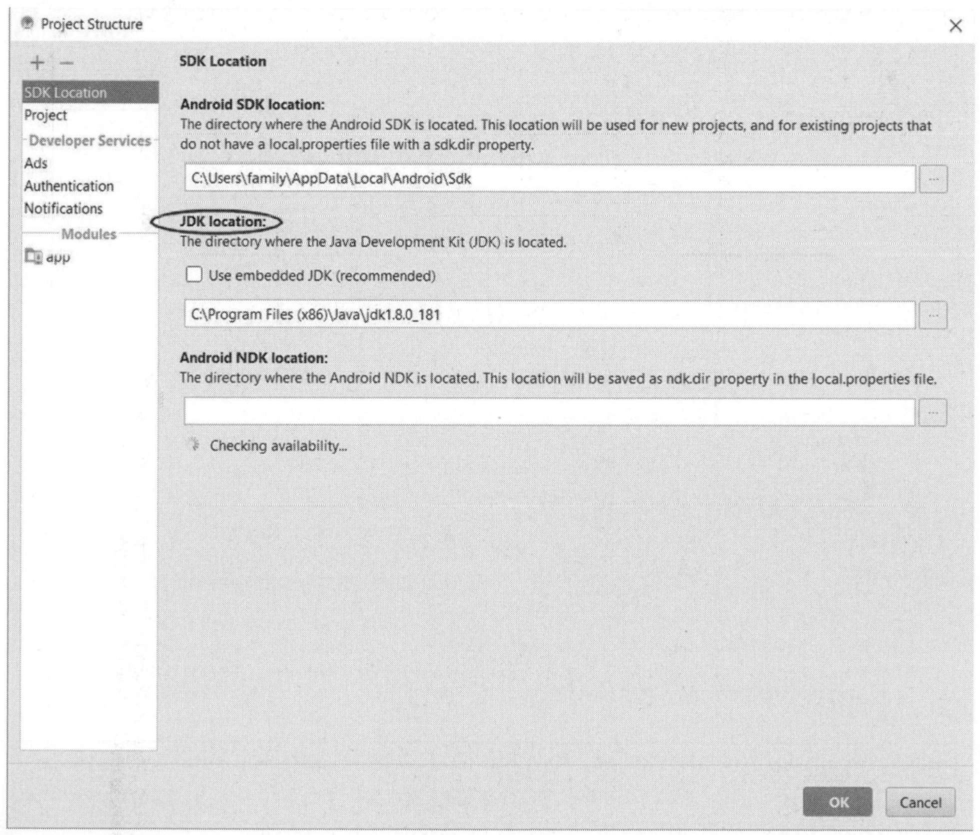

Figure 3-26 Changing the JDK compiler to an external version.

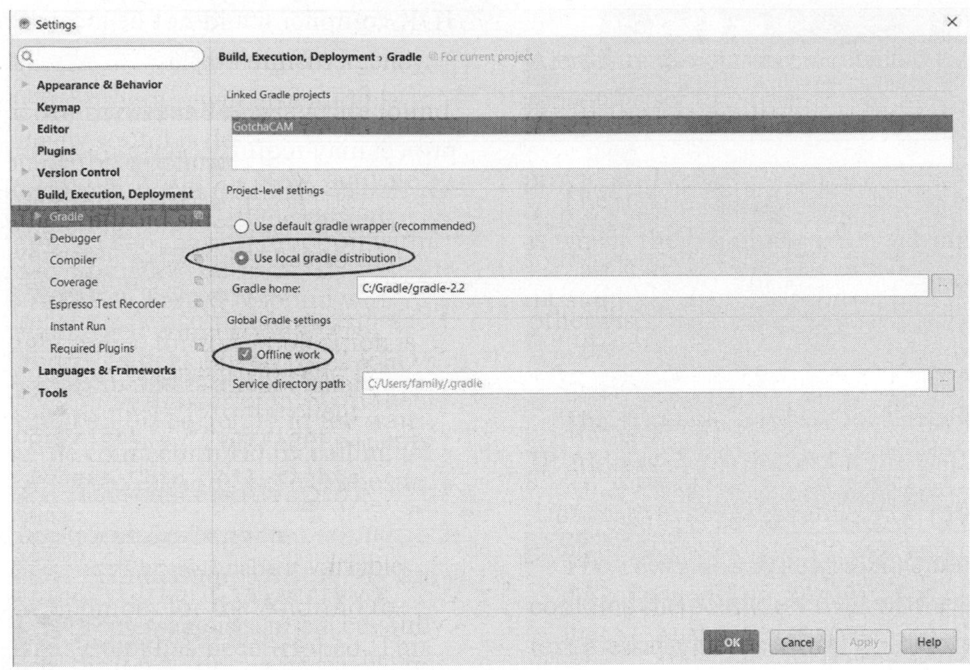

Figure 3-27 Using the local Gradle version 2.2 and working offline.

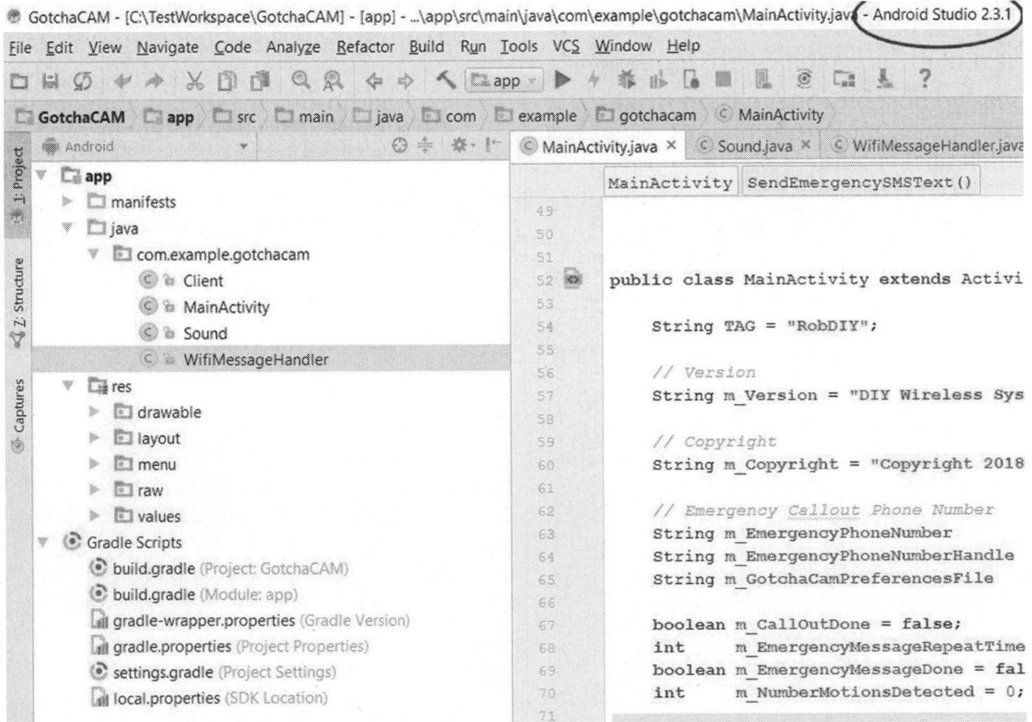

Figure 3-28 Successful importing of a project from Android Studio 1.5 to Android Studio 2.3.1.

Summary

This chapter gives the reader an introduction into the Android operating system and how to develop programs for this operating system. First some basic information about the Android operating system is presented and then information concerning Android cell phone service is discussed. Next, the system requirements for Android Studio that is used to develop programs for Android devices are covered. I then give an overview of the Android Studio program itself with emphasis on key features that you will need to build and deploy a program to your Android device. Next, I cover the Java programming language that is used to construct Android programs. I cover the basics of Java including data types, program flow control statements, classes, interfaces, and functions. I then cover the `Activity` class and the Android program's `Activity` life cycle. Next, a hands-on example is presented where I show the reader how to use Android Studio to create a new Android application from scratch, modify the program code, change the graphical user interface, and install and run the program on an Android device. Finally, I give an example on how to import an Android Studio project from one version of Android Studio to a newer version of Android Studio.

CHAPTER 4

Arduino with ESP-01 and Android Basic Wireless Framework

THIS CHAPTER DISCUSSES A BASIC wireless framework consisting of an Arduino using an ESP-01 module to provide Wi-Fi capability and an Android cell phone. I start off discussing the general overall framework of the system and how it works. I then cover the Android client source code for the application that serves as the user interface to the wireless network. Next, I discuss the server source code for the Arduino portion of the wireless framework. Finally, I guide the reader through a hands-on example of how to set up and operate the Arduino with ESP-01 and Android basic wireless framework.

The Arduino with ESP-01 and Android Basic Wireless Communication Framework Overview

The basic wireless communications framework consists of the Arduino Mega 2560 and ESP-01 Wi-Fi communications module presented earlier in this book and an Android cell phone. An overview of the Arduino, ESP8266, and Android wireless communications framework discussed in this chapter is shown in Figure 4-1.

This is the basic framework that will serve as a starting point for the examples involving the Arduino and ESP-01 Wi-Fi module. The Android cell phone communicates with the Arduino through the ESP-01 ESP8266-based module. A Wi-Fi connection is first established between the Android and the Arduino. After a Wi-Fi connection is established you can send commands to the Arduino from the Android, and receive responses to these commands and other alerts from the Arduino. Based on these responses and alerts the Android phone can call out to an emergency phone number or send a text message alert to another cell phone.

The Android application itself is called "Basic Framework v1.0" and is shown in Figure 4-2.

The emergency phone number for call outs and text messages can be set by the user and is "9876543210" in the above figure in the phone entry edit textbox. This same edit textbox is also used to set the name of the Android device that will be used in the basic wireless network. However, the main purpose of this phone entry edit textbox is to allow the user to send commands to the Arduino server by typing in the command into the textbox and then hitting the "Send Data" button. Debug messages and incoming text from the Arduino server are displayed in the edit textbox located directly below the phone entry textbox. On the right-hand side of the screen directly under the "Send Data" button you have a textbox that lists that state of the application. The first line is the name of the android phone that will be used in the basic wireless network. The second line is the emergency phone number that will receive call outs and text messages. The next line should be blank. The following line should be the state of the alarm system either "AlarmOn" if the alarm system is turned on or "AlarmOff" if the

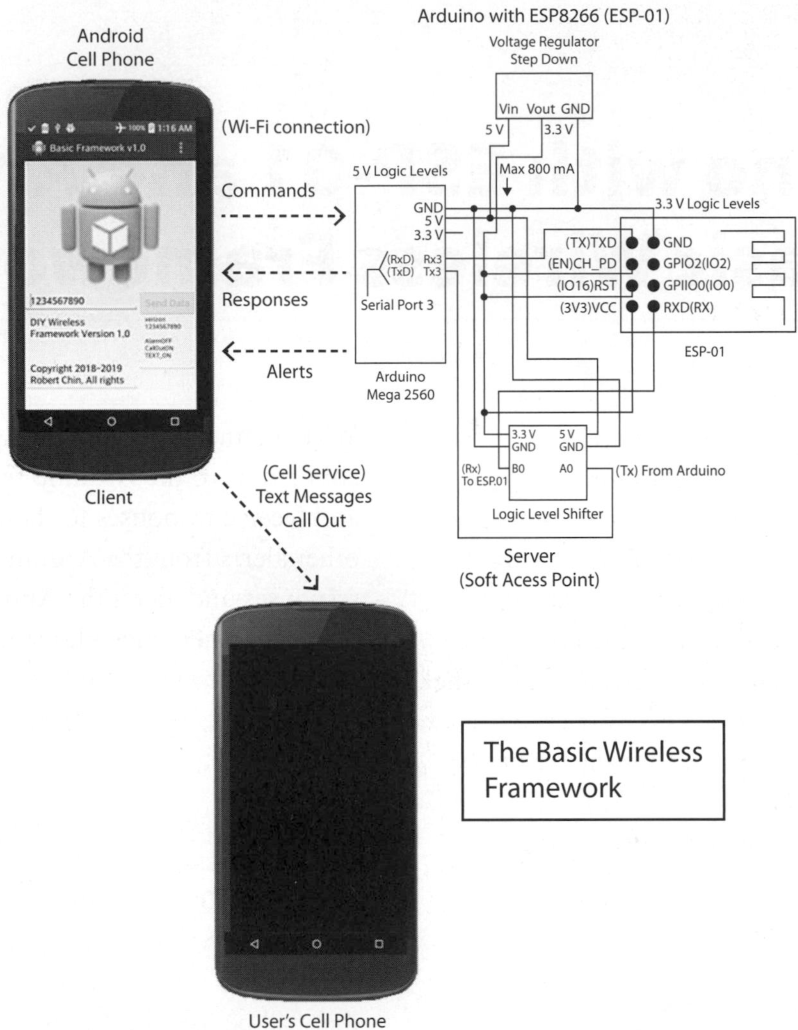

Figure 4-1 The Android, Arduino, ESP8266 basic wireless communications framework.

alarm system is turned off. The next line is either "CallOutON" if a call out to the emergency phone number is to be performed when the alarm is tripped or "CallOutOFF" if the call out feature is disabled. Finally, the last line is either "TEXT_ON" if an emergency text message is to be sent out when the alarm is tripped or "TEXT_OFF" if the emergency text message feature is disabled.

The Android Client Wireless Communication Framework Code

This section covers the code for the Android client portion of the basic wireless communications framework including code for the graphical user interface, menu system, and Android manifest.

The Android Manifest

The Android manifest file is an xml file that holds key information such as permissions and minimum operating system requirements. In order to use the Android Wi-Fi to connect to the Arduino device we need to add some permissions in the AndroidManifest.xml file. These permissions allow you to turn the Android's Wi-Fi on and off.

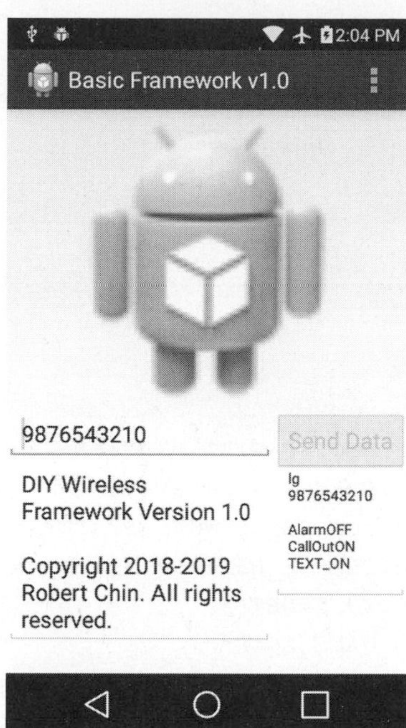

Figure 4-2 The Android program.

```
<uses-permission android:name="android.
permission.ACCESS_WIFI_STATE"/>
<uses-permission android:name="android.
permission.CHANGE_WIFI_STATE"/>
```

Permissions are also required to access and change the Android's connection state to a network such as when connecting to the ESP-01 module in Access Point mode.

```
<uses-permission android:name="android.
permission.ACCESS_NETWORK_STATE"/>
<uses-permission android:name="android.
permission.CHANGE_NETWORK_STATE"/>
```

Also, in order to send and receive data on the newly created network between the Arduino and the Android device you will need to give the Android permission to access the Internet.

```
<uses-permission android:name="android.
permission.INTERNET"/>
```

In order to keep the Android's Wi-Fi radio turned on all the time we need to add the wake lock permission.

```
<uses-permission android:name="android.
permission.WAKE_LOCK"/>
```

The other important item is the line:

```
android:windowSoftInputMode="adjustPan"
```

which sets the Android virtual keyboard so that when the keyboard is displayed a portion of the screen is scrolled upward so that the emergency phone text entry box is visible. The complete Android manifest file is shown in Listing 4-1 with the key permissions highlighted in bold print.

The Graphical User Interface (GUI) Layout

In the res/layout folder is the activity_main.xml file that holds the code for the application's GUI. You can double click on this file inside of Android Studio to bring up the GUI editor. You can click on each GUI component or click on the items under the component tree to the far right of the GUI editor to highlight each item that makes up the GUI. For example, click on the "phoneentry" item under the component tree and the corresponding GUI item should be highlighted. See Figure 4-3.

The code for the main GUI layout consists of

1. An ImageView object called imageView1 that can be used to hold a jpg image and display it on the Android screen. Currently this is not used but is there for possible future expansion.
2. A button object called TestMessageButton that has "Send Data" written on it and is used to send data to the Arduino server.
3. An EditText object called phoneentry that is located just to the left of the button and is used to set the emergency phone number, set the Android device name used on the wireless network, and to send data to the Arduino server.
4. An EditText object called debugmsg that is located under the phoneentry textbox

Listing 4-1 The Android Manifest

```
<?xml version="1.0" encoding="utf-8"?>
<manifest xmlns:android="http://schemas.android.com/apk/res/android"
    package="com.example.gotchacam"
    android:versionCode="1"
    android:versionName="1.0" >

    <uses-sdk
        android:minSdkVersion="8"
        android:targetSdkVersion="23" />
    <uses-permission android:name="android.permission.ACCESS_WIFI_STATE"/>
    <uses-permission android:name="android.permission.CHANGE_WIFI_STATE"/>
    <uses-permission android:name="android.permission.ACCESS_NETWORK_STATE"/>
    <uses-permission android:name="android.permission.CHANGE_NETWORK_STATE"/>
    <uses-permission android:name="android.permission.INTERNET"/>
    <uses-permission android:name="android.permission.WAKE_LOCK"/>
    <uses-permission android:name="android.permission.WRITE_EXTERNAL_STORAGE"/>
    <uses-permission android:name="android.permission.CALL_PHONE"/>
    <uses-permission android:name="android.permission.SEND_SMS"/>

    <application
        android:allowBackup="true"
        android:icon="@drawable/ic_launcher"
        android:label="@string/app_name"
        android:theme="@style/AppTheme" >
        <activity
            android:name="com.example.gotchacam.MainActivity"
            android:windowSoftInputMode="adjustPan"
            android:label="@string/app_name" >
            <intent-filter>
                <action android:name="android.intent.action.MAIN" />
                <category android:name="android.intent.category.LAUNCHER" />
            </intent-filter>
        </activity>
    </application>
</manifest>
```

and is used to display debug messages and text alerts and other data from the Arduino server.

5. An EditText object called textView1 that is located under the button and is used to display the Android device network name, the emergency phone number, and the state and options of the alarm system.

See Listing 4-2.

The Menu Items

In the res/menu project directory the file "activity_main.xml" holds the code for the basic wireless framework's menu system.

The menus for the Android program are as follows:

1. Call Out Setting Menu
2. Alarm Setting Menu

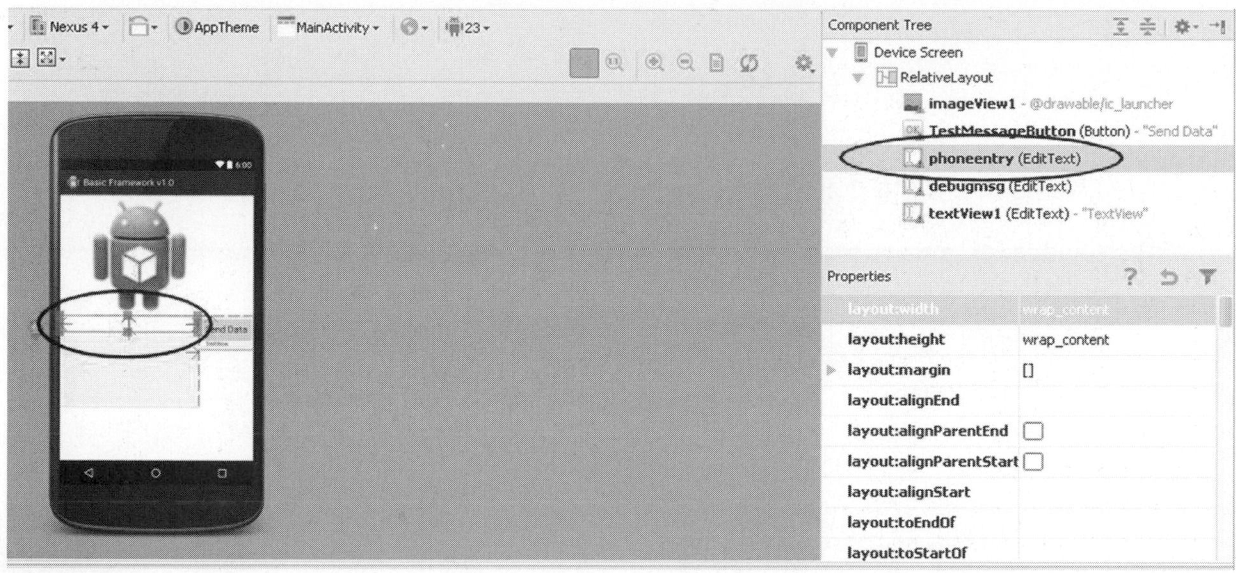

Figure 4-3 The Android Studio GUI editor.

Listing 4-2 The GUI Layout

```
<RelativeLayout xmlns:android="http://schemas.android.com/apk/res/android"
    xmlns:tools="http://schemas.android.com/tools"
    android:layout_width="match_parent"
    android:layout_height="match_parent"
    tools:context=".MainActivity" >

    <ImageView
        android:id="@+id/imageView1"
        android:layout_width="320dp"
        android:layout_height="240dp"
        android:layout_alignParentLeft="true"
        android:layout_alignParentTop="true"
        android:src="@drawable/ic_launcher" />

    <Button
        android:id="@+id/TestMessageButton"
        android:layout_width="wrap_content"
        android:layout_height="wrap_content"
        android:layout_alignParentRight="true"
        android:layout_alignTop="@+id/phoneentry"
        android:text="Send Data" />

    <EditText
        android:id="@+id/phoneentry"
        android:layout_width="wrap_content"
        android:layout_height="wrap_content"
        android:layout_alignParentLeft="true"
        android:layout_below="@+id/imageView1"
        android:layout_toLeftOf="@+id/TestMessageButton"
        android:ems="10"
```

```
            android:lines="1"
            android:maxLines="1"
            android:singleLine="true" />

        <EditText
            android:id="@+id/debugmsg"
            android:layout_width="wrap_content"
            android:layout_height="wrap_content"
            android:layout_alignParentLeft="true"
            android:layout_alignRight="@+id/phoneentry"
            android:layout_below="@+id/phoneentry"
            android:ems="10"
            android:lines="6"
            android:maxLines="6" />

        <EditText
            android:id="@+id/textView1"
            android:layout_width="wrap_content"
            android:layout_height="wrap_content"
            android:layout_alignLeft="@+id/TestMessageButton"
            android:layout_alignParentRight="true"
            android:layout_alignTop="@+id/debugmsg"
            android:ems="10"
            android:scrollHorizontally="true"
            android:scrollbarStyle="outsideOverlay"
            android:scrollbars="vertical"
            android:text="TextView"
            android:textSize="11dp" >

            <requestFocus />
        </EditText>

    </RelativeLayout>
```

3. Text Message Setting Menu
4. Save Android Device Menu

See Listing 4-3.

The Sound Class

The Sound class is a custom class that is responsible for playing back digital sounds.

The `m_SoundPool` variable holds a SoundPool class object that was created in the `MainActivity` class. The SoundPool object is the main interface to loading in, playing back, and managing our sounds.

```
private SoundPool m_SoundPool;
```

The `m_SoundIndex` variable holds the identification number of the sound. The number is the ID of the sound within `m_SoundPool`.

```
private int m_SoundIndex = -1;
```

The `m_LeftVolume` variable holds a value from 0 (no volume) to 1 (full volume) of the left channel of a stereo sound.

```
float m_LeftVolume = 1;
```

The `m_RightVolume` variable holds a value from 0 (no volume) to 1 (full volume) of the right channel of a stereo sound.

```
float m_RightVolume = 1;
```

Listing 4-3 The Menu Code

```
<menu xmlns:android="http://schemas.android.com/apk/res/android" >

    <item
        android:id="@+id/callout_settings"
        android:orderInCategory="100"
        android:showAsAction="never"
        android:title="Call Out Settings">
        <menu>
            <item android:id="@+id/callout_on" android:title="Turn On Emergency Call
Out"/>
            <item android:id="@+id/callout_off" android:title="Turn Off Emergency
Call Out"/>
        </menu>
    </item>

    <item android:id="@+id/alarm_settings" android:title="Alarm Settings">
        <menu>
            <item android:id="@+id/alarm_activate" android:title="Activate Alarm"/>
            <item android:id="@+id/alarm_deactivate" android:title="Deactivate Alarm"/>
        </menu>
    </item>

    <item android:id="@+id/textmess_settings" android:title="Text Message Settings">
        <menu>
            <item android:id="@+id/notext" android:title="No Text Messages" />
            <item android:id="@+id/text_only" android:title="Text Messages On"/>
        </menu>
    </item>

    <item android:id="@+id/savedevicename" android:title="Save Android Device Name">
    </item>
</menu>
```

The `m_Priority` variable holds the streaming priority of the sound with 0 being the lowest priority.

```
int m_Priority = 1;
```

If the `m_Loop` variable is set to -1 then the sound is set to loop continuously. If it is 0 then the sound is set to play one time.

```
int m_Loop = 0;
```

The `m_Rate` variable is the speed that the sound is played back. A 1 setting is normal playback speed. The range is from 0.5 to 2.0. The variable is initialized to 1.

```
float m_Rate = 1;
```

The `m_StreamID` variable holds the identification number of the sound stream that is currently playing the actual sound.

```
int m_StreamID = 0;
```

The `Sound` class constructor creates a `Sound` class object by loading in and initializing the sound designated by the `ResourceID` input parameter and putting it inside the existing `SoundPool` class object designated by the `Pool` input parameter. The `Activity` class object also must be input as the `iContext` parameter.

More specifically:

1. The global class variable m_SoundPool is set to the Pool input parameter.
2. The sound is loaded into the m_SoundPool class object by calling the m_SoundPool.load(iContext, ResourceID, 1) function. This function also returns the index number for the newly created sound in the SoundPool object. This index is then assigned to m_SoundIndex.

See Listing 4-4.

The PlaySound() function plays the sound through the Android device's speakers.

PlaySound(true) plays the sound in a continuous infinite loop. A PlaySound(false) plays the sound one time only.

More specifically the function does the following:

1. If the m_Loop variable is equal to –1 then this sound is set to continuously play, so there is no need to play it again so return without doing anything further.
2. If the input parameter Loop is set to true then set m_Loop to –1 to indicate this sound needs to be looped continuously. Otherwise, set m_Loop to 0.
3. Starts playing the sound effect by calling the m_SoundPool.play(m_SoundIndex, m_LeftVolume, m_RightVolume, m_Priority, m_Loop, m_Rate) function. The m_SoundIndex parameter indicates the sound within the sound pool that should be played back. The m_LeftVolume and m_RightVolume indicate the volume levels of the sound's left and right channels. The next parameters indicate the priority, whether to loop the sound or not, and the speed at which to play the sound back. The ID number of the stream that is playing back the sound is returned by the function and stored in the m_StreamID class variable.

See Listing 4-5.

The StopSound() function stops the sound playing by

1. Calling the m_SoundPool.stop(m_StreamID) function with the ID number of the stream associated with this playing sound.
2. Setting the m_Loop variable to 0 to indicate that this sound is not to be looped.

See Listing 4-6.

Listing 4-4 The Sound Constructor

```
Sound(Context iContext, SoundPool
Pool, int ResourceID)
{
  m_SoundPool = Pool;
  m_SoundIndex = m_SoundPool.
load(iContext, ResourceID, 1);
}
```

Listing 4-5 The PlaySound() Function

```
void PlaySound(boolean Loop)
{
  //public final int play (int
soundID, float leftVolume, float
rightVolume, int priority, int loop,
float rate)
  // Sound already playing
  if (m_Loop == -1)
  {
    return;
  }
  if (Loop)
  {
    m_Loop = -1;
  }
  else
  {
    m_Loop = 0;
  }
  m_StreamID = m_SoundPool.play(m_
SoundIndex, m_LeftVolume, m_
RightVolume, m_Priority, m_Loop, m_
Rate);
}
```

Listing 4-6 The `StopSound()` Function

```
void StopSound()
{
  m_SoundPool.stop(m_StreamID);
  m_Loop = 0;
}
```

Listing 4-7 The `GetSoundPool()` Function

```
SoundPool GetSoundPool()
{
 return m_SoundPool;
}
```

The `GetSoundPool()` function returns `m_SoundPool` which is a reference to the sound pool that is being used by this `Sound` class object. See Listing 4-7.

The `WifiMessageHandler` Class

The `WifiMessageHandler` class processes the data returned from the microcontroller based upon the last command issued by the Android controller. The incoming data is in either text format or binary format.

Text Data

The text data format consists of numbers and letters followed by a newline character which is "\n" that indicates the end of the text data. See Figure 4-4.

We specify the exact character to indicate the end of text using the variable `m_EndData`.

```
char m_EndData     = '\n';
```

The text data is read in by calling the `ReceiveTextData()` function

```
boolean ReceiveTextData(int NumberBytes,
byte[] Message)
```

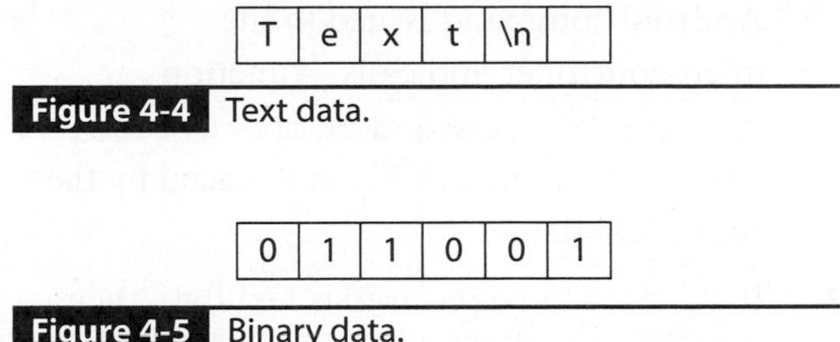

Figure 4-4 Text data.

Figure 4-5 Binary data.

Binary Data

The binary data is a series of 1's and 0's and the length of the returned data must be known when issuing an Android command that expects binary data returned from the microcontroller. See Figure 4-5.

The data length in bytes of the incoming binary data is held in the `m_DataIncomingLength` variable that is declared as:

```
int m_DataIncomingLength = 0;
```

The binary data is read in using the `ReceiveBinaryData()` function that is declared as follows:

```
boolean ReceiveBinaryData(int
NumberBytes, byte[] Message)
```

Class Overview

The `ReceiveMessage()` function is the main entry point to this class and is called from the `Client` class. This section gives you a general overview of how this class works and how you can add your own customizations to this class.

The general procedure to handle data that is being sent from the microcontroller in response to an Android command is:

1. The `Client` class object calls the `ReceiveMessage()` function with the data received from the microcontroller and the number of bytes that data consists of.
2. The `ReceiveMessage()` function processes the data based on the last

Android command issued to the microcontroller and calls a function `ProcessXXXXCommand(NumberBytes, Message)` where XXXX is replaced by the command name.

3. If the data to be received is text data then the `ReceiveTextData()` function is used to process the data.
4. If the data to be received is binary data then the `ReceiveBinaryData()` function is used to process the data.
5. Once all the text or binary data is received then you need to set a variable in the `MainActivity` class to indicate that the Android command has received a response from the microcontroller. You do this through a command such as `m_MainActivity.SetXXXXFinished()` where the XXXX is replaced by the command that has just received a response from the microcontroller.
6. In order to change the part of the Android user interface that the command updates you need to call `m_MainActivity.runOnUiThread(m_MainActivity)`. This executes the `run()` function in the `MainActivity` class where the `MainActivity` class's user interface objects can be accessed.

See Figure 4-6.

The `m_Command` is a `String` variable that holds an alphanumeric representation of the Android command that the Android expects to receive a response to. The `m_Command` variable is set using the `SetCommand()` function and is initialized to "GetTextData".

See Listing 4-8.

The `m_MainActivity` variable holds a reference to the `MainActivity` class object that created this `WifiMessageHandler` object.

```
MainActivity m_MainActivity;
```

The `m_USAscii` variable holds the character set to use when converting incoming bytes into alphanumeric text.

```
Charset     m_USAscii     = null;
```

The `m_Data` variable holds text data.

```
String      m_Data        = "";
```

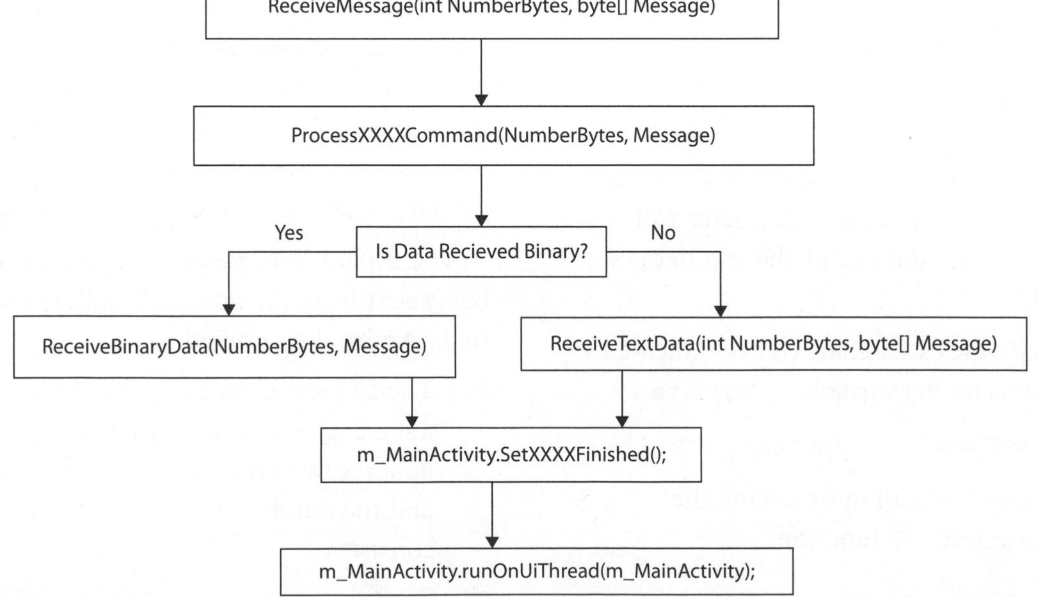

Figure 4-6 `WifiMessageHandler` class flowchart.

Listing 4-8 The `SetCommand()` Function

```
String m_Command = "GetTextData";
void SetCommand(String Command)
{
  m_Command = Command;
}
```

The `m_DataIncomingLength` variable holds the length in bytes of the data expected to be sent by the microcontroller.

```
int  m_DataIncomingLength = 0;
```

The `m_DataByte` array is an array of bytes of size `m_DataSize` that holds the binary data that will be sent by the microcontroller.

```
byte[]       m_DataByte    = new byte[m_
DataSize];
```

The `m_DataSize` variable holds the size of the binary data buffer used to store incoming binary data and is set to 700,000 bytes long.

```
int  m_DataSize     = 700000;
```

The `m_DataIndex` variable holds the index into the `m_DataByte` array and indicates the next available empty position in the array that can hold a new value.

```
int         m_DataIndex   = 0;
```

The `SetDataReceiveLength()` function is called from the `MainActivity` class and sets the length of the expected incoming binary data in bytes.

```
void SetDataReceiveLength(int length) {
m_DataIncomingLength = length;}
```

The `GetBinaryData()` function returns a reference to `m_DataByte` which holds binary data sent from the microcontroller.

```
byte[] GetBinaryData(){return m_
DataByte;}
```

The `GetBinaryDataLength()` function returns the length in bytes of the data in the `m_DataByte` array.

```
int GetBinaryDataLength() { return m_
DataIndex;}
```

The `m_Freeze` variable is true if the `Client` thread that calls this class has been halted and false otherwise. The volatile modifier on this variable means that value can be changed from another thread.

```
volatile boolean m_Freeze = false;
```

The `Freeze()` function sets the m_Freeze variable to true which halts this class from processing any more incoming data from the microcontroller.

```
void Freeze() {m_Freeze = true;}
```

Listing 4-9 `GetStringData()` Function

```
String GetStringData()
{
  String temp = m_Data;
  m_Data = "";

  return temp;
}
```

The `GetStringData()` Function

The `GetStringData()` function returns the text data sent from the microcontroller that is stored in `m_Data`. The `m_Data` variable is also reset to the null string. See Listing 4-9.

The `ResetData()` Function

The `ResetData()` function initializes the text and binary data structures that are used to receive data from the microcontroller by:

1. Setting `m_Data` which holds the incoming text data to `""`.
2. Setting `m_DataIndex` that holds the position of the next available byte in the incoming binary data buffer to 0 that is the beginning of the buffer.
3. Setting `m_DataIncomingLength` that holds the length in bytes of the expected binary data from microcontroller to 0.
4. Erasing the existing data in the binary data buffer `m_DataByte` by writing 0 to every array location.

See Listing 4-10.

Listing 4-10 The `ResetData()` Function

```
void ResetData()
{
 // Reset String Data
 m_Data = "";

 // Reset Binary Data
 m_DataIndex = 0;
 m_DataIncomingLength = 0;

 // Erase old data in array
 for (int i = 0 ; i < m_DataSize; i++)
 {
  m_DataByte[i] = 0;
 }
}
```

The `WifiMessageHandler` Class Constructor

The class constructor initializes the class object by:

1. Assigning the global variable `m_MainActivity` to the `Activity` class object that will use this handler.
2. Assigns the US ASCII character set to `m_USAscii` by calling `Charset.forName("US-ASCII")`. The `m_USAscii` variable is used to convert incoming text data into US English characters.

See Listing 4-11.

Listing 4-11 The Constructor

```
WifiMessageHandler(MainActivity
iActivity)
{
 m_MainActivity = iActivity;

 // Set Charset to US ASCII
translation
 m_USAscii = Charset.forName
("US-ASCII");
}
```

The `ReceiveTextData()` Function

The `ReceiveTextData()` function reads in the data from the Message byte array input parameter and returns true if a complete text data message has been received and false otherwise.

Specifically, the function does the following:

1. Reads in the valid data from the `Message` array based on the `NumberBytes` input parameter.
2. Converts this data into a `ByteBuffer` object using the `ByteBuffer.wrap(temp)` function with temp being the valid data from step 1.
3. Converts the `ByteBuffer` object into a `CharBuffer` object by calling the `m_USAscii.decode(bb)` function with the input parameter `bb` being the `ByteBuffer` obtained in step 2.
4. Converts the `CharBuffer` object obtained in the previous step into a `String` object by calling the `TempCharBuffer.toString()` function on the `CharBuffer` object.
5. Adds the `String` obtained from step 4 to the current text data message which is stored in `m_Data`.
6. Next, we check to see if the end of the text message has been reached. We first find the array index value of the text end message character that is stored in `m_EndData` by calling the `m_Data.indexOf(m_EndData)` function.
7. If the returned value is greater than or equal to 0 then we have found the end of text message marker and this text message is complete.

See Listing 4-12.

The `ReceiveBinaryData()` Function

The `ReceiveBinaryData()` function reads in the incoming binary data from the microcontroller.

Listing 4-12 The `ReceiveTextData()` Function

```
boolean ReceiveTextData(int NumberBytes, byte[] Message)
{
    boolean EndTextDataFound = false;
    byte[] temp = new byte[NumberBytes];
     // Capture readable data
     for (int i = 0; i < NumberBytes; i++)
     {
       temp[i] = Message[i];
     }
     ByteBuffer bb = ByteBuffer.wrap(temp);

     // Convert from bytes to printable characters
     CharBuffer TempCharBuffer = m_USAscii.decode(bb);

     // Debug output Number of bytes incoming and characters
     m_Message = TempCharBuffer.toString();
     Log.d("ZombieCopter Text Recieve NumberBytes " , NumberBytes + "");
     Log.d("ZombieCopter Text Recieve Partial Message",  m_Message);

     // Add Partial Message to Complete Data
     m_Data += m_Message;

     // Check if end of data has been reached
     int EndOfData = m_Data.indexOf(m_EndData);
     if (EndOfData >= 0)
     {
       // All Text Data Has been Read in
       EndTextDataFound = true;
     }
     return EndTextDataFound;
}
```

The function does the following:

1. Reads in the incoming binary data from the `Message` byte array and stores it in the `m_DataByte` array. The `m_DataIndex` variable that indicates the next empty position in the `m_DataByte` array is incremented.
2. If `m_DataIndex` is greater than or equal to the expected length of the incoming binary message which is held in `m_DataIncomingLength` then the incoming binary message has completed. Otherwise, more data needs to be read in from the microcontroller.

See Listing 4-13.

The `ReceiveMessage()` Function

The `ReceiveMessage()` function is called whenever data from the microcontroller is received by the Android using Wi-Fi.

The function does the following:

1. If `m_Freeze = true` then do not process any further input from the microcontroller.
2. If the Android command that is waiting on a response from the microcontroller is "GetImageSize" then process the command by calling the `ProcessGetImageSizeCommand(NumberBytes, Message)`

Listing 4-13 The `ReceiveBinaryData()` Function

```
boolean ReceiveBinaryData(int
NumberBytes, byte[] Message)
{
 boolean Finished = false;

 // Add incoming Binary data to bytes
data array
 for (int i = 0; i < NumberBytes; i++)
 {
  m_DataByte[m_DataIndex] = Message[i];
  m_DataIndex++;
 }

 // Check to see if all binary data
has been received
 if (m_DataIndex >= m_DataIncomingLength)
 {
  Finished = true;
 }
 return Finished;
}
```

function with the parameters `Message` which contains an array of bytes from the microcontroller and `NumberBytes` which contains the number of valid bytes in the array. This command retrieves the size of the image that was just taken and the size is in text format.

3. If the Android command is "GetImageData" then process the incoming data by calling the `ProcessGetImageDataCommand(NumberBytes, Message)` function which reads in the image data from the microcontroller.
4. If the Android command is "GetTextData" then call the `ProcessReceiveTextDataCommand(NumberBytes, Message)` function. This reads in incoming text data from the microcontroller.
5. If `m_Command` which is the Android command that is waiting on a response from the microcontroller does not match any of the above then print out an error message by calling the `Log()` function.

See Listing 4-14.

Listing 4-14 The `ReceiveMessage()` Function

```
void ReceiveMessage(int NumberBytes,
byte[] Message)
{
 // Check to see if this thread has
been frozen by the main activity
thread
 if (m_Freeze == true)
 {
  return;
 }

 // Process Incoming Data
 // Assume incoming data is
associated with the current m_Command
variable
 // and process the Message
accordingly.
 if (m_Command == "GetImageSize")
 {
  ProcessGetImageSizeCommand(Number
 Bytes, Message);
 }
 else
 if (m_Command == "GetImageData")
 {
  ProcessGetImageDataCommand(Number
 Bytes, Message);
 }
 else
 if (m_Command == "GetTextData")
 {
  ProcessReceiveTextDataCommand(Number
 Bytes, Message);
 }
 else
 {
  Log.e("WIFI Handler" , "Error -
 Command for Data Receive Not
 Found!!!!!!");
 }
}
```

The ProcessGetImage SizeCommand() Function

The ProcessGetImageSizeCommand() processes the size of the image that was just captured by the microcontroller's camera.

The ProcessGetImageSizeCommand() function does the following:

1. Reads in incoming data assuming that the data is a text string and is terminated by the character held in m_EndData by calling the ReceiveTextData(NumberBytes, Message) function.
2. If the incoming data has all been read in then:
    1. If this thread has not been halted then continue execution of this function.
    2. Set up the MainActivity's user interface to be updated by calling the m_MainActivity.TakePhotoCommandCallback() function.
    3. The MainActivity's user interface is actually updated by calling the m_MainActivity.runOnUiThread(m_MainActivity) function which executes the MainActivity class's run() function.

See Listing 4-15.

Listing 4-15 The ProcessGetImage SizeCommand() Function

```
void ProcessGetImageSizeCommand(int
NumberBytes, byte[] Message)
{
 boolean FinishedReceivingText =
ReceiveTextData(NumberBytes, Message);
  if (FinishedReceivingText)
  {
   if (m_Freeze == false)
   {
     // Process Remote Directory
    Command
     m_MainActivity.
    TakePhotoCommandCallback();

     // Update the User Interface
     m_MainActivity.runOnUiThread
    (m_MainActivity);
   }
  }
}
```

The ProcessGetImage DataCommand() Function

The ProcessGetImageDataCommand() function processes the actual incoming image data that was taken by the camera.

The ProcessGetImageDataCommand() function does the following:

1. Starts to receive the incoming binary data from the microcontroller by calling the ReceiveBinaryData(NumberBytes, Message) function.
2. If the binary data has finished being read in and the client thread has not been halted then set up the MainActivity class to process the data by calling the m_MainActivity.TakePhotoCommandCallback() function.
3. Performs the actual update to the MainActivity's user interface by calling the m_MainActivity.runOnUiThread(m_MainActivity) function.

See Listing 4-16.

The ProcessReceiveText DataCommand() Function

The ProcessReceiveTextDataCommand() function processes incoming text data from the microcontroller and illustrates how you can read in generic text data from the microcontroller and have it processed by the MainActivity class.

Listing 4-16 The `ProcessGetImageDataCommand()` Function

```
void ProcessGetImageDataCommand
(int NumberBytes, byte[] Message)
{
   boolean FinishedReceivingData =
  ReceiveBinaryData(NumberBytes,
  Message);
  if (FinishedReceivingData)
  {
   if (m_Freeze == false)
   {
     // Process Take Picture Command
     m_MainActivity.
    TakePhotoCommandCallback();

     // Update the User Interface
     m_MainActivity.runOnUiThread
    (m_MainActivity);
   }
  }
}
```

The function does the following:

1. Reads in text data from the microcontroller by calling the `ReceiveTextData(NumberBytes, Message)` function.
2. If all the text data has been read in and the execution of this class has not been halted then prepare the `MainActivity's` user interface to be updated by calling the `m_MainActivity.SetRecieveTextDataCallbackFinished()` function.
3. The actual update in the `MainActivity` class using the newly read in text data is done by calling the `m_MainActivity.runOnUiThread(m_MainActivity)` function which calls the `MainActivity` class's `run()` function.

See Listing 4-17.

The `Client` Class

The `Client` class is a custom class I created that extends the `Thread` class which is a standard built-in Android class.

Listing 4-17 The `ProcessReceiveTextDataCommand()` Function

```
void ProcessReceiveTextDataCommand
(int NumberBytes, byte[] Message)
{
   boolean FinishedReceivingText =
  ReceiveTextData(NumberBytes,
  Message);
   if (FinishedReceivingText)
   {
     if (m_Freeze == false)
     {
       // Process Receive Text Data
      Command
       m_MainActivity.SetRecieve
      TextDataCallbackFinished();

       // Update the User Interface
       m_MainActivity.runOnUiThread
      (m_MainActivity);
     }
   }
}
```

```
public class Client extends Thread
```

The important thing to take notice of here is that this class is a thread which executes separately from the `MainActivity` thread that controls the GUI. Also, in some versions of Android all Wi-Fi activities must be in a thread other than the `MainActivity` thread. The reason being that Wi-Fi activity can block other functions such as processing user input so that the user would incorrectly believe that the application has stopped working.

The `TAG` variable is a `String` that is used in combination with the `Log()` function to indicate that it is the `Client` class that `Log()` debug output print statements belong to.

```
String TAG = "CLIENT";
```

The key class to manage and control the Android's Wi-Fi is called the `WifiManager` class represented by the variable `m_WifiManager`.

```
WifiManager m_WifiManager = null;
```

In order to keep the Android's Wi-Fi on continuously even when the screen goes dark we need to use the WifiLock class represented by the m_WifiLock variable.

```
WifiLock    m_WifiLock = null;
```

In order to retrieve information about the Android's current Wi-Fi connection we need to use the WifiInfo class which is represented by the variable m_WifiInfo.

```
WifiInfo    m_WifiInfo = null;
```

The MacAddress of the Android device is held in the variable m_MacAddress.

```
String m_MacAddress = "";
```

The speed of the current Wi-Fi connection is held in the variable m_LinkSpeed.

```
int m_LinkSpeed = 0;
```

In order to determine if the Android's Wi-Fi is actually connected to the microcontroller we need to use the ConnectivityManager class.

```
ConnectivityManager m_
ConnectivityManager =    null;
```

The m_WifiConnectInfo variable holds information relating to the state of the Android's Wi-Fi connection with the microcontroller.

```
NetworkInfo m_WifiConnectInfo = null;
```

The access point's SSID which is the name for the access point located on the ESP-01 is held in the m_SSID variable as a String object.

```
String m_SSID = "";
```

The IP address of the server to connect to is held in the m_ServerIPAddressString variable.

```
String   m_ServerIPAddressString = "";
```

The port number for the server that is running on the microcontroller device is held in the m_ServerPortNumber variable.

```
int m_ServerPortNumber = 0;
```

The IP address of the Android phone that serves as the client received from the WifiManager class is stored here.

```
int m_ClientIPAddress = -1;
```

The m_ClientIPAddressString holds the IP address that was stored in m_ClientIPAddress in String format instead of integer format.

```
String m_ClientIPAddressString = "";
```

The m_ClientIPByteArray variable holds the Android's IP address in byte array format.

```
byte[] m_ClientIPByteArray = null;
```

The m_Socket variable holds the Android Socket object that will be used to connect to the microcontroller's TCP server.

```
Socket m_Socket = null;
```

The m_IsConnected variable is true if the Android client is connected to the server and false otherwise.

```
volatile boolean m_IsConnected = false;
```

The m_SocketOutputStream variable is an OutputStream class object that is used to send data from the Android to the microcontroller device.

```
OutputStream  m_SocketOutputStream  = null;
```

The m_SocketInputStream variable is an InputStream class object that is used to read in data from the microcontroller device.

```
InputStream m_SocketInputStream = null;
```

The m_WifiMessageHandler variable handles the incoming data from the microcontroller based upon the command that was initially sent to the microcontroller device from the Android device.

```
WifiMessageHandler m_WifiMessageHandler
= null;
```

The m_MainActivity variable holds a reference to the MainActivity class object that created this Client object instance.

```
MainActivity m_MainActivity = null;
```

The m_Freeze variable stops the execution of the thread if it is true. This variable is declared volatile so that a different thread can set this variable and this thread will see this variable change to the new value.

```
volatile boolean m_Freeze = false;
```

The `run()` Function

The `run()` function is executed when the `start()` function is called on the thread.

The `run()` function sets up the Wi-Fi, creates the client socket on the Android side, handles the connection to the microcontroller, and continually reads in and processes incoming Wi-Fi data from the microcontroller device. The `run()` function does this by:

1. Calling the `InitWifiManager()` function to initialize the Wi-Fi manager class object on the Android device.
2. Continuously checking to see if the Android has been connected to a Wi-Fi network. If it is not then check to see if the thread should be stopped. If the thread should be stopped then return from the `run()` function and stop execution. Otherwise, suspend program execution for 1000 milliseconds and then repeat step 2.
3. Retrieve the connection information for the currently connected Wi-Fi connection by calling the `GetWifiConnectionInfo()` function.
4. If the information was successfully retrieved from step 3 then call the `DisplayWifiConnectionInfoDebug()` function to display the Wi-Fi connection information on the Android program's debug window. Otherwise, exit out the `run()` function and print out an error using the `Log()` function.
5. Create a client socket and try to connect this to the server IP address and server port on the microcontroller by calling the `CreateConnectSocket()` function.
6. If the Android and microcontroller are connected by Wi-Fi then start to read in incoming data. Otherwise exit out the `run()` function.
7. In order to read in data from the microcontroller the program:
    1. Creates an array of bytes called "buffer" of length 1024.
    2. Creates an integer variable called "bytes" that holds the number of bytes that are read from the Wi-Fi connection.
    3. Reads in the incoming data from the Wi-Fi connection by calling the `m_SocketInputStream.read(buffer)` function with the buffer created in step 1. The function returns the number of bytes read and stores this in the bytes variable.
    4. Calls the `m_WifiMessageHandler.ReceiveMessage(bytes, buffer)` function which processes the incoming data using the `m_WifiMessageHandler` variable. The two input parameters are the number of bytes read in from step 3 and the buffer created in step 1.
    5. If there is an error reading in the data using the `read()` command in step 3 then an exception is thrown and the run() function is exited.

See Listing 4-18.

The `InitWifiManager()` Function

The `InitWifiManger()` function initializes the Android's Wi-Fi by:

1. Creating a new `WifiManager` object by calling the `m_MainActivity.getSystemService(MainActivity.WIFI_SERVICE)` function and setting the result to the `m_WifiManager` variable.
2. Creating a new `ConnectivityManager` object by calling the `m_MainActivity.`

Listing 4-18 The `run()` Function

```
@Override
public void run()
{
 // Init Wifi
 InitWifiManager();

 // Check for Network Connectivity
 while (!IsConnectedToWifi())
 {
   // If thread has been frozen then
stop execution and return from run()
function
   if (m_Freeze)
   {
     return;
   }

   // while wifi is not connected
then do nothing.
   try {
     sleep(1000);
   } catch (InterruptedException e) {
     // TODO Auto-generated catch
   block
     Log.e(TAG, "Sleep Command FAILED
   !!!!");
     e.printStackTrace();
   }
 }

 if (GetWifiConnectionInfo())
 {
  DisplayWifiConnectionInfoDebug();
 }
 else
 {
  // Get Wifi Connection Info Failed.
   Log.e(TAG, "Getting Wifi
  Connection Info Failed !!!");
   return;
 }

 // Try to Connect Socket to Server
 CreateConnectSocket();

 // If connection is not successfull
  then return
 if (!m_IsConnected)
 {
  return;
 }

 // Continuously read in data from
microcontroller
 byte[] buffer = new byte[1024];  //
buffer store for the stream
 int bytes; // number bytes returned
from read()

 // Keep listening to the InputStream
until an exception occurs
 while (m_IsConnected)
 {
  try
  {
   // Read from the InputStream
   bytes = m_SocketInputStream.
  read(buffer);

   // Send the obtained bytes to the
  UI activity
   m_WifiMessageHandler.
  ReceiveMessage(bytes, buffer);
  }
  catch (IOException e)
  {
   Log.e(TAG, "Error Reading Wifi
  DATA ... ERROR = " + e.toString());
   break;
  }
 }
}
```

`getSystemService(MainActivity.CONNECTIVITY_SERVICE)` function and setting the result to the m_ConnectivityManager variable.

3. If the Android's Wi-Fi is enabled then set a debug message to be printed out on the Android's debug message window indicating that the Wi-Fi is on.
4. If the Android's Wi-Fi is not enabled then turn on the Android's Wi-Fi by calling the

m_WifiManager.setWifiEnabled(true) function with the parameter true and then set a debug message to be displayed in the debug message window indicating that the Wi-Fi is being turned on.

5. If this thread instance has not been stopped by setting m_Freeze to true then update the program's user interface by calling the m_MainActivity.runOnUiThread(m_MainActivity) function. This calls the MainActivity class's run() function which will update the user interface data with the changes from the Client class.

See Listing 4-19.

Listing 4-19 The InitWifiManager() Function

```
void InitWifiManager()
{
 m_WifiManager = (WifiManager)
m_MainActivity.getSystemService
(MainActivity.WIFI_SERVICE);
 m_ConnectivityManager =
(ConnectivityManager)m_MainActivity.
getSystemService(MainActivity.
CONNECTIVITY_SERVICE);
 if (m_WifiManager.isWifiEnabled())
 {
  m_MainActivity.
 AddDebugMessageThread("Wifi is on
 ..." + "\n");
 }
 else
 {
  m_WifiManager.setWifiEnabled(true);
  m_MainActivity.
 AddDebugMessageThread("Wifi is NOT
 on. Turning on Wifi ..." + "\n");
 }
 if (m_Freeze == false)
 {
  // Update the User Interface
  m_MainActivity.runOnUiThread(m_
 MainActivity);
 }
}
```

The IsConnectedToWifi() Function

The IsConnectedToWifi() function returns true if the Android is connected to the server through Wi-Fi and returns false otherwise.

This function does the following:

1. First gets the current Wi-Fi network information through the connectivity manager by using the getNetworkInfo(ConnectivityManager.TYPE_WIFI) function which retrieves the current Wi-Fi connection information.
2. Returns the Wi-Fi connection state by calling the m_WifiConnectInfo.isConnected() function and returning the result.

See Listing 4-20.

The GetWifiConnectionInfo() Function

The GetWifiConnectionInfo() function gets information regarding the Android's Wi-Fi connection to the microcontroller device.

This function does the following:

1. If the Android's Wi-Fi is enabled then:
    1. The connection information structure containing all the network connection information is retrieved by calling the m_WifiManager.getConnectionInfo() function and is stored in the m_WifiInfo variable.

Listing 4-20 The IsConnectedToWifi() Function

```
boolean IsConnectedToWifi()
{
 m_WifiConnectInfo =
m_ConnectivityManager.getNetworkInfo
(ConnectivityManager.TYPE_WIFI);
 return m_WifiConnectInfo.
isConnected();
}
```

2. The speed of the Wi-Fi connection is retrieved by calling the `m_WifiInfo.getLinkSpeed()` function.
3. The IP address of the Android is retrieved by calling the `m_WifiInfo.getIpAddress()` function.
4. The access point's name is retrieved by calling the `m_WifiInfo.getSSID()` function.
5. The Android device's MAC address is retrieved by calling the `m_WifiInfo.getMacAddress()` function.
6. If the Wi-Fi connection information has been retrieved then true is returned otherwise false is returned.

See Listing 4-21.

Listing 4-21 The `GetWifiConnectionInfo()` Function

```
boolean GetWifiConnectionInfo()
{
 boolean result = false;
 if (m_WifiManager.isWifiEnabled())
 {
  //Return dynamic information about
 the current Wi-Fi connection, if any
 is active.
  m_WifiInfo = m_WifiManager.
 getConnectionInfo();

  // Get Wifi Connection Info
  m_LinkSpeed =m_WifiInfo.
 getLinkSpeed();
  m_ClientIPAddress = m_WifiInfo.
 getIpAddress();
  m_SSID = m_WifiInfo.getSSID();
  m_MacAddress = m_WifiInfo.
 getMacAddress();

  result = true;
 }
 return result;
}
```

The `DisplayWifiConnectionInfoDebug()` Function

The `DisplayWifiConnectionInfoDebug()` function displays the current Wi-Fi connection information inside the debug window of the Android application.

The `DisplayWifiConnectionInfoDebug()` function does the following:

1. Converts the Android's client IP address from a number to a `String` object by calling the `ConvertClientIPAddressToString(m_ClientIPAddress)` function with the numerical version of the IP address retrieved previously in the `GetWifiConnectionInfo()` function.
2. Sets up for the printing of the client IP address to the debug window.
3. Sets up for the printing of the static IP address of the microcontroller to the debug window.
4. Gets the IP address associated with the microcontroller server by calling the `InetAddress.getAllByName(m_ServerIPAddressString)` function with the `String` value of the server's IP address and setting the IA variable to the first element of the returned array. The IA variable now holds the Internet address of the server in the form of an `InetAddress` class object.
5. If the IA variable is not null then set up for the printing of the server IP address.
6. Sets up for the printing of the access point's SSID to the debug window.
7. Sets up for the printing of the link speed and the Android's mac address to the debug window.
8. If the thread has not been halted by setting the `m_Freeze` variable to true then do the actual update of the debug window by calling the `m_MainActivity.`

runOnUiThread(m_MainActivity) function which executes the run() function in the MainActivity class.

See Listing 4-22.

The ConvertClientIPAddressToString() Function

The ConvertClientIPAddressToString() function converts an IP address that is in integer format into one that is in String format and then returns it.

The ConvertClientIPAddressToString() function does the following:

1. If the input IP address is in little endian format then reverse the bytes in the number to big endian format. Big endian format is where the most significant byte in the number is stored in the lowest memory

Listing 4-22 The DisplayWifiConnectionInfoDebug() Function

```
void DisplayWifiConnectionInfoDebug()
{
 m_ClientIPAddressString = ConvertClientIPAddressToString(m_ClientIPAddress);
 m_MainActivity.AddDebugMessageThread("Android Client IP Address: " + m_
ClientIPAddressString + "\n");

 // AP Info
 m_MainActivity.AddDebugMessageThread("AP/ microcontroller Static IP Address: " +
m_ServerIPAddressString + "\n");

 InetAddress IA = null;
 try {
  IA = InetAddress.getAllByName(m_ServerIPAddressString)[0];
 } catch (UnknownHostException e) {
  m_MainActivity.AddDebugMessageThread(TAG + ": Can not get Server InetAddress from
 getAllByName(), Error = " + e.toString()  +  "\n");
  e.printStackTrace();
 }
 if (IA != null)
 {
  m_MainActivity.AddDebugMessageThread("Access Point Name: " + IA.getHostName() +
 "\n");
 }
  m_MainActivity.AddDebugMessageThread("Access Point SSID: " + m_SSID + "\n");
  // Connection Info
  m_MainActivity.AddDebugMessageThread("Link Speed: " + m_LinkSpeed + "\n");
  m_MainActivity.AddDebugMessageThread("Android Client MacAddress: " + m_MacAddress
 + "\n");

 if (m_Freeze == false)
 {
  // Update the User Interface
  m_MainActivity.runOnUiThread(m_MainActivity);
 }
}
```

location and the least significant byte is stored in the highest memory location.

2. Converts the IP address into an array of bytes and store it in the `m_ClientIPByteArray` variable by calling the `BigInteger.valueOf(IPAddress).toByteArray()` function with the IP address as the parameter.
3. Gets the `InetAddress` class object representation of the IP address by calling the `InetAddress.getByAddress(m_ClientIPByteArray)` function with the byte array calculated in step 2. Then gets the `String` representation by calling the `getHostAddress()` function of the `InetAddress` class object.
4. If there is an exception then writes out error messages by calling the `Log()` function and the `m_MainActivity.AddDebugMessageThread()` to add an error message to the debug window.
5. Returns the `String` representation of the IP address that was input to the function.

See Listing 4-23.

Listing 4-23 The `ConvertClientIPAddressToString()` Function

```
String ConvertClientIPAddressToString(
int IPAddress)
{
 String result = "N/A";

 // Convert little-endian to big-
endian if needed
 if (ByteOrder.nativeOrder().
equals(ByteOrder.LITTLE_ENDIAN))
 {
  IPAddress = Integer.
 reverseBytes(IPAddress);
 }

 m_ClientIPByteArray = BigInteger.
valueOf(IPAddress).toByteArray();
 try {
  result = InetAddress.
 getByAddress(m_ClientIPByteArray).
 getHostAddress();
 } catch (UnknownHostException ex) {
  Log.e("CLIENT", "Unable to get
 host address from Network Byte
 Array...");
  m_MainActivity.
 AddDebugMessageThread(TAG + ":Unable
 to get host address from Network
 Byte Array...\n");
  return result;
 }
 return result;
}
```

The `CreateConnectSocket()` Function

The `CreateConnectSocket()` function attempts to connect the Android to the microcontroller device over a Wi-Fi connection by doing the following:

1. Setting up a message to be printed to the debug window notifying the user that the Android is going to attempt to connect to the microcontroller device.
2. If the thread has not been frozen then update the debug window by calling the `m_MainActivity.runOnUiThread()` function.
3. Calling the `ConnectSocket()` function to perform the actual Wi-Fi connection between the Android and the microcontroller. If true is returned then:
    1. Set up a message to be printed to the debug window that notifies the user that the Wi-Fi connection has been established.
    2. Create a Wi-Fi lock that prevents the Android's Wi-Fi from shutting off by calling the `CreateWifiLock()` function.

3. Set `m_IsConnected` to true to indicate that the Wi-Fi connection has now been established.
4. Set up a message to be printed to the debug window that notifies the user that the Wi-Fi connection has been established by calling the `m_MainActivity.WifiSocketConnectedMessage()` function.
4. If the `ConnectSocket()` function returns false then set up a message to be printed to the debug window that the connection to the microcontroller has failed.
5. If this thread is still active and not stopped then do the actual update to the debug window by calling the `m_MainActivity.runOnUiThread(m_MainActivity)` function.

See Listing 4-24.

Listing 4-24 The `CreateConnectSocket` Function

```
void CreateConnectSocket()
{
  // Connect the Socket to the destination machine at IPAddress on PortNumber
  m_MainActivity.AddDebugMessageThread(TAG + ": Client Thread is Attempting to
Connect to Server at '" + m_ServerIPAddressString + "' on port number " + m_
ServerPortNumber + "\n");
  if (m_Freeze == false)
  {
    m_MainActivity.runOnUiThread(m_MainActivity);
  }

  if (ConnectSocket())
  {
    m_MainActivity.AddDebugMessageThread(TAG + ": Client Thread is Connected to Server
  !!! \n");

    // Since Thread is now connected we need to lock the wifi so that it does not turn
  off.
    CreateWifiLock();
    m_IsConnected = true;
    m_MainActivity.WifiSocketConnectedMessage();
  }
  else
  {
    m_MainActivity.AddDebugMessageThread(TAG + ": Client Thread FAILED to Connect to
  Server !!! \n");
  }

  if (m_Freeze == false)
  {
    // Update the User Interface
    m_MainActivity.runOnUiThread(m_MainActivity);
  }
}
```

The `ConnectSocket()` Function

The `ConnectSocket()` function performs the actual Wi-Fi connection procedure of connecting the Android client thread to the microcontroller device's TCP server.

The `ConnectSocket()` function does the following:

1. Sets the result variable true if the connection was successful and false if it has failed to the default value of false.
2. Creates a new Socket object and assigns this to `m_Socket`.
3. Retrieves the optional Socket information on the "keep alive" status by calling the `m_Socket.getKeepAlive()` function. With TCP the "keep alive" function is optional and defaults to false. By default TCP sockets like we are using do not automatically time out and disconnect. A "keep alive" message is sent between devices to maintain the connection and to make sure the connection is operating correctly. If there is an exception generated then print out an error message using the `Log()` function.
4. The socket created in step 1 is bound to the local host which is the Android device on any available port number by calling the `m_Socket.bind(null)` function with null sent as a parameter. The null value indicates that the socket should be bound to any free port on the local host. If the socket binding fails then print out error messages.
5. Retrieves the IP address of the server running on the microcontroller and assigns it to the IA variable by calling the `InetAddress.getAllByName(m_ServerIPAddressString)` function and referencing the first element of the returned array. The input parameter is the server IP address in string format. If there is an error in getting the IP address then error messages are output.
6. Creates a new `InetSocketAddress` class variable called `SocketAddress` that holds the socket address of that target we want to connect to which includes the IP address of the server and the port number that the TCP server is using.
7. Connects to the server by calling the `m_Socket.connect(SocketAddress)` function with the input parameter of the `SocketAddress` variable created in step 6. The `m_Socket` variable was created in step 2. If there is an error then error messages are printed out.
8. Retrieves the `OutputStream` class object that we will use to send data to the microcontroller device by calling the `m_Socket.getOutputStream()` function and assigning the returned value to the `m_SocketOutputStream` variable.
9. Retrieves the `InputStream` class object that we will use to read data from the microcontroller device by calling the `m_Socket.getInputStream()` function and assigning the returned value to the `m_SocketInputStream` variable.
10. If both the `OutputStream` and `InputStream` objects are successfully retrieved then the result variable is set to true. Otherwise an exception is generated and error messages are printed out.
11. The result variable is returned which indicates the success or failure of the connection attempt.

See Listing 4-25.

The `CreateWifiLock()` Function

The `CreateWifiLock()` function locks the Android's Wi-Fi so that the Wi-Fi radio is always on.

Listing 4-25 The `ConnectSocket()` Function

```
boolean ConnectSocket()
{
 boolean result = false;
 InetAddress IA = null;
 InetSocketAddress      SocketAddress = null;

 // Create new socket
 m_Socket = new Socket();

 try {
  boolean keepalive = m_Socket.getKeepAlive();
  Log.e(TAG,"KeepAlive socket value = " + keepalive);
 } catch (SocketException e2) {
  // TODO Auto-generated catch block
  Log.e(TAG,"ERROR getting Keep Alive status of socket !!!");
  e2.printStackTrace();
 }

 // Bind socket to local host
 try {
  m_Socket.bind(null);
 } catch (IOException e) {
  Log.e(TAG,"Client Thread Socket Bind failed !!! Error = " + e.toString());
  m_MainActivity.AddDebugMessageThread(TAG + ": Client Thread Socket Bind failed !!!
 Error = " + e.toString() + "\n");
  e.printStackTrace();
  return result;
 }

 // Get the IP address of machine to connect to
 try {
  IA = InetAddress.getAllByName(m_ServerIPAddressString)[0];
 } catch (UnknownHostException e1) {
  Log.e(TAG,"Getting Server InetAddress by getAllByName() Failed ! ERROR = " +
 e1.toString());
  m_MainActivity.AddDebugMessageThread(TAG + ": Getting Server InetAddress by
 getAllByName() Failed ! ERROR = " + e1.toString() + "\n");
  e1.printStackTrace();
  return result;
 }
 // Create a new SocketAddress based on the InetAddress and port number of machine
to
```

```
  // connect to.
  if (IA != null)
  {
   SocketAddress = new InetSocketAddress(IA, m_ServerPortNumber);
  }

  // Connect the socket to the destination machine on the port number specified by
 the SocketAddress object.
  try {
   m_Socket.connect(SocketAddress);
  } catch (IOException e) {
   Log.e(TAG,"Socket connection FAILED! ERROR = " + e.toString());
   m_MainActivity.AddDebugMessageThread(TAG + ": Socket Connection FAILED ! ERROR = "
  + e.toString() + "\n");
   e.printStackTrace();
   return result;
  }

  // Get Output and input stream from Socket for writing and reading data
  try {
   m_SocketOutputStream = m_Socket.getOutputStream();
   m_SocketInputStream = m_Socket.getInputStream();
   result = true;
  } catch (IOException e) {
   Log.e(TAG,"Cannot get output/input streams from socket! ERROR = " + e.toString());
   m_MainActivity.AddDebugMessageThread(TAG + ": Cannot get output/input streams
  from socket! ERROR = " + e.toString());
   e.printStackTrace();
   return result;
  }
  return result;
 }
```

The `CreateWifiLock()` function does the following:

1. Gets a `WifiLock` object from the current `WifiManager` by calling the `m_WifiManager.createWifiLock(TAG)` function with the `String` that will identify this particular `WifiLock` object in debug messages.
2. Locks the Android's Wi-Fi by calling the `m_WifiLock.acquire()` function.
3. If a call to the `m_WifiLock.isHeld()` function returns true then set up a message to be added to the debug window that notifies the user that the Wi-Fi has been successfully locked. Otherwise, the message will notify the user that the Wi-Fi has not been successfully locked.

See Listing 4-26.

Listing 4-26 The `CreateWifiLock()` Function

```
void CreateWifiLock()
{
 // WIFI LOCK
 m_WifiLock = m_WifiManager.
createWifiLock(TAG);
 m_WifiLock.acquire();
 if (m_WifiLock.isHeld())
 {
  m_MainActivity.
 AddDebugMessageThread("WifiLock is
 aquired ..." + "\n");
 }
 else
 {
  m_MainActivity.
 AddDebugMessageThread("ERROR !
 WifiLock aquire FAILED !!!" + "\n");
 }
}
```

Listing 4-27 The `IsConnected()` Function

```
boolean IsConnected()
{
 return m_IsConnected;
}
```

The `IsConnected()` Function

The `IsConnected()` function returns true if there is a Wi-Fi connection between the Android and the microcontroller using this `Client` class. Otherwise, it returns false. See Listing 4-27.

The `Client` Class Constructor

The `Client` class constructor initializes key class member variables with input from the `MainActivity` class. See Listing 4-28.

The `Freeze()` Function

The `Freeze()` function halts the execution of the class by setting `m_Freeze` to true and calling the `m_WifiMessageHandler.Freeze()` function in the Wi-Fi message handler for this class. See Listing 4-29.

Listing 4-28 The `Client` Constructor

```
Client(String ServerStaticIP, int
 PortNumber, WifiMessageHandler
 iWifiMessageHandler, MainActivity
 iMainActivity)
{
 // Initialization
 m_MainActivity = iMainActivity;
 m_ServerPortNumber = PortNumber;
 m_WifiMessageHandler =
iWifiMessageHandler;
 m_ServerIPAddressString =
ServerStaticIP;
}
```

Listing 4-29 The `Freeze()` Function

```
void Freeze()
{
 m_Freeze = true;

 // Freeze WIFI Message Handler
 m_WifiMessageHandler.Freeze();
}
```

The `Write()` Function

The `Write()` function sends data in the form of bytes to the microcontroller over Wi-Fi by calling the `m_SocketOutputStream.write(bytes)` function with the input parameter bytes which is an array of type byte. See Listing 4-30.

The `Cancel()` Function

The `Cancel()` function prepares the Wi-Fi connection to be terminated by:

1. Halting the execution of this thread by setting `m_IsConnected` to false indicating

Listing 4-30 The `Write()` Function

```
public void write(byte[] bytes)
{
 try
 {
  m_SocketOutputStream.write(bytes);
 }
 catch (IOException e)
 {
  Log.e(TAG,"Error Writing Bytes
 to SocketOutputStream. Error = " +
 e.toString());
 }
}
```

that the Android is no longer connected to the microcontroller.

2. If a Wi-Fi lock exists then release the lock by calling the `m_WifiLock.release()` function.
3. Close the open client connection to the TCP server on the microcontroller device by calling the `m_Socket.close()` function. If there is an error then print out an error message.

See Listing 4-31.

Listing 4-31 The `Cancel()` Function

```
public void cancel()
{
 // Stop Reading from Socket
 m_IsConnected = false;

 // Release WifiLock
 if (m_WifiLock != null)
 {
  Log.e(TAG, "Releasing m_WifiLock ...");
  m_WifiLock.release();
 }

 try
 {
  // Close open client socket to
 TCP server
  m_Socket.close();
 }
 catch (IOException e)
 {
  Log.e(TAG,"Error closing Wifi
 Socket ...");
 }
}
```

The `MainActivity` Class

The `MainActivity` class contains the main code for the basic wireless framework and alarm system. The code to handle the user interface, the alarm system settings and logic, and the cell phone communication system are all contained in this class.

The `MainActivity` class extends the `Activity` class and implements the `Runnable` interface by implementing the `run()` function. The class is declared as:

```
public class MainActivity extends
Activity implements Runnable
```

The key functions in this class are:

- `onCreate()` – This function is called when the Android application is created and is used to initialize the basic wireless framework.
- `OnOptionsItemSelected()` – This function is called when the user touches items in the graphical user interface belonging to the application.
- `run()` – This function is called from another thread and is used to process incoming data and to update the graphical user interface after new information is loaded into the Android from the microcontroller.
- `onDestroy()` – This function is called when the application ends and is used to shut down the Wi-Fi connection and other items.

The `MainActivity` Class Variables

The `TAG String` object variable is used with the `Log()` function for debugging purposes to indicate that an error or debug message came from the `MainActivity` class.

```
String TAG = "DIYWireless";
```

The version of this application is held in the `m_Version String` variable.

```
String m_Version = "DIY Wireless
Framework Version 1.0 \n";
```

The copyright notice for this application is held in `m_Copyright` and is a text string.

```
String m_Copyright = "Copyright
2018-2019 Robert Chin. All rights
reserved.\n";
```

The `m_EmergencyPhoneNumber` variable holds the phone number for the Android to call if the alarm system has been tripped. This number is set by the user in the phone entry text box on the Android application.

```
String m_EmergencyPhoneNumber ="";
```

The `m_EmergencyPhoneNumberHandle` variable holds the shared preferences handle that is used to save and then load in the value of the emergency phone number.

```
String m_EmergencyPhoneNumberHandle =
"PhoneNumber";
```

The `m_GotchaCamPreferencesFile` holds all the saved preference handles and the associated data for the security application including the handle and data pair for the emergency phone number.

```
String m_GotchaCamPreferencesFile =
"GotchaCamRecords";
```

If `m_CallOutDone` is true then the required emergency callout has already been completed. Otherwise the value is false.

```
boolean m_CallOutDone = false;
```

The `m_EmergencyMessageRepeatTimes` variable holds the number of times to repeat the vocal alarm generated by Android's text to speech function. For example, the default number of times to repeat the phrase "Intruder Alert Intruder has been detected by the motion sensor" is set to 1 time.

```
int m_EmergencyMessageRepeatTimes = 1;
```

The `m_EmergencyMessageDone` variable is true if the required speech warning of an intruder alert has started to execute and false otherwise.

```
boolean m_EmergencyMessageDone = false;
```

The `m_ServerStaticIP` variable holds the IP address of the server on the ESP-01.

```
String m_ServerStaticIP = "192.168.4.1";
```

The `TextMessageSetting` enumeration contains the possible values of the emergency text message alert setting. The `TEXT_OFF` value means that no emergency text messages will be sent. The `TEXT_ON` value means that emergency text messages will be sent

```
enum TextMessageSetting
{
  TEXT_OFF,
  TEXT_ON
};
```

The `m_AlarmSet` value is true if the alarm system is on. The value is false otherwise.

```
boolean m_AlarmSet = false;
```

The `m_AlarmTripped` variable is true if the alarm has been tripped. It is false otherwise.

```
boolean m_AlarmTripped = false;
```

The `m_CallOutSet` variable is true if the alarm system is set to call out to the emergency phone number when the alarm has been tripped. If it is false then an intruder alert message just appears locally on the debug window of the Android application and a sound is played continuously and a text speech voice is played to announce that an intruder has been detected by the motion detector.

```
boolean m_CallOutSet = true;
```

The m_TextMessagesSetting holds the current status of the emergency text message alert setting. This value is defaulted to TEXT_ON which sends out an emergency text message to the emergency phone number when an intruder is detected.

```
TextMessageSetting m_TextMessagesSetting
= TextMessageSetting.TEXT_ON;
```

The m_TextMessageSent variable is true if an emergency text message has been sent after the alarm has been tripped and false otherwise.

```
boolean m_TextMessageSent = false;
```

The m_TTS variable is used for text to speech conversion so that the Android device can vocalize such things as intruder alerts and important system messages.

```
TextToSpeech m_TTS = null;
```

The m_PortNumber variable value is the port that the TCP server running on the ESP-01 is listening to for a connection request.

```
int m_PortNumber = 80;
```

The m_WifiMessageHandler variable processes the data being returned from the microcontroller.

```
WifiMessageHandler m_WifiMessageHandler
= null;
```

The m_ClientConnectThread is the class object that will be used to establish a Wi-Fi connection with the microcontroller as well as send commands to the microcontroller.

```
Client m_ClientConnectThread = null;
```

The m_ButtonActive variable is true if the “Send Data” button is active and able to be pressed. If it is false then the button is not active and cannot be pressed.

```
boolean m_ButtonActive = false;
```

The m_Button represents the “Send Data” button which the user presses to send data to the microcontroller.

```
Button        m_Button = null;
```

The m_InfoTextView TextView class object holds the name of the Android device, the emergency phone number, and other alarm system settings.

```
TextView m_InfoTextView = null;
```

The m_PhoneEntryView is a text entry box that allows the user to enter the emergency phone number.

```
EditText    m_PhoneEntryView = null;
```

The m_DebugMsgView represents the Debug Window on the Android app that displays the debug information for the application as well as alerts.

```
EditText    m_DebugMsgView   = null;
```

The m_DebugMsg string holds the text that will be displayed in the Debug Message Window.

```
String m_DebugMsg= "";
```

If m_RefreshMessageWindows is true then the m_DebugMsg string is printed to the debug message window.

```
boolean m_RefreshMessageWindows = false;
```

If the value of m_RecieveTextData CallBackDone is true then new data has just been received from the microcontroller.

```
boolean  m_RecieveTextDataCallBackDone  =
false;
```

The m_PingHandler variable is used to generate a ping to the server at a set interval. This is done to prevent the server from closing the connection to the client.

```
Handler m_PingHandler = new Handler();
```

The interval between pings is defined as 120 seconds.

```
long  m_PingInterval  =  1000  *  120;  //
Number milliseconds
```

The name of the Android device as defined for the wireless network is defined by default as "android."

```
String m_AndroidClientName = "android";
```

The handle that the android device name is stored under is "DeviceName."

```
String m_AndroidDeviceNameHandle =
"DeviceName";
```

The onCreate() Function

The onCreate() function in the MainActivity class is executed when the application is created and contains the code for the initialization of the basic framework. The @Override keyword before the function indicates that this function overrides the same function in a parent class.

The onCreate() function does the following:

1. Calls the parent function by calling the super.onCreate(savedInstanceState) function with the input parameter to the function savedInstanceState.
2. Sets the Android user interface layout by calling the setContentView(R.layout.activity_main) function with the resource ID for the main activity layout as the input parameter.
3. Fixes the screen orientation of the Android screen to the portrait orientation by calling the this.setRequestedOrientation(ActivityInfo.SCREEN_ORIENTATION_PORTRAIT) function. The screen orientation is maintained regardless of the orientation of the Android device during the execution of the application.
4. The computerized voice that is used to read out plain text for alerts and information is initialized by calling the TextToSpeechInit() function.
5. Initializes the sound pool that manages the playback and loading of the sound effects by calling the CreateSoundPool() function.
6. Initializes the application's sound effects by calling the CreateSound(this) function.
7. Finds the phone entry text editbox from the Android user interface by calling the (EditText) findViewById(R.id.phoneentry) function with the resource ID number for the phone entry text box. Assigns the returned value to m_PhoneEntryView.
8. Sets the phone entry text editbox to focusable by calling the m_PhoneEntryView.setFocusable(true) function with the true parameter. If an edit text box is focusable that means the user can click on it and then enter text into it using the Android keyboard.
9. Sets a function to execute when the phone entry edit text box is clicked by the user by calling the m_PhoneEntryView.setOnClickListener() function with a newly created View.OnClickListener() object which stops the notification sound effect from being played. This is done by calling the m_Alert2SFX.StopSound() function.
10. Loads in the emergency call out phone number that is stored on the Android's internal storage system by calling the LoadInPhoneNumber() function.
11. Sets the emergency phone number read in from step 10 into the phone text editbox by calling the m_PhoneEntryView.setText(m_EmergencyPhoneNumber) function with the phone number as a parameter.
12. Finds the debug message edit text window by calling the (EditText) findViewById(R.id.debugmsg) function with the resource ID of the debug message

window and assigns the return value to `m_DebugMsgView`.

13. Adds the application version number to the `m_DebugMsg String` object.
14. Adds the copyright message to the `m_DebugMsg String` object.
15. Adds the emergency phone number to the `m_DebugMsg String` object.
16. Sets the debug message window with the contents of the `m_DebugMsg String` object by calling the `m_DebugMsgView.setText(m_DebugMsg.toCharArray(), 0, m_DebugMsg.length())` function.
17. Disables the debug message window's ability to get the focus by calling the `m_DebugMsgView.setFocusable(false)` function with the false parameter. This means that the window cannot receive text input from the user by touching the window and entering the text using the Android keyboard.
18. Sets a new `OnClickListener` function for the `m_DebugMsgView` window that stops any notification sounds being played by calling the `m_Alert2SFX.StopSound()` function.
19. Sets a new `OnEditorActionListener` object for the phone entry edit text box by calling the `m_PhoneEntryView.setOnEditorActionListener()` function. The function does the following if the "done" key has been pressed:
    1. Sets the `m_EmergencyPhoneNumber` to the value the user entered in the phone entry edit text box.
    2. Next, it displays a short message called a toast by calling the `Toast()` function notifying the user that the emergency phone number has been changed.
    3. Next, it erases the current message being displayed in the debug window and adds a message that notifies the user that the emergency phone number has changed.
    4. Next, it hides the Android keyboard from the user by calling the `HideKeyboard()` function.
    5. Finally, it saves the phone number to the Android device's permanent storage by calling the `SavePhoneNumber()` function.
20. Finds the settings information text box by calling the `(TextView) findViewById(R.id.textView1)` function with the resource ID for the information text box. The returned value is assigned to `m_InfoTextView`.
21. Updates the settings information text box by calling the `UpdateCommandTextView()` function.
22. The settings information text box is disabled for editing by calling the `m_InfoTextView.setFocusable(false)` function with false as a parameter.
23. Finds the "Send Data" button from the Android's layout by calling the `(Button) findViewById(R.id.TestMessageButton)` function with the resource ID of the button as the input parameter. Assigns the result to `m_Button`.
24. Enables the "Send Data" button so the button can be activated by the user touching it by calling the `m_Button.setEnabled(true)` function with the true parameter.
25. Creates a new `OnClickListener()` object for the button by calling the `m_Button.setOnClickListener()` function. The `onClick()` function that is created sends the data in the phone entry text box to the microcontroller over Wi-Fi. The `onClick()`

function does the following when the user presses the "Send Data" button:

1. Plays a sound effect by calling the `m_Alert1SFX.PlaySound(false)` function with false as a parameter to indicate that this sound will not loop but be played only once.
2. Clears the Wi-Fi message handler of old stored data and prepares it to accept new data by calling the `m_WifiMessageHandler.ResetData()` function.
3. Sets the command type that will be used to process the incoming data from the server by calling the `m_WifiMessageHandler.SetCommand("GetTextData")` function with the "GetTextData" parameter which is the command type.
4. Gets the command in the form of text data from the phone entry text box and adds a newline character to the end of the string.
5. If the client thread is connected which means that `m_ClientConnectThread.IsConnected()` evaluates to true then disable the button and send the command to the server over Wi-Fi by calling the `m_ClientConnectThread.write(Command.getBytes())` function. The function's parameter converts the `Command String` object into an array of bytes before sending it over Wi-Fi to the server.

26. Creates the client socket connection thread and begins its execution that is designed to connect the Android to the server using Wi-Fi by calling the `CreateClientConnection(m_ServerStaticIP, m_PortNumber)` function. The first parameter is the IP address of the server device and the second parameter is the port number that the TCP server is listening to for a connection request.
27. Loads in the name that will be used for the Android device on the wireless network by calling the `LoadInAndroidDeviceName()` function. Updates the information window by calling the `UpdateCommandTextView()` function. The name of the Android device should be updated.
28. Creates a new `Runnable` object called r that will be used to handle the ping updates to the server. In the `run()` function of the object a ping is sent to the server by calling the `PingHandler()` function. The next iteration of the `run()` function is then scheduled by calling the `m_PingHandler.postDelayed(this, m_PingInterval)` function which delays the next execution of the `run()` function by `m_PingInterval` time.
29. The ping handler called `r` that was created in the previous step is initialized by scheduling it to run in `m_PingInterval` time in the future by calling the `m_PingHandler.postDelayed(r, m_PingInterval)` function.
30. Finally, the Android device screen is prevented from shutting down by calling the `getWindow().addFlags()` function with the `WindowManager.LayoutParams.FLAG_KEEP_SCREEN_ON` parameter.

See Listing 4-32.

The `LoadInAndroidDeviceName()` Function

The `LoadInAndroidDeviceName()` function loads in the device name that is saved in the Android preferences file for this application. See Listing 4-32b.

Listing 4-32 The `onCreate()` Function

```
@Override
protected void onCreate(Bundle savedInstanceState) {
 super.onCreate(savedInstanceState);
 setContentView(R.layout.activity_main);

 // Set Orientation for Portrait
 this.setRequestedOrientation(ActivityInfo.SCREEN_ORIENTATION_PORTRAIT);

 // Initialize Text to Speech engine
 TextToSpeechInit();

 // Init Sounds
 CreateSoundPool();
 CreateSound(this);

 // Initialize the Output Message Window
 m_PhoneEntryView = (EditText) findViewById(R.id.phoneentry);
 m_PhoneEntryView.setFocusable(true);
 m_PhoneEntryView.setOnClickListener(new View.OnClickListener() {
  public void onClick(View v) {
   m_Alert2SFX.StopSound();
  }
 });

 LoadInPhoneNumber();
 m_PhoneEntryView.setText(m_EmergencyPhoneNumber);

 // Initialize the Debug Message Window
 m_DebugMsgView = (EditText) findViewById(R.id.debugmsg);
 m_DebugMsg = m_Version + "\n";
 m_DebugMsg += m_Copyright + "\n";
 m_DebugMsg += "Emergency Phone Number: " + m_EmergencyPhoneNumber + "\n\n";
 m_DebugMsgView.setText(m_DebugMsg.toCharArray(), 0, m_DebugMsg.length());
 m_DebugMsgView.setFocusable(false);
 m_DebugMsgView.setOnClickListener(new View.OnClickListener() {
 public void onClick(View v) {
  m_Alert2SFX.StopSound();
  //Toast.makeText(MainActivity.this, "DebugMSGVIEW Touched!! ", Toast.LENGTH_LONG).
 show();
  }
 });

 m_PhoneEntryView.setOnEditorActionListener(new TextView.OnEditorActionListener() {
  @Override
  public boolean onEditorAction(TextView view, int actionId, KeyEvent event) {
   int result = actionId & EditorInfo.IME_MASK_ACTION;
```

```
switch(result) {
  case EditorInfo.IME_ACTION_DONE:
 // done stuff
  m_EmergencyPhoneNumber = m_PhoneEntryView.getText().toString();
  Log.e(TAG,"In setOnEditorActionListener:: Emergency PhoneNumber = " +
 m_EmergencyPhoneNumber);
  Toast.makeText(MainActivity.
 this, "Emergency Phone Number Changed to " + m_EmergencyPhoneNumber, Toast.LENGTH_
 LONG).show();
  m_DebugMsg = "";
  AddDebugMessage("Emergency Phone Number Changed to " + m_EmergencyPhoneNumber);
  HideKeyboard();
  SavePhoneNumber();
  UpdateCommandTextView();
  break;
  case EditorInfo.IME_ACTION_NEXT:
  // next stuff
   HideKeyboard();
   break;
   }
   return true;
  }
 });

// Initialize the Information Text window
m_InfoTextView = (TextView) findViewById(R.id.textView1);
UpdateCommandTextView();
m_InfoTextView.setFocusable(false);

// Initialize Test Message Button
m_Button = (Button) findViewById(R.id.TestMessageButton);
m_Button.setEnabled(true);
m_Button.setOnClickListener(new View.OnClickListener() {
public void onClick(View v) {
 // Perform action on click
 m_Alert1SFX.PlaySound(false);

 // Basic Framework
 String Command = m_PhoneEntryView.getText().toString();
 // Set Up Data Handler
 m_WifiMessageHandler.ResetData();

 // NEW for testing for ESP8266
 m_WifiMessageHandler.SetCommand("GetTextData");
 Command += "\n";
 if (m_ClientConnectThread != null) {
  if (m_ClientConnectThread.IsConnected()) {
  m_ButtonActive = false;
  m_Button.setEnabled(false);
```

```
      m_ClientConnectThread.write(Command.getBytes());
      m_DebugMsg += "Sending COMMAND = " + Command + "\n";
    } else {
      m_DebugMsg += "ERROR! ClientConnectedThread is not connected!! \n";
      Log.e(TAG, "ERROR! ClientConnectedThread is not connected!!");
      if (m_TTS != null) {
       m_TTS.speak("ERROR ClientConnectedThread is not connected", TextToSpeech.QUEUE_
      ADD, null);
      }
    }
     } else {
      m_DebugMsg += "ClientConnectThread is Null!! \n";
     }

      m_DebugMsgView.setText(m_DebugMsg.toCharArray(), 0, m_DebugMsg.length());
    }

  });

  // Create the Client Socket Connection Thread
  CreateClientConnection(m_ServerStaticIP, m_PortNumber);

  // Initialize Android
 Device Name
  LoadInAndroidDeviceName();
  UpdateCommandTextView();
  // Setup Ping Handler
  // Send Pings to server at
 regular intervals before timeout value
  Runnable r = new Runnable() {
   public void run(){
   PingHandler();
   m_PingHandler.postDelayed(this, m_PingInterval);
  }
 };
   m_PingHandler.postDelayed(r, m_PingInterval);

   // Keep screen on and Android from sleeping getWindow().addFlags(WindowManager.
  LayoutParams.FLAG_KEEP_SCREEN_ON);
 }
```

The `PingHandler()` Function

The `PingHandler()` function sends a ping command to the server by doing the following:

1. Builds the "ping" command as a `String` object terminated by a new line character.
2. Clears the current data structures that hold incoming data from the Wi-Fi connection by calling the `m_WifiMessageHandler.ResetData()` function.
3. Sets the type of data that is expected to be returned by calling the `m_WifiMessageHandler.`

Listing 4-32b The `LoadInAndroidDeviceName()` Function

```
void LoadInAndroidDeviceName()
{
 SharedPreferences settings =
this.getSharedPreferences(m_
GotchaCamPreferencesFile, 0);
 m_AndroidClientName = settings.
getString(m_AndroidDeviceNameHandle,
m_AndroidClientName);
}
```

`SetCommand("GetTextData")` function with the parameter shown. The ping command is expecting text data terminated by a new line character as a response.

4. If the client thread is connected to the server then send the ping command to the server by calling the `m_ClientConnectThread.write(Command.getBytes())` function. The `Command` variable is converted to bytes before being sent over.
5. Prints out debug messages indicating that a ping command was just sent out from the Android to the server.

See Listing 4-32c.

Listing 4-32c The `PingHandler()` Function

```
void PingHandler()
{
  // Send Ping if Connected
  String Command = "ping\n";

  // Set Up Data Handler
  m_WifiMessageHandler.ResetData();
  m_WifiMessageHandler.
SetCommand("GetTextData");
  if (m_ClientConnectThread != null)
  {
   if (m_ClientConnectThread.
  IsConnected())
   {
    // Send Ping to Server
    m_ClientConnectThread.
   write(Command.getBytes());

    // Print out Ping sent debug
   message
    AddDebugMessageFront("PingSent..."
   + "\n");
    Log.e(TAG, "PingHandler() Executed
   ....");
   }
  }
}
```

Two variables that are references to the `Sound` class are declared and initialized to null. These variables are `m_Alert1SFX` and `m_Alert2SFX`.

```
private Sound  m_Alert1SFX = null;
private Sound m_Alert2SFX = null;
```

The `CreateSound()` function creates new Sound class objects for use with our basic framework system.

The function does the following:

1. Creates a new `Sound` object using the `.wav` sound file located in the res\raw folder of the main workspace directory for this project. This is done by calling the `Sound` constructor with the resource ID for this sound file which is `R.raw.playershoot2`. This sound effect is played when the "Send Data" button is pressed.

Initializing and Using the Sound Class

The `Sound` class is used in the `MainActivity` class for such things like playing sounds effects. For example, this class is used for playing a sound effect that indicates an intruder has been detected.

The `SoundPool` class needs to be imported using the import keyword.

```
import android.media.SoundPool;
```

The `m_SoundPool` variable is declared as a `SoundPool` class object and is initialized to null.

```
private SoundPool   m_SoundPool = null;
```

2. Creates a new `Sound` object using the `.wav` sound file located in the res\raw folder of the main workspace directory for this project. This is done by calling the `Sound` constructor with the resource ID for this sound file which is `R.raw.explosion1`. This sound effect is played when the alarm has been tripped.

***Note:* The R class is a globally available built-in and automatically generated class that contains resource ids that link to the actual sound files which are wave files located in the res\raw directory of the Android project. You can record your own wave files, copy the files into this directory, and change these lines of code to change the sounds being played.**

See Listing 4-33.

The `StopSounds()` function stops all the sound effects from playing by calling the `StopSound()` function on each of the Sound class objects. See Listing 4-34.

The `CreateSoundPool()` function creates a new `SoundPool` object by:

1. Calling the `SoundPool` constructor `SoundPool(maxStreams, streamType, srcQuality)` with the following input parameters:

Listing 4-33 The `CreateSound()` Function

```
void CreateSound(Context iContext)
{
 m_Alert1SFX = new Sound(iContext,
m_SoundPool, R.raw.playershoot2);
 m_Alert2SFX = new Sound(iContext,
m_SoundPool, R.raw.explosion1);
}
```

Listing 4-34 The `StopSounds()` Function

```
void StopSounds()
{
 m_Alert1SFX.StopSound();
 m_Alert2SFX.StopSound();
}
```

   1. `maxStreams` – The maximum number of simultaneous streams for this `SoundPool` object. A value of 10 is assigned by default as the maximum number of simultaneous streams.
   2. `streamType` – The audio stream type as described in `AudioManager`. This is set to a value of `STREAM_MUSIC`.
   3. `srcQuality` – The sample-rate converter quality. Currently has no effect. Set to 0 for the default setting.

2. The newly created `SoundPool` class object is then assigned to `m_SoundPool`.

See Listing 4-35.

The `TextToSpeechInit()` Function

The `TextToSpeechInitI()` function initializes a new text to speech object variable `m_TTS` by setting the language to be spoken to English. The function does this by calling the `m_TTS.setLanguage(Locale.US)` function with the `Locale.US` value as an input parameter.

See Listing 4-36.

Listing 4-35 The `CreateSoundPool()` Function

```
void CreateSoundPool()
{
 int maxStreams = 10;
 int streamType = AudioManager.
STREAM_MUSIC;
 int srcQuality = 0;

 m_SoundPool = new
SoundPool(maxStreams, streamType,
srcQuality);
 if (m_SoundPool == null)
 {
  Log.e("Main Activity " ,
 "m_SoundPool creation failure!
 !!!!!!!!!!!!!!!!!!!!!!!!!!!!
 !!!!!!!!!!!!!!!!!!!!!!!");
 }
}
```

Listing 4-36 The `TextToSpeechInit()` Function

```
void TextToSpeechInit()
{
 m_TTS = new TextToSpeech(MainActivity.
this, new TextToSpeech.OnInitListener()
{
   @Override
   public void onInit(int status) {
    if(status == TextToSpeech.SUCCESS)
   {
     int result=m_TTS.
    setLanguage(Locale.US);
      if(result==TextToSpeech.LANG_
     MISSING_DATA ||
      result==TextToSpeech.LANG_NOT_
     SUPPORTED){
      Log.e("error", "This Language is
     not supported");
     }
    }
    else
     Log.e("error", "Initilization
    Failed!");
    }
 });
}
```

The `LoadInPhoneNumber()` Function

The `LoadInPhoneNumber()` function retrieves the emergency phone number from the shared preferences file by:

1. Retrieving the shared preferences settings for the Android application by calling the `this.getSharedPreferences(m_GotchaCamPreferencesFile, 0)` function with the filename that contains the settings for this application. The result is assigned to the settings variable.
2. Retrieving the string that represents the emergency phone number by calling the `settings.getString(m_EmergencyPhoneNumberHandle, "None")` function with the handle that the emergency phone number was stored under. The retrieved string is assigned to `m_EmergencyPhoneNumber`.

See Listing 4-37.

Listing 4-37 The `LoadInPhoneNumber()` Function

```
void LoadInPhoneNumber()
{
 SharedPreferences settings =
this.getSharedPreferences(m_
GotchaCamPreferencesFile, 0);
 m_EmergencyPhoneNumber =
settings.getString(m_Emergency
PhoneNumberHandle, "None");
}
```

The `HideKeyboard()` Function

The `HideKeyboard()` function hides the keyboard after the user enters the emergency phone number into the phone text editbox by:

1. Retrieving the Android's `InputMethodManager` class object by calling the `getSystemService(Context.INPUT_METHOD_SERVICE)` function with the `Context.INPUT_METHOD_SERVICE` parameter.
2. Hiding the Android's keyboard from view by calling the `hideSoftInputFromWindow(TargetView.getWindowToken(), 0)` function from the `InputMethodManager` class that was retrieved in step 1.

See Listing 4-38.

The `SavePhoneNumber()` Function

The `SavePhoneNumber()` function saves the emergency phone number as a shared preference by:

1. Retrieving the shared preferences settings for this application by calling

Listing 4-38 The `HideKeyboard` Function

```
void HideKeyboard()
{
 View TargetView = this.
getCurrentFocus();
 if (TargetView != null)
 {
  InputMethodManager imm =
 (InputMethodManager)getSystemService
 (Context.INPUT_METHOD_SERVICE);
  imm.hideSoftInputFromWindow(TargetV
 iew.getWindowToken(), 0);
 }
}
```

the `this.getSharedPreferences(m_GotchaCamPreferencesFile, 0)` function with the filename of the shared preferences. The returned value is assigned to the settings variable.

2. Retrieving the `Editor` class object for this group of shared preferences by calling the `settings.edit()` function and assigning the returned value to the editor variable.
3. Putting the emergency phone number string value into the shared preferences for this application by calling the `editor.putString(m_EmergencyPhoneNumberHandle, m_EmergencyPhoneNumber)` function. The first parameter is the handle that is used as an index to store and read back in the associated value. The second parameter is the emergency phone number.
4. Saving the changes to the shared preferences by calling the `editor.commit()` function.

See Listing 4-39.

Listing 4-39 The `SavePhoneNumber()` Function

```
void SavePhoneNumber()
{
 SharedPreferences settings =
this.getSharedPreferences(m_
GotchaCamPreferencesFile, 0);
 SharedPreferences.Editor editor =
settings.edit();

 editor.putString(m_
EmergencyPhoneNumberHandle,
m_EmergencyPhoneNumber);
 editor.commit();
}
```

The `UpdateCommandTextView()` Function

The `UpdateCommandTextView()` function updates the information settings window that displays the key settings for the application.

The function displays the following:

1. The Android device's name.
2. The emergency phone number to send call outs and text messages to.
3. The alarm system status which is either on or off.
4. The emergency phone call out setting of the alarm system.
5. The emergency text message alert setting of the alarm system.

See Listing 4-40.

The `CreateClientConnection()` Function

The `CreateClientConnection()` initiates the Wi-Fi connection from the Android to the server by:

1. Creating a new `WifiMessageHandler` class with the current `MainActivity` class object as a parameter and assigning the result to `m_WifiMessageHandler`.
2. Creating a new `Client` class object by calling the `Client(ServerStaticIP, PortNumber, m_WifiMessageHandler, this)` function. The first parameter

Listing 4-40 The `UpdateCommandTextView()` Function

```
void UpdateCommandTextView()
{
 String AlarmStatus = "";
 String CallOutStatus = "";

 // Get Current Camera Resolution and
set text view
 if (m_AlarmSet)
 {
  AlarmStatus = "AlarmON";
 }
 else
 {
  AlarmStatus = "AlarmOFF";
 }

 if (m_CallOutSet)
 {
  CallOutStatus = "CallOutON";
 }
 else
 {
  CallOutStatus = "CallOutOFF";
 }

 String Info = m_AndroidClientName +
"\n" +
 m_EmergencyPhoneNumber + "\n" + "\n"
+
 AlarmStatus + "\n" +
 CallOutStatus + "\n" +
 m_TextMessagesSetting + "\n";
 int length = Info.length();
 m_InfoTextView.setText(Info.
toCharArray(), 0, length);
}
```

is the IP of the server. The second parameter is the server port number. The third parameter is the Wi-Fi message handler for this application. The last parameter is the current instance of the `MainActivity` class. The new Client object is assigned to `m_ClientConnectThread`.

Listing 4-41 The `CreateClientConnection()` Function

```
void CreateClientConnection(String
ServerStaticIP, int PortNumber)
{
 m_WifiMessageHandler = new
WifiMessageHandler(this);
 m_ClientConnectThread = new
Client(ServerStaticIP, PortNumber,
m_WifiMessageHandler, this);
 if (m_ClientConnectThread != null)
 {
  AddDebugMessage(TAG + ": Starting
 ClientConnectThread ...\n");
  m_ClientConnectThread.start();
 }
}
```

3. If the `m_ClientConnectThread` is not null then print out a message to the debug window indicating that the Wi-Fi connection is being started and start the client by calling the `m_ClientConnectThread.start()` function.

See Listing 4-41.

The `onOptionsItemSelected()` Function

The `onOptionsItemSelected()` function is called when the user touches an element in the Android application's graphic user interface.

The `onOptionsItemSelected()` does the following:

1. If the user selects `R.id.alarm_activate` which is the menu item to activate the alarm then the `SendCommand("ACTIVATE_ALARM")` is called with the parameter shown to send a message that will activate the alarm on the microcontroller side, the `SetAlarmStatus(true)` function is called with the true parameter to turn on the security system. The `UpdateCommandTextView()` function

is also called to update the information settings editbox with the new status of the alarm system. A short message is also displayed to confirm the selection by calling the `Toast.makeText()` function.

2. If the user selects `R.id.alarm_deactivate` which is the menu item to deactivate the alarm then the `SendCommand("DEACTIVATE_ALARM")` function is called with the parameter shown to deactivate the alarm on the microcontroller side, the `SetAlarmStatus(false)` function is called with the false parameter to turn off the security system. The `UpdateCommandTextView()` function is called to update the information settings editbox with the new status of the alarm system. A message is also displayed confirming the user selection by calling the `Toast.makeText()` function.
3. If the user selects `R.id.callout_on` which is the menu item to turn on the emergency phone number call out then call the `SetCallOutStatus(true)` function with the true parameter to activate the call out feature. Update the information settings editbox by calling the `UpdateCommandTextView()` function. Display a message to the user that confirms the selection by calling the `Toast.makeText()` function.
4. If the user selects `R.id.callout_off` then the `SetCallOutStatus(false)` function is called with the false parameter which turns off the emergency phone call out feature. The `UpdateCommandTextView()` function is called to update the information settings editbox. Display a message to the user that confirms the selection by calling the `Toast.makeText()` function.
5. If the user selects to turn emergency text message alerts off which maps to the `R.id.notext` value then call the `SetTextMessageStatus(TextMessageSetting.TEXT_OFF)` function with the value to turn off text messaging. The `UpdateCommandTextView()` function updates the information setting editbox. A `Message` is displayed that confirms that text message alerts have been turned off by calling the `Toast.makeText()` function.
6. If the user turns on the emergency text message alerts which maps to the `R.id.text_only` value then call the `SetTextMessageStatus(TextMessageSetting.TEXT_ON)` with the `TEXT_ON` parameter to turn on the text messaging system. The information settings editbox is updated by calling the `UpdateCommandTextView()` function. A confirmation message indicating that emergency text messaging alerts has been turned on is displayed by calling the `Toast.makeText()` function.
7. If the user clicks the menu that saves the Android device name then the `R.id.savedevicename` is selected and the current text that is displayed in the phone entry text box is saved as the Android device name. The text in the phone entry box is then replaced with the emergency phone number.

See Listing 4-42.

The `SendCommand()` Function

The `SendCommand()` function sends a command to the microcontroller over a TCP connection by doing the following:

1. Builds a command based on the input string by adding a new line character to the end of the string.
2. Resets the data structures that hold the incoming Wi-Fi data from the microcontroller.
3. Sets the return data type to text data with a terminating new line character.

Listing 4-42 The `onOptionsItemSelected()` Function

```
@Override
public boolean onOptionsItemSelected(MenuItem item)
{
 // Handle item selection
 switch (item.getItemId()) {
 ////////////////////////// Alarm Activation
 case R.id.alarm_activate:
  SendCommand("ACTIVATE_ALARM");
  SetAlarmStatus(true);
  UpdateCommandTextView();
  Toast.makeText(this, "Alarm Activated !!!", Toast.LENGTH_LONG).show();
  return true;

 case R.id.alarm_deactivate:
  SendCommand("DEACTIVATE_ALARM");
  SetAlarmStatus(false);
  UpdateCommandTextView();
  Toast.makeText(this, "Alarm Deactivated !!!", Toast.LENGTH_LONG).show();
  return true;

  ////////////////////////// Alarm Emergency Phone Callout
  case R.id.callout_on:
   SetCallOutStatus(true);
   UpdateCommandTextView();
   Toast.makeText(this, "Call Out Activated !!!", Toast.LENGTH_LONG).show();
   return true;

  case R.id.callout_off:
   SetCallOutStatus(false);
   UpdateCommandTextView();
   Toast.makeText(this, "Call Out Deactivated !!!", Toast.LENGTH_LONG).show();
   return true;

  //////////////////////// Text messages
  case R.id.notext:
   SetTextMessageStatus(TextMessageSetting.TEXT_OFF);
   UpdateCommandTextView();
   Toast.makeText(this, "Text Message Alerts Turned Off !!!", Toast.LENGTH_LONG).
  show();
   return true;
  case R.id.text_only:
   SetTextMessageStatus(TextMessageSetting.TEXT_ON);
   UpdateCommandTextView();
   Toast.makeText(this, "Text Message Alerts Activated !!!", Toast.LENGTH_LONG).
  show();
   return true;
```

```
    case R.id.savedevicename:
     m_AndroidClientName = m_PhoneEntryView.getText().toString();
     SaveDeviceName();
     Toast.makeText(this, "Saving Device Name: " + m_AndroidClientName, Toast.LENGTH_
    SHORT).show();
     AddDebugMessageFront("Android Device Name Changed to: " + m_AndroidClientName +
    "\n");
     m_PhoneEntryView.setText(m_EmergencyPhoneNumber);
     UpdateCommandTextView();
     return true;

     default
     return super.onOptionsItemSelected(item);
   }
 }
```

4. Sends the actual data over a TCP connection by calling the `m_ClientConnectThread.write(Command.getBytes())` function. The command is converted to bytes before being sent over to the microcontroller.
5. Debugs messages that the command has been sent.

See Listing 4-42b.

Listing 4-42b The `SendCommand()` Function

```
void SendCommand(String cmd)
{
 // Add terminating character to
command
 String Command = cmd + "\n";

 // Set Up Data Handler
 m_WifiMessageHandler.ResetData();
 m_WifiMessageHandler.
SetCommand("GetTextData");
 if (m_ClientConnectThread != null)
 {
  if (m_ClientConnectThread.
 IsConnected())
  {
   // Send Data to Server
   m_ClientConnectThread.
  write(Command.getBytes());

   // Print out Ping sent debug
  message
   AddDebugMessageFront("Command
  Sent: " + Command);
   Log.e(TAG, "SendCommand() method
  called ....");
  }
 }
}
```

The `SaveDeviceName()` Function

The `SaveDeviceName()` function saves the name of the Android device held in the `m_AndroidClientName` variable to the `m_GotchaCamPreferencesFile` file. The Android device name is associated with the `m_AndroidDeviceNameHandle` value.

See Listing 4-42c.

The `SetAlarmStatus()` Function

The `SetAlarmStatus()` turns on or turns off the security system by doing the following:

1. If the input parameter to the function is true then turn on the alarm and notify the user through the Android's text to speech function.

Listing 4-42c `SaveDeviceName`

```
void SaveDeviceName()
{
 SharedPreferences settings =
this.getSharedPreferences
(m_GotchaCamPreferencesFile, 0);
 SharedPreferences.Editor editor =
settings.edit();

 editor.putString(m_
AndroidDeviceNameHandle, m_
AndroidClientName);
 editor.commit();
}
```

2. If the input parameter to the function is false then turn off the alarm and notify the user through the Android's text to speech function. Reset the variables related to the emergency phone call out. Reset the number of detected motions to 0. Reset the emergency message sent flag.

   See Listing 4-43.

Listing 4-43 The `SetAlarmStatus` Function

```
void SetAlarmStatus(boolean status)
{
 if (status == true)
 {
  m_AlarmSet = true;
  m_TTS.speak("Activateing Alarm",
 TextToSpeech.QUEUE_ADD, null);
 }
 else
 {
  m_AlarmSet = false;
  m_TTS.speak("DeActivating Alarm",
 TextToSpeech.QUEUE_ADD, null);

  // Reset Callout Variables
  m_CallOutDone = false;
  m_EmergencyMessageDone = false;
  m_AlarmTripped = false;

  // Reset number of detected motions
 to 0
  m_NumberMotionsDetected = 0;

  // Reset Emergency Message Sent
 flag
  m_TextMessageSent = false;
 }
}
```

The `SetCallOutStatus()` Function

The `SetCallOutStatus()` function does the following:

1. Activates the emergency cell phone call out if the input parameter status is true. Also gives a vocalized notification that the call out has been activated.
2. Deactivates the emergency cell phone call out if the input parameter status is false. Also gives a vocalized notification that the call out has been turned off.

   See Listing 4-44.

The `SetTextMessageStatus()` Function

The `SetTextMessageStatus()` function does the following:

1. Sets the emergency text message alert setting to the one provided by the user.
2. Notifies the user that the emergency text message alert setting has been changed by issuing a vocal alert using the Android text to speech system.

   See Listing 4-45.

The `run()` Function

The `run()` function is called from another separate process or thread such as the `Client` class or the `WifiMessageHandler` class. The `run()` function is actually executed by calling the `m_MainActivity.runOnUiThread(m_MainActivity)` function with the `MainActivity` class object as the input parameter.

Listing 4-44 The `SetCallOutStatus()` Function

```
void SetCallOutStatus(boolean status)
{
 if (status == true)
 {
  m_CallOutSet = true;
  m_TTS.speak("Activating Emergency
 Call Out", Text
 ToSpeech.QUEUE_ADD, null);
 }
 else
 {
  m_CallOutSet = false;
  m_TTS.speak("DeActivating Emergency
 Call Out", TextToSpeech.QUEUE_ADD,
 null);
 }
}
```

Listing 4-45 The `SetTextMessageStatus()` Function

```
void SetTextMessageStatus(TextMessageS
etting Setting)
{
 m_TextMessagesSetting = Setting;
 m_TTS.speak("Emergency Text Messages
Set to " + m_TextMessagesSetting,
TextToSpeech.QUEUE_ADD, null);
}
```

The `run()` function does the following:

1. If text data has been received and needs to be processed then call the `ReceiveDisplayTextData()` function.
2. If debug messages were added in other threads then update the debug message window to show these new messages and play a sound to notify the user that there are new messages in the debug window.
3. Sets the state of the "Send Data" button based on the value of `m_ButtonActive`.

See Listing 4-46.

Listing 4-46 The `run()` Method

```
public void run()
{
 // Recieve amd Display Text Data
 if (m_RecieveTextDataCallBackDone)
 {
  ReceiveDisplayTextData();
  m_RecieveTextDataCallBackDone =
 false;
  m_ButtonActive = true;
 }
 else
 if (m_RefreshMessageWindows)
 {
  // Refresh Window Messages
  m_DebugMsgView.setText(m_DebugMsg.
 toCharArray(), 0, m_DebugMsg.
 length());
  m_Alert2SFX.PlaySound(false);
  m_RefreshMessageWindows = false;
 }
 else
 {
  // Error
  Log.e(TAG, "RUN:: ERROR IN RUN()
 ... No callback executed !!!!");
 }

 // Set Take Photo Button active
state
 if (m_ButtonActive == true)
 {
  m_Button.setEnabled(true);
 }
 else
 {
  m_Button.setEnabled(false);
 }
}
```

The `ReceiveDisplayTextData()` Function

The `ReceiveDisplayTextData()` function reads in the text data that was just sent from the server to the Android device over Wi-Fi and does the following:

1. The incoming text data is retrieved by calling the `m_WifiMessageHandler`.

GetStringData() function and the value is stored in the Data variable.

2. The text is then stripped of whitespace characters by calling the Data.trim() function and the result is assigned to the DataTrimmed variable.
3. The text is then written to the LogCat window in the Android IDE by calling the Log() function.
4. The text is also written to the debug message window.
5. If the incoming data is "PING_OK" then the server has responded to the Android client's automatic sending of a ping signal. A sound effect is also played.
6. If the incoming data is "ALARM_TRIPPED" then the alarm controlled by the microcontroller has been tripped so call the ProcessAlarmTripped() method to process the event.

See Listing 4-47.

The ProcessAlarmTripped() Method

The ProcessAlarmTripped() method tests for and processes an alarm tripped event. If the Android alarm is set and the alarm has been tripped then add a message to the debug window alerting the user that the alarm has been tripped. The CheckProcessCallout() and CheckProcessTextMessages() are also called in order to perform the call out and text message to the emergency phone number if needed.

See Listing 4-48.

The CheckProcessCallout() Function

The CheckProcessCallout() function processes the emergency phone call out function

Listing 4-47 The ReceiveDisplayTextData() Function

```
void ReceiveDisplayTextData()
{
 String Data = m_WifiMessageHandler.
GetStringData();
 String DataTrimmed = Data.trim();

 // Debug Print Outs
 Log.e(TAG, "Returned Text Data = " +
DataTrimmed);
 m_DebugMsg = "";
 AddDebugMessage(TAG + ": Returned
Text Data = " + "'" + DataTrimmed +
"'" + "\n");
 // Process ping response from
server
 if (DataTrimmed.equals("PING_OK"))
 {
  m_Alert1SFX.PlaySound(false);
  Log.e(TAG,"PING_OK RECEIVED !!!!!!!
 !!!!!!!!!!!!!!!!!!!!!!!!!");
 }
 else if (DataTrimmed.equals("ALARM_
TRIPPED"))
 {
  m_AlarmTripped = true;
  ProcessAlarmTripped();
 }
}
```

Listing 4-48 The ProcessAlarmTripped() Method

```
void ProcessAlarmTripped()
{
 if (m_AlarmSet)
 {
  if (m_AlarmTripped)
  {
   AddDebugMessage("ALARM
  TRIPPED!!!!\n");
   CheckProcessCallout();
   CheckProcessTextMessages();
  }
 }
}
```

once the alarm has been tripped by doing the following:

1. If the emergency phone call out has been set by the user and has not been done then call the emergency phone number by calling the `CallEmergencyPhoneNumber()` function. Set the `m_CallOutDone` variable to true.
2. If the emergency phone call out has not been set by the user and the emergency message has not been done then use the Android's text to speech function to notify the user that the alarm has been tripped. A sound effect that loops indefinitely is played. The `m_EmergencyMessageDone` variable is set to true.

See Listing 4-49.

Listing 4-49 The `CheckProcessCallout()` Function

```
void CheckProcessCallout()
{
 // Check to see if callout is
active and not yet done
 if (m_CallOutSet && !m_
CallOutDone)
 {
  // Call emergency number and set
 call out done flag
  CallEmergencyPhoneNumber();
  m_CallOutDone = true;
 }
 else
 if (!m_CallOutSet && !m_
EmergencyMessageDone)
 {
  // Repeat message
  for (int i = 0; i < m_
 EmergencyMessageRepeatTimes ;
 i++)
  {
   m_TTS.speak("Intruder Alert
  Intruder has been detected
  by the motion sensor",
  TextToSpeech.QUEUE_ADD, null);
  }

  m_Alert2SFX.PlaySound(true);
  m_EmergencyMessageDone = true;
 }
}
```

The `CallEmergency PhoneNumber()` Function

The `CallEmergencyPhoneNumber()` function actually makes the call by the Android cell phone to the emergency phone number by:

1. Creating a string consisting of "tel:" and the emergency phone number without any trailing newline or other white space characters such as tab, space, etc.
2. Creating a new Intent class object which is set to call out to a phone number.
3. Setting the phone number in the intent object to the emergency phone number from step 1 by calling the `setData()` function.
4. Dialing this phone number by calling the `startActivity(intent)` function with the intent from step 2 as an input parameter.

See Listing 4-50.

The `CheckProcessText Messages()` Function

The `CheckProcessTextMessages()` function processes the emergency text message alert function when the alarm has been tripped by doing the following:

1. If the emergency text message alert has already been sent then exit the function.
2. If the emergency text message alert has not been sent and the user has requested that a text message be sent then send the alert by calling the `SendEmergencySMSText()`

Listing 4-50 The `CallEmergencyPhoneNumber()` Function

```
void CallEmergencyPhoneNumber()
{
 // Call emergency number so that
home owner can listen in to what
tripped the
 // intruder alarm
 String uri = "tel:" + m_
EmergencyPhoneNumber.trim();
 Intent intent = new Intent(Intent.
ACTION_CALL);
 intent.setData(Uri.parse(uri));
 startActivity(intent);
}
```

function. Set the `m_TextMessageSent` variable to true.

See Listing 4-51.

The `SendEmergencySMSText()` Function

The `SendEmergencySMSText()` function sends an SMS emergency text alert message by doing the following:

1. Creates the emergency text message and assigns it to the `EmergencyMessage` variable.
2. Sends an SMS text message by calling the `SendSMS(EmergencyMessage)` function with the text message created in step 1.

***Note:* The maximum number of characters for a text message is 160 characters and depending on the encoding for your particular cell phone network the number of available characters may be significantly less.**

See Listing 4-52.

The `SendSMS()` Function

The `SendSMS()` function does the actual job of sending an SMS text message to the emergency phone number.

Listing 4-51 The `CheckProcessTextMessages()` Function

```
void CheckProcessTextMessages()
{
 if (m_TextMessageSent)
 {
  return;
 }

 if (m_TextMessagesSetting ==
TextMessageSetting.TEXT_ON)
 {
  SendEmergencySMSText();
  m_TextMessageSent = true;
 }
}
```

Listing 4-52 The `SendEmergencySMSText()` Function

```
void SendEmergencySMSText()
{
 String EmergencyMessage =
"***Intruder Detected, Number Motions
Detected: " + m_NumberMotionsDetected;
 SendSMS(EmergencyMessage);
}
```

The function does the following:

1. Retrieves a `SmsManager` object by calling the `SmsManager.getDefault()` function. The returned object is assigned to the `TextMessageManager` variable.
2. Sends the actual text message to the emergency phone number by calling the `TextMessageManager.sendTextMessage(m_EmergencyPhoneNumber, null, Msg, null, null)` function. The first parameter is the phone number to send the text message to. The third parameter is the actual text message that is sent. The rest of the parameters are not needed so they are set to null.
3. A message is displayed that contains the text "Emergency Text Message

Listing 4-53 The SendSMS() Function

```
void SendSMS(String Msg)
{
 SmsManager TextMessageManager =
 SmsManager.getDefault();

 //TextMessageManager.sendTextMessa
 ge(destinationAddress, scAddress,
 text, sentIntent, deliveryIntent);
 TextMessageManager.sendTextMessage(m_
 EmergencyPhoneNumber, null, Msg,
 null, null);
 Toast.makeText(this, "Emergency
 Text Message Sent to " + m_
 EmergencyPhoneNumber,Toast.LENGTH_
 LONG).show();
 Toast.makeText(this, Msg,Toast.
 LENGTH_LONG).show();
}
```

Sent to" and then the number held in the `m_EmergencyPhoneNumber` variable.

4. A message is displayed that contains the text message that was sent.

See Listing 4-53.

The `AddDebugMessage()` Function

The `AddDebugMessage()` function adds a `String` text message to the current debug message and updates the debug message window to reflect the change. See Listing 4-54.

The `AddDebugMessageThread()` Function

The `AddDebugMessageThread()` function is called from another thread such as the `Client` class and adds a text string to the existing debug message. It also sets the debug message window to refresh when the `run()` function is executed by setting the `m_RefreshMessageWindows` to true. See Listing 4-55.

Listing 4-54 The AddDebugMessage() Function

```
public void AddDebugMessage(String
Message)
{
 m_DebugMsg += Message;
 m_DebugMsgView.setText(m_DebugMsg.
toCharArray(), 0, m_DebugMsg.
length());
}
```

Listing 4-55 The AddDebugMessageThread() Function

```
public void
AddDebugMessageThread(String Message)
{
 m_DebugMsg += Message;
 m_RefreshMessageWindows = true;
}
```

The `WifiSocketConnectedMessage()` Function

The `WifiSocketConnectedMessage()` function is called from the `Client` class when a Wi-Fi connection has been established with the server. The function notifies the user that the Wi-Fi connection is now active, sets the "Send Data" button to active, and sets up the debug and output message windows to be updated when the `run()` function is called. It also sets the name of the Android device on the wireless network by calling the `SetAndroidControllerName()` method.

See Listing 4-56.

Listing 4-56 The `WifiSocketConnectedMessage()` Function

```
void WifiSocketConnectedMessage()
{
 m_TTS.speak("Wifi Socket Is now
Connected", TextToSpeech.QUEUE_ADD,
null);
 Log.e(TAG , "WIfi Socket is Now
Connected ...");

 m_DebugMsg += "Wifi Socket Is Now
Connected ...\n";

 // Activate Take Photo Button
 m_ButtonActive = true;

 // Refresh Message Windows
 m_RefreshMessageWindows = true;

 // Android Client connected to
server
 // Set Name
 SetAndroidControllerName();
}
```

The `SetAndroidControllerName()` Function

The `SetAndroidControllerName()` function sends the command to change the Android device's name to the `m_AndroidClientName` on the wireless network. The command is in the form of "name=androiddevicename\n". See Listing 4-57.

Listing 4-57 The `SetAndroidControllerName()` Function

```
void SetAndroidControllerName()
{
 String Command = "name=" + m_
AndroidClientName +"\n";

 // Prepare to send data to
server
 m_WifiMessageHandler.
ResetData();
 m_WifiMessageHandler.
SetCommand("GetTextData");
 if (m_ClientConnectThread
!= null)
 {
  if (m_ClientConnectThread.
 IsConnected()) {
   // Send Ping to Server
   m_ClientConnectThread.
 write(Command.getBytes());

   // Print out Ping sent debug
 message
   Log.e(TAG, "Android Client Name Set
 ....");
  }
 }
}
```

The `onDestroy()` Function

The `onDestroy()` function is called when the application is being terminated and does the following:

1. Stops the playing of any sound effects that may be active by calling the `StopSounds()` function.
2. Shuts down the Android text to speech function by calling the `m_TTS.shutdown()` function.
3. Cancels the Wi-Fi connection by calling the `m_ClientConnectThread.cancel()` function.
4. If there is no Wi-Fi connection between the Android and the server then stop the execution of the `Client` class by calling the `m_ClientConnectThread.Freeze()` function.
5. Turns off the Android Wi-Fi by calling the `m_ClientConnectThread.m_WifiManager.setWifiEnabled(false)` function with false as the input parameter.
6. Discards the `Client` thread so that there are no more references to the

`Client` object and it can be erased from the Android's memory by the memory manager by setting the `m_ClientConnectThread` to null.

See Listing 4-58.

The Arduino with ESP-01 Server Wireless Communication Framework Code

This section discusses the Arduino- and ESP-01-based wireless server that the Android client device will connect to. The Android client software was discussed earlier in this chapter.

Program Data

The ClientInfo structure holds the information for each client TCP connection to the ESP-01 server. The ConnectionID is a number in String format between 0 and 3 that represents the connection ID that is assigned to a client when it connects to the server running on the ESP-01. The Name is a String variable that represents the name of the client and is set by the Android device and can be changed by the user.

```
struct ClientInfo
{
 String ConnectionID;
 String Name;
};
```

The `MAX_CLIENTS` variable holds the maximum number of clients that can connect to the ESP-01 server.

```
const int MAX_CLIENTS = 4;
```

The `Clients` array variable holds all the clients that are currently connected to the server.

```
ClientInfo Clients[MAX_CLIENTS];
```

The `ESP01ResponseString` variable holds the incoming data received from the ESP-01 module.

```
String ESP01ResponseString = "";
```

Listing 4-58 The `onDestroy()` Function

```
@Override
protected void onDestroy()
{
 Log.e(TAG, "In onDestroy() FUNCTION
!!!!!!!!");
 super.onDestroy();

 // Stop playing sound effects
 StopSounds();

 // Shutdown Text To Speech
 m_TTS.shutdown();

 // Shutdown Client Thread to Server
 if (m_ClientConnectThread != null)
 {
  if (m_ClientConnectThread.
 IsConnected())
  {
   // Close Client Thread
   Log.e(TAG, "Cancelling Client
  Connect Thread!!!!!!!!");
   m_ClientConnectThread
  .cancel();
  }
  else
  {
   // Thread not connected to server
  so freeze its execution
   m_ClientConnectThread.Freeze();
  }
 }

 // Turn off Wifi
 m_ClientConnectThread.m_WifiManager.
setWifiEnabled(false);

 // Discard Thread
 m_ClientConnectThread = null;
}
```

The `ssid` variable holds the name of the access point that will be created on the ESP-01.

```
String ssid   = "ESP-01";
```

The `pwd` variable holds the password for the access point.

```
String pwd    = "esp8266esp01";
```

The `chl` variable is the channel ID which is set to 1.

```
int chl = 1;
```

The `ecn` is the type of password protection used on the access point that is set to 2 which is `WPA_PSK`.

```
int ecn       = 2;
```

The `maxconn` variable sets the maximum number of connections allowed by the server at one time.

```
int maxconn = 4;
```

The `ssidhidden` variable is set to 0 to allow the `ssid` of the access point to be broadcast.

```
int ssidhidden = 0;
```

The `SystemStartupOK` variable is true if the Arduino/ESP-01 system has started up successfully.

```
boolean SystemStartupOK = true;
```

The `AlarmActivated` variable is true if the alarm system has been activated and is ready to be tripped by an event and false otherwise.

```
boolean AlarmActivated = false;
```

The `AlarmTripped` variable is true if the alarm system has actually been tripped by an event and false otherwise.

```
boolean AlarmTripped = false;
```

The `Notification` variable holds incoming text that is being read in from the ESP-01.

```
String Notification = "";
```

The `setup()` Function

The `setup()` function is called first and is used to initialize the Arduino and ESP-01 Wi-Fi module. The `setup()` function does the following:

1. Initializes the serial communication for debugging to 9600 baud.
2. Prints out an initial program start-up message to the Serial Monitor.
3. Initializes communications with the ESP-01 module located on Serial Port 3 to 115,200.
4. Halts the execution of the program for 1000 milliseconds or 1 second.
5. Initializes the pin connected to the built-in LED on the Arduino board to an output pin that will provide current to the LED by calling the `pinMode(LED_BUILTIN, OUTPUT)` function.
6. Sets the built-in LED to off by calling the `digitalWrite(LED_BUILTIN, LOW)` function.
7. Initializes the data structures that hold information regarding the clients that will be connected to the server by calling the `InitializeClients()` function.
8. Sets the ESP-01 module to accept multiple connections from stations by calling the `SendATCommand(F("AT+CIPMUX=1\r\n"))` function with the command `"AT+CIPMUX=1\r\n"`.
9. Sets the ESP-01 module to access point mode by calling the `SendATCommand(F("AT+CWMODE_CUR=2\r\n"))` function with the `"AT+CWMODE_CUR=2\r\n"` command.
10. Creates a server on the ESP-01 module that will be listening for client connections on port 80 by calling the `SendATCommand(F("AT+CIPSERVER=1,80\r\n"))` function with the `"AT+CIPSERVER=1,80\r\n"` command.
11. Builds a `String` variable `APConfigCommand` that will be used to configure the access point on the ESP-01.
12. Configures the access point by calling the `SendATCommand(APConfigCommand)` function with the command built in the previous step.
13. Delays the program execution for 1000 milliseconds or 1 second.
14. Retrieves the current access point configuration by calling the `SendATCommand(F("AT+CWSAP_CUR?\r\n"))` function.

15. Retrieves the current TCP timeout value by calling the `SendATCommand( F("AT+CIPSTO?\r\n"))` function.
16. If the Arduino/ESP-01 server system has started up without any errors then turn on the built-in LED light and print out a message to the Serial Monitor indicating that the system is active. Otherwise keep the LED light off and print out an error message to the Serial Monitor.

See Listing 4-59.

The `InitializeClients()` Function

The `InitializeClients()` function initializes the connection ID and the name of each client slot in the `Clients` array to the null string which is `""`. See Listing 4-60.

The `SendATCommand()` Function

The `SendATCommand()` function sends a command designated by the `Command` input parameter to the ESP-01 module that is attached to the serial communications port number 3 by calling the `Serial3.print(Command)` function. It then reads in a response from the ESP-01 module by calling the `ProcessESP01Response()` function. The result is then returned to the caller. Debug messages are also printed out to the Serial Monitor indicating the command being sent and the response from the ESP-01 module. See Listing 4-61.

The `ProcessESP01Response()` Function

The `ProcessESP01Response()` function reads in incoming data from the ESP-01 by doing the following:

1. Sets the `DelayTime` variable which holds the amount of time to wait for a response from the ESP-01 to 4000 milliseconds or 4 seconds.
2. While the time that has elapsed is less than the value of `DelayTime` do the following:
    1. If there is incoming data from the ESP-01 then read in the character and add it to the `ESP01ResponseString` variable.
    2. Update the amount of time that has elapsed since we started reading in the data.
3. Prints out the `ESP01ResponseString` to the Serial Monitor.
4. Checks to see if the `ESP01ResponseString` variable contains an "OK" which means that the previous command that was issued was successful.
5. Returns a true if an "OK" is found in the response from the ESP-01 and false otherwise.

See Listing 4-62.

The `loop()` Function

The `loop()` function handles the main program logic that does the following:

1. Processes the Wi-Fi notifications coming from the ESP-01 module by calling the `ProcessESP01Notifications()` function.
2. Processes the sensor input by calling the `ProcessSensor()` function.

See Listing 4-63.

The `ProcessESP01Notifications()` Function

The `ProcessESP01Notifications()` function processes the incoming notifications from the ESP-01 Wi-Fi module by doing the following:

1. If there is data to read in from the ESP-01 module then continue otherwise exit the function.
2. Reads in the next character available from the ESP-01 module and adds it to the `Notification String` variable.

Listing 4-59 The `setup()` Function

```
void setup()
{
 // Initialize Serial
 Serial.begin(9600);
 Serial.println(F("****** ESP-01 Arduino Mega 2560 Server Program ********"));
 Serial.println(F("*********** Basic Wireless Framework v1.0 ************"));
 Serial.println();
 Serial.println();
 Serial.println();
 Serial.println();

 // Initialize ESP-01 on Serial Port 3 on Arduino Mega 2560
 Serial3.begin(115200);
 delay(1000);

 // Initialize digital pin LED_BUILTIN as an output and turn off.
 pinMode(LED_BUILTIN, OUTPUT);
 digitalWrite(LED_BUILTIN, LOW);

 // Initialize Clients Data Structure
 InitializeClients();
 // Set up ESP-01 as an AP with Server using AT commands
 Serial.println();
 Serial.println();

 // Enable Multiple Connections
 if (SendATCommand(F("AT+CIPMUX=1\r\n")))
 {
  Serial.println(F("ESP-01 mode set to allow multiple connections."));
 }
 else
 {
  Serial.println(F("ERROR - ESP-01 FAILED to allow for multiple connections."));
  SystemStartupOK = false;
 }

 Serial.println();
 Serial.println();
 // Set AP Mode
 if (SendATCommand(F("AT+CWMODE_CUR=2\r\n")))
 {
  Serial.println(F("ESP-01 mode changed to Access Point Mode."));
 }
 else
 {
  Serial.println(F("ERROR - ESP-01 FAILED to change to Access Point Mode."));
  SystemStartupOK = false;
 }
```

```
Serial.println();
Serial.println();
// Start TCP Server
if (SendATCommand(F("AT+CIPSERVER=1,80\r\n")))
{
 Serial.println(F("Server Started on ESP-01."));
}
else
{
 Serial.println(F("ERROR - Server FAILED to start on ESP-01."));
 SystemStartupOK = false;
}

Serial.println();
Serial.println();

// Set AP Configuration SSID, Password, etc
String APConfigCommand = "AT+CWSAP_CUR=";
APConfigCommand += "\"" + ssid + "\"" + "," + "\"" + pwd + "\"" + ",";
APConfigCommand += String(chl) + "," + String(ecn) + "," + String(maxconn) + "," +
String(ssidhidden) + "\r\n";
if (SendATCommand(APConfigCommand))
{
 Serial.println(F("AP Configuration Set Successfullly."));
}
else
{
 Serial.println(F("ERROR - AP Configureation was not Successfully Set."));
 SystemStartupOK = false;
}
Serial.println();
Serial.println();
delay (1000);

// Retrieve the current AP configureation of the ESP-01
if (SendATCommand(F("AT+CWSAP_CUR?\r\n")))
{
 Serial.println(F("This is the curent configuration of the AP on the ESP-01."));
}
else
{
 Serial.println(F("ERROR - AP info about the ESP-01 failed to retrieve."));
 SystemStartupOK = false;
}

Serial.println();
Serial.println();

// Retrieve the current TCP timeout of the ESP-01
if (SendATCommand(F("AT+CIPSTO?\r\n")))
{
 Serial.println(F("This is the curent TCP timeout in seconds on the ESP-01."));
}
```

```
  else
  {
   Serial.println(F("ERROR - Failed to retrieve TCP Timout Info."));
   SystemStartupOK = false;
  }

 Serial.println();
 Serial.println();
 if (SystemStartupOK)
 {
  Serial.println(F("************** System Has Started *******************"));
  digitalWrite(LED_BUILTIN, HIGH);
 }
 else
 {
  Serial.println(F("***** ERROR(s) ******* ERROR(s) OCCURRED ON STARTUP *******"));
  digitalWrite(LED_BUILTIN, LOW);
 }
}
```

Listing 4-60 The `InitializeClients()` Function

```
void InitializeClients()
{
 for (int i = 0; i < MAX_CLIENTS; i++)
 {
  Clients[i].ConnectionID = "";
  Clients[i].Name = "";
 }
}
```

Listing 4-61 The `SendATCommand()` Function

```
boolean SendATCommand(String Command)
{
 boolean result = false;

 // Send command
 Serial.print(F("AT Command Sent: "));
 Serial.print(Command);
 Serial3.print(Command);

 // Read in response from ESP-01
 Serial.print(F("ESP-01 Response: "));
 result = ProcessESP01Response();

 return result;
}
```

Listing 4-62 The `ProcessESP01Response()` Function

```
boolean ProcessESP01Response()
{
 boolean result = false;
 char Out;
 unsigned long DelayTimeCounter = 0;
 unsigned long DelayTimeStart = 0;
 int DelayTime = 4000; // in
milliseconds

 ESP01ResponseString = "";

 DelayTimeStart = millis();
 while(DelayTimeCounter < DelayTime)
 {
  if (Serial3.available()>0)
  {
   Out = (char)Serial3.read();
   ESP01ResponseString += Out;
  }
  DelayTimeCounter = millis() -
 DelayTimeStart;
 }
 Serial.println(ESP01ResponseString);

 // Check result of command
 if (ESP01ResponseString.
indexOf("OK") >= 0)
```

```
 {
  // OK is found. Command was
 successful
  result = true;
 }

 return result;
}
```

Listing 4-63 The `loop()` Function

```
void loop()
{
 // Process Wifi Notifications
 ProcessESP01Notifications();

 // Process Sensor Input
 ProcessSensor();
}
```

3. If the `Notification` variable contains the word "CONNECT" then a new client has just connected and the `ProcessNewClientConnection(Notification)` function is called to process the new client connection. A debug message is printed out to the Serial Monitor and the `Notification` variable is reset.
4. If the `Notification` variable contains the word "CLOSED" then a client has just closed its connection with the TCP server so call the `ProcessClientDisconnect(Notification)` function to process the client disconnection. A debug message is also printed out to the Serial Monitor and the `Notification` variable is reset.
5. If the last character read in is a new line character then
   1. If the `Notification` variable contains the string "+IPD" then check to see if the `Notification` variable contains an incoming command by calling the `CheckAndProcessIPD(Notification)` function. A debug message is printed out to the Serial Monitor and the `Notification` variable is reset.
   2. If the `Notification` variable does not contain the string "+IPD" then print out the `Notification` string to the Serial Monitor and reset the `Notification` variable.

See Listing 4-64.

Listing 4-64 The `ProcessESP01Notifications()` Function

```
void ProcessESP01Notifications()
{
 char Out;
 if (Serial3.available()>0)
 {
  Out = (char)Serial3.read();
  Notification += Out;
  if (Notification.indexOf("CONNECT")
 >= 0)
  {
   // Check for client connecting
  disconnecting
   Serial.print(F("Notification: '"));
   Serial.print(Notification);
   Serial.println(F("'"));
   ProcessNewClientConnection(Notific
  ation);
   Notification = "";
  }
  else
  if (Notification.indexOf("CLOSED")
 >= 0)
  {
   // Connection was just closed
   Serial.print(F("Notification: '"));
   Serial.print(Notification);
   Serial.print(F("'"));
   ProcessClientDisconnect
  (Notification);
   Notification = "";
  }
  else
  if (Out == '\n')
  {
   if (Notification.indexOf("+IPD")
  >= 0)
```

```
    {
     // Client Data notification found
     Serial.print(F("Notification:
    '"));
     Serial.print(Notification);
     Serial.print(F("'"));
     CheckAndProcessIPD(Notification);
     Notification = "";
    }
    else
    {
     // Print out to Serial Monitor
    and Ignore new line
     // Reset for next input line of
    incoming data
     Serial.print(F("Notification:
    '"));
     Serial.print(Notification);
     Serial.print(F("'"));
     Notification = "";
    }
   }
  }
}
```

The `ProcessNewClient Connection()` Function

The `ProcessNewClientConnection()` function processes a new client connection to the server running on the ESP-01 by doing the following:

1. Finds the new connection's ID from the incoming data where the format is "`ConnectionID,CONNECT`".
2. Finds an empty slot in the `Clients` array that holds a list of the current active clients and allocate this empty slot to the new client by assigning the new client's connection ID to the `ConnectionID` and the `Name` fields of the empty slot. Prints out debug information to the Serial Monitor indicating that a new client has connected to the server.

See Listing 4-65.

Listing 4-65 The `ProcessNewClient Connection()` Function

```
void ProcessNewClientConnection
(String line)
{
 int index = 0;
 String NewConnection = "";
 boolean Done = false;

 // 0,CONNECT
 Serial.println(F("Processing New
Client Connection ..."));
 index = line.indexOf(',');
 NewConnection = line.
substring(0,index);
 for (int i = 0; (i < MAX_CLIENTS) &&
!Done; i++)
 {
  if (Clients[i].ConnectionID == "")
  {
   // Empty Slot Found
   Clients[i].ConnectionID =
  NewConnection;
   Clients[i].Name = NewConnection;
   Serial.print(F("NewConnection
  Name: "));
   Serial.println(Clients[i].Name);
   Serial.println();
   Done = true;
  }
 }
}
```

The `ProcessClient Disconnect()` Function

The `ProcessClientDisconnect()` function processes a client's disconnect from the TCP server by doing the following:

1. Finds the connection ID associated with the client that is disconnecting from the server.
2. Searches the `Clients` array for the client that is disconnecting by matching the connection ID. After the ID is found, resets the `ConnectionID` and `Name` fields

Listing 4-66 The ProcessClientDisconnect() Function

```
void ProcessClientDisconnect(String
line)
{
 int index = 0;
 String Connection = "";
 boolean Done = false;
 // 0,CLOSED
 index = line.indexOf(',');
 Connection = line.substring(0,index);

 for (int i = 0; (i < MAX_CLIENTS) &&
!Done; i++)
 {
  if (Clients[i].ConnectionID ==
 Connection)
  {
   // CLient Slot Found
   Serial.print(F(" Connection
  Removed Name: "));
   Serial.println(Clients[i].Name);
   Serial.println();
   Clients[i].ConnectionID = "";
   Clients[i].Name = "";
   Done = true;
  }
 }
}
```

to the null string which is `""`. Prints out the name of the client that is disconnecting to the Serial Monitor.

See Listing 4-66.

The `CheckAndProcessIPD()` Function

The `CheckAndProcessIPD()` function checks for and processes commands from the Android client by doing the following:

1. If the input line from the ESP-01 module contains the "+IPD" string then it is a command so continue with the function. Otherwise exit the function.
2. From the input `String` parameter to the function find out the connection ID, the data length, and the data. The format of the incoming data should be "+IPD,<Connection ID>,<Data Length>:<Data>". The comma and colon character separators are used in determining this information.
3. If the `Data` portion of the +IPD message contains "listall" then send the requesting client a list of all the clients that are connected to the access point by calling the `ListClients(ClientIndex)` function.
4. If the `Data` portion of the +IPD message contains `"name="` then change the name of the requesting client on the wireless network. The new name consists of the characters after the equal sign with any trailing white space characters such as a new line being stripped out. The client connection ID and the new client name is printed out to the Serial Monitor. The new name of the client is set by calling the `SetNameOfClient(ClientIndex, ClientName)` function. If the name has been changed successfully then send the client an "OK" message by calling the `SendTCPMessage(ClientIndex, "3", "OK\n")` function and printing a debug message to the Serial Monitor. If the name has not changed successfully then send the client a "FAILED" message by calling the `SendTCPMessage(ClientIndex, "7", "FAILED\n")` function and printing out a debug message indicating that the change has failed.
5. If the `Data` portion of the +IPD message contains `":"` then this command requests to send a message to a client based on its name. The format of this command is `"ClientName:MessageToClient"`. The client name is found before the colon separator and the message to be sent to the

client if found after the colon. If the `":"` is found then the following is done:

1. The client name and message are found.
2. The client's connection ID is found by calling the `GetConnectionIDOf Client(clientname)` function with the client name.
3. If the returned connection ID is not equal to "None" then send the message using TCP to the client using the connection ID by calling the `SendTCPMessage(destID, String(messagelength), message))` function.
4. If the TCP message is sent successfully then set the `MessageSent` variable to true otherwise set the variable to false.
5. If the message has been successfully sent then print out a debug message to the Serial Monitor and send an "OK" message to the client that requested the redirected message by calling the `SendTCPMessage(ClientIndex, String(3), "OK\n")` function with the included parameters.
6. If the message has NOT successfully sent then print out a debug message to the Serial Monitor and send an "ERROR" message to the client that requested the redirected message by calling the `SendTCPMessage(ClientIndex, String(6), "ERROR\n")` function with the parameters as shown.

6. If the `Data` portion of the +IPD message is equal to "ping\n" then the client has sent the server a ping command to let the server know that it is alive and active. A debug message is printed to the Serial Monitor and a response message of "PING_OK" is sent back to the client by calling the `SendTCPMessage(ClientIndex, String(8), "PING_OK\n")` function with the parameters as shown.
7. If the `Data` portion of the +IPD message is equal to "ACTIVATE_ALARM\n" then activate the alarm on the Arduino by setting the `AlarmActivated` variable to true, create an acknowledgment message for the client, and send this message to the client by calling the `SendTCPMessage(ClientIndex, returnstring)` with the parameters as shown.
8. If the `Data` portion of the +IPD message is equal to "DEACTIVATE_ALARM\n" then deactivate the alarm on the Arduino side by setting the `AlarmActivated` variable to false; create an acknowledgment message, and send this message back to the client by calling the `SendTCPMessage(ClientIndex, returnstring)` function with the parameters as shown.
9. If the `Data` portion of the +IPD message is equal to "testalarm\n" then test the alarm on the Android side by sending an "ALARM_TRIPPED\n" string back to the Android client. This is done by calling the `SendTCPMessage(ClientIndex, returnstring)` function with the parameters as shown.
10. If the `Data` portion of the +IPD message does not fit into any of the above cases then this is an unknown command coming from the client so create an unknown command error message and send this to the client by calling the function `SendTCPMessage(ClientIndex, returnstring)` with the parameters as shown.

See Listing 4-67.

The `ListClients()` Function

The `ListClients()` function sends a list of the clients that are connected to the ESP-01 server back to the requesting client by doing the following:

Listing 4-67 CheckAndProcessIPD

```
void CheckAndProcessIPD(String line)
{
 int index1=0;
 int index2=0;
 int index3=0;
 String ClientIndex = "";
 String DataLength = "";
 String Data = "";

 // Process +IPD incoming data
notification
 // +IPD,0,5:QVGA
 // +IPD,<Connection ID>,<Data
Length>:<Data>
 // Check for +IPD keyword
 if (line.indexOf("+IPD") >= 0)
 {
  // Incoming Data Dectected
  index1 = line.indexOf(',');
  index2 = line.
 indexOf(',',index1+1);
  index3 = line.indexOf(':');

  ClientIndex = line.
 substring(index1+1, index2);
  DataLength = line.
 substring(index2+1, index3);
  Data = line.substring(index3+1);

  Serial.print(F("ClientIndex: "));
  Serial.println(ClientIndex);
  Serial.print(F("DataLength: "));
  Serial.println(DataLength);
  Serial.print(F("Data: "));
  Serial.println(Data);

  String ResponseText = "RESPONSE-" +
 Data;
  int length = ResponseText.length();

  if (Data.indexOf("listall") >= 0)
  {
   ListClients(ClientIndex);
  }
  else
  if (Data.indexOf("name=") >= 0)
  {
   // Set name of client
   String ClientName = "";
   int nameindex = 0;

   nameindex = Data.indexOf("=");
   ClientName =  Data.
  substring(nameindex+1);
   ClientName.trim();

   Serial.print(F("Client Name Change
  Requested, ClientIndex: "));
   Serial.print(ClientIndex);
   Serial.print(F(", NewClientName: "));
   Serial.println(ClientName);

   if (SetNameOfClient(ClientIndex,
  ClientName))
   {
    Serial.println(F("Client Name
   Change Succeeded, OK..."));
    SendTCPMessage(ClientIndex, "3",
   "OK\n");
   }
   else
   {
    Serial.println(F("Client Name
   Change FAILED ..."));
    SendTCPMessage(ClientIndex, "7",
   "FAILED\n");
   }
  }
  else
  if (Data.indexOf(":") >= 0)
  {
   // Send data to specific client by
  name
   // ClientName:MessageToClient
   // Send message to destination
  client
   // Send confirmation or failure
  message to sending client
   boolean MessageSent   = false;
   int destseparator = Data.
  indexOf(":");
   String  clientname = Data.
  substring(0,destseparator);
   String  message   = Data.
  substring(destseparator+1);
```

```
    String destID = GetConnectionIDOfC
lient(clientname);
    int messagelength = message.
length();
    if (destID != "None")
    {
     if (SendTCPMessage(destID,
    String(messagelength), message))
     {
      MessageSent = true;
     }
     else
     {
      MessageSent = false;
     }
    }
    else
    {
     Serial.println(F("ERROR ...
    Destination Client ID NOT FOUND
    ...."));
     MessageSent = false;
    }
    // Check if message sent
successfully
    if (MessageSent)
    {
     // Send OK response message to
    requesting client
     Serial.println(F("Sending OK
    Response to Sending Client..."));
     SendTCPMessage(ClientIndex,
    String(3), "OK\n");
    }
    else
    {
     // Send ERROR response message to
    requesting client
     Serial.println(F("Sending ERROR
    Response to Sending Client..."));
     SendTCPMessage(ClientIndex,
    String(6), "ERROR\n");
    }
  }
  else
  if (Data == "ping\n")
  {
    // Incoming ping from client
    Serial.println(F("Ping Recieved
  from Client ..."));
    SendTCPMessage(ClientIndex,
  String(8), "PING_OK\n");
  }
  else
  if (Data == "ACTIVATE_ALARM\n")
  {
    AlarmActivated = true;
    String returnstring = "Alarm Has
  Been Activated ...\n";
    SendTCPMessage(ClientIndex,
  returnstring);
  }
  else
  if (Data == "DEACTIVATE_ALARM\n")
  {
    AlarmActivated = false;
    String returnstring = "Alarm Has
  Been DE-Activated ...\n";
    SendTCPMessage(ClientIndex,
  returnstring);
  }
  else
  if (Data == "testalarm\n")
  {
    String returnstring = "ALARM_
  TRIPPED\n";
      SendTCPMessage(ClientIndex,
  returnstring);
  }
  else
  {
    // Unknown Command
    String returnstring = "Unknown
  Command: " + Data;
    SendTCPMessage(ClientIndex,
  returnstring);
  }
 }
}
```

1. For each entry in the `Clients` array that contains a valid connection ID, adds the name of the client to the `ActiveClients` variable that will contain all the clients connected to the server.
2. Adds a new line character to the end of the `ActiveClients` variable to indicate that this is the end of the list.

Listing 4-68 The `ListClients()` Function

```
void ListClients(String
RequestingConnectionID)
{
 String ActiveClients = "Active
Clients: ";
 for (int i = 0; i < MAX_CLIENTS; i++)
 {
  if (Clients[i].ConnectionID != "")
  {
  // Client Active
  ActiveClients += Clients[i].Name +
 ", ";
  }
 }
 ActiveClients += "\n";
 // Send the list to the requesting
client.
 int length = ActiveClients.length();
 SendTCPMessage(RequestingConnection
ID, String(length), ActiveClients);
}
```

Listing 4-69 The `SetNameOfClient()` Function

```
boolean SetNameOfClient(String
ConnectionID, String Name)
{
 boolean result = false;
 boolean done = false;

 for (int i = 0; (i < MAX_CLIENTS) &&
!done; i++)
 {
  if (Clients[i].ConnectionID ==
 ConnectionID)
  {
   Clients[i].Name = Name;
   done = true;
   result = true;
  }
 }
 return result;
}
```

3. Sends the list to the requesting client by calling the `SendTCPMessage(RequestingConnectionID, String(length), ActiveClients)` function with the parameters as shown.

See Listing 4-68.

The `SetNameOfClient()` Function

The `SetNameOfClient()` function sets the name of the client in the `Clients` array based on the connection ID by doing the following:

1. Finds the entry in the `Clients` array where `ConnectionID` is the connection ID and sets the `Name` field to Name.
2. If the connection ID is found in the `Clients` array then return true, otherwise return false.

See Listing 4-69.

The `GetConnectionIDOfClient()` Function

The `GetConnectionIDOfClient()` function returns the connection ID associated with a client's name by doing the following:

1. For each member of the `Clients` array test each name against the `ClientName` variable input parameter.
2. If the names match then return the associated connection ID of the client.

See Listing 4-70.

The `SendTCPMessage()` Function

The `SendTCPMessage()` function sends a text message to a client using the client's connection ID by calling the `SendTCPMessage(ClientIndex, slength, Data)` function using the parameters as shown. See Listing 4-71.

Listing 4-70 The `GetConnectionIDOfClient()` Function

```
String GetConnectionIDOfClient(String
ClientName)
{
 String ID = "None";
 boolean done = false;

 // Get the connection ID of the
client
 for (int i = 0; (i < MAX_CLIENTS) &&
!done; i++)
 {
   if (Clients[i].Name == ClientName)
   {
    ID = Clients[i].ConnectionID;
    done = true;
   }
 }

 return ID;
}
```

Listing 4-71 The `SendTCPMessage()` Function

```
boolean SendTCPMessage(String
ClientIndex, String Data)
{
 boolean result = false;
 int length = Data.length();
 String slength = String(length);

 result = SendTCPMessage
(ClientIndex, slength, Data);
 return result;
}
```

The `SendTCPMessage()` Function

The `SendTCPMessage()` function sends a text message to a client by doing the following:

1. Sets the `TimeOutValue` variable which holds the amount of time to wait for incoming data from the ESP-01. This variable is set to 4000 milliseconds or 4 seconds.
2. Sends the command to initiate the start of data transfer to a client over Serial Port 3 to the ESP-01 module in the format `"AT+CIPSEND=Client Index,DataLength"` followed by a carriage return and a new line character `"\r\n"`.
3. Waits for the ESP-01 to return a value of `">"` to indicate that it is ready to send incoming data to the client. However, if the `">"` is not received within the value of the `TimeOutValue` time then stop waiting and note that the operation was timed out.
4. If there was a time out from step 3 then print out an error message to the Serial Monitor and exit the function.
5. Prints out debug info concerning the client connection ID, the data length, and the data that will be sent.
6. Sends the actual data to the ESP-01 module (connected to Serial Port 3) for transfer to the client by calling the `Serial3.print(Data)` function with the parameters shown.
7. Waits for an "OK" to be received from the ESP-01 module. If the elapsed time is greater than the time out time then stop waiting and note that the operation was timed out.
8. If a time out has occurred then print an error message to the Serial Monitor and exit the function and return false as a value.
9. Returns true as a value since the function was executed without any errors.

See Listing 4-72.

The `ProcessSensor()` Function

The `ProcessSensor()` function is a placeholder function that does not currently hold any executable code. This function is meant to hold project-specific code related to sensors such as code to update the status of the sensor, or to decide if an alarm that is active has been tripped by the sensor. See Listing 4-73.

Listing 4-72 The `SendTCPMessage()` Function

```
boolean SendTCPMessage(String
ClientIndex, String DataLength, String
Data)
{
 boolean result = true;
 boolean ReadyToSend = false;
 boolean OKReceived = false;
 char Out;
 int TimeOutValue = 4000;
 boolean TimedOut = false;
 unsigned int TimeStart  = 0;
 unsigned int TimePassed = 0;

 ESP01ResponseString = "";

 // Send Command to ESP-01: AT+CIPSEN
D=ClientIndex,DataLength
 // Wait for ">" from ESP-01
 // Send Actual Data to ESP-01
 // Wait for OK from ESP-01
indicating send was successful

 // Send AT Command to ESP-01
 Serial3.print(F("AT+CIPSEND="));
 Serial3.print(ClientIndex);
 Serial3.print(F(","));
 Serial3.print(DataLength);
 Serial3.print(F("\r\n"));

 // Wait until ESP-01 responds with a
'>' symbol
 TimeStart = millis();
 while (!ReadyToSend && !TimedOut)
 {
  // Read incoming data
  if (Serial3.available()>0)
  {
   Out = (char)Serial3.read();
   ESP01ResponseString += Out;
   if (Out == '>')
   {
    ReadyToSend = true;
   }
  }

  // Check for time out
  TimePassed = millis() - TimeStart;
  if (TimePassed > TimeOutValue)
  {
   TimedOut = true;
  }
 }

 if (TimedOut)
 {
  Serial.println(F(" ***********
 ERROR, SendTCPMessage() Timed Out On
 Waiting For > "));
  Serial.print(ESP01ResponseString);
  return false;
 }

 ESP01ResponseString = "";

 // Send Debug Info
 Serial.println(F("Sending Data
....")); 
 Serial.print(F("TCP Connection ID: "));
 Serial.print(ClientIndex);
 Serial.print(F(" , DataLength: "));
 Serial.print(DataLength);
 Serial.print(F(" , Data: "));
 Serial.println(Data);
 // Send data to ESP 01
 Serial3.print(Data);

 // Wait for OK from ESP-01
 TimeStart = millis();
 while (!OKReceived && !TimedOut)
 {
  if (Serial3.available()>0)
  {
   Out = (char)Serial3.read();
   ESP01ResponseString += Out;
   if (ESP01ResponseString.
  indexOf("OK") >= 0)
   {
    OKReceived = true;
   }
  }

  // Check for time out
  TimePassed = millis() - TimeStart;
  if (TimePassed > TimeOutValue)
  {
   TimedOut = true;
  }
 }
```

```
  // Check for Time Out
  if (TimedOut)
  {
   Serial.println(F(" ****** ERROR,
  SendTCPMessage() Timed Out On Wating
  for OK after Send ... "));
   Serial.print(ESP01ResponseString);
   return false;
  }

  return result;
}
```

Listing 4-73 The `ProcessSensor()` Function

```
void ProcessSensor()
{
 // Put Code to process your sensors
or other hardware in this function.
}
```

Hands-on Example: The Basic Arduino, ESP-01, and Android Wireless Communications Framework

This hands-on example demonstrates the basic wireless network framework for the Android, and Arduino with ESP-01 module using the software for the Android and Arduino discussed in this chapter.

Downloading the Software

My publisher McGraw-Hill should have made available for download the Arduino and Android source code for this chapter at https://www.mhprofessional.com or another web site that they control. If you don't see the source code on that web site please contact McGraw-Hill for the latest location for the book's downloads.

Parts List

- 1 Android cell phone
- 1 Arduino Mega 2560 microcontroller
- 1 5- to 3.3-V step-down voltage regulator
- 1 Logic level converter from 5 to 3.3 V
- 1 ESP-01 Wi-Fi module
- 1 Arduino development station such as a desktop or a notebook
- 1 Breadboard
- Wires both male-to-male and female-to-male to connect the components

Setting Up the Arduino Hardware

You can use the Arduino hardware setup you used in Chapter 2 for the hands-on example entitled "Hands-on Example: Using an ESP-01 with an Arduino Mega 2560" for this hands-on example. So if you have successfully completed that previous hands-on example then you don't need to build any additional Arduino hardware. The following instructions and diagram have been copied from the previous hands-on example in Chapter 2.

Step-Down Voltage Regulator Connections

1. Connect the Vin pin on the voltage regulator to the 5-V power pin on the Arduino Mega 2560.
2. Connect the Vout pin on the voltage regulator to a node on your breadboard represented by a horizontal row of empty slots. This is the 3.3-V power node that will supply the 3.3-V power to the ESP-01 module as well as provide the reference voltage for the logic level shifter.
3. Connect the GND pin on the voltage regulator to the GND pin on the Arduino or form a GND node similar to the 3.3-V power node in the previous step.

ESP-01 Wi-Fi Module Connections

1. Connect the Tx or TxD pin to the Rx3 pin on the Arduino Mega 2560. This is

the receive pin for the serial hardware port 3.

2. Connect the EN or CH_PD pin to the 3.3-V power node from the step-down voltage regulator.
3. Connect the IO16 or RST pin to the 3.3-V power node from the step-down voltage regulator.
4. Connect the 3V3 or VCC pin to the 3.3-V power node from the step-down voltage regulator.
5. Connect the GND pin to the ground pin or node.
6. Connect the Rx or RxD pin to pin B0 on the 3.3-V side of the logic level shifter. The voltage from this pin will be shifted from 5 V from the Arduino to 3.3 V for the ESP-01 module.

Logic Level Shifter

1. Connect the 3.3-V pin to the 3.3-V power node from the step-down voltage regulator.
2. Connect the GND pin on the 3.3-V side to the ground node for the Arduino.
3. Connect the B0 pin on the 3.3-V side to Rx or RxD pin on the ESP-01. This corresponds to the A0 pin on the 5-V side.
4. Connect the 5-V pin to the 5-V power pin on the Arduino.
5. Connect the GND pin on the 5-V side to the ground node for the Arduino.
6. Connect the A0 pin on the 5-V side to Tx3 pin on the Arduino. This pin corresponds to the B0 pin on the 3.3-V side.

See Figure 4-7.

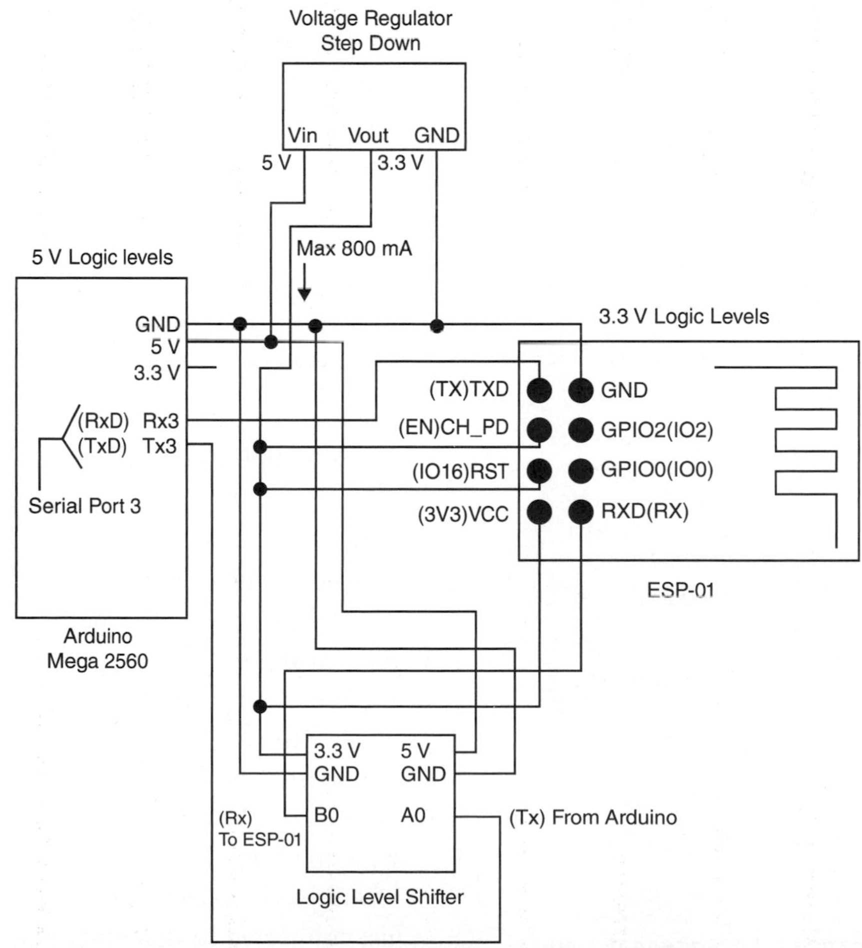

Figure 4-7 The Arduino with ESP-01 server setup (same as Figure 2-8 in Chapter 2).

Setting Up the Software

Unzip the source code for this chapter. You should see source code for the Android and the Arduino. Open the Android project in your Android Studio program. Note you may have to do some additional steps based on the version of Android Studio you are using. I used Android Studio version 1.5 for the project. You should be able to port this project version to later versions of Android Studio. See Chapter 3 for some tips on converting from one Android Studio version to another. Build the project and install it on your Android cell phone. Next load in the Arduino code for this chapter into the Arduino IDE. Connect your Arduino to your development system and upload the program to your Arduino Mega 2560.

Operating the Basic Wireless Framework System

Next, make sure your Arduino IDE is running. Unplug the Arduino, wait a few seconds, and then plug it back into the USB port of your development system. Start up the Serial Monitor. You should see the following start-up messages if the system initializes correctly.

```
****** ESP-01 Arduino Mega 2560 Server
Program ********
*********** Basic Wireless Framework
v1.0 *************

AT Command Sent: AT+CIPMUX=1
ESP-01 Response: AT+CIPMUX=1
OK
ESP-01 mode set to allow multiple
connections.

AT Command Sent: AT+CWMODE_CUR=2
ESP-01 Response: AT+CWMODE_CUR=2

OK

ESP-01 mode changed to Access Point
Mode.
AT Command Sent: AT+CIPSERVER=1,80
ESP-01 Response: AT+CIPSERVER=1,80
OK
Server Started on ESP-01.

AT Command Sent: AT+CWSAP_CUR="ESP-
01","esp8266esp01",1,2,4,0
ESP-01 Response: AT+CWSAP_CUR="ESP-
01","esp8266esp01",1,2,4,0
OK
AP Configuration Set Successfullly.

AT Command Sent: AT+CWSAP_CUR?
ESP-01 Response: AT+CWSAP_CUR?
+CWSAP_CUR:"ESP-
01","esp8266esp01",1,2,4,0
OK
This is the curent configuration of the
AP on the ESP-01.

AT Command Sent: AT+CIPSTO?
ESP-01 Response: AT+CIPSTO?
+CIPSTO:180
OK
This is the curent TCP timeout in
seconds on the ESP-01.

************** System Has Started
*******************
```

Next, on your Android device start up your Settings application. Click on the Wi-Fi section and turn on the Wi-Fi to see the available access points. Look for an access point called "ESP-01" which is the access point running on the ESP-01 module that is connected to your Arduino. Connect to that access point and use the password "esp8266esp01". Next, start up the Android application. The program should automatically connect to the TCP server running on the ESP-01 module. You should see something like the following on the Serial Monitor to indicate that the Android client has just connected to the server and the connection ID is 0.

```
Notification: '0,CONNECT'
Processing New Client Connection ...
NewConnection Name: 0
```

After the Android connects to the server it changes its name to the Android device name saved in the Android preferences file or if there is none then "android" by default. **(Note: To change the device name on the Android side enter the new name in the emergency phone number entry field and go to the Android menu items and select "Save Android Device Name." This new name will be sent to the server the next time the application is restarted.)** It sends a change device name command to the ESP-01 module. You should see something similar to the following:

```
Notification: '
'Notification: '
'Notification: '+IPD,0,8:name=lg
'ClientIndex: 0
DataLength: 8
Data: name=lg

Client Name Change Requested,
ClientIndex: 0, NewClientName: lg
Client Name Change Succeeded, OK...
Sending Data ....
TCP Connection ID: 0 , DataLength: 3,
Data: OK
```

In this case the Android device name was changed to "lg". Also, after the change is complete a status message is sent back to the client. In this case the name was successfully changed and an "OK" message was sent back to the Android device.

Next, type "listall" in the phone entry textbox on the Android client and press the "Send Data" button to send the command to the Arduino side. The listall command retrieves all the clients connected to the server and sends this list to the Android.

```
Notification: '
'Notification: '
'Notification: '+IPD,0,8:listall
'ClientIndex: 0
DataLength: 8
Data: listall

Sending Data ....
TCP Connection ID: 0 , DataLength: 21 ,
Data: Active Clients: lg,
```

In the above log from the Serial Monitor we can see the listall command coming in with a datalength of 8 which includes the new line character at the end of the listall command. Note that a new line character is always added to the end of any command sent to the Arduino. Then we see the Arduino sending a message back to the Android with a list of the connected clients which is "Active Clients: lg,". This text should also appear in the debug message window on your Android device.

Next, enter the "ping" command in the phone entry textbox and press the "Send Data" button. On the Arduino side you should see something like the following. The ping command is received with the +IPD notification line and a response is sent by the Arduino back to the client which is the message "PING_OK". This message should also show up in the debug message window on your Android device and a sound effect should be played to indicate that a ping reply has been received.

```
Notification: '
'Notification: '
'Notification: '+IPD,0,5:ping
'ClientIndex: 0
DataLength: 5
Data: ping

Ping Recieved from Client ...
Sending Data ....
TCP Connection ID: 0 , DataLength: 8,
Data: PING_OK
```

Next, let's send a command that will not be recognized by the Arduino such as the command "`qwerty`". On the Arduino side you should see the `qwerty` command coming in through a notification. Since there is currently no qwerty command defined on the Arduino side then an "`Unknown Command: qwerty`" message is sent back to the Android. On the Android in the debugmsg window you should see that message appear.

```
Notification: '
'Notification: '
'Notification: '+IPD,0,7:qwerty
'ClientIndex: 0
DataLength: 7
Data: qwerty

Sending Data ....
TCP Connection ID: 0 , DataLength: 24 ,
Data: Unknown Command: qwerty
```

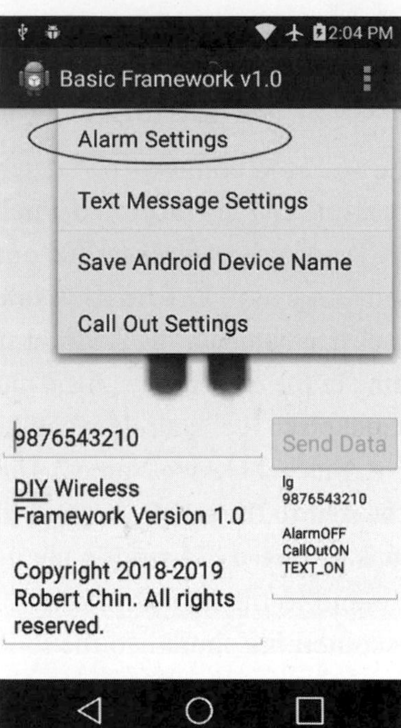

Figure 4-8 The Android basic framework menus.

Next, on the Android application bring up the menus. See Figure 4-8.

Under the Android menu selection "Alarm Settings" select the menu item to turn on the Alarm system which should be "Activate Alarm." See Figure 4-9.

Figure 4-9 Activating the alarm.

The Android then sends an "`ACTIVATE_ALARM`" message to the Arduino to indicate to the Arduino that it should activate the alarm system locally on its side. The following Arduino log shows the incoming message and the response message which is "`Alarm Has Been Activated ...`" that is sent back to the Android and should appear in the debug message window.

```
Notification: '
'Notification: '
'Notification: '+IPD,0,15:ACTIVATE_ALARM
'ClientIndex: 0
DataLength: 15
Data: ACTIVATE_ALARM

Sending Data ....
TCP Connection ID: 0 , DataLength: 29 ,
Data: Alarm Has Been Activated ...
```

Next, let's try to test out the alarm system. If you want to change the emergency telephone number then tap the phone entry textbox to bring up the keyboard input. Enter the new telephone number and press the "done" button or in the

Figure 4-10 Entering a new emergency phone number.

case shown in Figure 4-10 the "check" button. This is important because this is how the program knows when to actually change and save the phone number to the preferences file. If you just hit the back button the keyboard will disappear but the phone number will NOT be changed.

You can also change the emergency call out settings, and the emergency text message settings. If you turn off the emergency call out then an audible alarm will sound on the Android when the alarm is tripped and will continue indefinitely until you tap the debug msg window. Let's turn off the call out feature.

Next, let's give the command to test the actual alarm which is "`testalarm`." Enter "`testalarm`" in the phone entry textbox and hit the "Send Data" button. In the following log the Arduino receives the "`testalarm`" message and responds by sending the Android an "`ALARM_TRIPPED`" message. After receiving this message the Android should emit a sound effect indicating that the alarm has been tripped and there also should be an alert displayed in the debug message window.

```
Notification:
'Notification: '
'Notification: '+IPD,0,10:testalarm
'ClientIndex: 0
DataLength: 10
Data: testalarm

Sending Data ....
TCP Connection ID: 0 , DataLength: 14 ,
Data: ALARM_TRIPPED
```

Next, let's deactivate the alarm by selecting the "`Deactivate Alarm`" menu selection under the "Alarm Settings" menu. This sends a "`DEACTIVATE_ALARM`" message to the Arduino. The log below shows that the Arduino has received the message and has responded by sending a "`Alarm Has Been DE-Activated ...`" message back to the Android device.

```
Notification: '
'Notification: '
'Notification: '+IPD,0,17:DEACTIVATE_
ALARM
'ClientIndex: 0
DataLength: 17
Data: DEACTIVATE_ALARM

Sending Data ....
TCP Connection ID: 0 , DataLength: 32 ,
Data: Alarm Has Been DE-Activated ...
```

Summary

In this chapter I covered a basic wireless network system consisting of an Android cell phone and an Arduino with an ESP-01 module that provided Wi-Fi capability. I started off giving a general overview of the system including a diagram explaining its general operation. I then covered the Android source code for the client portion of the system that serves as the user interface to the wireless network. Next, I covered the Arduino source code for the server portion of the system. Finally, I guided the user through the setup, and operation of the actual Android and Arduino with ESP-01 wireless network system.

CHAPTER 5

Arduino with ESP8266 (ESP-01 Module) and Android Wireless Sensor and Remote Control Projects I

THIS CHAPTER COVERS WIRELESS SENSOR projects and wireless remote control projects using the Android and Arduino basic wireless framework discussed in Chapter 4. I first start off with a general system overview that gives a summary of how the Android- and Arduino-based wireless system works. Next, I cover an infrared motion detector, sensor-based alarm system that will detect the presence of humans and generate an alarm that can alert other users via cell phone text messages and phone calls. Next, a sound detector-based alarm system is discussed that can detect loud noises such as the breaking of glass caused by a forced entry into a home or business. Next, a distance sensor-based intruder alarm system is presented that senses the presence of an intruder similar to that of tripwire in that if the sensed distance is below a certain number that the user sets then an intruder alarm will be triggered. Next, a water leak alarm system is given that will trigger an alarm if the water sensor detects a wet environment that could be the start of a leak such as in that of a hot water heater used to warm the water in a home or office. Next, I show you how to control an LED remotely and wirelessly. I follow that up by showing the reader how to control an RGB LED that can produce a variety of colors remotely and wirelessly. Finally, I show the reader how to control a piezo buzzer remotely and wirelessly using the Android and Arduino wireless framework.

General System Overview

The system that is discussed in this chapter is based on the basic wireless framework presented in Chapter 4. This system consists of an Android cell phone, an Arduino Mega 2560 microcontroller, and an ESP-01 Wi-Fi module. The sensors or other hardware that is to be controlled is attached to the Arduino microcontroller. The Android cell phone acts as a user interface and a controller for the Arduino microcontroller and the attached sensors and other hardware. The Android sends commands such as "ACTIVATE_ALARM\n" and "DEACTIVATE_ALARM\n" to turn an alarm system located on the Arduino on and off. The Arduino processes these commands and may send a response such as "OK\n" or "ERROR\n" to acknowledge that the command was either processed successfully or failed. Alerts can be sent from the Arduino to the Android at any time which might trigger an emergency phone call out or text message from the Android cell phone. For example, a motion sensor that detects motion might send out an "ALARM_TRIPPED\n" or "ALARM_TRIPPED:3\n" message to the Android cell phone. Notifications can also be sent from the Arduino microcontroller to the Android that do not require any action to be taken but are displayed in a window on the Basic Wireless Framework application. For example, the messages "TEMP:80\n" and "HUMIDITY:50\n" could be notifications that relay the temperature and humidity information to the user.

See Figure 5-1.

The HC-SR501 Infrared Motion Detector

The HC-SR501 infrared motion detector senses the movement of humans and other objects. The specifications of the sensor are as follows:

- Operating voltage: 5 to 20 V
- Power consumption: 65 mA
- Logic levels: 3.3 V, 0 V
- Trigger methods: L – single trigger; H – repeat trigger
- Sensing range: Less than 120 degrees wide and 7 m maximum distance from the sensor
- Operating temperature: –15 to 70 degrees Centigrade

The sensor generally comes with a removable plastic dome as shown in Figure 5-2.

The dome can be removed to reveal the actual sensor and the pin labels. See Figure 5-3.

The output of the HC-SR501 sensor is 1 or high voltage when movement is detected and false or 0 or low voltage if no movement is detected.

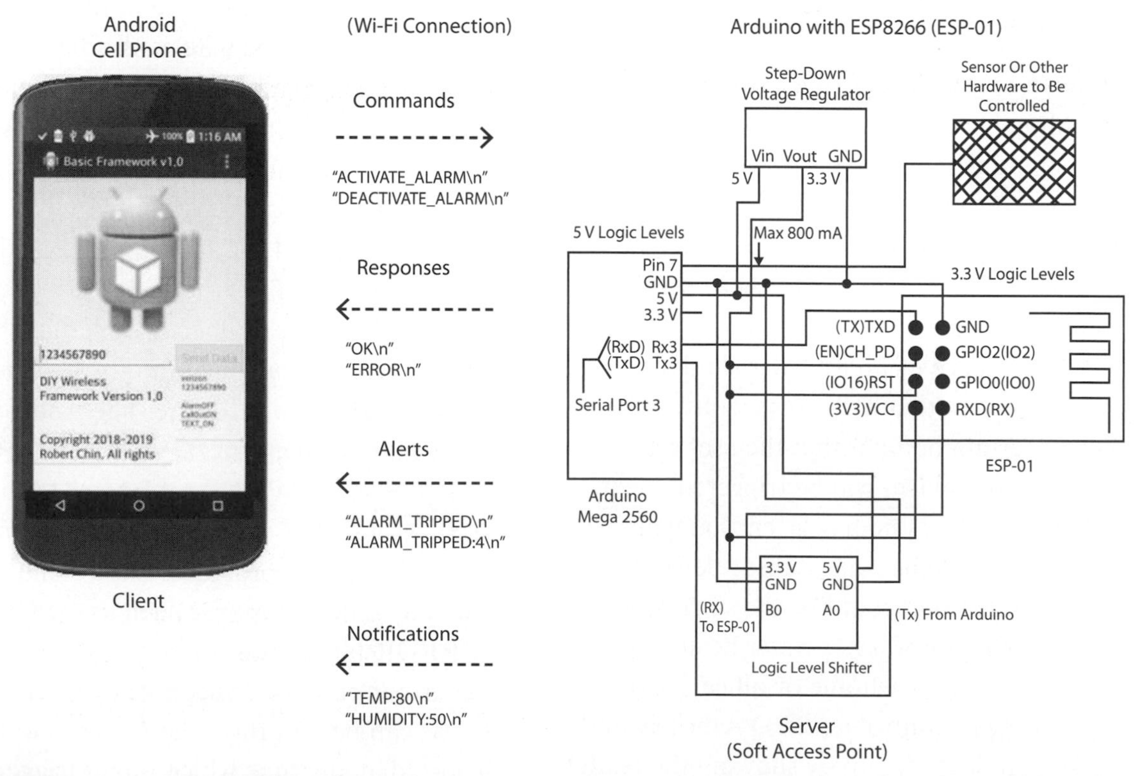

Figure 5-1 General system overview.

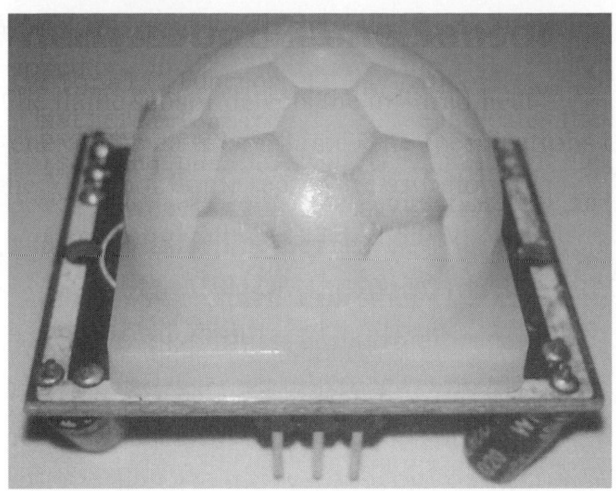

Figure 5-2 The HC-SR501 infrared motion sensor with the dome.

Figure 5-3 The HC-SR501 infrared motion sensor without the dome.

The sensor can be triggered in two ways based on a jumper located on the bottom of the sensor:

- Single trigger mode: When motion is detected then the output is 1 or high and then switches to 0 or low. Next, a time interval called the induction blocking time that is by default 2.5 seconds occurs where the sensor is inoperative. After the interval has passed, the sensor becomes operative and can once again detect motion.
- Repeat trigger mode: When human activity is detected then the output is 1 or high voltage and remains high as long as there is human activity present. After the human presence is out of range of the sensor, the output changes to 0 or low voltage.

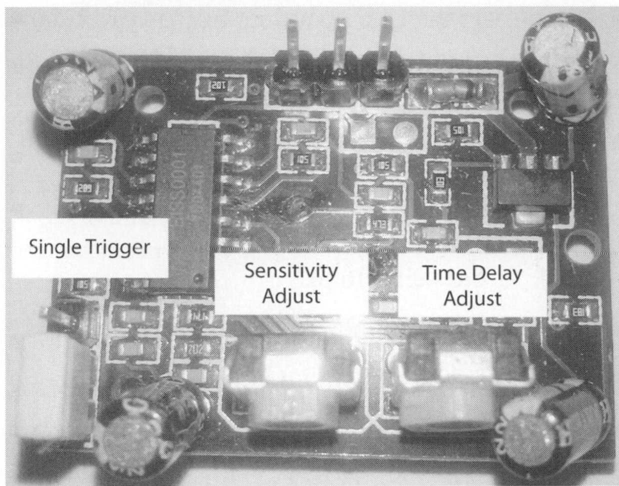

Figure 5-4 The bottom of the HC-SR501 infrared motion sensor set up for single trigger operation.

The distance at which the sensor can detect motion can be adjusted using the sensitivity adjust potentiometer. Turn the potentiometer clockwise to increase the distance at which motion can be detected. Turn the potentiometer counter clockwise to decrease the distance at which motion can be detected.

The induction delay interval can be adjusted by using the time delay adjust potentiometer. Turn the potentiometer clockwise to increase the delay and counter clockwise to decrease the delay.

Figure 5-4 shows the bottom view of an HC-SR501 motion sensor that is set for single trigger mode operation. In order to change the trigger mode to repeat trigger mode move the jumper to cover the other pin that is currently exposed. The sensitivity adjust and time delay adjust potentiometers are also shown.

Hands-on Example: The HC-SR501 Infrared Motion Detector Alarm System

This hands-on example shows you how to build and operate an infrared motion detection alarm system using the basic wireless framework that was discussed previously in Chapter 4 and the HC-SR501 sensor that we discussed in the previous section.

Parts List

- 1 Android cell phone
- 1 Arduino Mega 2560 microcontroller
- 1 5- to 3.3-V step-down voltage regulator
- 1 Logic level converter from 5 to 3.3 V
- 1 ESP-01 Wi-Fi module
- 1 Arduino development station such as a desktop or a notebook
- 1 Breadboard
- 1 HC-SR501 infrared motion detector
- 1 Package of wires both male-to-male and female-to-male to connect the components

Setting Up the Hardware

Step-Down Voltage Regulator Connections

1. Connect the Vin pin on the voltage regulator to the 5-V power pin on the Arduino Mega 2560.
2. Connect the Vout pin on the voltage regulator to a node on your breadboard represented by a horizontal row of empty slots. This is the 3.3-V power node that will supply the 3.3-V power to the ESP-01 module as well as provide the reference voltage for the logic level shifter.
3. Connect the GND pin on the voltage regulator to the GND pin on the Arduino or form a GND node similar to the 3.3-V power node in the previous step.

ESP-01 Wi-Fi Module Connections

1. Connect the Tx or TxD pin to the Rx3 pin on the Arduino Mega 2560. This is the receive pin for the serial hardware port 3.
2. Connect the EN or CH_PD pin to the 3.3-V power node from the step-down voltage regulator.
3. Connect the IO16 or RST pin to the 3.3-V power node from the step-down voltage regulator.
4. Connect the 3V3 or VCC pin to the 3.3-V power node from the step-down voltage regulator.
5. Connect the GND pin to the ground pin or node.
6. Connect the Rx or RxD pin to pin B0 on the 3.3-V side of the logic level shifter. The voltage from this pin will be shifted from 5 V from the Arduino to 3.3 V for the ESP-01 module.

Logic Level Shifter

1. Connect the 3.3-V pin to the 3.3-V power node from the step-down voltage regulator.
2. Connect the GND pin on the 3.3-V side to the ground node for the Arduino.
3. Connect the B0 pin on the 3.3-V side to Rx or RxD pin on the ESP-01. This corresponds to the A0 pin on the 5-V side.
4. Connect the 5-V pin to the 5-V power pin on the Arduino.
5. Connect the GND pin on the 5-V side to the ground node for the Arduino.
6. Connect the A0 pin on the 5-V side to Tx3 pin on the Arduino. This pin corresponds to the B0 pin on the 3.3-V side.

The HC-SR501 Infrared Motion Detector

1. Connect the VCC pin on the motion detector to the 5-V power node.
2. Connect the GND pin on the motion detector to the GND node.

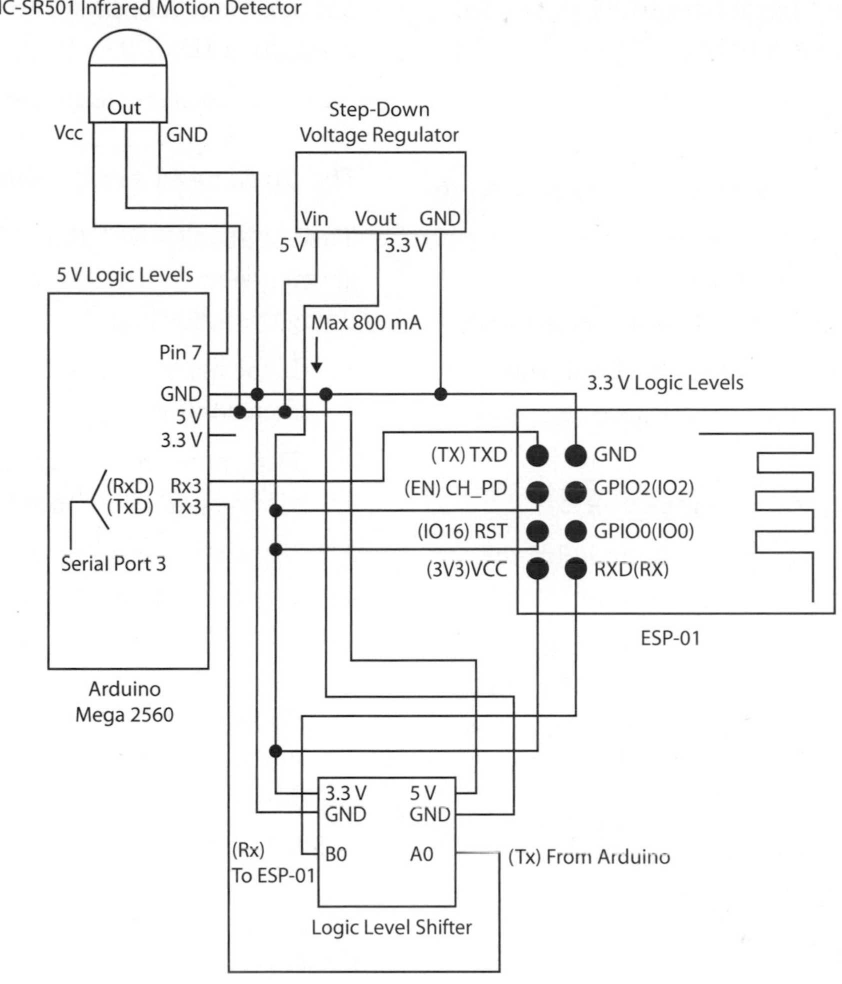

Figure 5-5 The HC-SR501 infrared motion detection system.

3. Connect the Out pin on the motion detector to Pin 7 on the Arduino.
4. Make sure that the jumper on the sensor is set to single trigger mode.

See Figure 5-5.

Setting Up the Arduino Software

For this hands-on example we build upon the basic Arduino server framework discussed in Chapter 4.

The `SensorPin` variable holds the Arduino pin number that the motion sensor output is connected to.

```
int SensorPin = 7;
```

The `NumberHits` variable keeps track of the number of times that the motion sensor has been tripped and is initialized to 0.

```
int NumberHits = 0;
```

The `SensorValue` variable holds the current output reading from the motion sensor. A 1 or high value indicates that movement was detected and a 0 or low value indicates that no movement was detected.

```
int SensorValue = 0;
```

The `PreviousSensorValue` variable holds the output value from the motion sensor that was read in previously. This variable is used to keep track of new alarm trips which are added to the `NumberHits` variable.

```
int PrevSensorValue = 0;
```

The `WaitTime` variable holds the time in milliseconds of the alarm wait state and is set to a default of 10 seconds.

```
unsigned long WaitTime = 1000 * 10;
// Set default wait time to 10 seconds.
```

The `WaitTimeStart` variable holds the time at which the alarm's wait state begins in milliseconds.

```
unsigned long WaitTimeStart = 0;
```

The `eAlarmState` enumeration holds the states of the alarm system which are:

1. `Alarm_WAIT`: The alarm is not active and is waiting for `WaitTime` milliseconds after which it changes to the `Alarm_ON` state.
2. `Alarm_ON`: The alarm is on and is capable of being tripped by a detected motion.
3. `Alarm_TRIPPED`: The sensor has detected a motion and the alarm has been tripped.
4. `Alarm_OFF`: The alarm is off and is not detecting motions.

```
enum eAlarmState
{
    Alarm_WAIT,
    Alarm_ON,
    Alarm_TRIPPED,
    Alarm_OFF
};
```

The `AlarmState` variable holds the current state of the alarm system and is set to a default value of off.

```
eAlarmState AlarmState = Alarm_OFF;
```

The `ConnectionIDForAlarm` variable holds the connection ID of the client that is setting the alarm and that will receive the alarm tripped notification if it occurs. The default is set to a connection ID of 0.

```
String ConnectionIDForAlarm = "0";
```

The `UpdateAlarm()` Function

The `UpdateAlarm()` function manages the alarm systems when it is in the wait state by doing the following:

1. If the alarm system is in the waiting state, then continue, otherwise exit the function.
2. Determine the amount of time that the alarm system has been in the waiting state.
3. If the amount of time is greater than or equal to the `WaitTime`, then turn the alarm on, send a message to the client associated with the alarm by calling the `SendAlarmNowActiveMessage()` function, and print out a debug message to the Serial Monitor.

See Listing 5-1.

The `SendAlarmNowActiveMessage()` Function

The `SendAlarmNowActiveMessage()` function sends out the message "`ALARM NOW ACTIVE...\n`" to the client that activated the alarm by calling the `SendTCPMessage(ConnectionIDForAlarm, returnstring)` function.

See Listing 5-2.

The `SendAlarmTrippedMessage()` Function

The `SendAlarmTrippedMessage()` function sends the message "`ALARM_TRIPPED:NumberHits\n`" to the client that activated the alarm where `NumberHits` is the number of motions that have been detected by the sensor. For example, a message of "`ALARM_TRIPPED:3\n`" means that the alarm has been tripped and the total number of trips is 3. The `SendTCPMessage(ConnectionIDForAlarm,`

Listing 5-1 The `UpdateAlarm()` Function

```
void UpdateAlarm()
{
 if (AlarmState == Alarm_WAIT)
 {
  // Alarm has been activated
  // Check to see if the wait delay
 has passed
  unsigned long ElapsedTime =
 millis() - WaitTimeStart;
  if (ElapsedTime >= WaitTime)
  {
  AlarmState = Alarm_ON;
  SendAlarmNowActiveMessage();
  Serial.println(F("ALARM WAIT STATE
  FINISHED ... Alarm Now ACTIVE
  ...."));
  }
 }
}
```

Listing 5-2 The `SendAlarmNowActiveMessage()` Function

```
void SendAlarmNowActiveMessage()
{
 // Send the Alarm tripped message
 String returnstring = "ALARM NOW
ACTIVE...\n";
 SendTCPMessage(ConnectionIDForAlarm,
returnstring);
}
```

`returnstring)` function is called to send the actual message.

See Listing 5-3.

Listing 5-3 The `SendAlarmTrippedMessage()` Function

```
void SendAlarmTrippedMessage()
{
 // Send the Alarm tripped message
 String returnstring = "ALARM_
TRIPPED:" + String(NumberHits) +
"\n";
 SendTCPMessage(ConnectionIDForAlarm,
returnstring);
}
```

The `ProcessSensor()` Function

The `ProcessSensor()` function reads in the input from the motion detection sensor and processes this input by doing the following:

1. If the alarm system is on or has been tripped, then continue with the function, otherwise exit the function.
2. Reads in the output value from the motion sensor by calling the `digitalRead(SensorPin)` function with the Arduino pin number that is attached to the motion sensor's output pin.
3. If the previous sensor value is 0 which means there was no motion detected and the current sensor value is 1 which means that motion was just detected, then do the following:
    1. Increase the variable that keeps track of the number of hits or motions that have been detected.
    2. Print out the sensor value and the total number of motions detected to the Serial Monitor.
    3. Send an alarm tripped message to the Android device by calling `SendAlarmTrippedMessage()` function.
4. If the conditions in step 3 were false then if the previous sensor value was 1 which means motion was detected and the current sensor value is 0 which means there is no motion detected, then print out debug messages to the Serial Monitor indicating that the sensor value has changed to 0. Reset the `PrevSensorValue` variable to 0 so that a new motion can be detected.

See Listing 5-4.

Listing 5-4 The `ProcessSensor()` Function

```
void ProcessSensor()
{
  // Put Code to process your sensors
 or other hardware in this function.
  if ((AlarmState == Alarm_ON) ||
 (AlarmState == Alarm_TRIPPED))
  {
    // Process Sensor Input
    SensorValue = digitalRead
   (SensorPin);
    if ((PrevSensorValue == 0) &&
   (SensorValue == 1))
    {
      // Motion Just Detected
      NumberHits++;
      Serial.print(F("SensorValue: "));
      Serial.print(SensorValue);
      Serial.print(F(" , Motion Detected
    ... NumberHits: "));
      Serial.println(NumberHits);
      PrevSensorValue = 1;

      // Send Alarm Tripped Message
      SendAlarmTrippedMessage();
    }
    else
    if ((PrevSensorValue == 1) &&
   (SensorValue == 0))
    {
      // Heat Source Moved out of range
      Serial.print(F("SensorValue: "));
      Serial.println(SensorValue);
      PrevSensorValue = 0;
    }
  }
}
```

Adding Code to the `setup()` Function

Code is added to the `setup()` function to initialize the pin mode for the Arduino pin connected to the output pin of the motion sensor to INPUT which means you will be able to read the voltage at that pin.

```
pinMode(SensorPin, INPUT);
```

Adding the Activate Alarm Command to the `CheckAndProcessIPD()` Function

If the incoming command from the client is the activate alarm command, then process the command by doing the following:

1. If the `"ACTIVATE_ALARM\n"` command is received by the Arduino, then continue with the block of code that processes the command.
2. If the alarm state is off, then:
    1. Set the number of motions detected to 0.
    2. Change the alarm state to the waiting state.
    3. Build a message indicating that the alarm has been activated and contains the wait time in seconds before the alarm is activated.
    4. Send the message to the client that has sent the activate alarm command.
    5. Set the `WaitTimeStart` variable to the current time since the Arduino was powered on by calling the `millis()` function which returns this time in milliseconds.
    6. Set the `ConnectionIDForAlarm` variable to the connection ID of the client that activated the alarm.
3. If the alarm state is not off, then send the message `"Alarm Already Active...\n"` to the client that tried to activate the alarm.

See Listing 5-5.

Adding the Deactivate Alarm Command to the `CheckAndProcessIPD()` Function

The code related to the deactivate alarm command is added to the `CheckAndProcessIPD()` function by doing the following:

1. If the incoming command from the Android is equal to `"DEACTIVATE_ALARM\n"`, then continue with the code in this block.

Listing 5-5 Adding the `Activate Alarm` Command

```
if (Data == "ACTIVATE_ALARM\n")
{
 if (AlarmState == Alarm_OFF)
 {
  NumberHits = 0;
  AlarmState = Alarm_WAIT;
  String returnstring = "Alarm
 Activated...WAIT TIME:" +
 String(WaitTime/1000) + "
 Seconds.\n";
  SendTCPMessage(ClientIndex,
 returnstring);
  WaitTimeStart = millis();
  ConnectionIDForAlarm = ClientIndex;
 }
 else
 {
  String returnstring = "Alarm
 Already Active...\n";
  SendTCPMessage(ClientIndex,
 returnstring);
 }
}
```

2. Set the alarm state to off.
3. Send the message `"Alarm Has Been DE-Activated ...\n"` to the client that sent the deactivate alarm command by calling the `SendTCPMessage(ClientIndex, returnstring)` function.

See Listing 5-6.

Listing 5-6 The `Deactivate Alarm` Command

```
if (Data == "DEACTIVATE_ALARM\n")
{
 AlarmState = Alarm_OFF;
   String returnstring = "Alarm Has
Been DE-Activated ...\n";
 SendTCPMessage(ClientIndex,
returnstring);
}
```

Adding the Hits Command to the `CheckAndProcessIPD()` Function

The code related to the hits command is added in to the `CheckAndProcessIPD()` function by doing the following:

1. If the incoming command from the Android client is equal to `"hits\n"` then build a text message indicating the total number of motions detected and then send this message to the Android client.

See Listing 5-7.

Listing 5-7 The `hits` Command

```
if (Data == "hits\n")
{
 // Return number of motions detected
to requesting client
 String returnstring = "Number
Motions Detected: " + String
(NumberHits) + "\n";
 SendTCPMessage(ClientIndex,
returnstring);
}
```

Adding the Reset Command to the `CheckAndProcessIPD()` Function

The code related to the reset command in the `CheckAndProcessIPD()` function does the following:

1. If the incoming command is equal to `"reset\n"` then do the following:
   1. Set the `NumberHits` variable to 0 to reset the total number of motions that have been detected by the sensor.
   2. Send a text message to the Android client that indicates that the total number of motions detected has been reset to 0.

See Listing 5-8.

Listing 5-8 The reset Command

```
if (Data == "reset\n")
{
   // Reset number hits to 0
   NumberHits = 0;
   String returnstring = "Number Hits
Reset to 0.\n";
   SendTCPMessage(ClientIndex,
returnstring);
}
```

Adding the Wait Command to the CheckAndProcessIPD() Function

The code implementing the wait command is added into the `CheckAndProcessIPD()` function by doing the following:

1. If the `"wait="` string is found within the command that is sent to the Arduino then process the wait command by doing the following:
    1. Find the wait time in milliseconds that is after the equal sign in the incoming command.
    2. Convert the string value of the wait time to its integer value and assign this value to the `WaitTime` variable. This becomes the new wait time when the alarm is reset and is activated again.
    3. Send a text message back to the Android client indicating that the wait time has been changed to the new value.

   See Listing 5-9.

Adding to the loop() Function

In the `loop()` function code was added which updates the alarm status by calling the `UpdateAlarm()` function.

See Listing 5-10.

Listing 5-9 The Wait Command

```
if (Data.indexOf("wait=") >= 0)
{
  int indexwait = Data.indexOf("=");
   String waitlength = Data.
substring(indexwait+1);
   WaitTime = waitlength.toInt();

   String returnstring = "";
   returnstring = returnstring +
F("WaitTime set to: ") + WaitTime/1000
+ F(" Seconds\n");
   SendTCPMessage(ClientIndex,
returnstring);
}
```

Listing 5-10 The loop() Function

```
void loop()
{
  // Process Wifi Notifications
  ProcessESP01Notifications();


  // Update Alarm Status
  UpdateAlarm();


  // Process Sensor Input
  ProcessSensor();
}
```

Setting Up the Android Software

Next, we will need to make some slight modifications to the basic Android wireless framework we covered in Chapter 4.

Modifying the ReceiveDisplayTextData() Function

The `ReceiveDisplayTextData()` function is called when the Android receives text data from the microcontroller. This function needs to be modified so that it can process the alert in the format "ALARM_TRIPPED:NumberHits\n" by doing the following:

1. Determines the number of hits or motions detected by the alarm and assigns that number to the m_TotalNumberEventsDetected variable.
2. Sets the alarm on the Android side to tripped by setting m_AlarmTripped = true.
3. Resets the m_TextMessageSent variable to false which will send out an emergency text message if the user has allowed for emergency text messages to be sent out.
4. Produces a sound effect that continually repeats to indicate that the alarm has been tripped by calling the m_Alert2SFX.PlaySound(true) function.
5. Processes the actual alarm trip by calling the ProcessAlarmTripped() function.

See Listing 5-11.

Listing 5-11 The ReceiveDisplayTextData() Function

```
void ReceiveDisplayTextData()
{
 String Data = m_WifiMessageHandler.
GetStringData();
 String DataTrimmed = Data.trim();

 // Debug Print Outs
 Log.e(TAG, "Returned Text Data = " +
DataTrimmed);
 m_DebugMsg = "";
 AddDebugMessage(TAG + ": Returned
Text Data = " + "'" + DataTrimmed +
"'" + "\n");

 // Process ping response from server
 if (DataTrimmed.equals("PING_OK"))
 {
  m_Alert1SFX.PlaySound(false);
  Log.e(TAG,"PING_OK RECEIVED !!!!!!
 !!!!!!!!!!!!!!!!!!!!!!!!!!");
 }
 else if (DataTrimmed.equals("ALARM_
TRIPPED"))
 {
  m_AlarmTripped = true;
  ProcessAlarmTripped();
 }
 else if (DataTrimmed.contains("ALARM_
TRIPPED:"))
 {
  // Process Alarm tripped message
 that contains the total number of
 hits
  int indexstart = DataTrimmed.
 indexOf(":");
  String NumberHits = DataTrimmed.
 substring(indexstart+1);

  m_TotalNumberEventsDetected =
 Integer.valueOf(NumberHits);
  m_AlarmTripped = true;
  m_TextMessageSent = false;

  m_Alert2SFX.PlaySound(true);
  ProcessAlarmTripped();
 }
}
```

The SendEmergencySMSText() Function

The SendEmergencySMSText() needs to be modified to work with the alarm system. The change is to the emergency text message that is sent out and the new text message reads "***Event Detected, Total Number of Events Detected: NumberEvents" where the NumberEvents is the m_TotalNumberEventsDetected value.

See Listing 5-12.

Operating the Motion Detection Alarm System

Make sure you download the Android project code called the "Basic Framework version 1.1" that is part of the source code for this chapter and install the application to your Android cell phone. Note that the Android APK install files for all the chapters are located in the

Listing 5-12 The `SendEmergencySMSText()` Function

```
void SendEmergencySMSText()
{
  //String EmergencyMessage =
"***Intruder Detected, Number Motions
Detected: " + m_NumberMotionsDetected;

  String EmergencyMessage =
  "***Event Detected, Total Number
  of Events Detected: " + m_
  TotalNumberEventsDetected;
  SendSMS(EmergencyMessage);
}
```

downloads section of Chapter 9. Also, download the Arduino motion sensor project code for this chapter and install that on your Arduino. Unplug your Arduino from your development system, wait a few seconds, plug it back in, and start up the Serial Monitor. You should see something like the following initialization information.

```
****** ESP-01 Arduino Mega 2560 Server
Program ********

***********  Motion Sensor Alarm System
*************

AT Command Sent: AT+CIPMUX=1
ESP-01 Response: AT+CIPMUX=1

OK

ESP-01 mode set to allow multiple
connections.

AT Command Sent: AT+CWMODE_CUR=2
ESP-01 Response: AT+CWMODE_CUR=2

OK

ESP-01 mode changed to Access Point
Mode.

AT Command Sent: AT+CIPSERVER=1,80
ESP-01 Response: AT+CIPSERVER=1,80

OK

Server Started on ESP-01.

AT Command Sent: AT+CWSAP_CUR="ESP-
01","esp8266esp01",1,2,4,0

ESP-01 Response: AT+CWSAP_CUR="ESP-
01","esp8266esp01",1,2,4,0

OK

AP Configuration Set Successfullly.

AT Command Sent: AT+CWSAP_CUR?
ESP-01 Response: AT+CWSAP_CUR?
+CWSAP_CUR:"ESP-01","esp8266esp01",
1,2,4,0

OK

This is the curent configuration of the
AP on the ESP-01.

AT Command Sent: AT+CIPSTO?
ESP-01 Response: AT+CIPSTO?
+CIPSTO:180

OK

This is the curent TCP timeout in
seconds on the ESP-01.

************** System Has Started
*******************
```

The Arduino server should now be up and running and the access point should be available on the ESP-01 module. On your Android device go to the Settings->Wifi section and turn on the Wi-Fi. Connect to the "ESP-01" access point which is the one running on the ESP-01 module. Start up the "Basic Framework v1.1" program which will automatically connect to the TCP server running on the ESP-01 module. After a TCP connection is made, the name of the Android client that is used on the wireless network is changed, and if successful then an "OK" message is sent back to the Android device. In the case below the name of the Android client is called "`lg`".

```
Notification: '0,CONNECT'
Processing New Client Connection ...
```

```
NewConnection Name: 0

Notification: '
'Notification: '
'Notification: '+IPD,0,8:name=lg
'ClientIndex: 0
DataLength: 8
Data: name=lg

Client Name Change Requested,
ClientIndex: 0, NewClientName: lg
Client Name Change Succeeded, OK...
Sending Data ....
TCP Connection ID: 0 , DataLength: 3 ,
Data: OK
```

Turn off the emergency call out on the Android side so that when the alarm is tripped an audible alarm will be generated instead of a phone call out.

Next, on the Android application menu select the "Alarm Settings" menu and click "`Activate Alarm`" to activate the alarm. The alarm setting will be changed locally on the Android device. An "`ACTIVATE_ALARM`" message will also be sent to the Arduino. The Arduino will then send a text message back to the Android indicating that the alarm has been activated and is in the wait state. The message will also show the wait time in seconds until the alarm can be tripped by motion.

```
Notification: '
'Notification: '
'Notification: '+IPD,0,15:ACTIVATE_
ALARM
'ClientIndex: 0
DataLength: 15
Data: ACTIVATE_ALARM

Sending Data ....
TCP Connection ID: 0 , DataLength: 40 ,
Data: Alarm Activated...WAIT TIME:10
Seconds.
```

After waiting for 10 seconds the alarm changes from the wait state to the `Alarm_ON` state where the alarm can now be tripped by a motion. The Arduino also sends a message which is "`ALARM NOW ACTIVE...`" to the Android. A debug message is also printed out to the Serial Monitor which is "`ALARM WAIT STATE FINISHED ... Alarm Now ACTIVE ....`" that indicates that the alarm can now be tripped.

```
Notification: '
' Sending Data ....
TCP Connection ID: 0 , DataLength: 20 ,
Data: ALARM NOW ACTIVE...

ALARM WAIT STATE FINISHED ... Alarm Now
ACTIVE ....
```

Next, trip the alarm by moving your hand in front of the motion sensor. You should see a message appear on the Serial Monitor indicating that a motion was detected and the total number of motions detected. A character string is also sent from the Arduino to the Android device that reads "`ALARM_TRIPPED:`" followed by the total number of motions that have been detected since the alarm was activated. You should also see the sensor value fall back to 0 which is printed to the Serial Monitor (assuming you have the sensor in single trigger mode).

```
Notification: '
'SensorValue: 1 , Motion Detected ...
NumberHits: 1
Sending Data ....

TCP Connection ID: 0 , DataLength: 16 ,
Data: ALARM_TRIPPED:1
Notification: '
'SensorValue: 0
```

Next, after waiting a few seconds to allow for the motion sensor to get past the automatic time delay (assuming single trigger mode) trigger the motion sensor again by moving your hand in front of it. Do this once or twice more to see that the alarm can be tripped multiple times which sends alerts to the Android cell phone each time.

```
SensorValue: 1 , Motion Detected ...
NumberHits: 2
Sending Data ....
TCP Connection ID: 0 , DataLength: 16 ,
Data: ALARM_TRIPPED:2
Notification: '
'SensorValue: 0
SensorValue: 1 , Motion Detected ...
NumberHits: 3
Sending Data ....
TCP Connection ID: 0 , DataLength: 16 ,
Data: ALARM_TRIPPED:3
Notification: '
'SensorValue: 0
```

Next, deactivate the alarm system by selecting the "Alarm Settings" menu item on the Android application and then selecting the "Deactivate Alarm" menu item. The alarm will be deactivated locally on the Android device. The Android will also send a message to the Arduino to deactivate the alarm on the Arduino's side which is the string "`DEACTIVATE_ALARM`". The Arduino will respond to this message by sending the string "`Alarm Has Been DE-Activated ...`" back to the Android device.

```
Notification: '
'Notification: '+IPD,0,17:DEACTIVATE_
ALARM
'ClientIndex: 0
DataLength: 17
Data: DEACTIVATE_ALARM
Sending Data ....
TCP Connection ID: 0 , DataLength: 32 ,
Data: Alarm Has Been DE-Activated ...
```

Next, type in the "`hits`" command into the phone entry textbox and hit the "Send Data" button to send this command to the Arduino. The Arduino should receive the "hits" command and then send back the number of motions that have been detected so far which in this example case is the string "`Number Motions Detected: 3`".

```
Notification: '
'Notification: '
'Notification: '+IPD,0,5:hits
'ClientIndex: 0
DataLength: 5
Data: hits
Sending Data ....
TCP Connection ID: 0 , DataLength: 27 ,
Data: Number Motions Detected: 3
```

Next, enter the command "`wait=30000`" into the phone entry textbox in the Android application and hit the "Send Data" button to send the command to the Arduino microcontroller. The Arduino should receive this command and change the wait time to 30,000 milliseconds or 30 seconds and send the character string "`WaitTime set to: 30 Seconds`" back to the Android.

```
Notification: '
'Notification: '
'Notification: '+IPD,0,11:wait=30000
'ClientIndex: 0
DataLength: 11
Data: wait=30000
Sending Data ....
TCP Connection ID: 0 , DataLength:
28 , Data: WaitTime set to:
30 Seconds
```

Next, activate the alarm on the Android side. This sends an activate message string to the Arduino that is "`ACTIVATE_ALARM`". The Arduino receives this message, activates the alarm, and sends a message that acknowledges that the alarm has been activated and gives the wait time which is "`Alarm Activated...WAIT TIME:30 Seconds`." Note that the wait time should have changed from 10 seconds to 30 seconds.

```
Notification: '
'Notification: '
'Notification: '+IPD,0,15:ACTIVATE_
ALARM
'ClientIndex: 0
DataLength: 15
Data: ACTIVATE_ALARM
Sending Data ....
TCP Connection ID: 0 , DataLength: 40 ,
Data: Alarm Activated...WAIT TIME:
30 Seconds.
```

Within about 30 seconds the alarm should be ready to detect motions. The Arduino should then send an "`ALARM NOW ACTIVE...`" message to the Android device as well as print a similar message to the Serial Monitor.

```
Notification: '
'Sending Data ....
TCP Connection ID: 0 , DataLength: 20 ,
Data: ALARM NOW ACTIVE...
ALARM WAIT STATE FINISHED ... Alarm Now
ACTIVE ....
```

This ends the basic demonstration of this alarm system.

The FC-04 Sound Sensor

The sound sensor detects sounds such as loud noises from breaking glass caused by an intruder attempting to get into a home through a closed window. The pins on my sound sensor are as follows:

- VCC: Power pin that you connect to the 3.3-V power supply.
- GND: Ground pin that you connect to the common ground of your circuit.
- OUT: 0 or LOW voltage if a sound is detected and 1 or HIGH voltage if no sound is detected.

Figure 5-6 The FC-04 sound sensor.

See Figure 5-6.

Hands-on Example: The Wireless Sound Sensor Alarm System

In this hands-on example, I present a wireless sound activated alarm system using the basic wireless framework previously presented in this book using an Android cell phone, and an Arduino Mega 2560 with an ESP8266-based ESP-01 module.

Parts List

- 1 Android cell phone
- 1 Arduino Mega 2560 microcontroller
- 1 5- to 3.3-V step-down voltage regulator
- 1 Logic level converter from 5 to 3.3 V
- 1 ESP-01 Wi-Fi module
- 1 Arduino development station such as a desktop or a notebook
- 1 Breadboard
- 1 FC-04 sound sensor
- 1 Package of wires both male-to-male and female-to-male to connect the components

Setting Up the Hardware

Step-Down Voltage Regulator Connections

1. Connect the Vin pin on the voltage regulator to the 5-V power pin on the Arduino Mega 2560.

2. Connect the Vout pin on the voltage regulator to a node on your breadboard represented by a horizontal row of empty slots. This is the 3.3-V power node that will supply the 3.3-V power to the ESP-01 module as well as provide the reference voltage for the logic level shifter.
3. Connect the GND pin on the voltage regulator to the GND pin on the Arduino or form a GND node similar to the 3.3-V power node in the previous step.

ESP-01 Wi-Fi Module Connections

1. Connect the Tx or TxD pin to the Rx3 pin on the Arduino Mega 2560. This is the receive pin for the serial hardware port 3.
2. Connect the EN or CH_PD pin to the 3.3-V power node from the step-down voltage regulator.
3. Connect the IO16 or RST pin to the 3.3-V power node from the step-down voltage regulator.
4. Connect the 3V3 or VCC pin to the 3.3-V power node from the step-down voltage regulator.
5. Connect the GND pin to the ground pin or node.
6. Connect the Rx or RxD pin to pin B0 on the 3.3-V side of the logic level shifter. The voltage from this pin will be shifted from 5 V from the Arduino to 3.3 V for the ESP-01 module.

Logic Level Shifter

1. Connect the 3.3-V pin to the 3.3-V power node from the step-down voltage regulator.
2. Connect the GND pin on the 3.3-V side to the ground node for the Arduino.
3. Connect the B0 pin on the 3.3-V side to Rx or RxD pin on the ESP-01. This corresponds to the A0 pin on the 5-V side.
4. Connect the 5-V pin to the 5-V power pin on the Arduino.
5. Connect the GND pin on the 5-V side to the ground node for the Arduino.
6. Connect the A0 pin on the 5-V side to Tx3 pin on the Arduino. This pin corresponds to the B0 pin on the 3.3-V side.

The Sound Sensor

1. Connect the VCC pin on the sound sensor to the 3.3-V node that contains the output from the 5- to 3.3-V step-down voltage regulator.
2. Connect the GND pin on the sound sensor to the ground node of the circuit.
3. Connect the OUT pin on the sound sensor to pin 7 on the Arduino.

See Figure 5-7.

Setting Up the Software

This alarm system is similar to the motion detection alarm system using the HC-SR501 motion sensor that was discussed in the previous hands-on example. Many variables and functions were unchanged from the previous example. In this section, I will only cover the new variables and other changes that were made to the infrared motion detector example that was previously discussed.

Setting Up the Arduino Software

The `SoundDetected` variable indicates the voltage level that the sound sensor outputs when a sound is detected. On my sound sensor

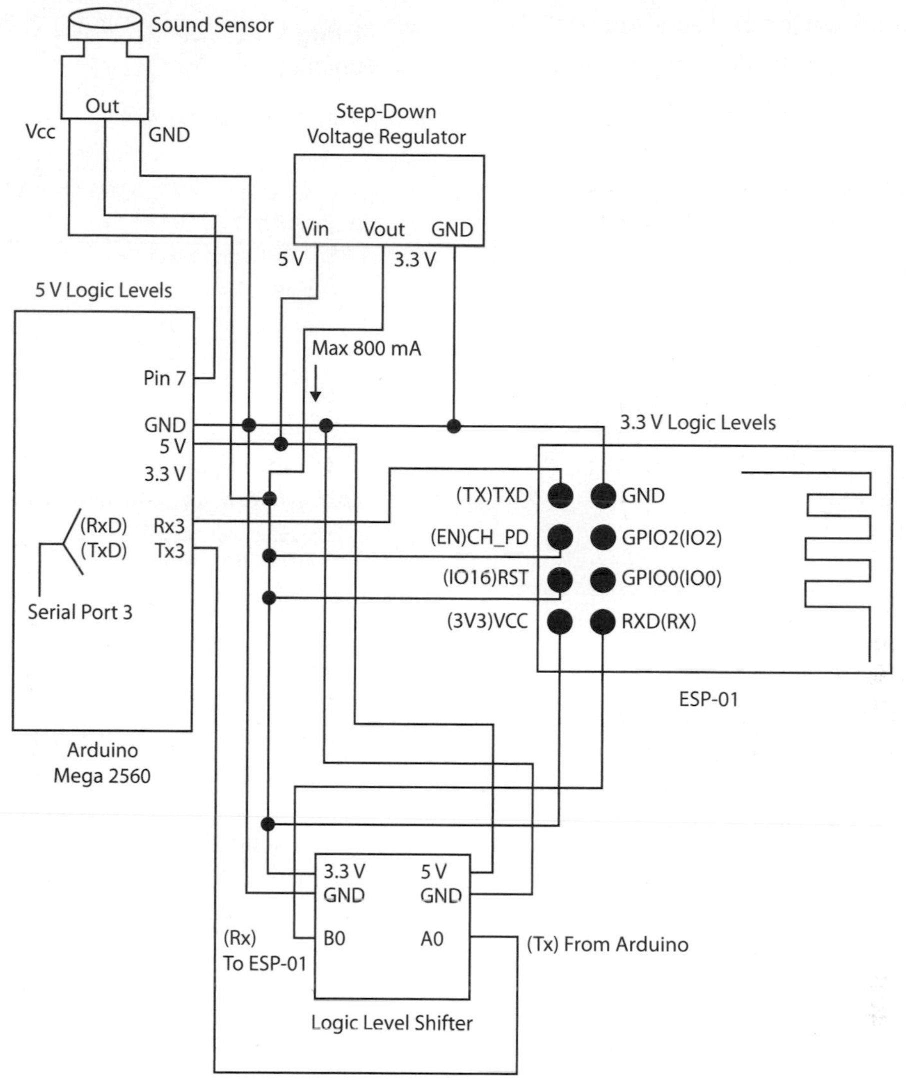

Figure 5-7 The sound sensor wireless alarm system.

when a sound is detected a 0 or LOW voltage is output.

```
int SoundDetected      = 0; // 0 or LOW
voltage indicates sound detected on my
sensor
```

The `SoundNotDetected` variable indicates the voltage level that the sound sensor outputs when no sound is detected. On my sound sensor when there is no sound detected the output is 1 or HIGH voltage.

```
int SoundNotDetected  = 1;
```

The `SensorValue` variable holds the current state of the sensor and is initialized to a value that means that no sound was detected.

```
int SensorValue = SoundNotDetected;
```

The `PrevSensorValue` variable holds the previous state of the sensor and is initialized to a value that means that no sound was detected.

```
int PrevSensorValue = SoundNotDetected;
```

The `MinDelayBetweenHits` variable controls the minimum delay between recording different hits from the sound sensor and sending the corresponding alerts to the Android device. This variable is set to a default value of 2500 milliseconds or 2.5 seconds.

```
unsigned long MinDelayBetweenHits =
2500; //
```

Set minimum time between registering new hits to 2.5 seconds.

The `MinDelayBetweenHitsStartTime` variable holds that time in milliseconds since the power of the Arduino that the last sound detection was registered. This time is initialized to 0 milliseconds.

```
unsigned long MinDelayBetweenHitsStart
Time = 0;
```

Adding to the ProcessSensor() Function

Some additional code was added to the `ProcessSensor()` function from the original code for the infrared motion detector in the previous hands-on example. Additional code was put in to make sure that new hits or the recognition of sounds as a new different sound would only occur at a minimum time period between sounds. This gives a more accurate view of the number of sounds that have occurred as well as reduces the number of extra alarm tripped messages that are sent to the Android device. The new code is shown in bold print in Listing 5-13.

Setting Up the Android Software

There are no changes you need to make to the Android application software for this example.

Operating the Wireless Sound Sensor Alarm System

Download the code for the "Android Basic Wireless Framework version 1.1" and the

Listing 5-13 The `ProcessSensor()` Function

```
void ProcessSensor()
{
 // Put Code to process your sensors
or other hardware in this function.
 if ((AlarmState == Alarm_ON) ||
(AlarmState == Alarm_TRIPPED))
 {
  // Process Sensor Input
  SensorValue = digitalRead
 (SensorPin);
  if ((PrevSensorValue == SoundNot
 Detected) && (SensorValue ==
 SoundDetected))
  {
   unsigned long
  ElapsedTime = millis() -
  MinDelayBetweenHitsStartTime;
   if (ElapsedTime >=
  MinDelayBetweenHits)
   {
    // Motion Just Detected
    NumberHits++;
    Serial.print(F("SensorValue: "));
    Serial.print(SensorValue);
    Serial.print(F(" , Motion
   Detected ... NumberHits: "));
    Serial.println(NumberHits);
    PrevSensorValue = SoundDetected;

    // Send Alarm Tripped Message
    SendAlarmTrippedMessage();

    // Time this hit occured
    MinDelayBetweenHitsStartTime =
   millis();
   }
  }
  else
  if ((PrevSensorValue ==
 SoundDetected) && (SensorValue ==
 SoundNotDetected))
  {
   // Heat Source Moved out of range
   Serial.print(F("SensorValue: "));
   Serial.println(SensorValue);
   PrevSensorValue = SoundNot
  Detected;
  }
 }
}
```

Arduino sound sensor project code and install the final programs on the Android and Arduino if you have not done this already. Note that the Android APK install files for all the chapters are located in the downloads section of Chapter 9. The code should be included in the code for this chapter on McGraw-Hill's web site.

After installing the sound sensor project code on the Arduino, unplug the Arduino's USB cable from your development system, wait a few seconds, and then plug the USB cable back into your computer and then bring up the Arduino's Serial Monitor. You should see something like the following:

```
****** ESP-01 Arduino Mega 2560 Server
Program ********
***********  Sound Sensor Alarm System
*************

AT Command Sent: AT+CIPMUX=1
ESP-01 Response: AT+CIPMUX=1

OK

ESP-01 mode set to allow multiple
connections.

AT Command Sent: AT+CWMODE_CUR=2
ESP-01 Response: AT+CWMODE_CUR=2

OK

ESP-01 mode changed to Access Point
Mode.

AT Command Sent: AT+CIPSERVER=1,80
ESP-01 Response: AT+CIPSERVER=1,80

OK

Server Started on ESP-01.

AT Command Sent: AT+CWSAP_CUR="ESP-
01","esp8266esp01",1,2,4,0
ESP-01 Response: AT+CWSAP_CUR="ESP-
01","esp8266esp01",1,2,4,0

OK

AP Configuration Set Successfullly.

AT Command Sent: AT+CWSAP_CUR?
ESP-01 Response: AT+CWSAP_CUR?

+CWSAP_CUR:"ESP-01","esp8266esp01",
1,2,4,0

OK

This is the curent configuration of the
AP on the ESP-01.

AT Command Sent: AT+CIPSTO?
ESP-01 Response: AT+CIPSTO?

+CIPSTO:180

OK

This is the curent TCP timeout in
seconds on the ESP-01.

************** System Has Started
*******************
```

Next, connect your Android to the Wi-Fi access point on the ESP-01 module called "ESP-01" using the password "esp8266esp01". Start up the Android basic framework application. The name of the Android device on the wireless network should be changed and then an "OK" message should be sent back to the Android device. In the example below the Android device's name on the wireless network is "`lg`".

```
Notification: '0,CONNECT'
Processing New Client Connection ...
NewConnection Name: 0

Notification: '
'Notification: '
'Notification: '+IPD,0,8:name=lg
'ClientIndex: 0
DataLength: 8
Data: name=lg

Client Name Change Requested,
ClientIndex: 0, NewClientName: lg
Client Name Change Succeeded, OK...
Sending Data ....
TCP Connection ID: 0 , DataLength: 3 ,
Data: OK
```

Next, turn off the emergency call out so that instead of the attempting to call an emergency phone number the program will give audible alarms when a new sound is detected by the sensor attached to the Arduino. Now, activate the alarm on the Android side. This should send an "ACTIVATE_ALARM" message to the Arduino. The Arduino should respond with an "Alarm Activated...WAIT TIME:10 Seconds." message to indicate that the alarm has been activated and is currently in the wait state.

```
Notification: '
'Notification: '
'Notification: '+IPD,0,15:ACTIVATE_
ALARM
'ClientIndex: 0
DataLength: 15
Data: ACTIVATE_ALARM

Sending Data ....
TCP Connection ID: 0 , DataLength:
40 , Data: Alarm Activated...WAIT
TIME:10 Seconds.
```

In around 10 seconds the alarm should be fully activated and ready to be tripped by a sound. The Arduino should send an "ALARM NOW ACTIVE..." message to your Android and a debug message should appear on the Serial Monitor indicating that the alarm is fully active now.

```
Notification: '
'Sending Data ....
TCP Connection ID: 0 , DataLength: 20 ,
Data: ALARM NOW ACTIVE...

ALARM WAIT STATE FINISHED ... Alarm Now
ACTIVE ....
```

Next, test out the alarm system by making a loud noise such as clapping your hands or shouting. The alarm should be tripped and an "ALARM_TRIPPED:1" message should be sent from the Arduino to the Android device. On the Android you should hear a sound effect and see the message in the debug message window. If you have the emergency email notification activated, then a text message will be sent to the emergency phone number that you have provided. Trip the alarm several more times waiting at least 2.5 seconds between each new sound. After each new alarm trip you should see the number of hits increasing. In the following example the alarm was tripped a total of four times.

```
Notification: '
'SensorValue: 0 , Motion Detected ...
NumberHits: 1
Sending Data ....
TCP Connection ID: 0 , DataLength: 16 ,
Data: ALARM_TRIPPED:1

SensorValue: 1
Notification: '
'SensorValue: 0 , Motion Detected ...
NumberHits: 2
Sending Data ....
TCP Connection ID: 0 , DataLength: 16 ,
Data: ALARM_TRIPPED:2

SensorValue: 1
Notification: '
'SensorValue: 0 , Motion Detected ...
NumberHits: 3
Sending Data ....
TCP Connection ID: 0 , DataLength: 16 ,
Data: ALARM_TRIPPED:3

SensorValue: 1
Notification: '
'SensorValue: 0 , Motion Detected ...
NumberHits: 4
Sending Data ....
TCP Connection ID: 0 , DataLength: 16 ,
Data: ALARM_TRIPPED:4

SensorValue: 1
```

Finally, deactivate the alarm from the Android application. A message will be sent from the Android to the Arduino which is "DEACTIVATE_ALARM" that will deactivate the alarm on the Arduino side. The Arduino will respond with an "`Alarm Has Been DE-Activated ...`" message that will be sent to the Android device.

```
Notification: '
'Notification: '
'Notification: '+IPD,0,17:DEACTIVATE_
ALARM
'ClientIndex: 0
DataLength: 17
Data: DEACTIVATE_ALARM
Sending Data ....
TCP Connection ID: 0 , DataLength: 32 ,
Data: Alarm Has Been DE-Activated ...
```

The HC-SR04 Distance Sensor

The HC-SR04 distance sensor can be used to measure distances from the sensor to an object. The sensor uses a sound pulse to determine the distance to an object. The sensor generates a sound pulse and detects the reflected sound or echo from that sound pulse when the sound is reflected by an object. The time between emitting the sound pulse and the echo is proportional to the distance to the object.

The specifications for the sensor are as follows:

- Working voltage: 5 V
- Working current: 15 mA
- Maximum range: 4 m
- Minimum range: 2 cm
- Measuring angle: 15 degrees

See Figure 5-8.

In order to find the distance to the nearest object in front of the HC-SR04 sensor, you have to do the following:

1. Set the Trig pin on the sensor from 0 or LOW to 1 or HIGH for 10 microseconds and then set the pin value back to 0 or LOW.
2. The distance sensor will then emit a series of sound pulses that will be reflected back to the sensor when an object has been hit by the sound waves.
3. Read in the pulse from the echo pin on the sensor. The time of the high portion of the pulse is proportional to the distance to the object that has been detected. The distance in inches to an object that is detected is the time in microseconds of the high portion of the pulse divided by 148.

Figure 5-8 The HC-SR04 distance sensor.

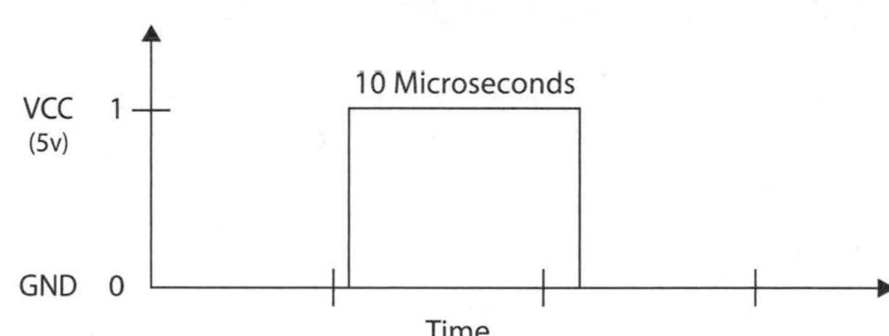

Figure 5-9 Triggering the sensor by setting the Trig pin high for 10 microseconds.

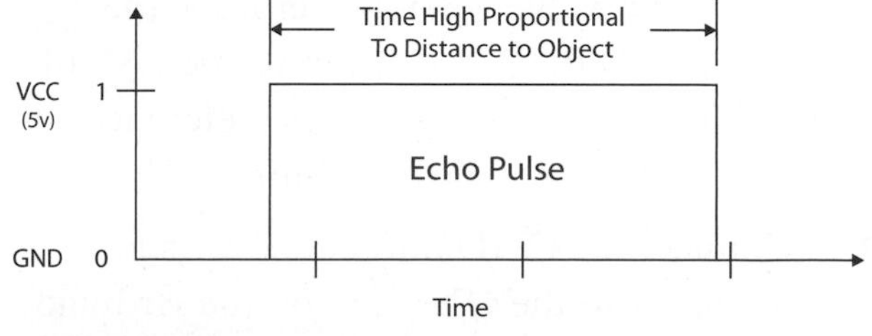

Figure 5-10 Reading the echo pulse.

Hands-on Example: HC-SR04 Distance Sensor Intruder Alarm System

In this hands-on example I show you how to build and operate an Android-/Arduino-based wireless alarm system using a distance sensor and an ESP-01 Wi-Fi module. This system is based on the Android/Arduino wireless basic framework covered previously.

Parts List

- 1 Android cell phone
- 1 Arduino Mega 2560 microcontroller
- 1 5- to 3.3-V step-down voltage regulator
- 1 Logic level converter from 5 to 3.3 V
- 1 ESP-01 Wi-Fi module
- 1 Arduino development station such as a desktop or a notebook
- 1 Breadboard
- 1 HC-SR04 ultrasonic distance sensor
- 1 Package of wires both male-to-male and female-to-male to connect the components

Setting Up the Hardware

Step-Down Voltage Regulator Connections

1. Connect the Vin pin on the voltage regulator to the 5-V power pin on the Arduino Mega 2560.
2. Connect the Vout pin on the voltage regulator to a node on your breadboard represented by a horizontal row of empty slots. This is the 3.3-V power node that will supply the 3.3-V power to the ESP-01 module as well as provide the reference voltage for the logic level shifter.
3. Connect the GND pin on the voltage regulator to the GND pin on the Arduino or form a GND node similar to the 3.3-V power node in the previous step.

ESP-01 Wi-Fi Module Connections

1. Connect the Tx or TxD pin to the Rx3 pin on the Arduino Mega 2560. This is the receive pin for the serial hardware port 3.
2. Connect the EN or CH_PD pin to the 3.3-V power node from the step-down voltage regulator.
3. Connect the IO16 or RST pin to the 3.3-V power node from the step-down voltage regulator.
4. Connect the 3V3 or VCC pin to the 3.3-V power node from the step-down voltage regulator.
5. Connect the GND pin to the ground pin or node.
6. Connect the Rx or RxD pin to pin B0 on the 3.3-V side of the logic level shifter. The voltage from this pin will be shifted from 5 V from the Arduino to 3.3 V for the ESP-01 module.

Logic Level Shifter

1. Connect the 3.3-V pin to the 3.3-V power node from the step-down voltage regulator.
2. Connect the GND pin on the 3.3-V side to the ground node for the Arduino.
3. Connect the B0 pin on the 3.3-V side to Rx or RxD pin on the ESP-01. This corresponds to the A0 pin on the 5-V side.
4. Connect the 5-V pin to the 5-V power pin on the Arduino.
5. Connect the GND pin on the 5-V side to the ground node for the Arduino.
6. Connect the A0 pin on the 5-V side to Tx3 pin on the Arduino. This pin corresponds to the B0 pin on the 3.3-V side.

HC-SR04 Ultrasonic Distance Sensor

1. Connect the VCC pin to the Arduino's 5-V node.

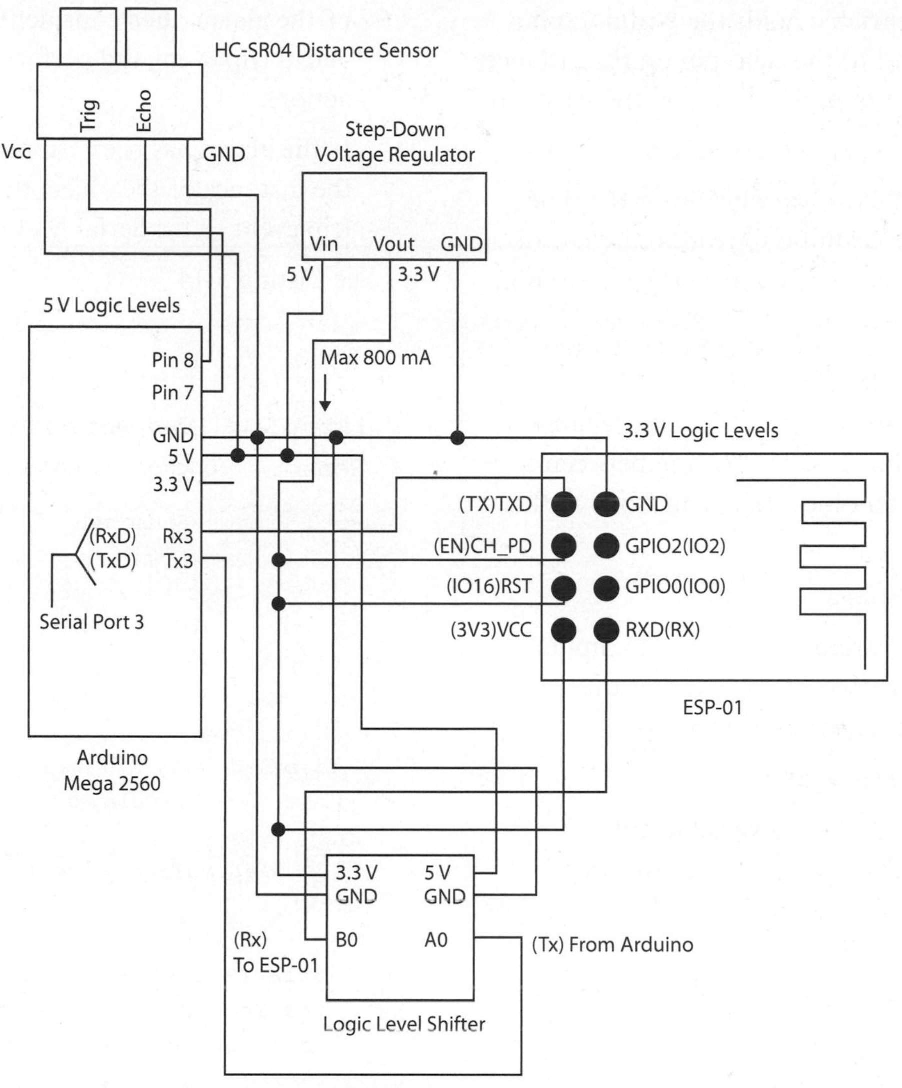

Figure 5-11 The distance sensor wireless alarm system.

2. Connect the GND pin to the ground node for the circuit.
3. Connect the Trig pin to pin 8 on the Arduino.
4. Connect the echo pin to pin 7 on the Arduino.

See Figure 5-11.

Setting Up the Software

The only changes in the software that are needed for this hands-on example are on the Arduino side. The Arduino code for this example is based on the code for the sound sensor detector presented in our last hands-on example. In this section I will only cover the code that is new to the distance sensor alarm project. Code that is the same as the sound sensor example will not be covered here.

The `TriggerPin` variable holds the Arduino pin that is connected to the trigger pin on the distance sensor that needs to be activated to generate a pulse. The pin is set to digital pin 8 on the Arduino.

```
int TriggerPin = 8; // output pin
```

The `EchoPin` variable holds the Arduino pin that is connected to the echo pin on the distance sensor and is set to digital pin 7 on the Arduino.

```
int EchoPin = 7; // input pin
```

The `DurationPulse` variable holds the time that it takes the Arduino to read in the return echo pulse that is reflected from objects in front of the distance sensor.

```
long DurationPulse = 0;
```

The `DistInches` variable holds the distance in inches that the sensor is reading between the sensor and an object that is in front of the sensor.

```
float DistInches = 0;
```

The `DistTrip` variable holds the maximum distance in inches where an object at this distance will cause the alarm to trip.

```
float DistTrip = 15;
```

The `MeasurementDelay` variable holds the delay in milliseconds between distance measurements. This is needed to ensure accurate readings from the distance sensor.

```
int MeasurementDelay = 80; // 80
milliseconds delay
```

The `ProcessSensor()` Function

The `ProcessSensor()` function is modified to incorporate additional code for the new distance sensor. The code does the following:

1. Before triggering the distance sensor to initiate a pulse makes sure that at least `MeasurementDelay` microseconds have passed since the last pulse.
2. Gets the distance in inches of any object that is in front of the distance sensor by calling the `GetDistanceInches()` function.
3. If the distance of the object in front of the distance sensor is less than the trip distance of the alarm, then consider this to be an alarm trip event and perform the required actions.
4. If the alarm has been tripped, then print out the distance of the object that produced the trip event to the Serial Monitor.

See Listing 5-14.

Listing 5-14 Modifying the `ProcessSensor()` Function for the Distance Sensor

```
void ProcessSensor()
{
 // Put Code to process your sensors
or other hardware in this function.
 if ((AlarmState == Alarm_ON) ||
(AlarmState == Alarm_TRIPPED))
 {
 unsigned long ElapsedTime =
millis() - MinDelayBetweenHits
StartTime;
 if (ElapsedTime >= MinDelayBetween
Hits)
   {
   delay(MeasurementDelay);
   DistInches = GetDistanceInches();
   if (DistInches < DistTrip)
    {
     // Motion Just Detected
     NumberHits++;
     Serial.print(F("Trip
    DistanceInches: "));
     Serial.print(DistInches);
     Serial.print(F(" , Motion
    Detected ... NumberHits: "));
     Serial.println(NumberHits);
     // Send Alarm Tripped Message
    SendAlarmTrippedMessage();
     // Time this hit occured
    MinDelayBetweenHitsStartTime =
    millis();
    }
   }
 }
}
```

The SensorSetup() Function

The SensorSetup() function initializes the distance sensor by doing the following:

1. Setting the pin TriggerPin on the Arduino to be an output pin that can produce a voltage level and provide current.
2. Setting the pin EchoPin on the Arduino to be an input pin that can read in the voltage at that pin.
3. Printing out the trip distance in inches to the Serial Monitor.

See Listing 5-15.

The MicrosecondsToInches() Function

The MicrosecondsToInches() function converts the duration of the echo pulse from the distance sensor into the distance in inches the object is from the distance sensor.

See Listing 5-16.

Listing 5-15 The SensorSetup()

```
void SensorSetup()
{
 pinMode(TriggerPin, OUTPUT);
 pinMode(EchoPin, INPUT);

 Serial.print(F("Trip Distance: "));
 Serial.print(DistTrip);
 Serial.println(" inches");
}
```

Listing 5-16 The MicrosecondsTo Inches() Function

```
float MicrosecondsToInches(long
Microseconds)
{
 // Distance in inches = uS/148
 return Microseconds/148.0;
}
```

Listing 5-17 The SendTriggerSignal() Function

```
void SendTriggerSignal()
{
 // Send Sound Pulse to detect object
 digitalWrite(TriggerPin, 0);
 delayMicroseconds(10);
 digitalWrite(TriggerPin, 1);
 delayMicroseconds(10);
 digitalWrite(TriggerPin, 0);
}
```

The SendTriggerSignal() Function

The SendTriggerSignal() function sends a 10-millisecond pulse to the distance sensor in order to trigger an ultrasonic pulse from the sensor that is used to determine the position of any object or objects in front of the sensor.

See Listing 5-17.

The GetDistanceInches() Function

The GetDistanceInches() function gets the distance of the object or objects in front of the distance sensor and returns this value in inches to the caller by doing the following:

1. Initiating a distance measurement by calling the SendTriggerSignal() to send the trigger signal to the sensor.
2. Waiting for and determining the time it takes to read in a pulse generated by the distance sensor on the echo pin by calling the pulseIn(EchoPin, HIGH) function.
3. Converting the time it took to read in the pulse from the sensor to the distance an object is from the sensor in inches by calling the MicrosecondsToInches (DurationPulse) function and returning this value.

See Listing 5-18.

Listing 5-18 The `GetDistanceInches()` Function

```
float GetDistanceInches()
{
  float retval = 0;

  SendTriggerSignal();

  // Read in return echo data
  DurationPulse = pulseIn(EchoPin,
HIGH);

  // Convert the returned echo pulse
into a distance
  retval = MicrosecondsToInches
(DurationPulse);

  return retval;
}
```

Adding the `trip` Command

The `trip` command lets the user assign the trip distance in inches using the Android basic wireless framework application. An object will trip the alarm if it is less than the trip distance from the sensor.

The format of the command is:

```
trip=distance
```

It is implemented in the `CheckAndProcessIPD()` function by doing the following:

1. If the "`trip=`" string is found in the incoming data from the client, then this is the trip command, so process it by doing the following:
    1. Find the index of the "`=`" sign.
    2. Find the string after the equal sign which should be a number.
    3. Convert the string to its equivalent number and assign it to the `DistTrip` variable that holds the trip distance for the alarm system.

Listing 5-19 Adding the `trip` Command

```
if (Data.indexOf("trip=") >= 0)
{
 int indextrip = Data.indexOf("=");
   String triplength = Data.
substring(indextrip+1);
   DistTrip = triplength.toInt();

   String returnstring = "";
   returnstring = returnstring +
F("DistTrip set to: ") + DistTrip +
F(" Inches\n");
   SendTCPMessage(ClientIndex,
returnstring);
}
```

    4. Create a return string indicating that the trip distance was set to the user specified distance.
    5. Send this string message to the requesting client by calling the `SendTCPMessage(ClientIndex, returnstring)` function.

See Listing 5-19.

Adding the `gettrip` Command

The `gettrip` command returns the alarm system's trip distance to the Android client and the new code is added in to the `CheckAndProcessIPD()` function. The trip distance is sent back to the requesting client by calling the `SendTCPMessage(ClientIndex, returnstring)` command.

See Listing 5-20.

Adding the `getdist` Command

The `getdist` command gets the current distance measurement from the distance sensor if the alarm system is active and returns this value to the Android device. This command is added into the `CheckAndProcessIPD()` function. The value is returned to the Android as a character string

Listing 5-20 Adding the `gettrip` Command

```
if (Data == "gettrip\n")
{
  // Return trip distance to
requesting client
   String returnstring = "Trip
Distance is: " + String(DistTrip) +
"\n";
   SendTCPMessage(ClientIndex,
returnstring);
}
```

Listing 5-21 Adding the `getdist` Command

```
if (Data == "getdist\n")
{
   String returnstring = "Current
Distance is: " + String(DistInches) +
" Inches...\n";
   SendTCPMessage(ClientIndex,
returnstring);
}
```

by calling the `SendTCPMessage(ClientIndex, returnstring)` function.

See Listing 5-21.

Operating the Intruder Alarm System

Download, compile, and install the Android Studio project "Basic Wireless Framework version 1.1" for this chapter on the Android. Do the same for the Arduino project related to the distance sensor for this chapter from McGraw-Hill's website. After installing the distance sensor alarm system program on your Arduino, bring up the Serial Monitor. You should see something like the following appear:

```
****** ESP-01 Arduino Mega 2560 Server
Program ********
****** HC-SR04 Ultra Sonic Distance
Sensor Alarm System  *******

Trip Distance: 15.00 inches

AT Command Sent: AT+CIPMUX=1
ESP-01 Response: AT+CIPMUX=1

OK

ESP-01 mode set to allow multiple
connections.

AT Command Sent: AT+CWMODE_CUR=2
ESP-01 Response: AT+CWMODE_CUR=2

OK

ESP-01 mode changed to Access Point
Mode.

AT Command Sent: AT+CIPSERVER=1,80
ESP-01 Response: AT+CIPSERVER=1,80

OK

Server Started on ESP-01.

AT Command Sent: AT+CWSAP_CUR="ESP-
01","esp8266esp01",1,2,4,0
ESP-01 Response: AT+CWSAP_CUR="ESP-
01","esp8266esp01",1,2,4,0

OK

AP Configuration Set Successfullly.

AT Command Sent: AT+CWSAP_CUR?
ESP-01 Response: AT+CWSAP_CUR?

+CWSAP_CUR:"ESP-01","esp8266esp01",1,
2,4,0

OK

This is the curent configuration of the
AP on the ESP-01.

AT Command Sent: AT+CIPSTO?
ESP-01 Response: AT+CIPSTO?

+CIPSTO:180

OK

This is the curent TCP timeout in
seconds on the ESP-01.

************** System Has Started
*******************
```

Connect to the Wi-Fi access point on the ESP-01 module that is named "ESP-01" and log in using the password "esp8266esp01". Start up the Android basic wireless framework version 1.1 which will automatically connect to the TCP server running on the ESP-01 module. Once the Android client is connected, the name of the device is changed to the one the user has selected or "android" by default. On my phone I set my Android device name to "`lg`". If the name change was successful, then an "OK" message is sent back to the Android device.

```
Notification: '0,CONNECT'
Processing New Client Connection ...
NewConnection Name: 0
Notification: '
'Notification: '
'Notification: '+IPD,0,8:name=lg
'ClientIndex: 0
DataLength: 8
Data: name=lg
Client Name Change Requested,
ClientIndex: 0, NewClientName: lg
Client Name Change Succeeded, OK...
Sending Data ....
TCP Connection ID: 0 , DataLength: 3 ,
Data: OK
```

Next, set the emergency call out to off. Activate the alarm system from the Android side. The Android should then send out an "`ACTIVATE_ALARM`" character string to the Arduino over the TCP connection. You should see this as an incoming notification on the Arduino side in the Serial Monitor. Once the Arduino activates the alarm on its side, it sends an "`Alarm Activated...WAIT TIME:10 Seconds.`" message to the Android client.

```
Notification: '
'Notification: '
'Notification: '+IPD,0,15:ACTIVATE_
ALARM
'ClientIndex: 0
DataLength: 15
Data: ACTIVATE_ALARM
Sending Data ....
TCP Connection ID: 0 , DataLength: 40 ,
Data: Alarm Activated...WAIT TIME:10
Seconds.
```

Wait around 10 seconds and you should see a notification in the Arduino's Serial Monitor indicating that the "`ALARM NOW ACTIVE...`" message has been sent to the Android device. The Arduino also prints out a message to the Serial Monitor indicating that the alarm is now fully active and is ready to be tripped.

```
Notification: '
'Sending Data ....
TCP Connection ID: 0 , DataLength: 20 ,
Data: ALARM NOW ACTIVE...
ALARM WAIT STATE FINISHED ... Alarm Now
ACTIVE ....
```

Next, trip the alarm several times to test out the system by putting an object in front of the distance sensor. Each alarm trip will display the distance at which the sensor detected an object in front of it. It should also send an alarm tripped message with the number of total alarm trips so far to the Android device. On the Android side each alarm trip should trigger a sound effect that plays continuously until you touch the debug message window to stop it. The incoming alarm tripped messages should also be shown in the debug message window. You should see something similar to the following in the Arduino's Serial Monitor:

```
Notification: '
'Trip DistanceInches: 11.97 , Motion
Detected ... NumberHits: 1
Sending Data ....
TCP Connection ID: 0 , DataLength: 16 ,
Data: ALARM_TRIPPED:1
Notification: '
'Trip DistanceInches: 11.79 , Motion
Detected ... NumberHits: 2
Sending Data ....
TCP Connection ID: 0 , DataLength: 16 ,
Data: ALARM_TRIPPED:2
Notification: '
'Trip DistanceInches: 14.86 , Motion
Detected ... NumberHits: 3
Sending Data ....
```

```
TCP Connection ID: 0 , DataLength: 16 ,
Data: ALARM_TRIPPED:3
```

Next, deactivate the alarm on the Android side. The Android should send a "DEACTIVATE_ALARM" message to the Arduino. The Arduino should receive the notification and deactivate the alarm. The Arduino should then send an "Alarm Has Been DE-Activated ..." message to the Android device.

```
Notification: '
'Notification: '
'Notification: '+IPD,0,17:DEACTIVATE_
ALARM
'ClientIndex: 0
DataLength: 17
Data: DEACTIVATE_ALARM
Sending Data ....
TCP Connection ID: 0 , DataLength: 32 ,
Data: Alarm Has Been DE-Activated ...
```

Next, let's test the gettrip command by entering that in the phone number textbox on the Android application and pressing the "Send Data" button. This command retrieves the current trip distance for the alarm system. The Arduino receives this notification and returns the value of the trip distance. You should see something like the following on the Arduino Serial Monitor. On the Android side you should see the text message "Trip Distance is: 15.00" show up in the debug message window.

```
Notification: '
'Notification: '
'Notification: '+IPD,0,8:gettrip
'ClientIndex: 0
DataLength: 8
Data: gettrip
Sending Data ....
TCP Connection ID: 0 , DataLength: 24 ,
Data: Trip Distance is: 15.00
```

Next, set the trip distance to 6 inches by typing "trip=6" into the phone number entry text box on the Android application and pressing "Send Data". The Arduino receives this command and sets the trip distance to 6 inches and sends back to the Android a confirmation message that the trip distance has been changed.

```
Notification: '
'Notification: '
'Notification: '+IPD,0,7:trip=6
'ClientIndex: 0
DataLength: 7
Data: trip=6
Sending Data ....
TCP Connection ID: 0 , DataLength: 29 ,
Data: DistTrip set to: 6.00 Inches
```

Next, turn on the alarm. Test the distance at which the distance sensor trips and it should be roughly 6 inches from the sensor. For example, the following is from the Serial Monitor where the exact trip distance is 5.69 inches.

```
Notification: '
'Trip DistanceInches: 5.69 , Motion
Detected ... NumberHits: 1
Sending Data ....
TCP Connection ID: 0 , DataLength: 16 ,
Data: ALARM_TRIPPED:1
```

Next, let's find out what distance the distance sensor is actually reading by using the getdist command. Send this command from your Android device. The Arduino should receive it as a notification and then send the current distance information back to the Android client. The information from the Serial Monitor output below indicates that the current distance being measured by the sensor is 46.32 inches.

```
Notification: '
'Notification: '
'Notification: '+IPD,0,8:getdist
'ClientIndex: 0
DataLength: 8
Data: getdist
Sending Data ....
TCP Connection ID: 0 , DataLength: 37 ,
Data: Current Distance is: 46.32
Inches...
```

This ends the demonstration of the distance sensor-based intruder alarm system.

The YL-38/YL-69 Water/ Moisture Detector

The YL-38/YL-69 water/moisture detector can measure the amount of moisture in soil or in the environment in general.

Its specifications are as follows:

- Operating voltage: 5 V
- Operating current: 3 mA
- Operating temperature: –40 to 60 degrees Centigrade

There are two parts of this sensor. The first part contains most of the electrical components and has pins to connect the device to the Arduino. This part is called the YL-38 component. See Figure 5-12.

The second part of the moisture sensor is the fork-like measuring probe. This probe is used to measure the amount of moisture in soil or the environment. See Figure 5-13.

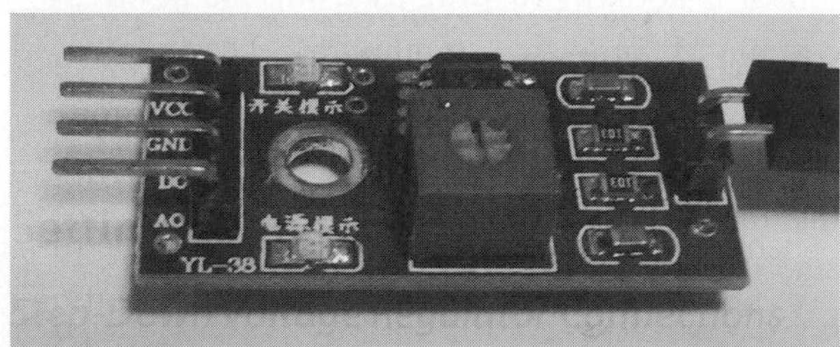

Figure 5-12 YL-38 the water/moisture detector main unit and connector.

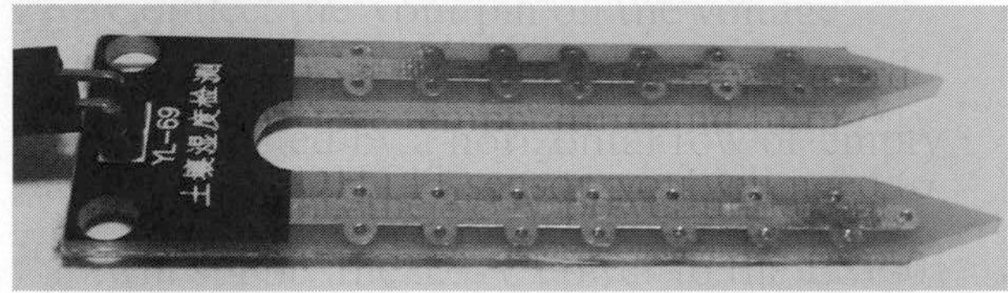

Figure 5-13 YL-69 water/moisture detector measuring probe.

Hands-on Example: The Water Detector Water Leak Wireless Alarm System

This hands-on example will demonstrate how to use the YL-38/YL-69 water/moisture detector in a wireless water leak alarm system. The alarm system uses the basic wireless framework consisting of an Android cell phone, and an Arduino Mega 2560 with an ESP8266-based ESP-01 Wi-Fi module.

Parts List

- 1 Android cell phone
- 1 Arduino Mega 2560 microcontroller
- 1 5- to 3.3-V step-down voltage regulator
- 1 Logic level converter from 5 to 3.3 V
- 1 ESP-01 Wi-Fi module
- 1 Arduino development station such as a desktop or a notebook
- 1 Breadboard
- 1 YL-38/YL-69 water/moisture detector
- 1 Package of wires both male-to-male and female-to-male to connect the components

Setting Up the Hardware

Step-Down Voltage Regulator Connections

1. Connect the Vin pin on the voltage regulator to the 5-V power pin on the Arduino Mega 2560.
2. Connect the Vout pin on the voltage regulator to a node on your breadboard represented by a horizontal row of empty slots. This is the 3.3-V power node that will supply the 3.3-V power to the ESP-01 module as well as provide the reference voltage for the logic level shifter.

3. Connect the GND pin on the voltage regulator to the GND pin on the Arduino or form a GND node similar to the 3.3-V power node in the previous step.

ESP-01 Wi-Fi Module Connections

1. Connect the Tx or TxD pin to the Rx3 pin on the Arduino Mega 2560. This is the receive pin for the serial hardware port 3.
2. Connect the EN or CH_PD pin to the 3.3-V power node from the step-down voltage regulator.
3. Connect the IO16 or RST pin to the 3.3-V power node from the step-down voltage regulator.
4. Connect the 3V3 or VCC pin to the 3.3-V power node from the step-down voltage regulator.
5. Connect the GND pin to the ground pin or node.
6. Connect the Rx or RxD pin to pin B0 on the 3.3-V side of the logic level shifter. The voltage from this pin will be shifted from 5 V from the Arduino to 3.3 V for the ESP-01 module.

Logic Level Shifter

1. Connect the 3.3-V pin to the 3.3-V power node from the step-down voltage regulator.
2. Connect the GND pin on the 3.3-V side to the ground node for the Arduino.
3. Connect the B0 pin on the 3.3-V side to Rx or RxD pin on the ESP-01. This corresponds to the A0 pin on the 5-V side.
4. Connect the 5-V pin to the 5-V power pin on the Arduino.
5. Connect the GND pin on the 5-V side to the ground node for the Arduino.
6. Connect the A0 pin on the 5-V side to Tx3 pin on the Arduino. This pin corresponds to the B0 pin on the 3.3-V side.

YL-38/YL-69 Water/Moisture Detector

1. Connect the VCC pin to the 5-V pin on the Arduino.
2. Connect the GND pin to the ground node for the circuit.
3. Connect the A0 pin to analog pin 7 on the Arduino.

See Figure 5-14.

Setting Up the Software

The code for this project only needs modifications on the Arduino side. The Android basic wireless framework version 1.1 can be used just like in the last hands-on example. The Arduino code is based on the code for the sound detector alarm hands-on example and has been slightly modified for use with the water detector sensor. I will cover only the changes made to the code related to the water detector sensor in this section.

I have experimentally determined some general values for the level of wetness for the sensor reading of the water detector. The lower the sensor value, the more moist the environment.

```
const int VERY_WET = 400;
const int WET      = 500;
const int DAMP     = 700;
const int DRY      = 850;
const int VERY_DRY = 950;
```

The `SensorPin` variable holds the analog pin on the Arduino that is connected to the water/moisture detector's output pin. The default value is analog pin A7 on the Arduino.

```
int SensorPin = A7;
```

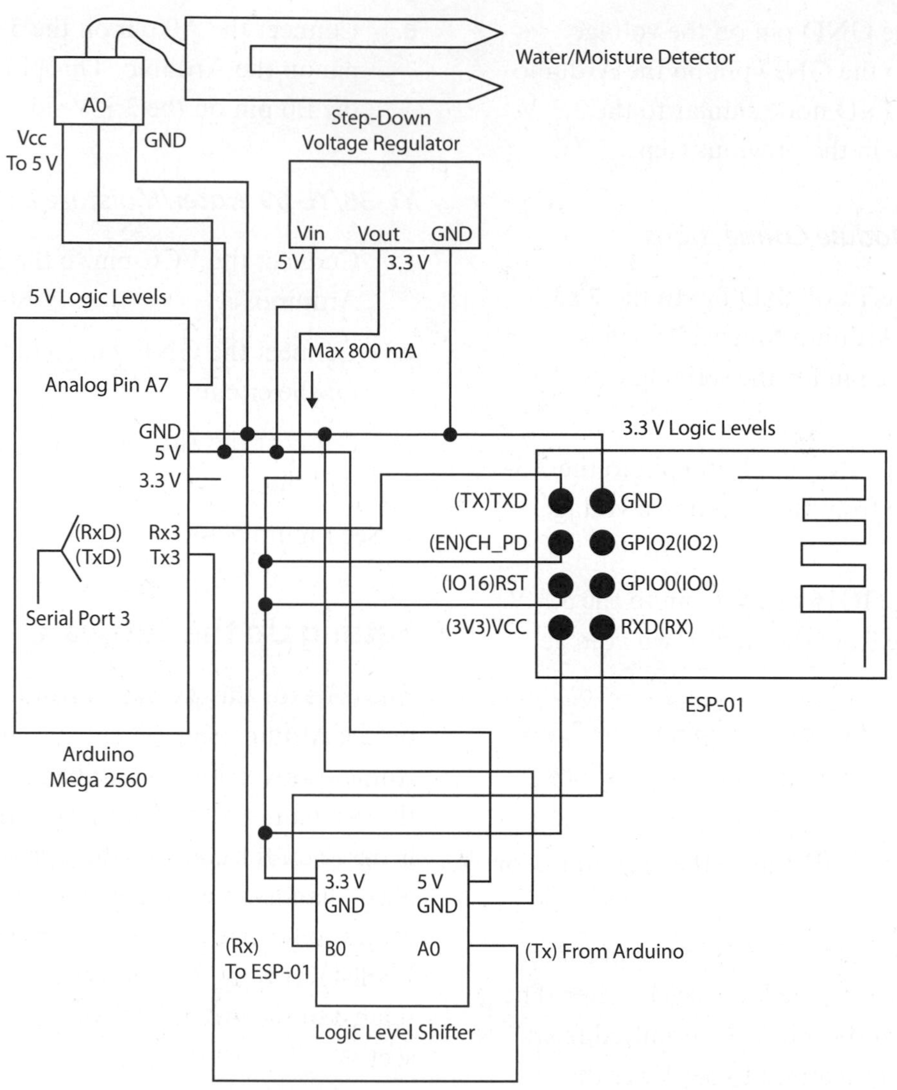

Figure 5-14 The water leak detector wireless alarm system.

The `SetupSensor()` Function

The `SetupSensor()` function initializes the water detector/moisture sensor by setting the `SensorPin` which is the Arduino pin connected to the output pin on the sensor to be able to read in voltages at that pin.

See Listing 5-22.

Listing 5-22 The `SetupSensor()` Function

```
void SetupSensor()
{
  // Set up Sensor
  pinMode(SensorPin, INPUT);
}
```

The `TestWaterLeak()` Function

The `TestWaterLeak()` function reads in the output from the water detector sensor and determines if the environment is considered wet. If the environment is wet, then there is a water leak, so return true. Otherwise return false.

See Listing 5-23.

The `ProcessSensor()` Function

The `ProcessSensor()` function is modified so that if the sensor detects a damp environment

Listing 5-23 The `TestWaterLeak()` Function

```
boolean TestWaterLeak()
{
 boolean result = false;

 SensorValue = analogRead(SensorPin);
 if (SensorValue < DAMP)
 {
  result = true;
 }
 return result;
}
```

then the alarm is tripped. That is if the `TestWaterLeak()` function is called and returns true, then a water leak has been detected.

See Listing 5-24.

Modifying the `setup()` Function

In the `setup()` function I added the code to call the `SetupSensor()` function which initializes the water leak sensor.

Operating the Water Leak Alarm System

This example uses the Android basic wireless framework version 1.1 and the water sensor Arduino project code for this chapter that should be downloadable from McGraw-Hill's website. Download the project files and install the programs on the Android and the Arduino. Note that the Android APK install files for all the chapters are located in the downloads section of Chapter 9.

Start up the Arduino's Serial Monitor and you should see something like the following:

```
****** ESP-01 Arduino Mega 2560 Server
Program ********
***********  Water Sensor Alarm System
*************
```

Listing 5-24 The `ProcessSensor()` Function

```
void ProcessSensor()
{
 // Put Code to process your
sensors or other hardware in this
function.
 if ((AlarmState == Alarm_ON) ||
(AlarmState == Alarm_TRIPPED))
 {
  unsigned long ElapsedTime =
 millis() - MinDelayBetweenHits
 StartTime;
  if (ElapsedTime >= MinDelay
 BetweenHits)
  {
   if (TestWaterLeak() == true)
   {
    // Motion Just Detected
    NumberHits++;
    Serial.print(F("Sensor
   Value: "));
    Serial.print(SensorValue);
    Serial.print(F(" , Water Leak
   Detected ... NumberHits: "));
    Serial.println(NumberHits);

    // Send Alarm Tripped Message
    SendAlarmTrippedMessage();

    // Time this hit occured
    MinDelayBetweenHitsStartTime =
   millis();
   }
  }
 }
}
```

```
AT Command Sent: AT+CIPMUX=1
ESP-01 Response: AT+CIPMUX=1

OK

ESP-01 mode set to allow multiple
connections.
```

```
AT Command Sent: AT+CWMODE_CUR=2
ESP-01 Response: AT+CWMODE_CUR=2

OK

ESP-01 mode changed to Access Point
Mode.

AT Command Sent: AT+CIPSERVER=1,80
ESP-01 Response: AT+CIPSERVER=1,80

OK

Server Started on ESP-01.

AT Command Sent: AT+CWSAP_CUR="ESP-
01","esp8266esp01",1,2,4,0
ESP-01 Response: AT+CWSAP_CUR="ESP-01",
"esp8266esp01",1,2,4,0

OK

AP Configuration Set Successfullly.

AT Command Sent: AT+CWSAP_CUR?
ESP-01 Response: AT+CWSAP_CUR?
+CWSAP_CUR:"ESP-
01","esp8266esp01",1,2,4,0

OK

This is the curent configuration of the
AP on the ESP-01.

AT Command Sent: AT+CIPSTO?
ESP-01 Response: AT+CIPSTO?
+CIPSTO:180

OK

This is the curent TCP timeout in
seconds on the ESP-01.

************** System Has Started
*******************
```

Connect your Android to the Wi-Fi access point “ESP-01” that is running on the ESP-01 module using the password “esp8266esp01”. Start up the basic wireless framework version 1.1 on the Android and this will connect to the TCP server running on the ESP-01 module. After connecting to the server, the Android will send the command to change the name of the device on the network. In my case the name of the Android device is changed to “`lg`”. If the change succeeded, then an “OK” string is sent back to the Android device. See the following log from the Serial Monitor:

```
Notification: '0,CONNECT'
Processing New Client Connection ...
NewConnection Name: 0

Notification: '
'Notification: '
'Notification: '+IPD,0,8:name=lg
'ClientIndex: 0
DataLength: 8
Data: name=lg

Client Name Change Requested,
ClientIndex: 0, NewClientName: lg
Client Name Change Succeeded, OK...
Sending Data ....
TCP Connection ID: 0 , DataLength: 3 ,
Data: OK
```

Next, turn off the emergency call out and activate the alarm on the Android side. The Android will send an “`ACTIVATE_ALARM`” string to the Arduino. The Arduino will activate the alarm and send an acknowledgment message back to the Android.

```
Notification: '
'Notification: '
'Notification: '+IPD,0,15:ACTIVATE_
ALARM
'ClientIndex: 0
DataLength: 15
Data: ACTIVATE_ALARM

Sending Data ....
TCP Connection ID: 0 , DataLength: 40 ,
Data: Alarm Activated...WAIT TIME:10
Seconds.
```

Wait for around 10 seconds for the alarm to fully activate. The Arduino should send a string to the Android device indicating that the alarm is now fully active. A message should also appear on the Arduino’s Serial Monitor indicating that alarm is now fully activated and can be tripped.

```
Notification: '
'Sending Data ....
TCP Connection ID: 0 , DataLength: 20 ,
Data: ALARM NOW ACTIVE...

ALARM WAIT STATE FINISHED ... Alarm Now
ACTIVE ....
```

Next, get a wet paper towel and wrap it around the water detector's measurement probe until the alarm is tripped. When the alarm is tripped the exact value that tripped the alarm is given. Any number less than DAMP or 700 will trip the alarm and send a message to the Android indicating that the alarm is tripped and the total number of alarm trips so far. See the log from the Serial Monitor below to see what values I got when I tripped my alarm using a wet paper towel.

```
Notification: '
'SensorValue: 699 , Water Leak Detected
... NumberHits: 1
Sending Data ....
TCP Connection ID: 0 , DataLength: 16 ,
Data: ALARM_TRIPPED:1

Notification: '
'SensorValue: 699 , Water Leak Detected
... NumberHits: 2
Sending Data ....
TCP Connection ID: 0 , DataLength: 16 ,
Data: ALARM_TRIPPED:2

Notification: '
'SensorValue: 551 , Water Leak Detected
... NumberHits: 3
Sending Data ....
TCP Connection ID: 0 , DataLength: 16 ,
Data: ALARM_TRIPPED:3

Notification: '
'SensorValue: 536 , Water Leak Detected
... NumberHits: 4
Sending Data ....
TCP Connection ID: 0 , DataLength: 16 ,
Data: ALARM_TRIPPED:4

Notification: '
'SensorValue: 699 , Water Leak Detected
... NumberHits: 5
Sending Data ....
TCP Connection ID: 0 , DataLength: 16 ,
Data: ALARM_TRIPPED:5
```

Next, deactivate the alarm on the Android side. A deactivate alarm message is sent to the Arduino. The Arduino deactivates the alarm and sends a message back to the Android indicating that the alarm has been deactivated.

```
Notification: '
'Notification: '
'Notification: '+IPD,0,17:DEACTIVATE_
ALARM
'ClientIndex: 0
DataLength: 17
Data: DEACTIVATE_ALARM

Sending Data ....
TCP Connection ID: 0 , DataLength: 32 ,
Data: Alarm Has Been DE-Activated ...
```

This ends the demonstration of the water leak detector wireless alarm system.

The Light Emitting Diode (LED)

The LED or light emitting diode emits light when a voltage is connected across it. The exact specifications vary depending on the exact type of LED but most will work with the Arduino which works on 5-V logic and can provide up to 20 mA of current to an electrical component. The LED has a positive terminal and a negative terminal. The positive terminal is longer than the negative terminal. The bottom rim of the LED casing should also be flat on the negative terminal side. See Figure 5-15.

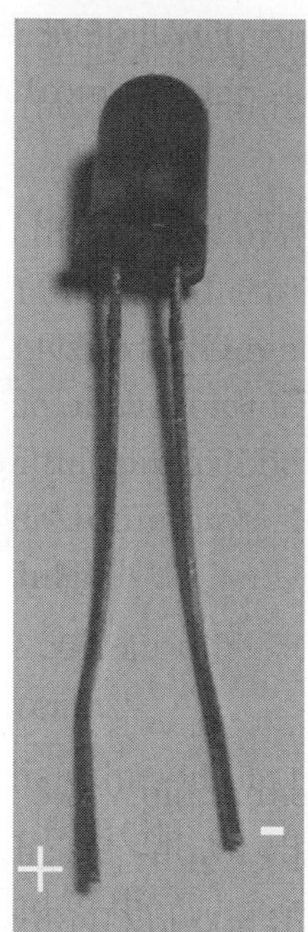

Figure 5-15 The LED.

Hands-on Example: The Remote Wireless Control of an LED

In this hands-on example I show you how to control an LED remotely through the Android and Arduino basic wireless framework using the ESP-01 Wi-Fi module. This example will demonstrate how to turn on an LED, turn off an LED, blink an LED, and set the blink parameters for an LED.

Parts List

- 1 Android cell phone
- 1 Arduino Mega 2560 microcontroller
- 1 5- to 3.3-V step-down voltage regulator
- 1 Logic level converter from 5 to 3.3 V
- 1 ESP-01 Wi-Fi module
- 1 Arduino development station such as a desktop or a notebook
- 1 Breadboard
- 1 LED
- 1 Resistor (OPTIONAL: Used to dim the LED a few hundred ohms such as 300 ohms should work)
- 1 Package of wires both male-to-male and female-to-male to connect the components

Setting Up the Hardware

Step-Down Voltage Regulator Connections

1. Connect the Vin pin on the voltage regulator to the 5-V power pin on the Arduino Mega 2560.
2. Connect the Vout pin on the voltage regulator to a node on your breadboard represented by a horizontal row of empty slots. This is the 3.3-V power node that will supply the 3.3-V power to the ESP-01 module as well as provide the reference voltage for the logic level shifter.
3. Connect the GND pin on the voltage regulator to the GND pin on the Arduino or form a GND node similar to the 3.3-V power node in the previous step.

ESP-01 Wi-Fi Module Connections

1. Connect the Tx or TxD pin to the Rx3 pin on the Arduino Mega 2560. This is the receive pin for the serial hardware port 3.
2. Connect the EN or CH_PD pin to the 3.3-V power node from the step-down voltage regulator.
3. Connect the IO16 or RST pin to the 3.3-V power node from the step-down voltage regulator.
4. Connect the 3V3 or VCC pin to the 3.3-V power node from the step-down voltage regulator.

5. Connect the GND pin to the ground pin or node.
6. Connect the Rx or RxD pin to pin B0 on the 3.3-V side of the logic level shifter. The voltage from this pin will be shifted from 5 V from the Arduino to 3.3 V for the ESP-01 module.

Logic Level Shifter

1. Connect the 3.3-V pin to the 3.3-V power node from the step-down voltage regulator.
2. Connect the GND pin on the 3.3-V side to the ground node for the Arduino.
3. Connect the B0 pin on the 3.3-V side to Rx or RxD pin on the ESP-01. This corresponds to the A0 pin on the 5-V side.
4. Connect the 5-V pin to the 5-V power pin on the Arduino.
5. Connect the GND pin on the 5-V side to the ground node for the Arduino.
6. Connect the A0 pin on the 5-V side to Tx3 pin on the Arduino. This pin corresponds to the B0 pin on the 3.3-V side.

The LED

1. Connect digital pin 7 on the Arduino to one end of a 330-ohm resistor.
2. Connect the other end of the 330-ohm resistor to the positive terminal of the LED.
3. Connect the negative terminal of the LED to the common ground node for the system.

See Figure 5-16.

Setting Up the Software

The Arduino code has been modified from the basic framework code discussed in Chapter 4. This section will cover the modifications to this basic framework.

The `LEDPin` variable holds the Arduino pin that is connected to the positive terminal of the LED and will be used to supply the voltage. The default is set to Arduino digital pin 7.

```
int LEDPin = 7;
```

The `LEDOn` variable is true if the LED is on and false otherwise. The default value is set to false.

```
boolean LEDOn = false;
```

The `Blink` variable is true if the LED blink mode is on and false otherwise. The default value is set to false.

```
boolean Blink = false;
```

The `LEDOffTime` holds the time in milliseconds that the LED should be off during the blink cycle and is set by default to 1000 milliseconds or 1 second.

```
int LEDOffTime = 1000;
```

The `LEDOnTime` variable holds the time in milliseconds that the LED should be on during the blink cycle and is set by default to 1000 milliseconds or 1 second.

```
int LEDOnTime = 1000;
```

The `BlinkStartTime` variable holds the time in milliseconds of the beginning of a blink on period or blink off period and is used to keep track of the amount of time that the LED has been on or off during blink mode.

```
unsigned long BlinkStartTime = 0;
```

The `SetupSensorAndHardware()` Function

The `SetupSensorAndHardware()` function sets up the LED hardware by setting the pin mode for the LEDPin on the Arduino which is connected to the LED to be an output pin that provides voltage and current. See Listing 5-25.

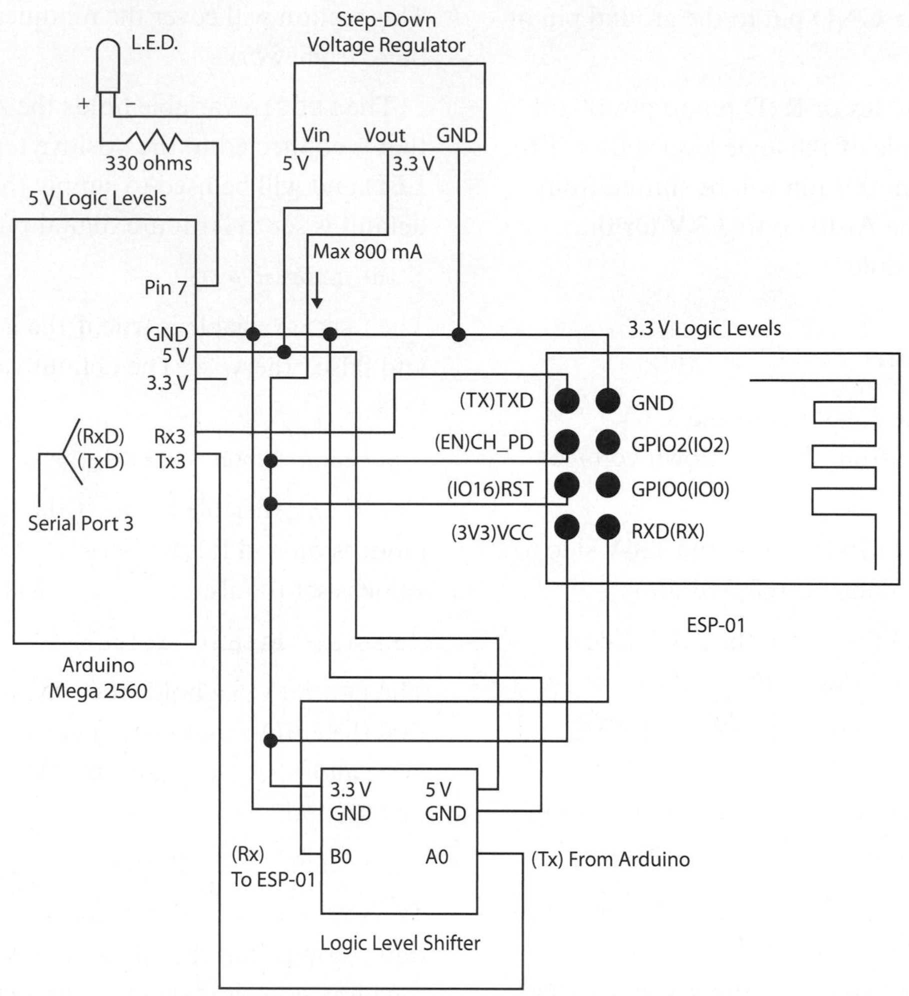

Figure 5-16 The remote wireless LED controller system.

Listing 5-25 The `SetupSensorAndHardware()` Function

```
void SetupSensorAndHardware()
{
 pinMode(LEDPin, OUTPUT);
}
```

The `PerformLEDBlink()` Function

The `PerformLEDBlink()` updates the blink status of the LED by doing the following:

1. Calculating the elapsed time since the start of the last blink period either an LED on period or LED off period.
2. If the LED is on, then do the following:
    1. If the elapsed time is greater than the target LED blink on time, then shut off the LED, set the `LEDOn` variable to false, and reset the `BlinkStartTime` variable to the current time.
3. If the LED is off, then do the following:
    1. If the elapsed time is greater than the target LED blink off time then turn on the LED, set the `LEDOn` variable to true, and reset the `BlinkStartTime` variable to the current time.

See Listing 5-26.

Listing 5-26 The `PerformLEDBlink()` Function

```
void PerformLEDBlink()
{
 unsigned long ElapsedTime = millis()
- BlinkStartTime;
 if (LEDOn)
 {
  // LED is on
  if (ElapsedTime > LEDOnTime)
  {
   // if LED is on and LED on time
  has passed then shut LED off
   digitalWrite(LEDPin, LOW);
   LEDOn = false;
   BlinkStartTime = millis();
  }
 }
 else
 {
  // LED is off
  if (ElapsedTime > LEDOffTime)
  {
   digitalWrite(LEDPin,HIGH);
   LEDOn = true;
   BlinkStartTime = millis();
  }
 }
}
```

The `ProcessSensorAndHardware()` Function

The `ProcessSensorAndHardware()` function processes the LED blink function if active by calling the `PerformLEDBlink()` function.

See Listing 5-27.

Modifying the `setup()` Function

The sensors and other hardware such as the LED are initialized in the `setup()` function by calling the `SetupSensorAndHardware()` function.

`Adding the ledon Command`

Listing 5-27 The `ProcessSensorAndHardware()` Function

```
void ProcessSensorAndHardware()
{
 // Put Code to process your sensors
or other hardware in this function.
 if (Blink)
 {
  PerformLEDBlink();
 }
}
```

Listing 5-28 Adding the `ledon` Command

```
if (Data == "ledon\n")
{

 LEDOn = true;
 digitalWrite(LEDPin, HIGH);

 String returnstring = "LED turned
ON...\n";
 SendTCPMessage(ClientIndex,
returnstring);
}
```

The code for the "`ledon`" command is added into the `CheckAndProcessIPD()` function. The code turns on the LED and sends back a text string to the Android device indicating that the LED has been turned on.

See Listing 5-28.

Adding the `ledoff` Command

The "`ledoff`" command was added to the `CheckAndProcessIPD()` function. The command turns off the LED and sends a string message back to the Android device indicating that the LED has been turned off.

See Listing 5-29.

Adding the `blink` Command

The "`blink`" command was added to the `CheckAndProcessIPD()` function. The `blink`

Listing 5-29 The `ledoff` Command

```
if (Data == "ledoff\n")
{
 LEDOn = false;
 digitalWrite(LEDPin, LOW);

 String returnstring = "LED turned
OFF...\n";
 SendTCPMessage(ClientIndex,
returnstring);
}
```

Listing 5-30 Adding the `blink` Command

```
if (Data == "blink\n")
{
 Blink = true;
 BlinkStartTime = millis();

 String returnstring = "LED Blinking
Started...\n";
 SendTCPMessage(ClientIndex,
returnstring);
}
```

command initializes and turns on the LED blinking feature and sends a text string back to the Android device acknowledging that the LED blinking feature has been turned on.

See Listing 5-30.

Adding the `noblink` Command

The code for the "`noblink`" command is added to the `CheckAndProcessIPD()` function. The `noblink` code turns off the blink function, sets the LED to off, and sends a message to the Android device indicating that the blink function has been turned off.

See Listing 5-31.

Listing 5-31 The `noblink` Command

```
if (Data == "noblink\n")
{
 Blink = false;
 LEDOn = false;
 digitalWrite(LEDPin, LOW);

 String returnstring = "LED
Blinking Stopped...\n";
 SendTCPMessage(ClientIndex,
returnstring);
}
```

Adding the `ledtimeoff` Command

The code for the "`ledtimeoff`" command is added to the `CheckAndProcessIPD()` function. The "`ledtimeoff`" command sets the amount of time in milliseconds that the LED blink cycle stays off. The code finds the actual string value of the off time following the equals sign and assigns the numerical equivalent to the `LEDOffTime` variable. A string message is then sent back to the Android acknowledging that the time has been changed.

See Listing 5-32.

Listing 5-32 The `ledtimeoff` Command

```
if (Data.indexOf("ledtimeoff=") >= 0)
{
 int indextime = Data.indexOf("=");
 String timelength = Data.
substring(indextime+1);
 LEDOffTime = timelength.toInt();

 String returnstring = "";
 returnstring = returnstring +
F("LEDOffTime set to: ") + LEDOffTime
+ F(" MilliSeconds\n");
 SendTCPMessage(ClientIndex,
returnstring);
}
```

Adding the `ledtimeon` Command

The code for the `ledtimeon` command is added to the `CheckAndProcessIPD()` function. This code sets the time that the LED is on in the

Listing 5-33 The `ledtimeon` Command

```
if (Data.indexOf("ledtimeon=") >= 0)
{
  int indextime = Data.indexOf("=");
  String timelength = Data.
substring(indextime+1);
  LEDOnTime = timelength.toInt();

  String returnstring = "";
  returnstring = returnstring +
F("LEDOnTime set to: ") + LEDOnTime +
F(" MilliSeconds\n");
  SendTCPMessage(ClientIndex,
returnstring);
}
```

Listing 5-34 Modifying the `loop()` function

```
void loop()
{
  // Process Wifi Notifications
  ProcessESP01Notifications();

  // Process Sensor Input
  ProcessSensorAndHardware();
}
```

blink cycle. The actual time is first determined as the string after the equals sign, converted to its numeric representation, and then assigned to the `LEDOnTime` variable. A character string is sent back to the Android stating that the time has been changed.

See Listing 5-33.

Modifying the `loop()` Function

The `loop()` function is also modified to rename the `ProcessSensor()` function to the more descriptive `ProcessSensorAndHardware()` function which includes processing the LED blink cycle.

See Listing 5-34.

Operating the Remote Wireless LED Control System

Download the Android basic wireless framework version 1.1 for this chapter and the Arduino code for the LED project from McGraw-Hill's website. Compile and install the programs. Note that the Android APK install files for all the chapters are located in the downloads section of Chapter 9. Start up the Serial Monitor on the Arduino and you should get the following start-up screen:

```
****** ESP-01 Arduino Mega 2560 Server
Program ********
*********** LED Control Program
*************

AT Command Sent: AT+CIPMUX=1
ESP-01 Response: AT+CIPMUX=1

OK

ESP-01 mode set to allow multiple
connections.

AT Command Sent: AT+CWMODE_CUR=2
ESP-01 Response: AT+CWMODE_CUR=2

OK

ESP-01 mode changed to Access Point
Mode.

AT Command Sent: AT+CIPSERVER=1,80
ESP-01 Response: AT+CIPSERVER=1,80

OK

Server Started on ESP-01.

AT Command Sent: AT+CWSAP_CUR="ESP-
01","esp8266esp01",1,2,4,0
ESP-01 Response: AT+CWSAP_CUR="ESP-
01","esp8266esp01",1,2,4,0

OK

AP Configuration Set Successfullly.

AT Command Sent: AT+CWSAP_CUR?
ESP-01 Response: AT+CWSAP_CUR?
+CWSAP_CUR:"ESP-01",
"esp8266esp01",1,2,4,0
```

```
OK
This is the curent configuration of the
AP on the ESP-01.
AT Command Sent: AT+CIPSTO?
ESP-01 Response: AT+CIPSTO?
+CIPSTO:180
OK
This is the curent TCP timeout in
seconds on the ESP-01.
************** System Has Started
*******************
```

Connect the Android to the "ESP-01" access point using the password "esp8266esp01". Start up the basic wireless framework app on the Android side and this will connect to the TCP server running on the ESP-01 module. After connecting to the server the name of the Android device will be changed on the wireless network. If the name change was successful, then an "OK" response is sent by the Arduino to the Android device.

```
Notification: '0,CONNECT'
Processing New Client Connection ...
NewConnection Name: 0
Notification: '
'Notification: '
'Notification: '+IPD,0,8:name=lg
'ClientIndex: 0
DataLength: 8
Data: name=lg
Client Name Change Requested,
ClientIndex: 0, NewClientName: lg
Client Name Change Succeeded, OK...
Sending Data ....
TCP Connection ID: 0 , DataLength: 3 ,
Data: OK
```

Next, enter and send the command "`ledon`" from the Android application to the Arduino. This will turn the LED light on. You should see the LED light turn on as well as a return message from the Arduino to the Android stating that the LED is now turned on.

```
Notification: '
'Notification: '
'Notification: '+IPD,0,6:ledon
'ClientIndex: 0
DataLength: 6
Data: ledon
Sending Data ....
TCP Connection ID: 0 , DataLength: 17 ,
Data: LED turned ON...
```

Next, type in and send the "`ledoff`" command from the Android application. After receiving the message the Arduino will respond with an acknowledgment stating that the LED has been turned off.

```
Notification: '
'Notification: '
'Notification: '+IPD,0,7:ledoff
'ClientIndex: 0
DataLength: 7
Data: ledoff
Sending Data ....
TCP Connection ID: 0 , DataLength: 18 ,
Data: LED turned OFF...
```

Next, enter and send the "`blink`" command in the Android application to start the LED blinking cycle. Once the Arduino receives the "`blink`" command, it will send back a message to the Android device stating that the blinking cycle has started.

```
Notification: '
'Notification: '
'Notification: '+IPD,0,6:blink
'ClientIndex: 0
DataLength: 6
Data: blink

Sending Data ....
TCP Connection ID: 0 , DataLength: 24 ,
Data: LED Blinking Started...
```

You should be seeing the LED blink in intervals of 1 second off and then 1 second on that will repeat indefinitely.

Next, enter and send a "`ledtimeon=50`" command to change the time the LED is on during a blink cycle to 50 milliseconds instead of 1 second or 1000 milliseconds. The Arduino should receive this command and change the rate of blinking. The Arduino should also send back a message to the Android indicating that the blink on time has been changed.

```
Notification: '
'Notification: '
'Notification: '+IPD,0,13:ledtimeon=50
'ClientIndex: 0
DataLength: 13
Data: ledtimeon=50
Sending Data ....
TCP Connection ID: 0 , DataLength: 34 ,
Data: LEDOnTime set to: 50 MilliSeconds
```

Next, let's change the blink off time to 50 milliseconds by sending the "`ledtimeoff=50`" command to the Arduino using the Android application. The Arduino will receive the command, change the rate of blinking, and send a response message back to the Android.

```
Notification: '
'Notification: '
'Notification: '+IPD,0,14:ledtimeoff=50
'ClientIndex: 0
DataLength: 14
Data: ledtimeoff=50
Sending Data ....
TCP Connection ID: 0 , DataLength:
35 , Data: LEDOffTime set to: 50
MilliSeconds
```

Finally, turn off the LED blinking feature by entering "`noblink`" into the Android application and sending this command to the Arduino. The Arduino will process the incoming command and turn off the blinking LED and then send a message back to the Android stating that the LED blinking has been deactivated.

```
Notification: '
'Notification: '
'Notification: '+IPD,0,8:noblink
'ClientIndex: 0
DataLength: 8
Data: noblink
Sending Data ....
TCP Connection ID: 0 , DataLength: 24 ,
Data: LED Blinking Stopped...
```

This ends the demonstration of the wireless LED control system.

The RGB LED (Common Ground Version)

The RGB LED can output red, green, and blue colors as well as combinations of these colors. The RGB LED has four terminals which are red, green, blue, and ground terminals. See Figure 5-17.

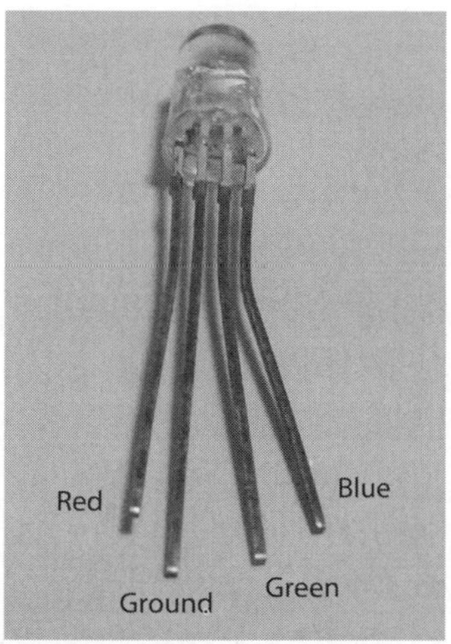

Figure 5-17 The RGB LED.

Hands-on Example: The Remote Wireless RGB LED Control System

In this hands-on example I show you how to control an RGB LED remotely and wirelessly using the Android/Arduino basic wireless framework. In this example you will be able to remotely control the red, green, and blue values of the RGB LED and create a variety of composite LED colors.

Parts List

- 1 Android cell phone
- 1 Arduino Mega 2560 microcontroller
- 1 5- to 3.3-V step-down voltage regulator
- 1 Logic level converter from 5 to 3.3 V
- 1 ESP-01 Wi-Fi module
- 1 Arduino development station such as a desktop or a notebook
- 1 Breadboard
- 1 RGB LED (Common Ground Version)
- 4 330-ohm resistors (recommended)
- 1 Package of wires both male-to-male and female-to-male to connect the components

Setting Up the Hardware

Step-Down Voltage Regulator Connections

1. Connect the Vin pin on the voltage regulator to the 5-V power pin on the Arduino Mega 2560.
2. Connect the Vout pin on the voltage regulator to a node on your breadboard represented by a horizontal row of empty slots. This is the 3.3-V power node that will supply the 3.3-V power to the ESP-01 module as well as provide the reference voltage for the logic level shifter.
3. Connect the GND pin on the voltage regulator to the GND pin on the Arduino or form a GND node similar to the 3.3-V power node in the previous step.

ESP-01 Wi-Fi Module Connections

1. Connect the Tx or TxD pin to the Rx3 pin on the Arduino Mega 2560. This is the receive pin for the serial hardware port 3.
2. Connect the EN or CH_PD pin to the 3.3-V power node from the step-down voltage regulator.
3. Connect the IO16 or RST pin to the 3.3-V power node from the step-down voltage regulator.
4. Connect the 3V3 or VCC pin to the 3.3-V power node from the step-down voltage regulator.
5. Connect the GND pin to the ground pin or node.
6. Connect the Rx or RxD pin to pin B0 on the 3.3-V side of the logic level shifter. The voltage from this pin will be shifted from 5 V from the Arduino to 3.3 V for the ESP-01 module.

Logic Level Shifter

1. Connect the 3.3-V pin to the 3.3-V power node from the step-down voltage regulator.
2. Connect the GND pin on the 3.3-V side to the ground node for the Arduino.
3. Connect the B0 pin on the 3.3-V side to Rx or RxD pin on the ESP-01. This corresponds to the A0 pin on the 5-V side.

4. Connect the 5-V pin to the 5-V power pin on the Arduino.
5. Connect the GND pin on the 5-V side to the ground node for the Arduino.
6. Connect the A0 pin on the 5-V side to Tx3 pin on the Arduino. This pin corresponds to the B0 pin on the 3.3-V side.

RGB LED

1. Connect the Red terminal to one end of a resistor (around 330 ohm recommended). Connect the other end of the resistor to digital pin 7 on the Arduino.
2. Connect the Ground terminal to one end of a resistor (around 330 ohm recommended). Connect the other end of the resistor to the ground node for the circuit.
3. Connect the Green terminal to one end of a resistor (around 330 ohm recommended). Connect the other end of the resistor to digital pin 8 on the Arduino.
4. Connect the Blue terminal to on end of a resistor (around 330 ohm recommended). Connect other end of the resistor to digital pin 9 on the Arduino.

***Note:* The resistors are used here to dim the brightness of the LED by lowering the voltage across the LED. You may be able to safely use the LED without the resistors depending on the characteristics of the specific LED. Just remember that the Arduino can only output about 20 mA of current safely on the input/output pins such as the ones connected to the Red, Green, and Blue terminals of the RGB LED and that the Arduino outputs 5 V on the input/output pins.**

See Figure 5-18.

Setting Up the Software

The software for this hands-on example uses the previous software for the regular LED example. The software for the Android side does not change. Only the software for the Arduino is modified from the software used in the previous LED hands-on example. This section will cover those modifications.

The `RedPin` variable holds the Arduino digital pin number that is connected to the Red terminal lead on the RGB LED.

```
int RedPin = 7;
```

The `GreenPin` variable holds the Arduino digital pin number that is connected to the Green terminal lead on the RGB LED.

```
int GreenPin = 8;
```

The `BluePin` variable holds the Arduino digital pin number that is connected to the Blue terminal lead on the RGB LED.

```
int BluePin = 9;
```

The `RGBColor` structure holds the color values for the LED.

```
struct RGBColor
{
 int Red;
 int Green;
 int Blue;
};
```

The `LEDColor` variable holds the color value of the LED in terms of red, green, and blue color values that range from 0 through 255.

```
RGBColor LEDColor;
```

The `Initialize LEDColor()` Function

The `InitializeLEDColor()` function initializes the color of the RGB LED to white which means the red, green, and blue values of the LED are all set to 255.

See Listing 5-35.

The `SetRGBLED()` Function

The `SetRGBLED()` function sets the color of the RGB LED to the red, green, and blue values from the input parameters. See Listing 5-36.

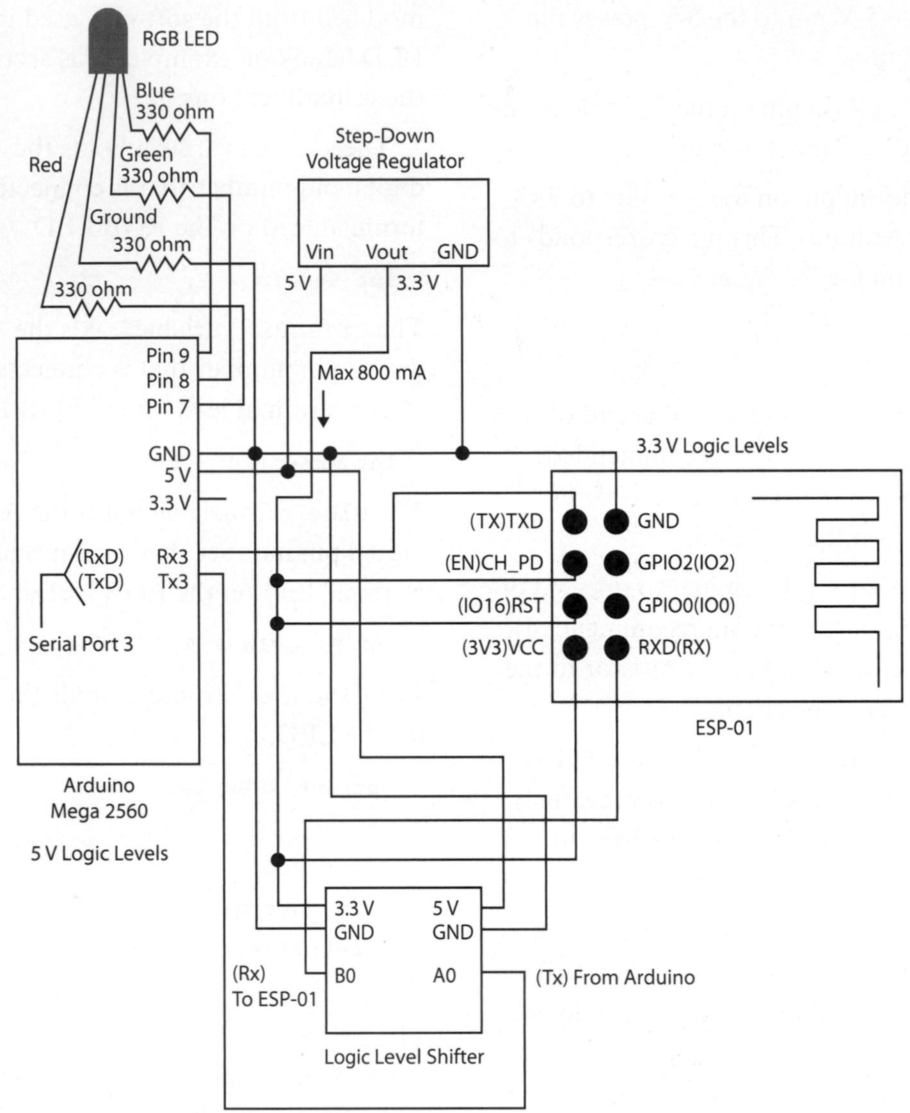

Figure 5-18 The remote wireless RGB LED control system.

Listing 5-35 The `InitializeLEDColor()` Function

```
void InitializeLEDColor()
{
  LEDColor.Red = 255;
  LEDColor.Green = 255;
  LEDColor.Blue = 255;
}
```

Listing 5-36 The `SetRGBLED()` Function

```
void SetRGBLED(int Red, int Green, int
Blue)
{
  analogWrite(RedPin, Red);
  analogWrite(GreenPin, Green);
  analogWrite(BluePin, Blue);
}
```

The `SetupSensorAndHardware()` Function

The `SetupSensorAndHardware()` function initializes the RGB LED by doing the following:

1. Calling the `InitializeLEDColor()` function to set the LED color to white.
2. Turning off the LED by setting the red, green, and blue components of the LED to 0.

Listing 5-37 The `SetupSensorAndHardware()` Function

```
void SetupSensorAndHardware()
{
 // No need to call pinMode for
analogWrite() function
 InitializeLEDColor();

 // Turn off LED
 SetRGBLED(0, 0, 0);
}
```

See Listing 5-37.

Modifying the `PerformLEDBlink()` Function

The `PerformLEDBlink()` function is modified from the previous one in the regular LED example by:

1. Turning off the LED by setting all the color components to 0.
2. Turning on the LED by setting the color components to the current color values held in the `LEDColor` variable.

See Listing 5-38. The code modifications are highlighted in bold print.

The `ProcessSensorAndHardware()` Function

The `ProcessSensorAndHardware()` function performs the LED blink feature if the blink mode is turned on by calling the `PerformLEDBlink()` function.

See Listing 5-39.

Modifying the `ledon` Command

The code for the "`ledon`" command for the regular LED is modified. To turn on the RGB LED we now have to set the current red, green, and blue color components. See Listing 5-40. The code modifications are in bold print.

Listing 5-38 The `PerformLEDBlink()` Function

```
void PerformLEDBlink()
{
 unsigned long ElapsedTime = millis()
- BlinkStartTime;
 if (LEDOn)
 {
  // LED is on
  if (ElapsedTime > LEDOnTime)
  {
   // if LED is on and LED on time
  has passed then shut LED off
   SetRGBLED(0, 0, 0);
   LEDOn = false;
   BlinkStartTime = millis();
  }
 }
 else
 {
  // LED is off
  if (ElapsedTime > LEDOffTime)
  {
   SetRGBLED(LEDColor.Red, LEDColor.
  Green, LEDColor.Blue);
   LEDOn = true;
   BlinkStartTime = millis();
  }
 }
}
```

Listing 5-39 The `ProcessSensorAndHardware()` Function

```
void ProcessSensorAndHardware()
{
 // Put Code to process your sensors
or other hardware in this function.
 if (Blink)
 {
  PerformLEDBlink();
 }
}
```

Modifying the `ledoff` Command

The code for the "`ledoff`" was modified from the previous hands-on example for the regular

Listing 5-40 Modifying the `ledon` Command

```
if (Data == "ledon\n")
{
 LEDOn = true;
 SetRGBLED(LEDColor.Red, LEDColor.
Green, LEDColor.Blue);

 String returnstring = "LED turned
ON...\n";
 SendTCPMessage(ClientIndex,
returnstring);
}
```

Listing 5-41 Modifying the `ledoff` Command

```
if (Data == "ledoff\n")
{
 LEDOn = false;
 SetRGBLED(0, 0, 0);

 String returnstring = "LED turned
OFF...\n";
 SendTCPMessage(ClientIndex,
returnstring);
}
```

LED. The RGB LED is turned off by setting all the color components of the LED to 0. See Listing 5-41. The modified code is in bold print.

Modifying the `noblink` Command

The code for the "`noblink`" command from the previous hands-on example for the regular LED was modified. The modified code now turns off the RGB LED by setting all its color components to 0. See Listing 5-42. The modified code is in bold print.

Listing 5-42 Modifying the `noblink` Command

```
if (Data == "noblink\n")
{
 Blink = false;
 LEDOn = false;

 // turn off LED
 SetRGBLED(0, 0, 0);

 String returnstring = "LED Blinking
Stopped...\n";
 SendTCPMessage(ClientIndex,
returnstring);
}
```

The `red` Command

The "`red`" command is added to the `CheckAndProcessIPD()` function and sets the red component of the RGB LED by doing the following:

1. Finding the value of the red component, converting the string representation into numerical format, and assigning it to the `LEDColor.Red` variable.
2. Changing the RGB LED to reflect the new red component value by calling the `SetRGBLED(LEDColor.Red, LEDColor.Green, LEDColor.Blue)` function with the newest component values.
3. Sending a text string back to the Android device indicating that the red component was changed.

See Listing 5-43.

Adding the `green` Command

The "`green`" command is added to the `CheckAndProcessIPD()` function and sets the green component of the RGB LED by doing the following:

1. Finding the value of the green component, converting the string representation into numerical format, and assigning it to the `LEDColor.Green` variable.

Listing 5-43 The `red` Command

```
if (Data.indexOf("red=") >= 0)
{
 int indexcolor = Data.indexOf("=");
 String color = Data.
substring(indexcolor+1);
 LEDColor.Red = color.toInt();

 SetRGBLED(LEDColor.Red, LEDColor.
Green, LEDColor.Blue);

 String returnstring = "";
 returnstring = returnstring + F("Red
component set to: ") + LEDColor.Red +
F("\n");
 SendTCPMessage(ClientIndex,
returnstring);
}
```

Listing 5-44 The `green` Command

```
if (Data.indexOf("green=") >= 0)
{
 int indexcolor = Data.indexOf("=");
 String color = Data.
substring(indexcolor+1);
 LEDColor.Green = color.toInt();

 SetRGBLED(LEDColor.Red, LEDColor.
Green, LEDColor.Blue);

 String returnstring = "";
 returnstring = returnstring +
F("Green component set to: ") +
LEDColor.Green + F("\n");
 SendTCPMessage(ClientIndex,
returnstring);
}
```

2. Changing the RGB LED to reflect the new green component value by calling the `SetRGBLED(LEDColor.Red, LEDColor.Green, LEDColor.Blue)` function with the newest component values.
3. Sending a text string back to the Android device indicating that the green component was changed.

See Listing 5-44.

The `blue` Command

The "`blue`" command is added to the `CheckAndProcessIPD()` function and sets the blue component of the RGB LED by doing the following:

1. Finding the value of the blue component, converting the string representation into numerical format, and assigning it to the `LEDColor.Blue` variable.
2. Changing the RGB LED to reflect the new blue component value by calling the `SetRGBLED(LEDColor.Red, LEDColor.Green, LEDColor.Blue)` function with the newest component values.
3. Sending a text string back to the Android device indicating that the blue component was changed.

See Listing 5-45.

Listing 5-45 The `blue` Command

```
if (Data.indexOf("blue=") >= 0)
{
 int indexcolor = Data.indexOf("=");
 String color = Data.substring
(indexcolor+1);
 LEDColor.Blue = color.toInt();

 SetRGBLED(LEDColor.Red, LEDColor.
Green, LEDColor.Blue);

 String returnstring = "";
 returnstring = returnstring +
F("Blue component set to: ") +
LEDColor.Blue + F("\n");
 SendTCPMessage(ClientIndex,
returnstring);
}
```

Listing 5-46 The `getcolor` Command

```
if (Data == "getcolor\n")
{
 String returnstring = "";
 returnstring = returnstring + F("LED
Color(R,G,B): ") +
   LEDColor.Red + F(",") +
   LEDColor.Green + F(",") +
   LEDColor.Blue + F("\n");
 SendTCPMessage(ClientIndex,
returnstring);
}
```

The `getcolor` Command

The "`getcolor`" command retrieves the color component information for the RGB LED and sends it to the Android device. See Listing 5-46.

Operating the RGB LED Control System

Download and install the Android basic wireless framework version 1.1 project on your Android device if you haven't done so already. Note that the Android APK install files for all the chapters are located in the downloads section of Chapter 9. Download and install the RGB LED Arduino project code on your Arduino. Start up the Arduino Serial Monitor and you should see the following start-up sequence:

```
****** ESP-01 Arduino Mega 2560 Server
Program ********
*********** RGB LED Control Program
*************

AT Command Sent: AT+CIPMUX=1
ESP-01 Response: AT+CIPMUX=1

OK

ESP-01 mode set to allow multiple
connections.

AT Command Sent: AT+CWMODE_CUR=2
ESP-01 Response: AT+CWMODE_CUR=2

OK

ESP-01 mode changed to Access Point
Mode.

AT Command Sent: AT+CIPSERVER=1,80
ESP-01 Response: AT+CIPSERVER=1,80

OK

Server Started on ESP-01.

AT Command Sent: AT+CWSAP_CUR="ESP-
01","esp8266esp01",1,2,4,0
ESP-01 Response: AT+CWSAP_CUR="ESP-
01","esp8266esp01",1,2,4,0

OK

AP Configuration Set Successfullly.

AT Command Sent: AT+CWSAP_CUR?
ESP-01 Response: AT+CWSAP_CUR?
+CWSAP_CUR:"ESP-
01","esp8266esp01",1,2,4,0

OK

This is the curent configuration of the
AP on the ESP-01.

AT Command Sent: AT+CIPSTO?
ESP-01 Response: AT+CIPSTO?
+CIPSTO:180

OK

This is the curent TCP timeout in
seconds on the ESP-01.

************** System Has Started
*******************
```

Next, connect your Android device to the "ESP-01" Wi-Fi access point using the password "esp8266esp01". Start up the Android basic wireless framework version 1.1. The application will connect to the TCP server running on the ESP-01 and change the name of the android device on the wireless network. In my case the name was changed to "`lg`". After the name was changed an "OK" string was sent back to the Android device.

```
Notification: '0,CONNECT'
Processing New Client Connection ...
```

```
NewConnection Name: 0
Notification: '
'Notification: '
'Notification: '+IPD,0,8:name=lg
'ClientIndex: 0
DataLength: 8
Data: name=lg
Client Name Change Requested,
ClientIndex: 0, NewClientName: lg
Client Name Change Succeeded, OK...
Sending Data ....
TCP Connection ID: 0 , DataLength: 3 ,
Data: OK
```

Next, issue the "ledon" command from the Android application to turn on the RGB LED. The Arduino should receive the command and send a string back to the Android indicating that the LED has been turned on. The LED should be white in color.

```
Notification: '
'Notification: '
'Notification: '+IPD,0,6:ledon
'ClientIndex: 0
DataLength: 6
Data: ledon
Sending Data ....
TCP Connection ID: 0 , DataLength: 17 ,
Data: LED turned ON...
```

Next, let's get the exact red, green, and blue colors of the LED which should all be 255 (which is the default value upon startup) by sending the "getcolor" command to the Arduino. You should see the response "LED Color(R,G,B): 255,255,255" show up in the Android application's debug message window which indicates that all the color components are 255.

```
Notification: '
'Notification: '
'Notification: '+IPD,0,9:getcolor
'ClientIndex: 0
DataLength: 9
Data: getcolor
Sending Data ....
TCP Connection ID: 0 , DataLength: 30 ,
Data: LED Color(R,G,B): 255,255,255
```

Next, turn off the LED by sending the Arduino the "ledoff" command. The LED should turn off and the message "LED turned OFF..." should appear in the debug message window of the Android application.

```
Notification: '
'Notification: '
'Notification: '+IPD,0,7:ledoff
'ClientIndex: 0
DataLength: 7
Data: ledoff
Sending Data ....
TCP Connection ID: 0 , DataLength: 18 ,
Data: LED turned OFF...
```

Next, turn on the blinking feature of the LED by issuing the "blink" command. The Arduino should receive the blink command and send back a message acknowledging that the blinking has started.

```
Notification: '
'Notification: '
'Notification: '+IPD,0,6:blink
'ClientIndex: 0
DataLength: 6
Data: blink
Sending Data ....
TCP Connection ID: 0 , DataLength: 24 ,
Data: LED Blinking Started...
```

Next, change the blue color component of the RGB LED to 0 by typing in "blue=0" into the phone entry text box in the Android application and pressing the "Send Data" button to send that command to the Arduino. The Arduino should send back an acknowledgment message to the Android device. You should see the color of the blinking LED change.

```
Notification: '
'Notification: '
```

```
'Notification: '+IPD,0,7:blue=0
'ClientIndex: 0
DataLength: 7
Data: blue=0

Sending Data ....
TCP Connection ID: 0 , DataLength: 25 ,
Data: Blue component set to: 0
```

You can also set the color component values for the green and red values of the RGB LED in the same way. Try and experiment with the values for the red, green, and blue components of the LED to produce a variety of composite colors.

Figure 5-19 The CEM-1203(42) piezo buzzer.

The CEM-1203(42) Piezo Buzzer

The piezo buzzer is a common electrical component used with microcontrollers like the Arduino to generate sound. The specific model we will use here is the CEM-1203(42). The specifications of this buzzer are listed below.

- Loudness: 85 dB
- Operating temperature: –20 through 60 degrees Centigrade
- Operating current: 35 mA
- Operating voltage: 3.5 V

The buzzer has two terminals: one positive and the other negative. See Figure 5-19.

Hands-on Example: The Remote Wireless Piezo Buzzer Control System

In this hands-on example I show you how to remotely and wirelessly control a piezo buzzer using an Android, and an Arduino equipped with an ESP8266-based ESP-01 Wi-Fi module. Both the Android and Arduino will be using software that is part of the basic wireless framework that was presented in Chapter 4.

Parts List

- 1 Android cell phone
- 1 Arduino Mega 2560 microcontroller
- 1 5- to 3.3-V step-down voltage regulator
- 1 Logic level converter from 5 to 3.3 V
- 1 ESP-01 Wi-Fi module
- 1 Arduino development station such as a desktop or a notebook
- 1 Breadboard
- 1 CEM-1203(42) piezo buzzer
- 1 Package of wires both male-to-male and female-to-male to connect the components

Setting Up the Hardware

Step-Down Voltage Regulator Connections

1. Connect the Vin pin on the voltage regulator to the 5-V power pin on the Arduino Mega 2560.
2. Connect the Vout pin on the voltage regulator to a node on your breadboard represented by a horizontal row of empty slots. This is the 3.3-V power node that will supply the 3.3-V power to the ESP-01

module as well as provide the reference voltage for the logic level shifter.

3. Connect the GND pin on the voltage regulator to the GND pin on the Arduino or form a GND node similar to the 3.3-V power node in the previous step.

ESP-01 Wi-Fi Module Connections

1. Connect the Tx or TxD pin to the Rx3 pin on the Arduino Mega 2560. This is the receive pin for the serial hardware port 3.
2. Connect the EN or CH_PD pin to the 3.3-V power node from the step-down voltage regulator.
3. Connect the IO16 or RST pin to the 3.3-V power node from the step-down voltage regulator.
4. Connect the 3V3 or VCC pin to the 3.3-V power node from the step-down voltage regulator.
5. Connect the GND pin to the ground pin or node.
6. Connect the Rx or RxD pin to pin B0 on the 3.3-V side of the logic level shifter. The voltage from this pin will be shifted from 5 V from the Arduino to 3.3 V for the ESP-01 module.

Logic Level Shifter

1. Connect the 3.3-V pin to the 3.3-V power node from the step-down voltage regulator.
2. Connect the GND pin on the 3.3-V side to the ground node for the Arduino.
3. Connect the B0 pin on the 3.3-V side to Rx or RxD pin on the ESP-01. This corresponds to the A0 pin on the 5-V side.
4. Connect the 5-V pin to the 5-V power pin on the Arduino.
5. Connect the GND pin on the 5-V side to the ground node for the Arduino.
6. Connect the A0 pin on the 5-V side to Tx3 pin on the Arduino. This pin corresponds to the B0 pin on the 3.3-V side.

Piezo Buzzer

1. Connect the positive terminal of the buzzer to digital pin 7 on the Arduino.
2. Connect the negative terminal of the buzzer to the ground node for the circuit.

See Figure 5-20.

Setting Up the Software

The new code for this hands-on example consists of the Arduino code that was modified from the regular LED control example. The same Android program which is the basic wireless framework program version 1.1 is used here to control the piezo buzzer. Only the modifications to the previous Arduino LED control program will be discussed in this section.

The `BuzzerPin` variable holds the Arduino digital pin number that is connected to the positive terminal of the piezo buzzer.

```
int BuzzerPin = 7;
```

The `Freq` variable holds the frequency in hertz of the tone that the buzzer will produce when active and is set by default to 500 hertz.

```
int Freq = 500;
```

The `BuzzerOn` variable is true if the buzzer is currently producing a sound and false otherwise.

```
boolean BuzzerOn = false;
```

The `SirenOn` variable is true if the buzzer is currently in siren mode.

```
boolean SirenOn = false;
```

The `SirenOffTime` variable holds the time in milliseconds that the buzzer should be off during siren mode and has a default value of 1000 milliseconds or 1 second.

```
int SirenOffTime = 1000;
```

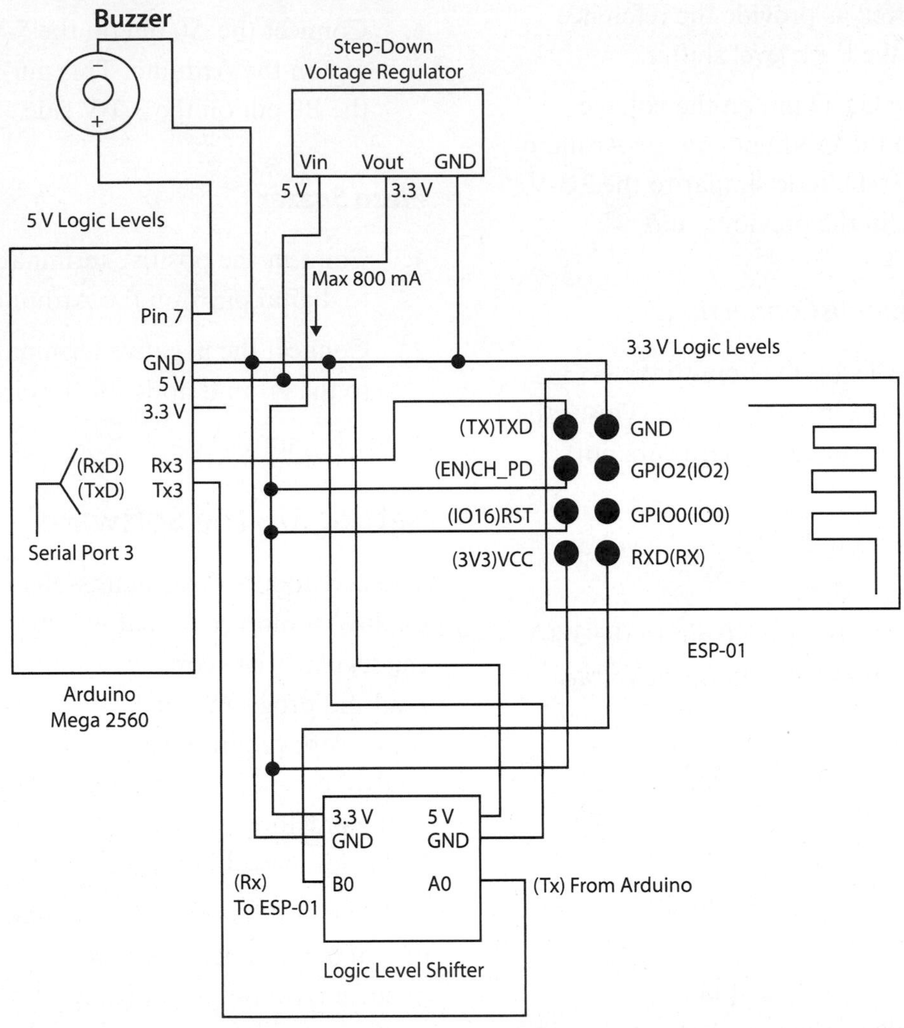

Figure 5-20 The remote wireless piezo buzzer control system.

The `SirenOnTime` variable holds the time in milliseconds that the buzzer should be on during siren mode and has a default value of 1000 milliseconds or 1 second.

```
int SirenOnTime = 1000;
```

The `SirenStartTime` variable holds the start time of the siren on time and the siren off time. It is used to keep track of the duration of the siren on time and the siren off time.

```
unsigned long SirenStartTime = 0;
```

The `SetupSensorAndHardware()` Function

The `BuzzerPin` pin on the Arduino which is connected to the piezo buzzer is set as an output pin to provide voltage and current to the buzzer. See Listing 5-47.

Listing 5-47 The `SetupSensorAndHardware()` Function

```
void SetupSensorAndHardware()
{
  pinMode(BuzzerPin, OUTPUT);
}
```

The `PerformBuzzerSiren()` Function

The `PerformBuzzerSiren()` function cycles the buzzer from on to off by doing the following:

1. Finding the elapsed time since the buzzer's cycle was last changed.

Listing 5-48 The `PerformBuzzerSiren()` Function

```
void PerformBuzzerSiren()
{
 unsigned long ElapsedTime = millis()
 - SirenStartTime;
 if (BuzzerOn)
 {
   // Buzzer is on
   if (ElapsedTime > SirenOnTime)
   {
    // if Buzzer is on and Buzzer on
   time has passed then shut Buzzer
   off
    noTone(BuzzerPin);
    BuzzerOn = false;
    SirenStartTime = millis();
   }
 }
 else
 {
   // Buzzer is off
   if (ElapsedTime > SirenOffTime)
   {
    tone(BuzzerPin, Freq);
    BuzzerOn = true;
    SirenStartTime = millis();
   }
 }
}
```

2. If the buzzer is on and the elapsed time is greater than the time the siren mode needs to be active then turn off the buzzer and reset the siren cycle timer variable.
3. If the buzzer was off in step 2 instead of on and if the elapsed time is greater than the time the siren mode needs to be inactive then turn on the buzzer and reset the siren cycle timer variable.

See Listing 5-48.

The `ProcessSensorAndHardware()` Function

If the buzzer siren is on, then the `ProcessSensorAndHardware()` function processes the buzzer's siren mode by calling the `PerformBuzzerSiren()` function. See Listing 5-49.

Listing 5-49 The `ProcessSensorAndHardware()` Function

```
void ProcessSensorAndHardware()
{
  // Put Code to process your sensors
or other hardware in this function.
 if (SirenOn)
 {
   PerformBuzzerSiren();
 }
}
```

All of the following commands are added into the `CheckAndProcessIPD()` function which processes incoming notifications from the Android device.

The `buzzeron` Command

The "`buzzeron`" command turns on the buzzer and sends a character string back to the Android indicating that the buzzer is now on and producing a tone. See Listing 5-50.

Listing 5-50 The buzzeron Command

```
if (Data == "buzzeron\n")
{
 BuzzerOn = true;
 tone(BuzzerPin, Freq);

 String returnstring = "Buzzer turned
ON...\n";
 SendTCPMessage(ClientIndex,
returnstring);
}
```

The `buzzeroff` Command

The "`buzzeroff`" command turns off the buzzer by calling the `noTone()` function and then sends a character string back to the Android device

Listing 5-51 The `buzzeroff` Command

```
if (Data == "buzzeroff\n")
{
 BuzzerOn = false;
 noTone(BuzzerPin);

 String returnstring = "Buzzer turned
OFF...\n";
 SendTCPMessage(ClientIndex,
returnstring);
}
```

Listing 5-52 The `siren` Command

```
if (Data == "siren\n")
{
 SirenOn = true;
 SirenStartTime = millis();

 String returnstring = "Buzzer Siren
Started...\n";
 SendTCPMessage(ClientIndex,
returnstring);
}
```

indicating that the buzzer has been turned off. See Listing 5-51.

The `siren` Command

The `siren` command initializes and activates the buzzer's siren mode and sends a character string back to the Android device indicating that the buzzer's siren mode has been activated. See Listing 5-52.

The `nosiren` Command

The "`nosiren`" command turns off the siren mode, turns off the buzzer, and sends a character string back to the Android indicating that the siren mode has been turned off. See Listing 5-53.

Listing 5-53 The `nosiren` Command

```
if (Data == "nosiren\n")
{
 SirenOn = false;
 BuzzerOn = false;
 noTone(BuzzerPin);

 String returnstring = "Buzzer Siren
Stopped...\n";
 SendTCPMessage(ClientIndex,
returnstring);
}
```

Listing 5-54 The `sirentimeoff` Command

```
if (Data.indexOf("sirentimeoff=") >= 0)
{
 int indextime = Data.indexOf("=");
 String timelength = Data.
substring(indextime+1);
 SirenOffTime = timelength.toInt();

 String returnstring = "";
 returnstring = returnstring
+ F("SirenOffTime set to: ") +
SirenOffTime + F(" MilliSeconds\n");
 SendTCPMessage(ClientIndex,
returnstring);
}
```

The `sirentimeoff` Command

The `sirentimeoff` command sets the time interval that the buzzer will be off in siren mode and sends a message back to the Android indicating that this time was changed. See Listing 5-54.

The `sirentimeon` Command

The `sirentimeon` command sets the time interval that the buzzer will be active during siren mode and sends a message back to the Android device indicating that this time was changed. See Listing 5-55.

Listing 5-55 The `sirentimeon` Command

```
if (Data.indexOf("sirentimeon=") >= 0)
{
 int indextime = Data.indexOf("=");
 String timelength = Data.
substring(indextime+1);
 SirenOnTime = timelength.toInt();

 String returnstring = "";
 returnstring = returnstring
+ F("SirenOnTime set to: ") +
SirenOnTime + F(" MilliSeconds\n");
 SendTCPMessage(ClientIndex,
returnstring);
}
```

Listing 5-56 The `freq` Command

```
if (Data.indexOf("freq=") >= 0)
{
 int indexfreq = Data.indexOf("=");
 String freqhz = Data.substring
(indexfreq+1);
 Freq = freqhz.toInt();

 String returnstring = "";
 returnstring = returnstring +
F("Buzzer Frequency set to: ") + Freq
+ F(" Hertz\n");
 SendTCPMessage(ClientIndex,
returnstring);
}
```

The `freq` Command

The `freq` command sets the frequency of the tone generated by the buzzer and sends back a message to the Android device indicating that the buzzer's frequency was changed. See Listing 5-56.

Operating the Remote Wireless Piezo Buzzer Control System

Download and install the Android basic wireless framework version 1.1 on your Android cell phone if you have not done so already. Download and install the Arduino buzzer project on your Arduino Mega. Start up the Arduino's Serial Monitor and the following should show up:

```
****** ESP-01 Arduino Mega 2560 Server
Program ********
*********** Piezo Buzzer Control
Program *************

AT Command Sent: AT+CIPMUX=1
ESP-01 Response: AT+CIPMUX=1

OK

ESP-01 mode set to allow multiple
connections.

AT Command Sent: AT+CWMODE_CUR=2
ESP-01 Response: AT+CWMODE_CUR=2

OK

ESP-01 mode changed to Access Point
Mode.

AT Command Sent: AT+CIPSERVER=1,80
ESP-01 Response: AT+CIPSERVER=1,80

OK

Server Started on ESP-01.

AT Command Sent: AT+CWSAP_CUR="ESP-
01","esp8266esp01",1,2,4,0
ESP-01 Response: AT+CWSAP_CUR="ESP-
01","esp8266esp01",1,2,4,0

OK

AP Configuration Set Successfullly.

AT Command Sent: AT+CWSAP_CUR?
ESP-01 Response: AT+CWSAP_CUR?
+CWSAP_CUR:"ESP-01",
"esp8266esp01",1,2,4,0

OK

This is the curent configuration of the
AP on the ESP-01.

AT Command Sent: AT+CIPSTO?
ESP-01 Response: AT+CIPSTO?
+CIPSTO:180

OK
```

```
This is the curent TCP timeout in
seconds on the ESP-01.
************** System Has Started
*******************
```

Connect your Android phone to the "ESP-01" access point running on the ESP-01 module using the password "esp8266esp01". Start up the Android basic wireless framework program and it will connect to the TCP server running on the ESP-01 module, and change the Android's device name on the wireless network. The Arduino will send back a message to the Android indicating that the name change was successful.

```
Notification: '0,CONNECT'
Processing New Client Connection ...
NewConnection Name: 0

Notification: '
'Notification: '
'Notification: '+IPD,0,8:name=lg
'ClientIndex: 0
DataLength: 8
Data: name=lg

Client Name Change Requested,
ClientIndex: 0, NewClientName: lg
Client Name Change Succeeded, OK...
Sending Data ....
TCP Connection ID: 0 , DataLength: 3 ,
Data: OK
```

Next, turn on the buzzer by sending the "`buzzeron`" command to the Arduino from the Android application. Once the Arduino receives the command, it turns on the buzzer and then sends a message back to the Android indicating that the buzzer has been turn on. You should now hear a sound being generated by the buzzer.

```
Notification: '
'Notification: '
'Notification: '+IPD,0,9:buzzeron
'ClientIndex: 0
DataLength: 9
Data: buzzeron
Sending Data ....
TCP Connection ID: 0 , DataLength: 20 ,
Data: Buzzer turned ON...
```

Next, send a "`buzzeroff`" command to the Arduino to turn off the buzzer. The Arduino should receive the command and turn off the buzzer. The Arduino should then send a message back to the Android indicating that the buzzer has been turned off. The tone that the buzzer was generating should have stopped.

```
Notification: '
'Notification: '
'Notification: '+IPD,0,10:buzzeroff
'ClientIndex: 0
DataLength: 10
Data: buzzeroff

Sending Data ....
TCP Connection ID: 0 , DataLength: 21 ,
Data: Buzzer turned OFF...
```

Next, turn on the siren by sending a "`siren`" command to the Arduino from the Android device. The Arduino should receive the command, turn on the buzzer's siren mode, and then send a message back to the Android indicating that the siren mode has been turned on. You should hear the buzzer generate a tone for 1 second and then shut off for 1 second in a continuous cycle.

```
Notification: '
'Notification: '
'Notification: '+IPD,0,6:siren
'ClientIndex: 0
DataLength: 6
Data: siren

Sending Data ....
TCP Connection ID: 0 , DataLength: 24 ,
Data: Buzzer Siren Started...
```

Next, set the siren's on time to a short interval such as 100 milliseconds by sending a "`sirentimeon=100`" command to the Arduino from the Android application. The Arduino will receive the command and change the siren's time on interval to 100 milliseconds. A message is also sent from the Arduino back to the Android indicating that the time interval has been changed. You should hear the length of the tone generated by the buzzer shorten significantly.

```
Notification: '
'Notification: '
'Notification:
'+IPD,0,16:sirentimeon=100
'ClientIndex: 0
DataLength: 16
Data: sirentimeon=100

Sending Data ....
TCP Connection ID: 0 , DataLength:
37 , Data: SirenOnTime set to: 100
MilliSeconds
```

Next, change the siren's off time to 100 milliseconds by sending the "`sirentimeoff=100`" command to the Arduino. The Arduino will change the time interval and send a message back to the Android indicating that the time interval was changed. You should hear the time delay between tones shorten considerably.

```
Notification: '
'Notification: '
'Notification:
'+IPD,0,17:sirentimeoff=100
'ClientIndex: 0
DataLength: 17
Data: sirentimeoff=100

Sending Data ....
TCP Connection ID: 0 , DataLength:
38 , Data: SirenOffTime set to: 100
MilliSeconds
```

Next, change the pitch of the tone generated by the buzzer by changing the frequency. Send the "`freq=50`" command to the Arduino to lower the tone considerably. As frequency increases so does the pitch of the tone generated by the buzzer.

```
Notification: '
'Notification: '
'Notification: '+IPD,0,8:freq=50
'ClientIndex: 0
DataLength: 8
Data: freq=50

Sending Data ....
TCP Connection ID: 0 , DataLength: 34 ,
Data: Buzzer Frequency set to: 50 Hertz
```

Finally, turn off the siren by sending the "`nosiren`" command to the Arduino.

```
Notification: '
'Notification: '
'Notification: '+IPD,0,8:nosiren
'ClientIndex: 0
DataLength: 8
Data: nosiren

Sending Data ....
TCP Connection ID: 0 , DataLength: 24 ,
Data: Buzzer Siren Stopped...
```

Summary

In this chapter I covered many different projects involving remote and wireless sensors and remote and wireless control over various electronics hardware. The sensor projects I covered included an infrared motion detector alarm system, a glass break sound alarm, a trip wire style alarm system using a distance sensor, and a water leak alarm system that can be used to detect flooding from a damaged hot water heater. The hardware control projects included controlling an LED, controlling an RGB LED, and controlling a piezo buzzer.

CHAPTER 6

Arduino with ESP8266 (ESP-01 Module) and Android Wireless Sensor and Remote Control Projects II

In this chapter I cover more hands-on examples of the Arduino- and Android-based wireless and remote control system that I presented earlier. An Android cell phone will be used as a wireless remote controller that will communicate over Wi-Fi with an Arduino equipped with an ESP8266-based ESP-01 module which provides the Wi-Fi connectivity. The Arduino will be connected to various sensors that can be controlled or monitored by the Android device. The sensors that will be covered in this chapter will be the reed switch, the flame sensor, the tilt switch, the TMP36 temperature sensor, the photo resistor which detects light, the DHT11 temperature and humidity sensor, and the ArduCAM OV2640 Mini camera.

The Reed Switch Magnetic Field Sensor

The reed switch magnetic field sensor detects the presence of a magnetic field.

The specifications of the reed switch sensor that I used are the following:

- VCC pin: Connect to an operating voltage of 3.3 V
- GND pin: Connect to the ground node of the circuit
- D0 pin: Outputs a 0 or LOW voltage if a magnetic field is detected or 1 or HIGH voltage if no magnetic field is detected
- Power LED: On when the sensor is receiving power
- Magnet detected LED: On when a magnet is detected by the sensor

***Note:* I was not able to find a specific model number for this sensor, so I had to experimentally determine the exact behavior of the unit.**

The glass tube portion of the sensor is the one that actually detects the magnetic field. See Figure 6-1.

Hands-on Example: The Wireless Reed Switch Door Entry Alarm System

In this hands-on example I show you how to build and operate a wireless door entry alarm system based on the reed switch and the Android/Arduino basic wireless framework discussed previously in this book. In order to use a reed switch to detect the opening of a door you would place the reed switch on one side of the door and a magnet on the other side of the door and have them align with each other when the

Figure 6-1 The reed switch magnetic field sensor.

door is closed. When the door is closed the reed switch will detect the magnet. When the door is open the magnet and reed switch will move apart from one another causing the door entry alarm system to be triggered.

Parts List

- 1 Android cell phone
- 1 Arduino Mega 2560 microcontroller
- 1 5- to 3.3-V step-down voltage regulator
- 1 Logic level converter from 5 to 3.3 V
- 1 ESP-01 Wi-Fi module
- 1 Arduino development station such as a desktop or a notebook
- 1 Breadboard
- 1 Reed switch magnetic sensor
- 1 Package of wires both male-to-male and female-to-male to connect the components

Setting Up the Hardware

Step-Down Voltage Regulator Connections

1. Connect the Vin pin on the voltage regulator to the 5-V power pin on the Arduino Mega 2560.
2. Connect the Vout pin on the voltage regulator to a node on your breadboard represented by a horizontal row of empty slots. This is the 3.3-V power node that will supply the 3.3-V power to the ESP-01 module as well as provide the reference voltage for the logic level shifter.
3. Connect the GND pin on the voltage regulator to the GND pin on the Arduino or form a GND node similar to the 3.3-V power node in the previous step.

ESP-01 Wi-Fi Module Connections

1. Connect the Tx or TxD pin to the Rx3 pin on the Arduino Mega 2560. This is the receive pin for the serial hardware port 3.
2. Connect the EN or CH_PD pin to the 3.3-V power node from the step-down voltage regulator.
3. Connect the IO16 or RST pin to the 3.3-V power node from the step down voltage regulator.
4. Connect the 3V3 or VCC pin to the 3.3-V power node from the step down voltage regulator.
5. Connect the GND pin to the ground pin or node.
6. Connect the Rx or RxD pin to pin B0 on the 3.3-V side of the logic level shifter. The voltage from this pin will be shifted from 5 V from the Arduino to 3.3 V for the ESP-01 module.

Logic Level Shifter

1. Connect the 3.3-V pin to the 3.3-V power node from the step-down voltage regulator.
2. Connect the GND pin on the 3.3-V side to the ground node for the Arduino.
3. Connect the B0 pin on the 3.3-V side to Rx or RxD pin on the ESP-01. This corresponds to the A0 pin on the 5-V side.
4. Connect the 5-V pin to the 5-V power pin on the Arduino.
5. Connect the GND pin on the 5-V side to the ground node for the Arduino.

6. Connect the A0 pin on the 5-V side to Tx3 pin on the Arduino. This pin corresponds to the B0 pin on the 3.3-V side.

The Reed Switch Magnetic Sensor

1. Connect the VCC pin to the 3.3-V power node from the voltage regulator.
2. Connect the GND pin to the common ground node for the circuit.
3. Connect the D0 pin to digital pin 7 on the Arduino.

See Figure 6-2.

Setting Up the Software

The Arduino software in this hands-on example is based on the code for the sound sensor alarm system that was presented in the previous chapter. I will discuss in this section only the code that has been modified.

The `MagnetDetected` variable indicates the voltage level that is output on pin D0 when the reed switch detects a magnet. The default value is set to 0 or LOW voltage.

```
int MagnetDetected      = 0;
```

The `MagnetNotDetected` variable indicates the voltage level that is output on pin D0 when

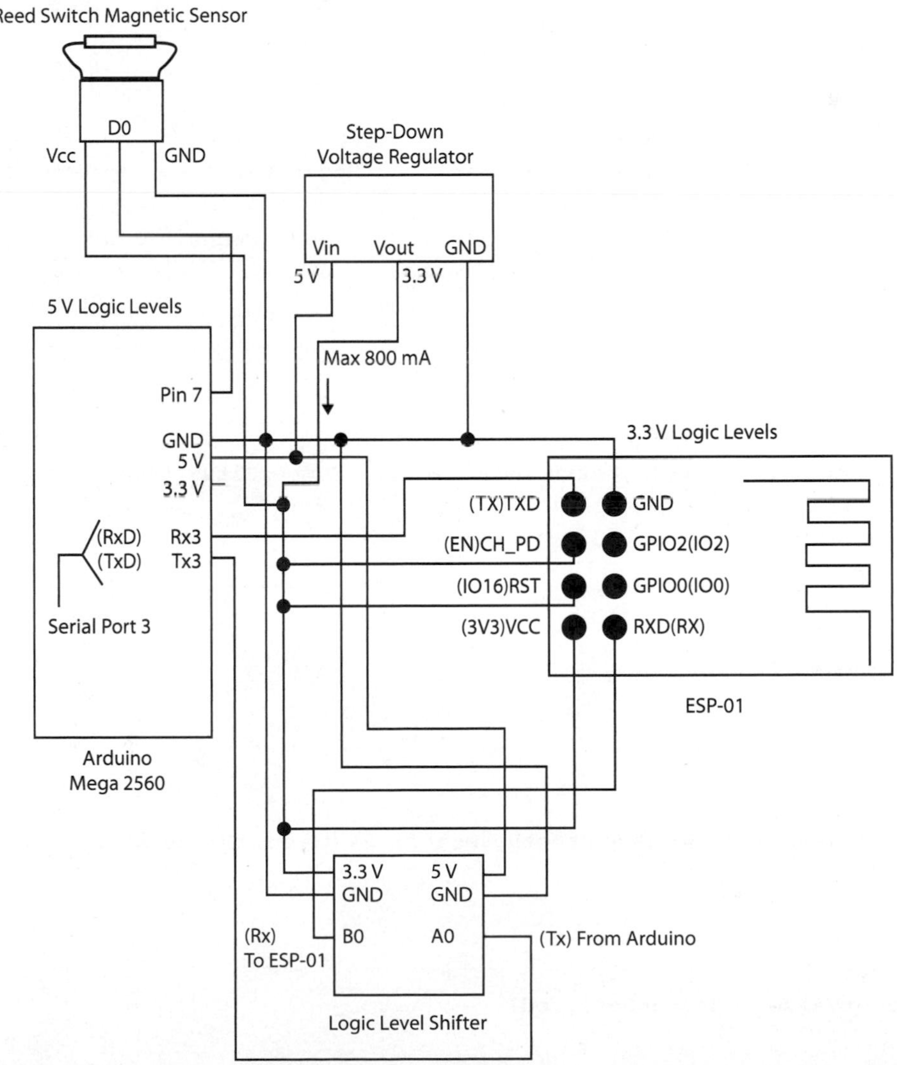

Figure 6-2 The reed switch magnetic door entry alarm system.

the reed switch does not detect a magnet. The default value is set to 1 or HIGH voltage.

```
int MagnetNotDetected  = 1;
```

The `SensorValue` holds the current sensor output value read in from the reed switch and is initialized to a reading that indicates a magnet was not detected.

```
int SensorValue = MagnetNotDetected;
```

The `PrevSensorValue` holds the reading from the reed switch that was obtained in the previous iteration and is set to a default value that indicates that a magnet was not detected.

```
int PrevSensorValue = MagnetNotDetected;
```

The `ProcessSensor()` Function

The `ProcessSensor()` function is modified for the reed switch. The alarm is now tripped when the previous iteration detected a magnet and the current reading does not detect a magnet. This means that the door has just opened. See Listing 6-1. The code changes are in bold print.

Listing 6-1 The `ProcessSensor()` Function

```
void ProcessSensor()
{
 // Put Code to process your sensors or other hardware in this function.
 if ((AlarmState == Alarm_ON) || (AlarmState == Alarm_TRIPPED))
 {
  // Process Sensor Input
  SensorValue = digitalRead(SensorPin);
  if ((PrevSensorValue == MagnetDetected) && (SensorValue == MagnetNotDetected))
  {
   unsigned long ElapsedTime = millis() - MinDelayBetweenHitsStartTime;
   if (ElapsedTime >= MinDelayBetweenHits)
   {
    // Motion Just Detected
    NumberHits++;
    Serial.print(F("SensorValue: "));
    Serial.print(SensorValue);
    Serial.print(F(" , Door Entry Detected ... NumberHits: "));
    Serial.println(NumberHits);
    PrevSensorValue = MagnetNotDetected;

    // Send Alarm Tripped Message
    SendAlarmTrippedMessage();

    // Time this hit occured
    MinDelayBetweenHitsStartTime = millis();
   }
  }
  else
  if ((PrevSensorValue == MagnetNotDetected) && (SensorValue == MagnetDetected))
  {
   // Door Closed
   Serial.print(F("SensorValue: "));
   Serial.println(SensorValue);
   PrevSensorValue = MagnetDetected;
  }
 }
}
```

Operating the System

Download and install the Android basic wireless framework version 1.1 from the last chapter if you have not done so already. Note that the Android APK install files for all the chapters are located in the downloads section of Chapter 9. Also, download and install the reed switch Arduino project on your Arduino. Bring up the Arduino's Serial Monitor and you should see the following initialization screen:

```
****** ESP-01 Arduino Mega 2560 Server
Program ********
***********  Reed Switch Sensor Alarm
System  *************

AT Command Sent: AT+CIPMUX=1
ESP-01 Response: AT+CIPMUX=1

OK

ESP-01 mode set to allow multiple
connections.

AT Command Sent: AT+CWMODE_CUR=2
ESP-01 Response: AT+CWMODE_CUR=2

OK

ESP-01 mode changed to Access Point
Mode.

AT Command Sent: AT+CIPSERVER=1,80
ESP-01 Response: AT+CIPSERVER=1,80

no change

OK

Server Started on ESP-01.

AT Command Sent: AT+CWSAP_CUR="ESP-01",
"esp8266esp01",1,2,4,0
ESP-01 Response: AT+CWSAP_CUR="ESP-01",
"esp8266esp01",1,2,4,0

OK

AP Configuration Set Successfullly.

AT Command Sent: AT+CWSAP_CUR?
ESP-01 Response: AT+CWSAP_CUR?

+CWSAP_CUR:"ESP-01","esp8266esp01",
1,2,4,0

OK

This is the curent configuration of the
AP on the ESP-01.

AT Command Sent: AT+CIPSTO?
ESP-01 Response: AT+CIPSTO?

+CIPSTO:180

OK

This is the curent TCP timeout in
seconds on the ESP-01.

******** System Has Started ********
```

Connect your Android device to the "ESP-01" access point located on the ESP-01 module using the password "esp8266esp01". Next, start up the Android basic framework application. This will connect the Android to the server running on the Arduino/ESP-01 hardware. Once the Android is connected to the TCP server the name of the Android device is changed on the wireless network and an "OK" response is sent back to the Android device.

```
Notification: '0,CONNECT'
Processing New Client Connection ...
NewConnection Name: 0
Notification: '
'Notification: '
'Notification: '+IPD,0,8:name=lg
'ClientIndex: 0
DataLength: 8
Data: name=lg
Client Name Change Requested,
ClientIndex: 0, NewClientName: lg
Client Name Change Succeeded, OK...
Sending Data ....
TCP Connection ID: 0 , DataLength: 3 ,
Data: OK
```

Next, turn off the emergency call out on the Android side. This will enable the Android to play back a sound effect when an alarm tripped message is received. Now, activate the alarm on the Android side. This will send a text character string "`ACTIVATE_ALARM`" to the Arduino. The

Arduino will receive it, activate the alarm, and send back a message indicating that the alarm was activated and is in wait mode for the next 10 seconds.

```
Notification: '
'Notification: '
'Notification: '+IPD,0,15:ACTIVATE_ALARM
'ClientIndex: 0
DataLength: 15
Data: ACTIVATE_ALARM

Sending Data ....
TCP Connection ID: 0 , DataLength: 40
, Data: Alarm Activated...WAIT TIME:10
Seconds.
```

Next, place a magnet next to the reed switch near the glass tube. If you have the same model of reed switch sensor I have then the LED that indicates the presence of a magnet should light up. This situation is the same as when the door is closed. After about 10 seconds the alarm should now be fully active and ready to be tripped. The Arduino will send a message to the Android indicating that the alarm is now active. A debug message will also be printed to the Serial Monitor indicating that the alarm is now fully active. The current sensor value is also printed out to the Serial Monitor which is 0 that indicates that a magnet has been detected. This would indicate that the door is closed.

```
Notification: '
'Sending Data ....
TCP Connection ID: 0 , DataLength: 20 ,
Data: ALARM NOW ACTIVE...

ALARM WAIT STATE FINISHED ... Alarm Now
ACTIVE ....
SensorValue: 0
```

Next, since the alarm is now active let's try to trip the alarm by moving the magnet away from the reed switch. This is similar to opening a door so that the reed switch and magnet that were once close together and aligned along opposite sides of the door move far apart to indicate an opened door. The alarm should trip and send an alarm tripped message to the Android device along with the total number of times that alarm has been tripped.

```
Notification: '
'SensorValue: 1 , Door Entry Detected
... NumberHits: 1
Sending Data ....
TCP Connection ID: 0 , DataLength: 16 ,
Data: ALARM_TRIPPED:1
```

Next, after the alarm is tripped quickly place the magnet near the reed switch again. The alarm system should register this as a door closed state and print out the sensor value of 0 to indicate that a magnet was detected. Now, pull the magnet away from the reed switch to trip the alarm again. This should be the second time it is tripped. A text string will be sent to the Android to indicate a second alarm trip event. A debug message should also appear on the Serial Monitor indicating a door entry event was detected.

```
Notification: '
'SensorValue: 0
SensorValue: 1 , Door Entry Detected ...
NumberHits: 2
Sending Data ....
TCP Connection ID: 0 , DataLength: 16 ,
Data: ALARM_TRIPPED:2
```

Finally, deactivate the alarm on the Android side. This will send a deactivate alarm text string to the Arduino. The Arduino will deactivate the alarm and send back a text string to the Android indicating the alarm has been deactivated.

```
Notification: '
'Notification: '
'Notification: '+IPD,0,17:DEACTIVATE_
ALARM
'ClientIndex: 0
DataLength: 17
Data: DEACTIVATE_ALARM

Sending Data ....
TCP Connection ID: 0 , DataLength: 32 ,
Data: Alarm Has Been DE-Activated ...
```

The Ywrobot Flame Sensor

The flame sensor detects the infrared heat generated by a fire or flame. The specifications of the Ywrobot flame sensor are as follows:

- Operating voltage: 3.3 and 5.0 V
- Light detection range: 760 to 1100 nm (red light through infrared light)
- Distance detection: 30 to 800 mm (31.496 inches)
- Flame LED indicator light: Lights up when a flame is detected

See Figure 6-3.

Figure 6-3 The flame sensor made by Ywrobot.

Hands-on Example: The Wireless Flame Sensor Fire Alarm System

In this hands-on example I show you how to build and operate a wireless fire alarm system based on a Ywrobot flame sensor, an Android cell phone, an Arduino, and the ESP8266-based ESP-01 module. This alarm system will be tripped when infrared heat is detected such as from a flame or fire.

Parts List

- 1 Android cell phone
- 1 Arduino Mega 2560 microcontroller
- 1 5- to 3.3-V step-down voltage regulator
- 1 Logic level converter from 5 to 3.3 V
- 1 ESP-01 Wi-Fi module
- 1 Arduino development station such as a desktop or a notebook
- 1 Breadboard
- 1 Ywrobot flame sensor (other brand of flame sensors ok but you may have to modify the code)
- 1 Incandescent flashlight, match, lighter, or candle that can produce infrared light
- 1 Package of wires both male-to-male and female-to-male to connect the components

Setting Up the Hardware

Step-Down Voltage Regulator Connections

1. Connect the Vin pin on the voltage regulator to the 5-V power pin on the Arduino Mega 2560.
2. Connect the Vout pin on the voltage regulator to a node on your breadboard represented by a horizontal row of empty slots. This is the 3.3-V power node that will supply the 3.3-V power to the ESP-01 module as well as provide the reference voltage for the logic level shifter.
3. Connect the GND pin on the voltage regulator to the GND pin on the Arduino or form a GND node similar to the 3.3-V power node in the previous step.

ESP-01 Wi-Fi Module Connections

1. Connect the Tx or TxD pin to the Rx3 pin on the Arduino Mega 2560. This is the receive pin for the serial hardware port 3.
2. Connect the EN or CH_PD pin to the 3.3-V power node from the step-down voltage regulator.
3. Connect the IO16 or RST pin to the 3.3-V power node from the step-down voltage regulator.

4. Connect the 3V3 or VCC pin to the 3.3-V power node from the step-down voltage regulator.
5. Connect the GND pin to the ground pin or node.
6. Connect the Rx or RxD pin to pin B0 on the 3.3-V side of the logic level shifter. The voltage from this pin will be shifted from 5 V from the Arduino to 3.3 V for the ESP-01 module.

Logic Level Shifter

1. Connect the 3.3-V pin to the 3.3-V power node from the step-down voltage regulator.
2. Connect the GND pin on the 3.3-V side to the ground node for the Arduino.
3. Connect the B0 pin on the 3.3-V side to Rx or RxD pin on the ESP-01. This corresponds to the A0 pin on the 5-V side.
4. Connect the 5-V pin to the 5-V power pin on the Arduino.
5. Connect the GND pin on the 5-V side to the ground node for the Arduino.
6. Connect the A0 pin on the 5-V side to Tx3 pin on the Arduino. This pin corresponds to the B0 pin on the 3.3-V side.

Flame Sensor

1. Connect the VCC pin to the 3.3-V power node from the voltage regulator.
2. Connect the GND pin to the ground node of the circuit.
3. Connect the DOUT pin to digital pin 7 on the Arduino.

See Figure 6-4.

Setting Up the Software

The Arduino code for the flame sensor alarm system has been modified from the previous reed switch-based alarm system hands-on example. I will only cover the code that has changed from the previous version.

The `FlameDetected` variable holds the voltage level that indicates that a flame has been detected by the flame sensor. The default value is set to 0 or LOW voltage.

```
int FlameDetected = 0;
```

The `FlameNotDetected` variable holds the voltage level that indicates that a flame has not been detected by the flame sensor. The default value is set to 1 or HIGH voltage.

```
int FlameNotDetected  = 1;
```

The `SensorValue` variable holds the value read in from the output pin of the flame sensor and is set by default to a value that indicates that no flame was detected.

```
int SensorValue = FlameNotDetected;
```

The `ProcessSensor()` Function

The `ProcessSensor()` processes the sensor input by doing the following:

1. If the alarm is on or has already been tripped, then continue with the function, otherwise exit the function.
2. Reads in the output from the flame sensor.
3. If the sensor detects a flame, then do the following:
    1. Find the elapsed time since the last flame was detected.
    2. If the elapsed time is greater than the minimum time, then do the following:
        1. Increase the record of the number of hits by 1.
        2. Print out the flame sensor value that was just read in, a message indicating that a flame was detected and the current number of hits.
        3. Send an alarm tripped message to the Android device.

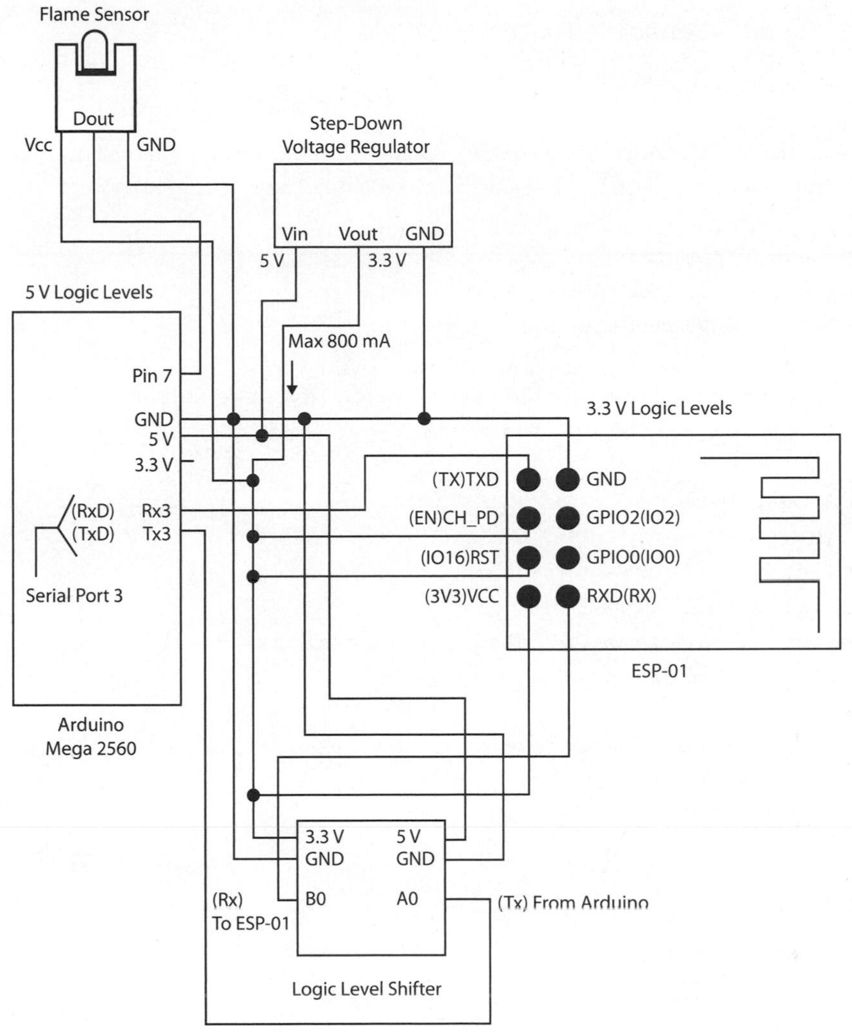

Figure 6-4 The flame sensor fire wireless fire alarm system.

4. Set the MinDelayBetween HitsStartTime variable that is used to determine the elapsed time since the last hit to the current time.

See Listing 6-2.

Operating the Wireless Flame Sensor Alarm System

Download and install the basic wireless framework 1.1 from the previous chapter on your Android device if you have not done so already. Note that the Android APK install files for all the chapters are located in the downloads section of Chapter 9. Also, download and install the flame sensor project for this chapter to your Arduino. Start up the Arduino's Serial Monitor and you should see something like the following:

```
****** ESP-01 Arduino Mega 2560 Server
Program ********
*********** Flame Sensor Alarm System
*************

AT Command Sent: AT+CIPMUX=1
ESP-01 Response: AT+CIPMUX=1

OK

ESP-01 mode set to allow multiple
connections.

AT Command Sent: AT+CWMODE_CUR=2
ESP-01 Response: AT+CWMODE_CUR=2

OK
```

Listing 6-2 The `ProcessSensor()` Function

```
void ProcessSensor()
{
  // Put Code to process your sensors or other hardware in this function.
  if ((AlarmState == Alarm_ON) || (AlarmState == Alarm_TRIPPED))
  {
    // Process Sensor Input
    SensorValue = digitalRead(SensorPin);
    if (SensorValue == FlameDetected)
    {
     unsigned long ElapsedTime = millis() - MinDelayBetweenHitsStartTime;
     if (ElapsedTime >= MinDelayBetweenHits)
     {
       // Motion Just Detected
       NumberHits++;
       Serial.print(F("SensorValue: "));
       Serial.print(SensorValue);
       Serial.print(F(" , Flame Detected ... NumberHits: "));
       Serial.println(NumberHits);

       // Send Alarm Tripped Message
       SendAlarmTrippedMessage();

       // Time this hit occured
       MinDelayBetweenHitsStartTime = millis();
     }
    }
  }
}
```

```
ESP-01 mode changed to Access Point
Mode.
AT Command Sent: AT+CIPSERVER=1,80
ESP-01 Response: AT+CIPSERVER=1,80

OK

Server Started on ESP-01.

AT Command Sent: AT+CWSAP_CUR="ESP-
01","esp8266esp01",1,2,4,0
ESP-01 Response: AT+CWSAP_CUR="ESP-
01","esp8266esp01",1,2,4,0

OK

AP Configuration Set Successfullly.

AT Command Sent: AT+CWSAP_CUR?
ESP-01 Response: AT+CWSAP_CUR?

+CWSAP_CUR:"ESP-
01","esp8266esp01",1,2,4,0

OK

This is the curent configuration of the
AP on the ESP-01.

AT Command Sent: AT+CIPSTO?
ESP-01 Response: AT+CIPSTO?

+CIPSTO:180

OK

This is the curent TCP timeout in
seconds on the ESP-01.

******** System Has Started ********
```

2. Connect the Vout pin on the voltage regulator to a node on your breadboard represented by a horizontal row of empty slots. This is the 3.3-V power node that will supply the 3.3-V power to the ESP-01 module as well as provide the reference voltage for the logic level shifter.
3. Connect the GND pin on the voltage regulator to the GND pin on the Arduino or form a GND node similar to the 3.3-V power node in the previous step.

ESP-01 Wi-Fi Module Connections

1. Connect the Tx or TxD pin to the Rx3 pin on the Arduino Mega 2560. This is the receive pin for the serial hardware port 3.
2. Connect the EN or CH_PD pin to the 3.3-V power node from the step-down voltage regulator.
3. Connect the IO16 or RST pin to the 3.3-V power node from the step-down voltage regulator.
4. Connect the 3V3 or VCC pin to the 3.3-V power node from the step-down voltage regulator.
5. Connect the GND pin to the ground pin or node.
6. Connect the Rx or RxD pin to pin B0 on the 3.3-V side of the logic level shifter. The voltage from this pin will be shifted from 5 V from the Arduino to 3.3 V for the ESP-01 module.

Logic Level Shifter

1. Connect the 3.3-V pin to the 3.3-V power node from the step-down voltage regulator.
2. Connect the GND pin on the 3.3-V side to the ground node for the Arduino.
3. Connect the B0 pin on the 3.3-V side to Rx or RxD pin on the ESP-01. This corresponds to the A0 pin on the 5-V side.
4. Connect the 5-V pin to the 5-V power pin on the Arduino.
5. Connect the GND pin on the 5-V side to the ground node for the Arduino.
6. Connect the A0 pin on the 5-V side to Tx3 pin on the Arduino. This pin corresponds to the B0 pin on the 3.3-V side.

Sunfounder Tilt Switch Sensor

1. Connect the VCC pin to the 5-V pin on the Arduino.
2. Connect the GND pin to the ground node of the circuit.
3. Connect the SIG pin to digital pin 7 on the Arduino.

See Figure 6-6.

Setting Up the Software

The Arduino code for this hands-on example is based on the code for the sound sensor alarm in the previous chapter. In this section I will cover only the code modifications that are specific to the tilt sensor alarm system.

The `TiltDetected` variable holds the voltage value that indicates that the tilt sensor has detected a tilt. The default value is set to 0 or LOW voltage.

```
int TiltDetected = 0;
```

The `TiltNotDetected` variable holds the voltage value that indicates that the tilt sensor does not detect a tilt. The default value is set to 1 or HIGH voltage.

```
int TiltNotDetected  = 1;
```

The `SensorValue` variable holds the current status of the tilt sensor and is initialized to a value indicating that a tilt was not detected by the tilt sensor.

```
int SensorValue = TiltNotDetected;
```

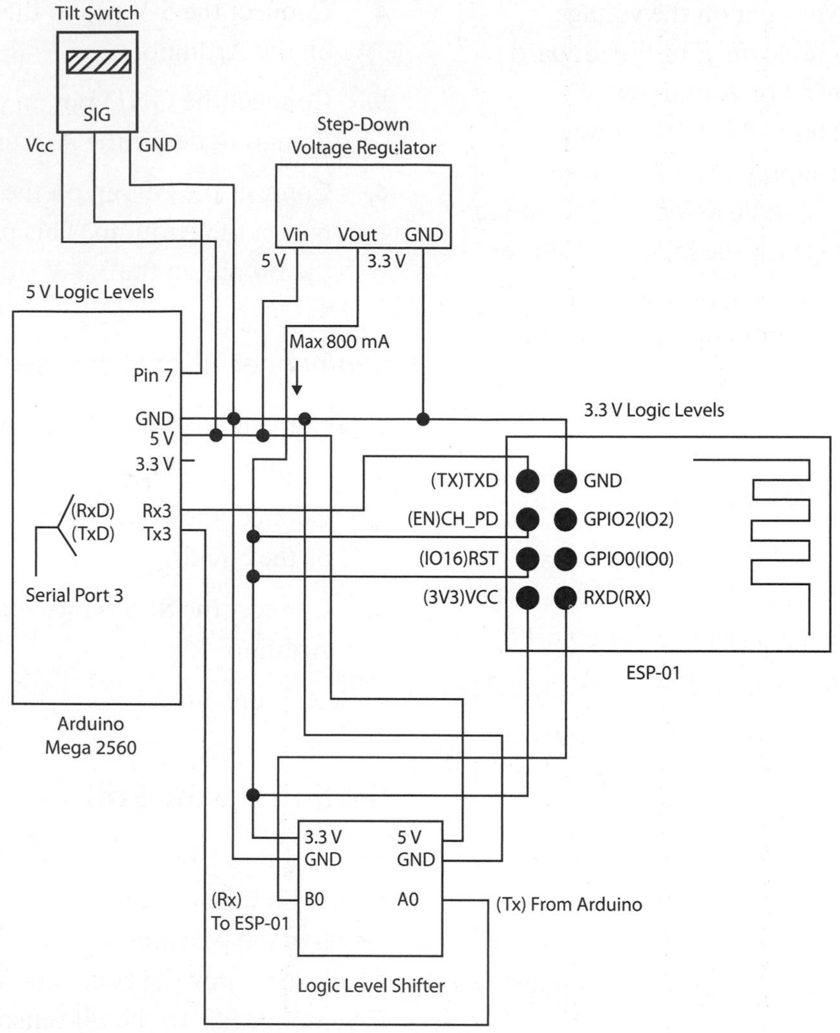

Figure 6-6 The tilt/vibration wireless alarm system.

The `PrevSensorValue` variable holds the status of the tilt sensor from the previous reading and is initialized to a value that indicates a tilt was not detected.

```
int PrevSensorValue = TiltNotDetected;
```

The `ProcessSensor()` Function

The `ProcessSensor()` function reads in and processes the data from the tilt sensor. The code is mostly the same as with the previous sound sensor. Instead of a sound being detected and processed a tilt is being detected and processed. The general idea is the same and thus the code is generally the same. The modifications are shown in bold print. See Listing 6-3.

Operating the Tilt/Vibrate Wireless Alarm System

Download and install the basic wireless framework version 1.1 from the last chapter on your Android cell phone if you have not done so already. Note that the Android APK install files for all the chapters are located in the downloads section of Chapter 9. Also, download and install on your Arduino the tilt sensor alarm system Arduino program from this chapter. Start up the Arduino's Serial Monitor and the following start-up screen should come up:

```
****** ESP-01 Arduino Mega 2560 Server
Program ********
```

Listing 6-3 The `ProcessSensor()` Function

```
void ProcessSensor()
{
 // Put Code to process your sensors or other hardware in this function.
 if ((AlarmState == Alarm_ON) || (AlarmState == Alarm_TRIPPED))
 {
  // Process Sensor Input
  SensorValue = digitalRead(SensorPin);
  if ((PrevSensorValue == TiltNotDetected) && (SensorValue == TiltDetected))
  {
   unsigned long ElapsedTime = millis() - MinDelayBetweenHitsStartTime;
   if (ElapsedTime >= MinDelayBetweenHits)
   {
    // Motion Just Detected
    NumberHits++;
    Serial.print(F("SensorValue: "));
    Serial.print(SensorValue);
    Serial.print(F(" , Motion Detected ... NumberHits: "));
    Serial.println(NumberHits);
    PrevSensorValue = TiltDetected;

    // Send Alarm Tripped Message
    SendAlarmTrippedMessage();

    // Time this hit occured
    MinDelayBetweenHitsStartTime = millis();
   }
  }
  else
  if ((PrevSensorValue == TiltDetected) && (SensorValue == TiltNotDetected))
  {
   // Tilted back
   Serial.print(F("SensorValue: "));
   Serial.println(SensorValue);
   PrevSensorValue = TiltNotDetected;
  }
 }
}
```

```
*********** Tilt Sensor Alarm System
*************

AT Command Sent: AT+CIPMUX=1
ESP-01 Response: AT+CIPMUX=1

OK

ESP-01 mode set to allow multiple
connections.

AT Command Sent: AT+CWMODE_CUR=2
ESP-01 Response: AT+CWMODE_CUR=2

OK

ESP-01 mode changed to Access Point
Mode.

AT Command Sent: AT+CIPSERVER=1,80
ESP-01 Response: AT+CIPSERVER=1,80
```

```
OK

Server Started on ESP-01.

AT Command Sent: AT+CWSAP_CUR="ESP-
01","esp8266esp01",1,2,4,0
ESP-01 Response: AT+CWSAP_CUR="ESP-
01","esp8266esp01",1,2,4,0

OK

AP Configuration Set Successfullly.

AT Command Sent: AT+CWSAP_CUR?
ESP-01 Response: AT+CWSAP_CUR?

+CWSAP_CUR:"ESP-
01","esp8266esp01",1,2,4,0

OK

This is the curent configuration of the
AP on the ESP-01.

AT Command Sent: AT+CIPSTO?
ESP-01 Response: AT+CIPSTO?

+CIPSTO:180

OK

This is the curent TCP timeout in
seconds on the ESP-01.

******** System Has Started ********
```

Next, connect your Android to the "ESP-01" access point that is running on the ESP-01 module using the password "esp8266esp01". Start up the Android basic wireless framework version 1.1 application. This will connect the Android to the TCP server running on the ESP-01 module. The Android will send a name change command to the Arduino which will change the name of the Android on the wireless network and send back an "OK" text string verifying the name change.

```
Notification: '0,CONNECT'
Processing New Client Connection ...
NewConnection Name: 0

Notification: '
'Notification: '
'Notification: '+IPD,0,8:name=lg
'ClientIndex: 0
DataLength: 8
Data: name=lg

Client Name Change Requested,
ClientIndex: 0, NewClientName: lg
Client Name Change Succeeded, OK...
Sending Data ....
TCP Connection ID: 0 , DataLength: 3 ,
Data: OK
```

Next, turn off the emergency call out and emergency email text features on the Android application. Then, activate the alarm on the Android side. This will send a text string to the Arduino that activates the alarm. The Arduino will respond with a text string indicating that the alarm has been activated and the number of seconds till the alarm is fully active and ready to be tripped.

```
Notification: '
'Notification: '
'Notification: '+IPD,0,15:ACTIVATE_ALARM
'ClientIndex: 0
DataLength: 15
Data: ACTIVATE_ALARM

Sending Data ....
TCP Connection ID: 0 , DataLength: 40 ,
Data: Alarm Activated...WAIT TIME:10
Seconds.
```

In around 10 seconds the alarm will become fully active and send a text string to the Android device indicating this active status. A message is also printed to the Arduino's Serial Monitor indicating that the alarm is now fully active and can be tripped.

```
Notification: '
'Sending Data ....
TCP Connection ID: 0 , DataLength: 20 ,
Data: ALARM NOW ACTIVE...

ALARM WAIT STATE FINISHED ... Alarm Now
ACTIVE ....
```

Next, let's trip that alarm by setting the tilt sensor on its side and hitting it against a hard surface. This will trigger the alarm. An alarm tripped text string will be sent to the Android with the total number of trips detected so far. Wait a few seconds for the alarm trip delay to pass and then trip that alarm a few more times.

```
Notification: '
'SensorValue: 0 , Motion Detected ...
NumberHits: 1
Sending Data ....
TCP Connection ID: 0 , DataLength: 16 ,
Data: ALARM_TRIPPED:1

SensorValue: 1
Notification: '
'SensorValue: 0 , Motion Detected ...
NumberHits: 2
Sending Data ....
TCP Connection ID: 0 , DataLength: 16 ,
Data: ALARM_TRIPPED:2

Notification: '
'SensorValue: 1
SensorValue: 0 , Motion Detected ...
NumberHits: 3
Sending Data ....
TCP Connection ID: 0 , DataLength: 16 ,
Data: ALARM TRIPPED:3

SensorValue: 1
```

Finally, deactivate the alarm using the Android application and exit the Android application by pressing the back button. The connection will be closed and the Android's name on the wireless network will be removed from the list of current clients.

```
Notification: '
'Notification: '
'Notification: '+IPD,0,17:DEACTIVATE_ALARM
'ClientIndex: 0
DataLength: 17
Data: DEACTIVATE_ALARM

Sending Data ....
TCP Connection ID: 0 , DataLength: 32 ,
Data: Alarm Has Been DE-Activated ...

Notification: '
'Notification: '0,CLOSED' Connection
Removed Name: lg

Notification: '
'
```

TMP36 Temperature Sensor

The TMP36 temperature sensor provides an output voltage that is linearly proportional to the Celsius (Centigrade) temperature. The TMP36 sensor is shown in Figure 6-7.

There are 3 pins on the TMP36. The +Vs pin should be connected to the 3.3-V pin on the Arduino. The center Vout pin outputs voltage that represents the measured temperature in Celsius. The GND pin should be connected to ground. The view of these pins from the bottom of the sensor is shown in Figure 6-8.

The relationship between output voltage and temperature for the TMP36 is linear. That means that it can be modeled by the equation of a line.

The equation of a line is

$$y = mx + b$$

where y is the output voltage from the TMP36 sensor, x is the temperature in Celsius, m is the slope of the line, and b is the y intercept value.

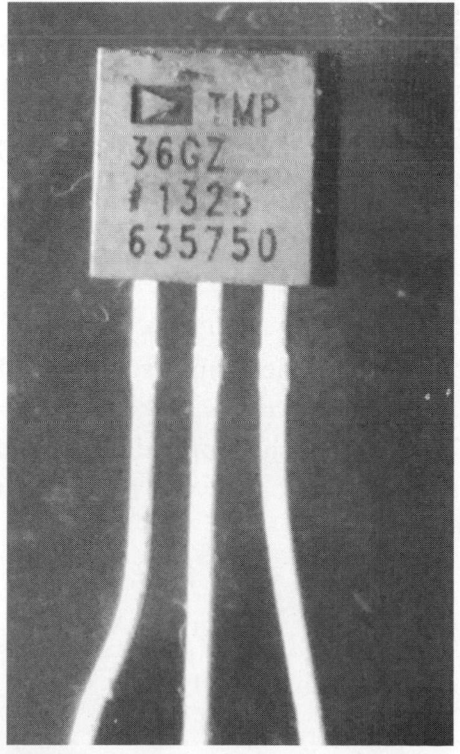

Figure 6-7 The TMP36 temperature sensor.

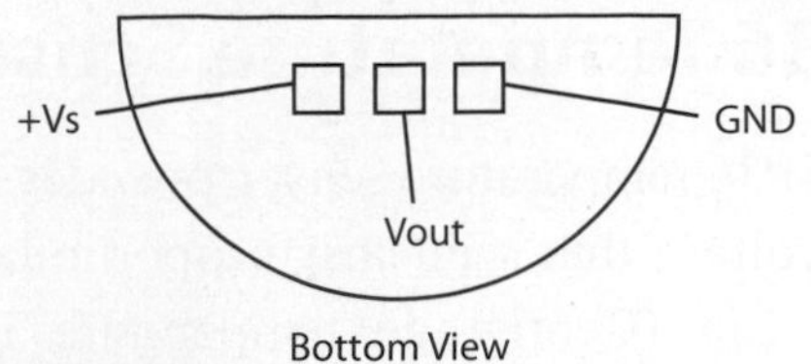

Figure 6-8 The TMP36 pinouts.

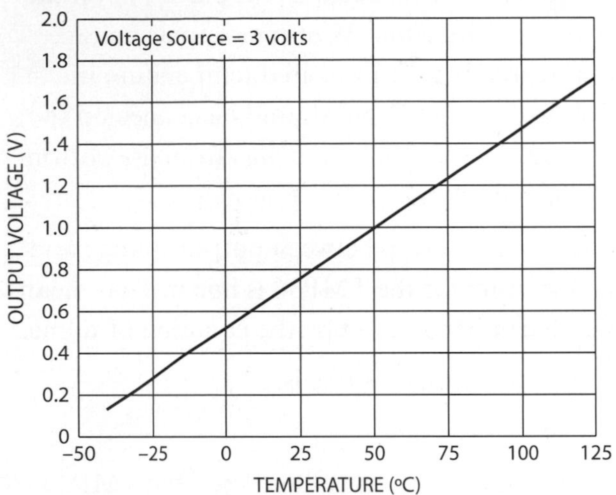

Figure 6-9 TMP36 voltage versus temperature output graph.

From the data sheet for the TMP36 from Analog Devices we can see in Figure 6-9, which shows a graph of the voltage output versus temperature, that the y intercept value is roughly 0.5.

The slope of a line is the change between the y values divided by the change in x values between two points on a line or:

m = (y2 – y1)/(x2 – x1)

Two points on the line from Figure 6-9 or Figure 6 (note that Figure 6 refers to the figure in the actual TMP36 data sheet provided by Analog Devices) in the Analog Devices data sheet for the TMP36 sensor are:

pt1(0, 0.5)

pt2(50, 1)

The slope is then calculated as:

m = (1 – 0.5)/(50 – 0)

m = 0.5/50 = 0.01 = 1/100

The equation of the line where y is the output voltage and x is temperature is as follows:

y = x/100 + 0.5

In order to find x which is temperature based on y which is the output voltage we need to solve the equation for the x variable as follows:

y – 0.5 = x/100

(y – 0.5) * 100 = x = Temperature in Celsius

Thus we have the final equation that calculates the temperature in Celsius based on the output voltage from the TMP36 sensor as follows:

Temp = (Vout – 0.5) * 100

Hands-on Example: The TMP36 Wireless Temperature Monitoring and Alarm System

In this hands-on example I show you how to build and operate a TMP36-based wireless temperature monitoring and alarm system. You will be able to remotely retrieve the current temperature and set a temperature range where any temperature outside that range will trigger the alarm.

Parts List

- 1 Android cell phone
- 1 Arduino Mega 2560 microcontroller
- 1 5- to 3.3-V step-down voltage regulator
- 1 Logic level converter from 5 to 3.3 V
- 1 ESP-01 Wi-Fi module
- 1 Arduino development station such as a desktop or a notebook
- 1 Breadboard
- 1 TMP36 temperature sensor
- 1 Package of wires both male-to-male and female-to-male to connect the components

Setting Up the Hardware

Step-Down Voltage Regulator Connections

1. Connect the Vin pin on the voltage regulator to the 5-V power pin on the Arduino Mega 2560.
2. Connect the Vout pin on the voltage regulator to a node on your breadboard represented by a horizontal row of empty slots. This is the 3.3-V power node that will supply the 3.3-V power to the ESP-01 module as well as provide the reference voltage for the logic level shifter.
3. Connect the GND pin on the voltage regulator to the GND pin on the Arduino or form a GND node similar to the 3.3-V power node in the previous step.

ESP-01 Wi-Fi Module Connections

1. Connect the Tx or TxD pin to the Rx3 pin on the Arduino Mega 2560. This is the receive pin for the serial hardware port 3.
2. Connect the EN or CH_PD pin to the 3.3-V power node from the step-down voltage regulator.
3. Connect the IO16 or RST pin to the 3.3-V power node from the step-down voltage regulator.
4. Connect the 3V3 or VCC pin to the 3.3-V power node from the step-down voltage regulator.
5. Connect the GND pin to the ground pin or node.
6. Connect the Rx or RxD pin to pin B0 on the 3.3-V side of the logic level shifter. The voltage from this pin will be shifted from 5 V from the Arduino to 3.3 V for the ESP-01 module.

Logic Level Shifter

1. Connect the 3.3-V pin to the 3.3-V power node from the step-down voltage regulator.
2. Connect the GND pin on the 3.3-V side to the ground node for the Arduino.
3. Connect the B0 pin on the 3.3-V side to Rx or RxD pin on the ESP-01. This corresponds to the A0 pin on the 5-V side.
4. Connect the 5-V pin to the 5-V power pin on the Arduino.
5. Connect the GND pin on the 5-V side to the ground node for the Arduino.
6. Connect the A0 pin on the 5-V side to Tx3 pin on the Arduino. This pin corresponds to the B0 pin on the 3.3-V side.

TMP36 Sensor

1. Connect the Vs pin to the 3.3-V power node.
2. Connect the GND pin to the ground node for the circuit.
3. Connect the Vout pin to analog pin 7 on the Arduino.

See Figure 6-10.

Setting Up the Software

The Arduino software for this example is based on the previous code for the tilt sensor alarm. I will cover only the new code specific to this hands-on example in this section.

The `UnitsToVolts` variable converts the raw value from an analog read to the corresponding voltage value.

```
const float UnitsToVolts = 5.0/1024.0;
// 0.0049 volts (4.9 mV) per unit
```

The `MinTemp` variable holds the minimum safe temperature in the user-defined temperature range. A temperature reading below this value will trigger the alarm. The default value is set to 40 degrees Fahrenheit.

```
int MinTemp = 40; // Degrees in
Farenheit
```

The `MaxTemp` variable holds the maximum safe temperature in the user-defined temperature

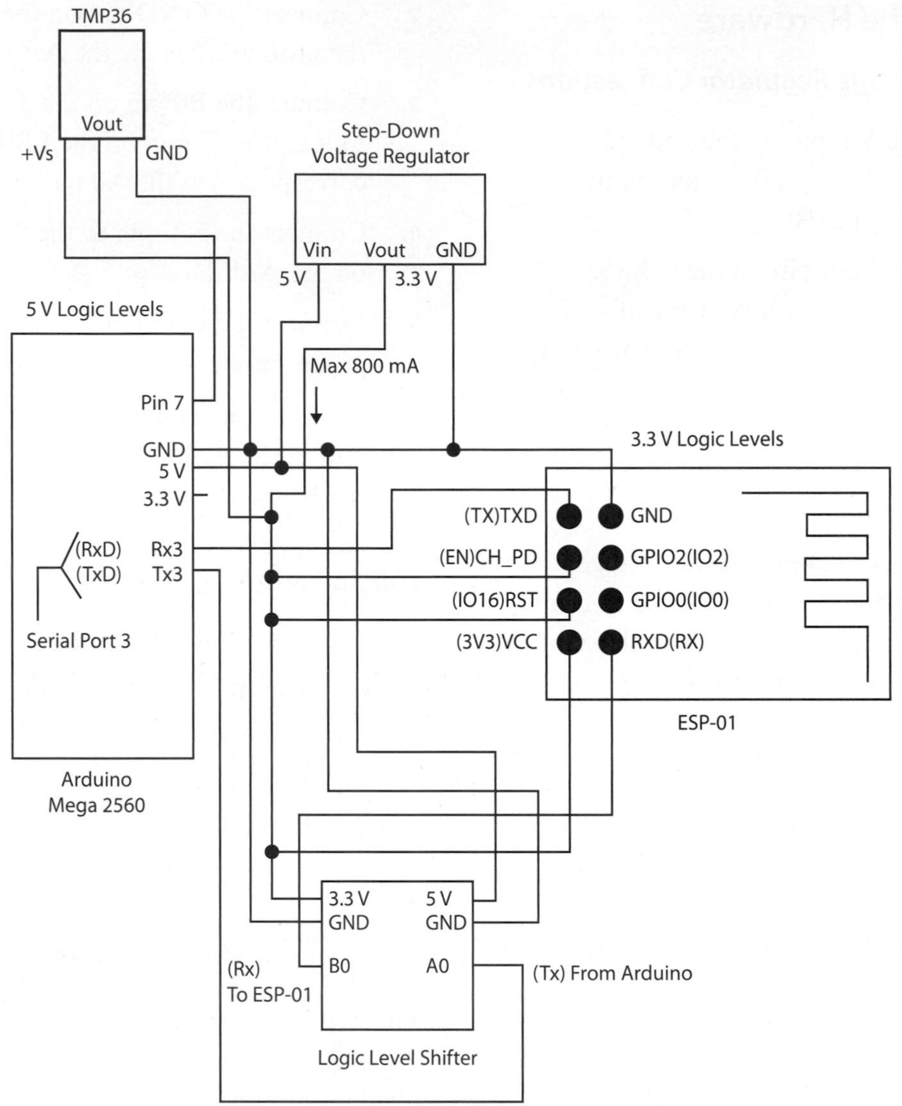

Figure 6-10 The TMP36 temperature monitoring and alarm system.

range. A temperature reading above this value will trigger the alarm. The default value is set to 90 degrees Fahrenheit.

```
int MaxTemp = 90; // Degrees in
Farenheit
```

The `SensorPin` variable holds the Arduino pin that is connected to the output pin on the TMP36 sensor. The default pin is set to analog pin 7 on the Arduino.

```
int SensorPin = A7;
```

The `MinDelayBetweenReadings` variable holds the minimum time in milliseconds between successive readings of the TMP35 temperature sensor. The default value is set to 2500 milliseconds or 2.5 seconds.

```
unsigned long MinDelayBetweenReadings =
2500;
```

The `MinDelayBetweenReadingsStartTime` variable holds the time of the last temperature reading in milliseconds.

```
unsigned long MinDelayBetween
ReadingsStartTime = 0;
```

The `GetTempCelsius()` Function

The `GetTempCelsius()` function converts the voltage reading from the TMP36 sensor into the corresponding Celsius temperature based on the formula previously determined which is Temperature = (VoltageOut – 0.5) * 100. See Listing 6-4.

The `GetTempFarenheit()` Function

The `GetTempFarenheit()` function converts the voltage reading from the TMP36 sensor to the corresponding Fahrenheit value. See Listing 6-5.

The `ReadTMP36F()` Function

The `ReadTMP36F()` function retrieves the temperature value output by the TMP36 temperature sensor by doing the following:

1. Reading in the analog value from the TMP36 sensor's output pin.
2. Converting this analog value to the corresponding voltage value that is output by the TMP36 sensor.

Listing 6-4 The `GetTempCelsius()` Function

```
float GetTempCelsius(float Voltage)
{
  float temp = 0;

  temp = (Voltage-0.5)*100;
  return temp;
}
```

Listing 6-5 The `GetTempFarenheit()` Function

```
float GetTempFarenheit(float Voltage)
{
  float temp = 0;
  float CTemp =
  GetTempCelsius(Voltage);
  temp = CTemp*(9.0/5.0)+32.0;
  return temp;
}
```

3. Converting the voltage value that is output by the TMP36 to the corresponding Fahrenheit temperature value by calling the `GetTempFarenheit(VoltageValue)` function.
4. Returning the temperature value to the calling function.

See Listing 6-6.

The `ReadTMP36C()` Function

The `ReadTMP36C()` function retrieves the Celsius value from the TMP36 sensor by doing the following:

1. Reading in the analog value from the TMP36's output pin.
2. Converting this analog value to the corresponding voltage value.
3. Converting the voltage value to the corresponding Celsius temperature value by calling the GetTempCelsius(VoltageValue) function.
4. Returning the temperature value to the function's caller.

See Listing 6-7.

The `ProcessSensor()` Function

The `ProcessSensor()` function processes the TMP36 sensor by doing the following:

1. If the alarm is on or the alarm has been tripped, then continue with the execution of the function, otherwise exit the function.
2. Determine the elapsed time since the last time the TMP36 sensor was read.
3. If the elapsed time is greater than the minimum time between successive readings, then do the following:
    1. Read in the temperature from the TMP36 by calling the ReadTMP36F() function.

Listing 6-6 The `ReadTMP36F()` Function

```
float ReadTMP36F()
{
 float VoltageValue = 0.0f;
 float Temp = 0.0f;

 SensorValue = analogRead(SensorPin);
 VoltageValue = SensorValue *
UnitsToVolts;
 Temp =
GetTempFarenheit(VoltageValue);

 return Temp;
}
```

Listing 6-7 The `ReadTMP36C()` Function

```
float ReadTMP36C()
{
 float VoltageValue = 0.0f;
 float Temp = 0.0f;

 SensorValue = analogRead(SensorPin);
 VoltageValue = SensorValue *
UnitsToVolts;
 Temp = GetTempCelsius(VoltageValue);

 return Temp;
}
```

2. If the temperature is less than MinTemp or the temperature is greater than MaxTemp, then do the following:
   1. Increase the number of hits by 1.
   2. Print out debug information to the Arduino Serial Monitor.
   3. Send out a text string to the Android indicating that the alarm has been tripped and the number of times so far it has been tripped.
3. Set the MinDelayBetweenReadings-StartTime variable to the current time since the temperature from the TMP36 sensor was just read.

See Listing 6-8.

The `tempf` Command

The `tempf` command gets the current temperature in Fahrenheit from the TMP36 sensor and sends it back to the Android device using the TCP connection. See Listing 6-9.

The `tempc` Command

The `tempc` command retrieves the current temperature in Celsius from the TMP36 sensor and sends it to the Android device using the TCP connection. See Listing 6-10.

The `mintemp` Command

The `mintemp` command sends the minimum temperature for the alarm system to the Android device using the TCP connection. See Listing 6-11.

The `maxtemp` Command

The `maxtemp` command sends the maximum temperature for the alarm system to the Android device using the TCP connection. See Listing 6-12.

The `mintemp assign` Command

The `mintemp assign` command which is of the format "`mintemp=tempvalue`" assigns a value to the minimum temperature for the alarm system. The temperature value is the value after the equals sign. A text string is sent back to the Android device acknowledging the change. See Listing 6-13.

The `maxtemp assign` Command

The `maxtemp assign` command which is of the format "`maxtemp=tempvalue`" assigns a temperature value to the maximum temperature for the alarm system. The temperature value is located after the equals sign. A text string acknowledging the change is sent back to the Android device over the TCP connection. See Listing 6-14.

Listing 6-8 The `ProcessSensor()` Function

```
void ProcessSensor()
{
 // Put Code to process your sensors or other hardware in this function.
 if ((AlarmState == Alarm_ON) || (AlarmState == Alarm_TRIPPED))
 {
  unsigned long ElapsedTime = millis() - MinDelayBetweenReadingsStartTime;
  if (ElapsedTime >= MinDelayBetweenReadings)
  {
   float Temp = ReadTMP36F();
   if ((Temp < MinTemp) || (Temp > MaxTemp))
   {
    // Temp Outside Safe Range Just Detected
    NumberHits++;
    Serial.print(F("Current Temp (Farenheit): "));
    Serial.print(Temp);
    Serial.print(F(" , Temperature Exceeded Safe Range ... NumberHits: "));
    Serial.println(NumberHits);

    // Send Alarm Tripped Message
    SendAlarmTrippedMessage();
   }
   MinDelayBetweenReadingsStartTime = millis();
  }
 }
}
```

Listing 6-9 The `tempf` Command

```
if (Data == "tempf\n")
{
 float Temp = ReadTMP36F();
 String returnstring = "";
 returnstring = returnstring +
F("TEMP (Farenheit): ") + String(Temp)
+ "\n";
 SendTCPMessage(ClientIndex,
returnstring);
}
```

Listing 6-10 The `tempc` Command

```
if (Data == "tempc\n")
{
 float Temp = ReadTMP36C();
 String returnstring = "";
 returnstring = returnstring +
F("TEMP (Celsius): ") + String(Temp)
+ "\n";
 SendTCPMessage(ClientIndex,
returnstring);
}
```

Operating the TMP36 Wireless Temperature Monitoring and Alarm System

This example uses the same Android program from the previous chapter which was the basic wireless framework version 1.1. Note that the Android APK install files for all the chapters are located in the downloads section of Chapter 9. Download the TMP36-related Arduino program for this chapter and install it on your Arduino. Start up the Serial Monitor. The start-up sequence should be similar to the following:

```
****** ESP-01 Arduino Mega 2560 Server
Program ********
```

Listing 6-11 The `mintemp` Command

```
if (Data == "mintemp\n")
{
 String returnstring = "";
 returnstring = returnstring +
 F("MinTemp (Farenheit): ") + MinTemp
 + "\n";
 SendTCPMessage(ClientIndex,
 returnstring);
}
```

Listing 6-12 The `maxtemp` Command

```
if (Data == "maxtemp\n")
{
 String returnstring = "";
 returnstring = returnstring +
 F("MaxTemp (Farenheit): ") + MaxTemp
 + "\n";
 SendTCPMessage(ClientIndex,
 returnstring);
}
```

Listing 6-13 The `mintemp assign` Command

```
if (Data.indexOf("mintemp=") >= 0)
{
 int indextemp = Data.indexOf("=");
 String mintemp = Data.
 substring(indextemp+1);
 MinTemp = mintemp.toInt();

 String returnstring = "";
 returnstring = returnstring +
 F("MinTemp set to: ") + MinTemp +
 F(" Deg Farenheit\n");
 SendTCPMessage(ClientIndex,
 returnstring);
}
```

```
***********  TMP36 Sensor Alarm System
*************

AT Command Sent: AT+CIPMUX=1
ESP-01 Response: AT+CIPMUX=1

OK
```

Listing 6-14 The `maxtemp assign` Command

```
if (Data.indexOf("maxtemp=") >= 0)
{
 int indextemp = Data.indexOf("=");
 String maxtemp = Data.
substring(indextemp+1);
 MaxTemp = maxtemp.toInt();
 String returnstring = "";
 returnstring = returnstring +
F("MaxTemp set to: ") + MaxTemp +
F(" Deg Farenheit\n");
 SendTCPMessage(ClientIndex,
returnstring);
}
```

```
ESP-01 mode set to allow multiple
connections.

AT Command Sent: AT+CWMODE_CUR=2
ESP-01 Response: AT+CWMODE_CUR=2

OK

ESP-01 mode changed to Access Point
Mode.

AT Command Sent: AT+CIPSERVER=1,80
ESP-01 Response: AT+CIPSERVER=1,80

OK

Server Started on ESP-01.

AT Command Sent: AT+CWSAP_CUR="ESP-
01","esp8266esp01",1,2,4,0
ESP-01 Response: AT+CWSAP_CUR="ESP-
01","esp8266esp01",1,2,4,0

OK

AP Configuration Set Successfullly.

AT Command Sent: AT+CWSAP_CUR?
ESP-01 Response: AT+CWSAP_CUR?

+CWSAP_CUR:"ESP-
01","esp8266esp01",1,2,4,0

OK

This is the curent configuration of the
AP on the ESP-01.
```

```
AT Command Sent: AT+CIPSTO?
ESP-01 Response: AT+CIPSTO?

+CIPSTO:180

OK

This is the curent TCP timeout in
seconds on the ESP-01.


******** System Has Started ********
```

Connect your Android cell phone to the ESP-01 Wi-Fi access point using the password shown in the Arduino source code. Start up the Android application. The Android will connect to the TCP server running on the ESP-01 module and send the change name command to change the name of the Android device on the wireless network. In my case the name was changed to "`lg`". After changing the name the Arduino sends an "OK" string back to the Android.

```
Notification: '0,CONNECT'
Processing New Client Connection ...
NewConnection Name: 0

Notification: '
'Notification: '
'Notification: '+IPD,0,8:name=lg
'ClientIndex: 0
DataLength: 8
Data: name=lg

Client Name Change Requested,
ClientIndex: 0, NewClientName: lg
Client Name Change Succeeded, OK...
Sending Data ....
TCP Connection ID: 0 , DataLength: 3 ,
Data: OK
```

Next, send a tempf command to the Arduino using the Android application. This command will be processed by the Arduino and a text string will be sent back to the Android indicating the current temperature in Fahrenheit.

```
Notification: '
'Notification: '
'Notification: '+IPD,0,6:tempf
'ClientIndex: 0
DataLength: 6
Data: tempf

Sending Data ....
TCP Connection ID: 0 , DataLength: 24 ,
Data: TEMP (Farenheit): 68.56
```

Next, let's get the current temperature in Celsius by sending a "`tempc`" command from the Android application. The Arduino will receive the tempc command and respond by sending a text string with the current temperature in Celsius back to the Android device.

```
Notification: '
'Notification: '
'Notification: '+IPD,0,6:tempc
'ClientIndex: 0
DataLength: 6
Data: tempc

Sending Data ....
TCP Connection ID: 0 , DataLength: 22 ,
Data: TEMP (Celsius): 20.31
```

Next, let's get the current minimum temperature for the alarm system by issuing a "`mintemp`" command from the Android application. This sends this command to the Arduino which processes it and responds with a text string indicating the minimum temperature which should be 40 degrees Fahrenheit.

```
Notification: '
'Notification: '
'Notification: '+IPD,0,8:mintemp
'ClientIndex: 0
DataLength: 8
Data: mintemp

Sending Data ....
TCP Connection ID: 0 , DataLength: 24 ,
Data: MinTemp (Farenheit): 40
```

Next, let's get the current maximum temperature for the alarm system by sending the "`maxtemp`" command from the Android application to the Arduino. The Arduino will receive the command and send back a text string that contains the maximum temperature for the alarm system which should be 90 degrees Fahrenheit.

```
Notification: '
'Notification: '
'Notification: '+IPD,0,8:maxtemp
'ClientIndex: 0
DataLength: 8
Data: maxtemp

Sending Data ....
TCP Connection ID: 0 , DataLength: 24 ,
Data: MaxTemp (Farenheit): 90
```

Next, turn off the emergency call out and the emergency text messages on the Android application. Activate the alarm on the Android side. This will send an activate alarm command to the Arduino. The Arduino will receive this command, activate the alarm, and send back a text string indicating that the alarm was activated and put into wait mode for the next 10 seconds.

```
Notification: '
'Notification: '
'Notification: '+IPD,0,15:ACTIVATE_ALARM
'ClientIndex: 0
DataLength: 15
Data: ACTIVATE_ALARM

Sending Data ....
TCP Connection ID: 0 , DataLength: 40
, Data: Alarm Activated...WAIT TIME:10
Seconds.
```

After approximately 10 seconds an alarm active text string should be sent to your Android device, the alarm will be fully activated and ready to be tripped, and another text message is printed to the Serial Monitor indicating that the wait state has finished.

```
Notification: '
'Sending Data ....
TCP Connection ID: 0 , DataLength: 20 ,
Data: ALARM NOW ACTIVE...

ALARM WAIT STATE FINISHED ... Alarm Now
ACTIVE ....
```

The alarm will trip if the temperature is less than 40 degrees or greater than 90 degrees Fahrenheit. The current temperature for me is around 70 degrees Fahrenheit.

Next, let's set the minimum temperature for the alarm system to above the current temperature in order to trip the alarm. In my case I set the minimum temperature to 75 degrees using the "`mintemp=75`" command. The Arduino receives the command, sets the new minimum temperature, and sends back a text string indicating that the change has been made.

```
Notification: '
'Notification: '
'Notification: '+IPD,0,11:mintemp=75
'ClientIndex: 0
DataLength: 11
Data: mintemp=75
Sending Data ....
TCP Connection ID: 0 , DataLength: 33 ,
Data: MinTemp set to: 75 Deg Farenheit
```

The alarm should now begin to trip every 2.5 seconds. For each alarm trip the current temperature is displayed on the Serial Monitor along with a message indicating that the current temperature has exceeded the safe range. The total number of times that the alarm has been tripped since the alarm was activated is also displayed.

```
Notification: '
'Current Temp (Farenheit): 68.56 ,
Temperature Exceeded Safe Range ...
NumberHits: 1
Sending Data ....
TCP Connection ID: 0 , DataLength: 16 ,
Data: ALARM_TRIPPED:1
Notification: '
'Current Temp (Farenheit): 67.68 ,
Temperature Exceeded Safe Range ...
NumberHits: 2
Sending Data ....
TCP Connection ID: 0 , DataLength: 16 ,
Data: ALARM_TRIPPED:2
Notification: '
'Current Temp (Farenheit): 68.56 ,
Temperature Exceeded Safe Range ...
NumberHits: 3
Sending Data ....
TCP Connection ID: 0 , DataLength: 16 ,
Data: ALARM_TRIPPED:3
```

```
Notification: '
'Current Temp (Farenheit): 67.68 ,
Temperature Exceeded Safe Range ...
NumberHits: 4
Sending Data ....
TCP Connection ID: 0 , DataLength: 16 ,
Data: ALARM_TRIPPED:4
Notification: '
'Current Temp (Farenheit): 68.56 ,
Temperature Exceeded Safe Range ...
NumberHits: 5
Sending Data ....
TCP Connection ID: 0 , DataLength: 16 ,
Data: ALARM_TRIPPED:5
Notification: '
'Current Temp (Farenheit): 67.68 ,
Temperature Exceeded Safe Range ...
NumberHits: 6
Sending Data ....
TCP Connection ID: 0 , DataLength: 16 ,
Data: ALARM_TRIPPED:6
```

Next, deactivate the alarm using the Android application.

```
Notification: '
'Notification: '
'Notification: '+IPD,0,17:DEACTIVATE_ALARM
'ClientIndex: 0
DataLength: 17
Data: DEACTIVATE_ALARM
Sending Data ....
TCP Connection ID: 0 , DataLength: 32 ,
Data: Alarm Has Been DE-Activated ...
```

Next, set the minimum temperature to a value that would not trigger the alarm such as 50 in my case. Send the command "`mintemp=50`" to the Arduino from the Android application.

```
Notification: '
'Notification: '
'Notification: '+IPD,0,11:mintemp=50
'ClientIndex: 0
DataLength: 11
Data: mintemp=50
Sending Data ....
TCP Connection ID: 0 , DataLength: 33 ,
Data: MinTemp set to: 50 Deg Farenheit
```

Next, activate the alarm using the Android application.

```
Notification: '
'Notification: '
'Notification: '+IPD,0,15:ACTIVATE_ALARM
'ClientIndex: 0
DataLength: 15
Data: ACTIVATE_ALARM
Sending Data ....
TCP Connection ID: 0 , DataLength: 40
, Data: Alarm Activated...WAIT TIME:10
Seconds.
```

In about 10 seconds the alarm will be fully active and ready to be tripped.

```
Notification: '
'Sending Data ....
TCP Connection ID: 0 , DataLength: 20 ,
Data: ALARM NOW ACTIVE...
ALARM WAIT STATE FINISHED ... Alarm Now
ACTIVE ....
```

Next, let's change the maximum temperature for the alarm system so that the alarm will be tripped. In my case any maximum temperature that is lower than around 67 will trip the alarm. So I used the "`maxtemp=50`" to set the maximum temperature to 50 degrees. Any temperature reading above 50 degrees will cause the alarm to trip.

```
Notification: '
'Notification: '
'Notification: '+IPD,0,11:maxtemp=50
'ClientIndex: 0
DataLength: 11
Data: maxtemp=50
Sending Data ....
TCP Connection ID: 0 , DataLength: 33 ,
Data: MaxTemp set to: 50 Deg Farenheit
```

The alarm should start tripping in roughly 2.5-second intervals. The temperature at which each alarm trip occurred is displayed along with the total number of alarm trips.

```
Notification: '
'Current Temp (Farenheit): 68.56 ,
Temperature Exceeded Safe Range ...
NumberHits: 1
Sending Data ....
TCP Connection ID: 0 , DataLength: 16 ,
Data: ALARM_TRIPPED:1
```

```
Notification: '
'Current Temp (Farenheit): 68.56 ,
Temperature Exceeded Safe Range ...
NumberHits: 2
Sending Data ....
TCP Connection ID: 0 , DataLength: 16 ,
Data: ALARM_TRIPPED:2

Current Temp (Farenheit): 68.56 ,
Temperature Exceeded Safe Range ...
NumberHits: 3
Sending Data ....
TCP Connection ID: 0 , DataLength: 16 ,
Data: ALARM_TRIPPED:3

Notification: '
'Current Temp (Farenheit): 68.56 ,
Temperature Exceeded Safe Range ...
NumberHits: 4
Sending Data ....
TCP Connection ID: 0 , DataLength: 16 ,
Data: ALARM_TRIPPED:4

Notification: '
'Current Temp (Farenheit): 68.56 ,
Temperature Exceeded Safe Range ...
NumberHits: 5
Sending Data ....
TCP Connection ID: 0 , DataLength: 16 ,
Data: ALARM_TRIPPED:5
```

Finally, deactivate the alarm using the Android application.

```
'Notification: '
'Notification: '+IPD,0,17:DEACTIVATE_
ALARM
'ClientIndex: 0
DataLength: 17
Data: DEACTIVATE_ALARM

Sending Data ....
TCP Connection ID: 0 , DataLength: 32 ,
Data: Alarm Has Been DE-Activated ...
```

The Photo Resistor

The photo resistor detects the presence of light and acts like a variable resistor in the presence or absence of light. When the resistance of the photo resistor changes, the voltage across the resistor also changes. Based on this property we can determine the amount of light present in the surrounding environment according to the voltage level. A photo resistor has a unique appearance with a square wave looking pattern on the top of the resistor. See Figure 6-11.

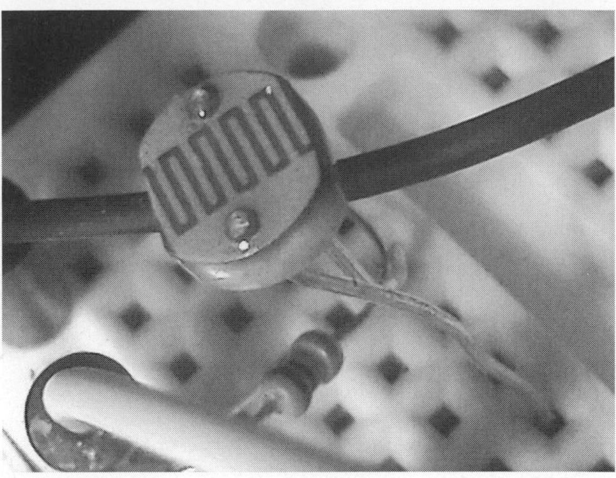

Figure 6-11 The photo resistor.

Recall that the voltage across a resistor is equal to the current multiplied by the resistance such that:

Voltage = Current * Resistance

or

$$V = IR$$

We can determine the amount of light in the environment by measuring the voltage across a resistor that is connected in series to a photo resistor. To measure the voltage we use the Arduino's analog pin to read in the voltage amount that is converted to a number between 0 and 1023 for an input voltage in the range of 0 to 5 V. See Figure 6-12.

Hands-on Example: The Wireless Remote Controlled Light Activated LED

In this hands-on example I show you how to use a photo resistor to control the blinking of an

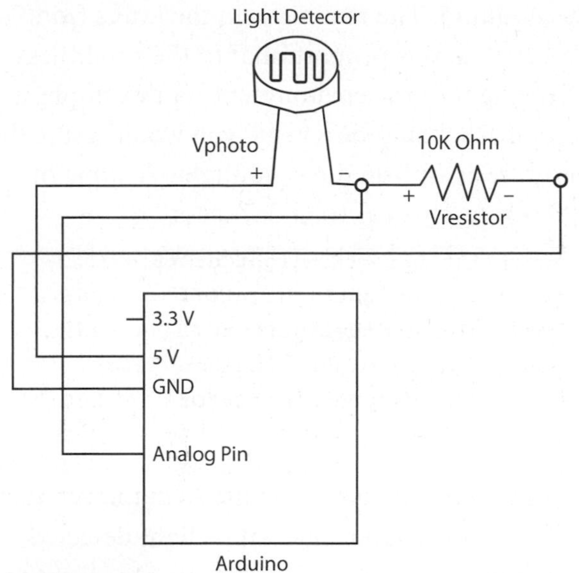

Figure 6-12 Determining the amount of light in the environment using a photo resistor and Arduino.

LED. When the amount of light is low enough, then the LED is allowed to blink if it is set to blinking mode. With this basic framework you can expand it to include your own custom projects such as light activated LED flares and light activated camping lights and beacons.

Parts List

- 1 Android cell phone
- 1 Arduino Mega 2560 microcontroller
- 1 5- to 3.3-V step-down voltage regulator
- 1 Logic level converter from 5 to 3.3 V
- 1 ESP-01 Wi-Fi module
- 1 Arduino development station such as a desktop or a notebook
- 1 Breadboard
- 1 LED
- 1 Photo resistor
- 1 330-ohm resistor (recommended)
- 1 10K-ohm resistor (recommended)
- 1 Package of wires both male-to-male and female-to-male to connect the components

Setting Up the Hardware

Step-Down Voltage Regulator Connections

1. Connect the Vin pin on the voltage regulator to the 5-V power pin on the Arduino Mega 2560.
2. Connect the Vout pin on the voltage regulator to a node on your breadboard represented by a horizontal row of empty slots. This is the 3.3-V power node that will supply the 3.3-V power to the ESP-01 module as well as provide the reference voltage for the logic level shifter.
3. Connect the GND pin on the voltage regulator to the GND pin on the Arduino or form a GND node similar to the 3.3-V power node in the previous step.

ESP-01 Wi-Fi Module Connections

1. Connect the Tx or TxD pin to the Rx3 pin on the Arduino Mega 2560. This is the receive pin for the serial hardware port 3.
2. Connect the EN or CH_PD pin to the 3.3-V power node from the step-down voltage regulator.
3. Connect the IO16 or RST pin to the 3.3-V power node from the step-down voltage regulator.
4. Connect the 3V3 or VCC pin to the 3.3-V power node from the step-down voltage regulator.
5. Connect the GND pin to the ground pin or node.
6. Connect the Rx or RxD pin to pin B0 on the 3.3-V side of the logic level shifter. The voltage from this pin will be shifted from 5 V from the Arduino to 3.3 V for the ESP-01 module.

Logic Level Shifter

1. Connect the 3.3-V pin to the 3.3-V power node from the step-down voltage regulator.

2. Connect the GND pin on the 3.3-V side to the ground node for the Arduino.
3. Connect the B0 pin on the 3.3-V side to Rx or RxD pin on the ESP-01. This corresponds to the A0 pin on the 5-V side.
4. Connect the 5-V pin to the 5-V power pin on the Arduino.
5. Connect the GND pin on the 5-V side to the ground node for the Arduino.
6. Connect the A0 pin on the 5-V side to Tx3 pin on the Arduino. This pin corresponds to the B0 pin on the 3.3-V side.

Photo Resistor

1. Connect one end of the photo resistor to the 5-V power node on the Arduino.
2. Connect the other end of the photo resistor to a node that is connected to both analog pin 7 and one end of a resistor.
3. Connect the other end of the resistor to the ground node for the circuit.

LED

1. Connect the positive terminal of the LED to one end of a resistor (few hundred ohms recommended to dim brightness if desired).
2. Connect the other end of the resistor to digital pin 7.
3. Connect the negative terminal of the LED to the ground node for the circuit.

See Figure 6-13.

Setting Up the Software

The Arduino code for this hands-on example is based on the code for the LED example from Chapter 5. In this section I will cover the changes made to the LED Arduino code.

To measure the amount of light we perform an `analogRead()` on the analog A7 pin on the Arduino. The reading is in the range from 0 to 1023 and is proportional to the brightness of the light in the environment. For example, a value 1023 would be a value you would get if the photo resistor is in direct sunlight. A value of 100 is roughly near total darkness.

```
const int DirectSunlightCutOff = 1023;
const int BrightLightCutOff   = 900;
const int MediumLightCutOff   = 525;
const int LowLightCutOff   = 310;
const int VeryLowLightCutOff   = 200;
const int NightCutOff = 100;
```

The `SensorPin` variable is the Arduino pin that is used to determine amount of light detected by the photo resistor. The default value is set to analog pin 7 on the Arduino.

```
int SensorPin = A7;
```

The SensorValue variable holds the value read in from the Sensor Pin and is initialized to 0.

```
int SensorValue = 0;
```

The `MaxLightActive` variable holds the maximum value of the light intensity where the LED will still turn on. A value greater than the `MaxLightActive` value will shut off the LED. The default value is set to a low level of light which is 310.

```
int MaxLightActive = LowLightCutOff;
```

The `SetupSensorAndHardware()` Function

The `SetupSensorAndHardware()` function has been modified to set the `SensorPin` as an `INPUT` pin so that a voltage level can be read in. See Listing 6-15.

Listing 6-15 The `SetupSensorAndHardware()` Function

```
void SetupSensorAndHardware()
{
 pinMode(LEDPin, OUTPUT);
 pinMode(SensorPin, INPUT);
}
```

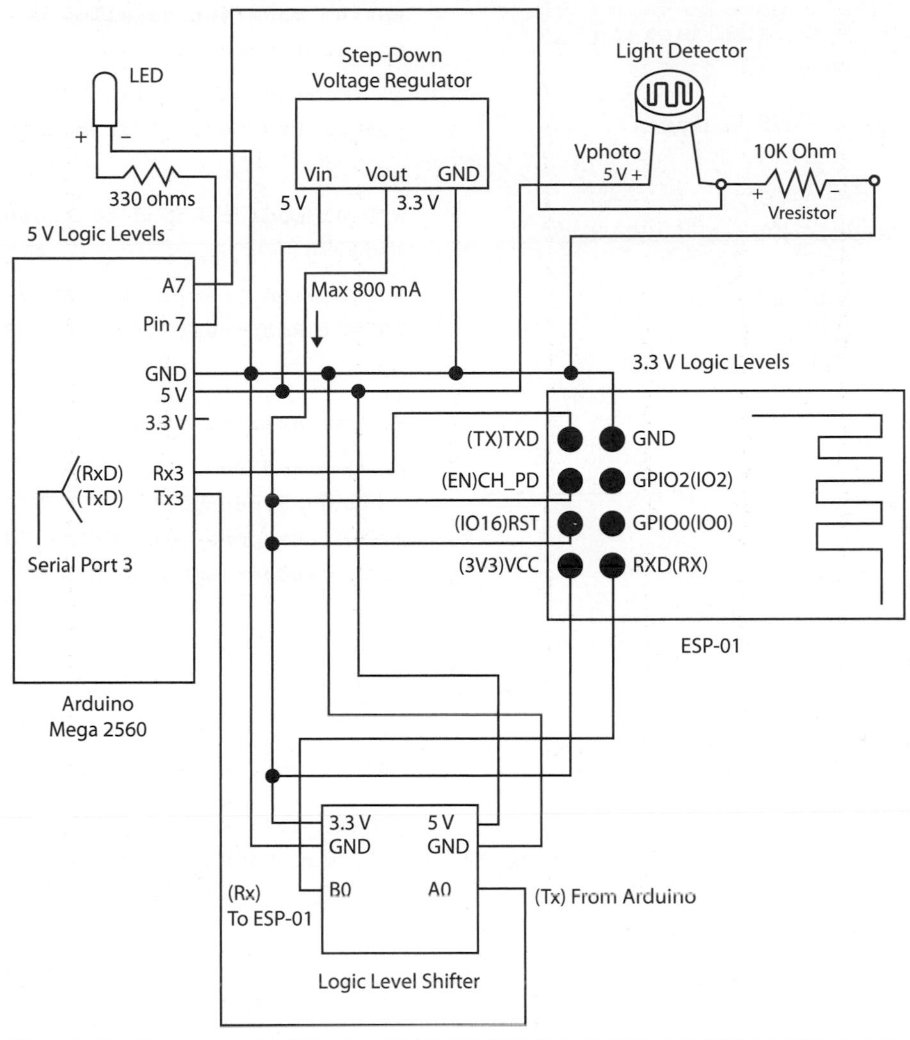

Figure 6-13 The light activated wireless LED control system.

The `ProcessSensorAndHardware()` Function

The `ProcessSensorAndHardware()` function controls the blinking of the LED by doing the following:

1. Finding the intensity of light in the environment.
2. If this value is greater than the maximum intensity of light for the operation of the LED, then turn off the LED and exit the function.
3. If the blink feature of the LED is turned on, then perform the blink function by calling the `PerformLEDBlink()` function.

See Listing 6-16. The additional code is highlighted in bold print.

The Light Level Command

The "`ll`" command retrieves the light intensity of the environment from the photo resistor and sends it to the Android device over the TCP connection. See Listing 6-17.

Operating the System

This example uses the Android basic framework version 1.1 application from a previous chapter that you should have already installed on your Android cell phone. Note that the Android

Listing 6-16 The `ProcessSensorAndHardware()` Function

```
void ProcessSensorAndHardware()
{
 SensorValue = analogRead(SensorPin);
 if (SensorValue > MaxLightActive)
 {
  digitalWrite(LEDPin, LOW);
  LEDOn = false;
  return;
 }
 // Put Code to process your sensors
 or other hardware in this function.
 if (Blink)
 {
  PerformLEDBlink();
 }
}
```

Listing 6-17 The `ll` Command

```
if (Data == "ll\n")
{
 SensorValue = analogRead(SensorPin);

 String returnstring = "";
 returnstring = returnstring + "Light
Level: " + SensorValue + "\n";
 SendTCPMessage(ClientIndex,
returnstring);
}
```

APK install files for all the chapters are located in the downloads section of Chapter 9. The new software you will need to install is the Arduino code related to the photo resistor from this chapter. After installing the program to the Arduino, start up the Serial Monitor. You should see something like the following:

```
****** ESP-01 Arduino Mega 2560 Server
Program ********
*********** Photo Resistor LED Control
Program *************

AT Command Sent: AT+CIPMUX=1
ESP-01 Response: AT+CIPMUX=1

OK

ESP-01 mode set to allow multiple
connections.

AT Command Sent: AT+CWMODE_CUR=2
ESP-01 Response: AT+CWMODE_CUR=2

OK

ESP-01 mode changed to Access Point
Mode.

AT Command Sent: AT+CIPSERVER=1,80
ESP-01 Response: AT+CIPSERVER=1,80

OK

Server Started on ESP-01.

AT Command Sent: AT+CWSAP_CUR="ESP-
01","esp8266esp01",1,2,4,0
ESP-01 Response: AT+CWSAP_CUR="ESP-
01","esp8266esp01",1,2,4,0

OK

AP Configuration Set Successfullly.

AT Command Sent: AT+CWSAP_CUR?
ESP-01 Response: AT+CWSAP_CUR?

+CWSAP_CUR:"ESP-
01","esp8266esp01",1,2,4,0

OK

This is the curent configuration of the
AP on the ESP-01.

AT Command Sent: AT+CIPSTO?
ESP-01 Response: AT+CIPSTO?

+CIPSTO:180

OK

This is the curent TCP timeout in
seconds on the ESP-01.

******** System Has Started ********
```

Next, connect your Android device to the access point running on the ESP-01 module. Start up the Android application. The application will connect to the TCP server running on the ESP-01 module and change the name of the Android device on the wireless network to the one set by the user which in my case is "`lg`". The Arduino will send back an "OK" text string to confirm that the name change has taken place.

```
Notification: '0,CONNECT'
Processing New Client Connection ...
NewConnection Name: 0

Notification: '
'Notification: '
'Notification: '+IPD,0,8:name=lg
'ClientIndex: 0
DataLength: 8
Data: name=lg

Client Name Change Requested,
ClientIndex: 0, NewClientName: lg
Client Name Change Succeeded, OK...
Sending Data ....
TCP Connection ID: 0 , DataLength: 3 ,
Data: OK
```

Next, let's use the light level command to get the current level of light in the environment. Type in "`ll`" into the phone entry text box in the android application and send this command to the Arduino. The Arduino will receive this command and send back the current light level that the photo resistor is detecting. In my case I issued the light level command twice and received values of 163 and 167 which indicate a very low level of light.

```
Notification: '
'Notification: '
'Notification: '+IPD,0,3:ll
'ClientIndex: 0
DataLength: 3
Data: ll

Sending Data ....
TCP Connection ID: 0 , DataLength: 17 ,
Data: Light Level: 163

Notification: '
'Notification: '
'Notification: '+IPD,0,3:ll
'ClientIndex: 0
DataLength: 3
Data: ll

Sending Data ....
TCP Connection ID: 0 , DataLength: 17 ,
Data: Light Level: 167
```

Next, let's turn on the LED blinking function by sending a "`blink`" command to the Arduino using the Android application. The Arduino should receive the blink command, turn on LED blinking, and send a response text to the Android indicating that the blinking has started.

```
Notification: '
'Notification: '
'Notification: '+IPD,0,6:blink
'ClientIndex: 0
DataLength: 6
Data: blink

Sending Data ....
TCP Connection ID: 0 , DataLength: 24 ,
Data: LED Blinking Started...
```

Next, shine a light on or near the photo resistor. The light should illuminate the area surrounding the photo resistor causing the LED to shut off and stop blinking if the level of light is greater than low level light which is around 310. Issue a light level command to determine the intensity of the light detected by the photo resistor. I did this twice and found that the light levels were 547 and 559 which are both above the cutoff level of 310.

```
Notification: '
'Notification: '
'Notification: '+IPD,0,3:ll
'ClientIndex: 0
DataLength: 3
Data: ll

Sending Data ....
TCP Connection ID: 0 , DataLength: 17 ,
Data: Light Level: 547

Notification: '
'Notification: '
'Notification: '+IPD,0,3:ll
'ClientIndex: 0
DataLength: 3
Data: ll

Sending Data ....
TCP Connection ID: 0 , DataLength: 17 ,
Data: Light Level: 559
```

Next, turn off the light. The LED should begin to blink again. Issue another light level command. The light level should again be below the light cutoff value which in my case was 161.

```
Notification: '
'Notification: '
```

```
'Notification: '+IPD,0,3:11
'ClientIndex: 0
DataLength: 3
Data: 11
Sending Data ....
TCP Connection ID: 0 , DataLength: 17 ,
Data: Light Level: 161
```

Also note that you may have to bend the photo resistor away from the LED so that the light from the LED does not affect the measured light value.

Figure 6-14 The Ywrobot DHT11 temperature/humidity sensor.

The DHT11 Temperature/Humidity Sensor

The DHT11 is a digital sensor that measures temperature and humidity. See Figure 6-14.

Operational Information

The DHT11 can operate on both 3.3 and 5 V. The data from the sensor consists of 40 bits or 5 bytes of continuous data. The data consists of the integer part of the relative humidity followed by the decimal part of the relative humidity, followed by the integer part of the temperature in Celsius, followed by the decimal part of the temperature, and finally the checksum. The checksum is the last 8 bits of the sum of all the previous 4 bytes of data. See Figure 6-15.

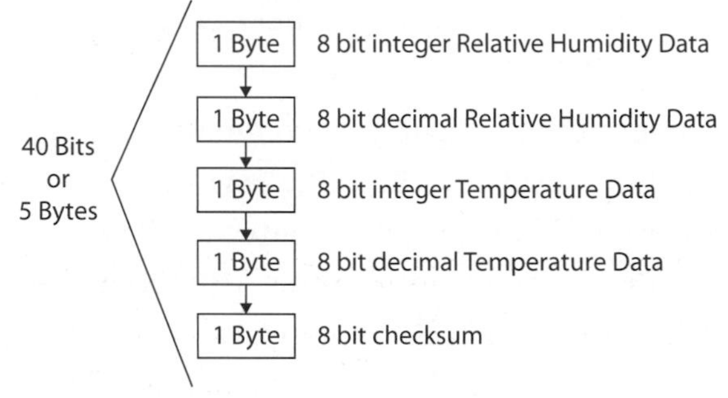

Figure 6-15 The DHT11 humidity/temperature data structure.

In addition, the data comes in with the most significant bit from each byte first. So a good graphical representation of the incoming data stream would be Figure 6-16.

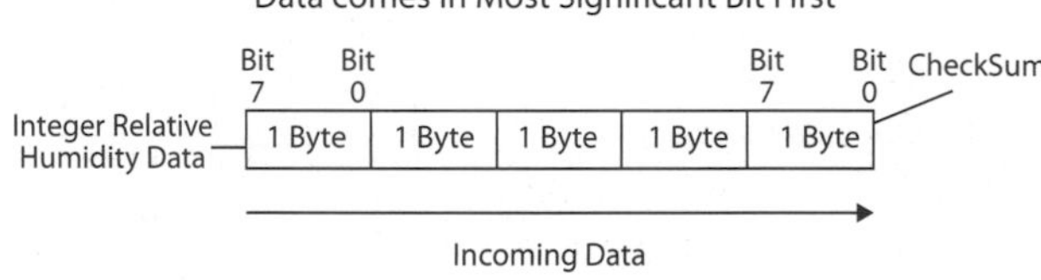

Figure 6-16 The DHT11 incoming data order.

Requesting Data from the DHT11 Sensor

In order to retrieve temperature and humidity data from the DHT11 sensor, you will need to do the following:

1. Send a Request to Retrieve Data from the DHT11 – Have your microcontroller such as your Arduino set the voltage on the sensor's data pin to GND or 0 for at least 18 milliseconds. Then set the voltage on the data pin to HIGH or 1 and hold this for 20 to 40 microseconds. This will signal the DHT11 sensor that it needs to provide temperature and humidity data. See Figure 6-17.
2. Read in the response from the DHT11 sensor to make sure that the DHT11 has acknowledged the data request. The DHT11 sets the voltage low on the data pin for 80 microseconds and then sets the voltage high

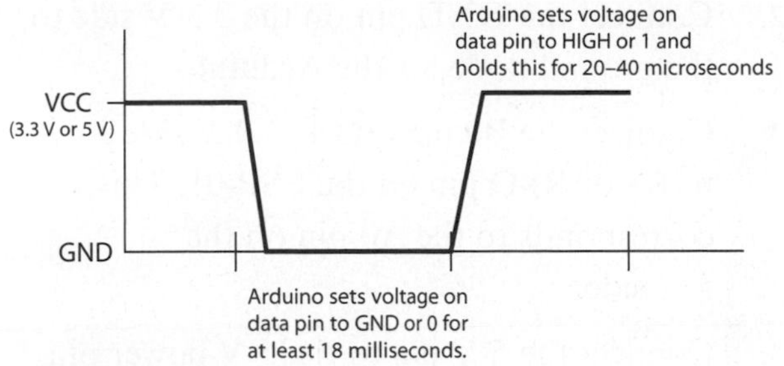

Figure 6-17 Requesting a data reading from the DHT11.

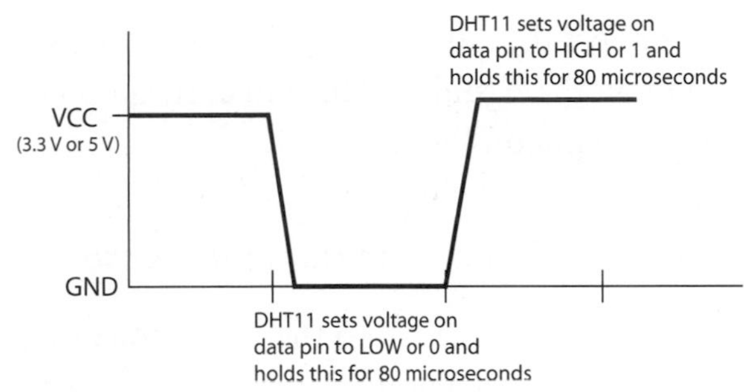

Figure 6-18 The DHT11's response to the data request.

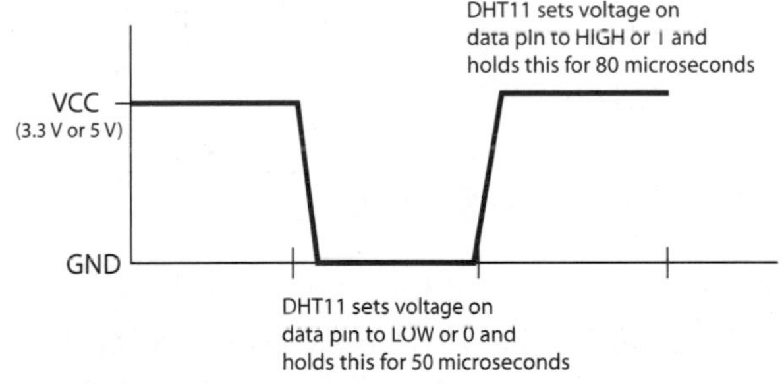

Figure 6-19 The DHT11 sending a "0" data bit.

on the data pin for 80 microseconds. See Figure 6-18.

3. Read in the data which is 5 bytes total one bit at a time. A 0 bit is indicated by the DH11 setting the voltage low on the data pin and holding it there for 50 microseconds followed by setting the voltage high and holding it there for 26 to 28 microseconds. See Figure 6-19.

A 1 data bit is generated by the DHT11 by setting the voltage low on the data pin for 50 microseconds and then setting the voltage high on the data pin for 70 microseconds. See Figure 6-20.

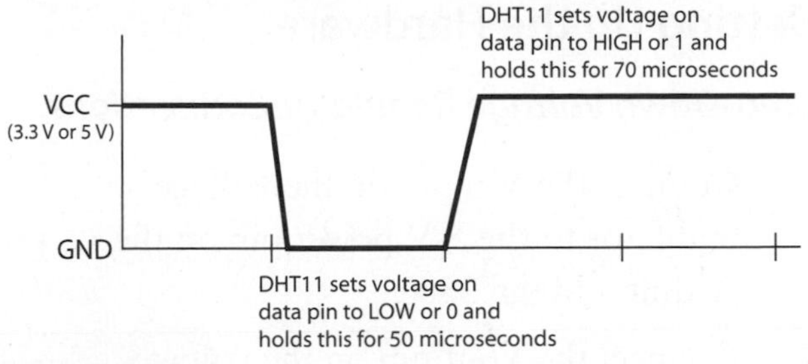

Figure 6-20 The DHT11 sending a "1" data bit.

Hands-on Example: The Wireless DHT11 Temperature/ Humidity Remote Monitoring and Alarm System

In this hands-on example I show you how to build a wireless temperature and humidity monitoring and alarm system that can be accessed remotely. The temperature and humidity sensor is the DHT11 sensor that operates using the Arduino's digital pins instead of Arduino's analog pins as in the hands-on example involving the TMP36 sensor previously discussed in this chapter.

Parts List

- 1 Android cell phone
- 1 Arduino Mega 2560 microcontroller
- 1 5- to 3.3-V step-down voltage regulator
- 1 Logic level converter from 5 to 3.3 V
- 1 ESP-01 Wi-Fi module
- 1 Arduino development station such as a desktop or a notebook
- 1 Breadboard
- 1 DHT11 temperature and humidity sensor
- 1 Package of wires both male-to-male and female-to-male to connect the components

Setting Up the Hardware

Step-Down Voltage Regulator Connections

1. Connect the Vin pin on the voltage regulator to the 5-V power pin on the Arduino Mega 2560.
2. Connect the Vout pin on the voltage regulator to a node on your breadboard represented by a horizontal row of empty slots. This is the 3.3-V power node that will supply the 3.3-V power to the ESP-01 module as well as provide the reference voltage for the logic level shifter.
3. Connect the GND pin on the voltage regulator to the GND pin on the Arduino or form a GND node similar to the 3.3-V power node in the previous step.

ESP-01 Wi-Fi Module Connections

1. Connect the Tx or TxD pin to the Rx3 pin on the Arduino Mega 2560. This is the receive pin for the serial hardware port 3.
2. Connect the EN or CH_PD pin to the 3.3-V power node from the step-down voltage regulator.
3. Connect the IO16 or RST pin to the 3.3-V power node from the step-down voltage regulator.
4. Connect the 3V3 or VCC pin to the 3.3-V power node from the step-down voltage regulator.
5. Connect the GND pin to the ground pin or node.
6. Connect the Rx or RxD pin to pin B0 on the 3.3-V side of the logic level shifter. The voltage from this pin will be shifted from 5 V from the Arduino to 3.3 V for the ESP-01 module.

Logic Level Shifter

1. Connect the 3.3-V pin to the 3.3-V power node from the step-down voltage regulator.
2. Connect the GND pin on the 3.3-V side to the ground node for the Arduino.
3. Connect the B0 pin on the 3.3-V side to Rx or RxD pin on the ESP-01. This corresponds to the A0 pin on the 5-V side.
4. Connect the 5-V pin to the 5-V power pin on the Arduino.
5. Connect the GND pin on the 5-V side to the ground node for the Arduino.
6. Connect the A0 pin on the 5-V side to Tx3 pin on the Arduino. This pin corresponds to the B0 pin on the 3.3-V side.

DHT11 Temperature and Humidity Sensor

1. Connect the Vcc pin to the 3.3-V node from the voltage regulator.
2. Connect the GND pin to the ground node for the circuit.
3. Connect the DAT pin to digital pin 7 on the Arduino.

See Figure 6-21.

Setting Up the Software

The Arduino code for this example is based on the code for the TMP36 hands-on example presented earlier in this chapter. In this section I will cover only the new code that is specific to the new DHT11 temperature and humidity sensor.

The `SensorPin` variable holds the Arduino pin number that is connected to the DHT11's DAT pin which is used to both trigger and read in temperature and humidity readings. The default value is set to digital pin 7 on the Arduino.

```
int SensorPin = 7;
```

The `NumEntries` variable holds the number of bytes of an incoming temperature/humidity reading and is set to 5 bytes.

```
const int NumEntries = 5;
```

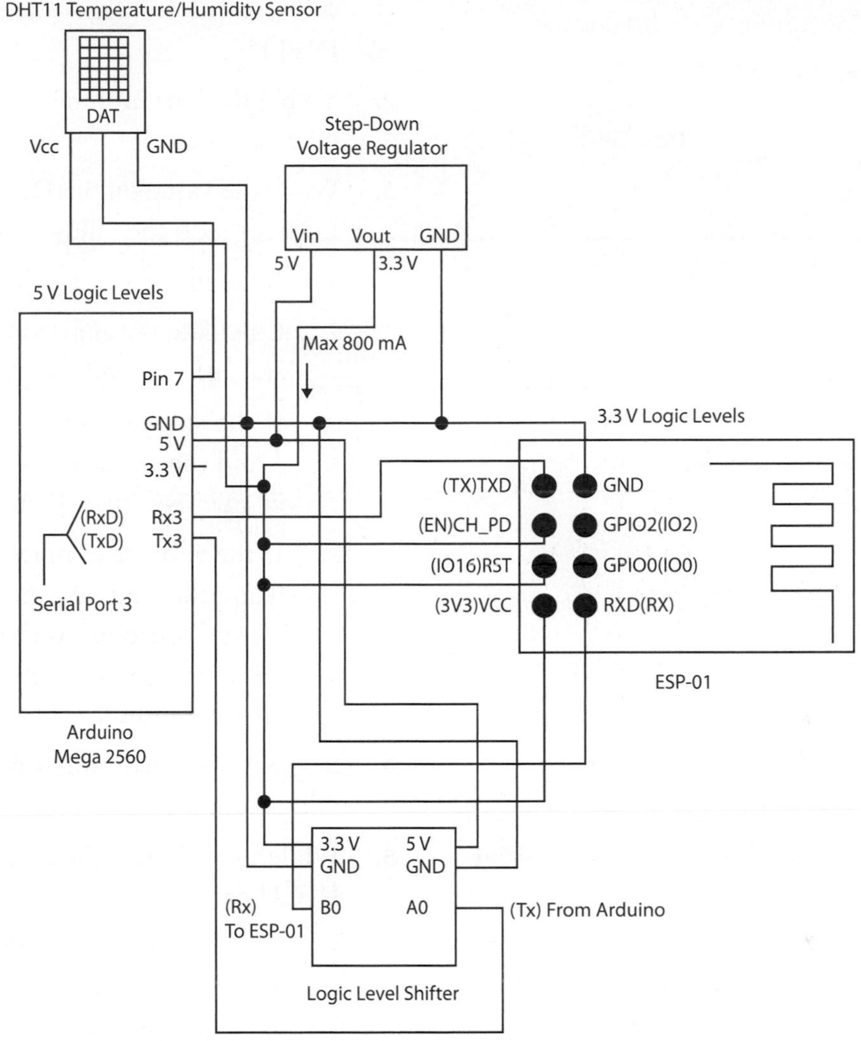

Figure 6-21 The wireless DHT11 temperature/humidity remote monitoring and alarm system.

The `TotalDataBits` variable holds the total number of bits in an incoming temperature/humidity reading and it is set to 40 bits.

```
const int TotalDataBits = 40;
```

The incoming data from the DHT11 is held in the `Data` array with each element in the array representing 1 byte from the temperature/humidity data.

```
byte Data[NumEntries];
```

The `TempC` variable holds the temperature reading in Celsius.

```
float TempC = 0;
```

The `TempF` variable holds the temperature reading in Fahrenheit.

```
float TempF = 0;
```

The `Humidity` variable holds the humidity reading from the DHT11.

```
float Humidity = 0;
```

The `ResetData()` Function

The `ResetData()` function sets all of the elements in the `Data` array to 0. See Listing 6-18.

The `StartSignalToDHT11()` Function

The `StartSignalToDHT11()` function requests data from the DHT11 by doing the following:

1. Setting the pin on the Arduino that is connected to the data pin on the DHT11 to

Listing 6-18 The `ResetData()` Function

```
void ResetData()
{
 for(int i = 0; i < NumEntries; i++)
 {
  Data[i] = 0;
 }
}
```

Listing 6-19 The `StartSignalToDHT11()` Function

```
void StartSignalToDHT11()
{
 pinMode(SensorPin, OUTPUT);

 digitalWrite(SensorPin, LOW);
 delay(18);
 digitalWrite(SensorPin, HIGH);
 delayMicroseconds(40);
}
```

be an output pin so that a voltage signal can be output to the DHT11. This is done using the `pinMode()` function.

2. Setting the voltage on the pin connected to the data pin on the DHT11 to LOW or 0 using the `digitalWrite()` function.
3. Holding the LOW voltage value on the pin by suspending program execution for 18 milliseconds by calling the `delay()` function.
4. Setting the voltage on the data pin to HIGH by calling the `digitalWrite()` function.
5. Holding the voltage on the data pin HIGH by suspending the execution of the program for 40 microseconds.

See Listing 6-19.

The `ReadDHT11StartSignal()` Function

The `ReadDHT11StartSignal()` function reads in the DHT11 response to the start signal sent by the Arduino by doing the following:

1. Setting the Arduino pin connected to the data pin on the DHT11 as an input pin that will read the voltages generated from the DHT11.
2. Saving the current time since start up in microseconds.
3. While the value on the DHT11 data pin is LOW or 0 do the following:
   1. Read in the value of the data pin.
   2. Calculate the elapsed time since the start of the while loop.
   3. If the value read in from the data pin is 1 and the elapsed time is greater than 70 microseconds, then exit the loop.
   4. If the value read in from the data pin is 0 and the elapsed time is greater than 90 microseconds, then exit the function by returning false which indicates an error condition.
4. Saving the current time since start up in microseconds.
5. While the value on the DHT11 data pin is HIGH or 1 do the following:
   1. Read in the value of the data pin.
   2. Calculate the elapsed time since the start of the while loop.
   3. If the value of the data pin is LOW or 0 and the elapsed time is greater than 70 microseconds. then exit the while loop.
   4. If the value of the data pin is HIGH or 1 and the elapsed time is greater than 90 microseconds, then return false which indicates an error condition.
6. Returning true if the DHT11 responded correctly or false if there was an error condition.

See Listing 6-20.

The `ConvertCelsius2Fahrenheit()` Function

The `ConvertCelsius2Fahrenheit()` function converts a temperature in Celsius into a temperature in Fahrenheit. See Listing 6-21.

Listing 6-20 The `ReadDHT11StartSignal()` Function

```
boolean ReadDHT11StartSignal()
{
 // Returns true if DHT sends a low
 voltage for 80us and
 // then a high voltage for 80us then
 goes low.
 // Returns false otherwise.

 boolean result = true;
 boolean done = false;
 unsigned long StartTime = 0;
 unsigned long CurrentTime = 0;
 unsigned long ElapsedTime = 0;

 // Change pin to read in data from
 DHT11
 pinMode(SensorPin, INPUT);

 // Wait until value goes from low to
 high after 80us or times out
 StartTime = micros();
 done = false;
 while(!done)
 {
  // Read in pin value
  SensorValue =
  digitalRead(SensorPin);
  CurrentTime = micros();
  ElapsedTime = CurrentTime -
  StartTime;
  if (SensorValue == 1)
  {
   if (ElapsedTime > 70)
   {
    done = true;
   }
  }
  else
  {
   // Check Timeout
   if(ElapsedTime > 90)
   {
   // Response Takes Too Long
   return false;
   }
  }
 }
 // Wait until value goes from high
 for 80us to low or times out.
 StartTime = micros();
 done = false;
 while(!done)
 {
  // Read in pin value
  SensorValue =
  digitalRead(SensorPin);
  CurrentTime = micros();
  ElapsedTime = CurrentTime -
  StartTime;
  if (SensorValue == 0)
  {
   if (ElapsedTime > 70)
   {
    done = true;
   }
  }
  else
  {
   // Check Timeout
   if(ElapsedTime > 90)
   {
    // Response Takes Too Long
    return false;
   }
  }
 }
 return result;
}
```

Listing 6-21 The `ConvertCelsius2Fahrenheit()` Function

```
float ConvertCelsius2Fahrenheit(float
TempC)
{
 float temp = 0;

 temp = TempC *(9.0/5.0) + 32.0;
 return temp;
}
```

The `ReadDataEntryDHT11()` Function

The `ReadDataEntryDHT11()` function reads in the 40 bits or 5 bytes of incoming data from the

DHT11 by doing the following:

1. For each of the 40 incoming bits do the following:
    1. Read in the data from the DHT11's data pin until a value of 1 or HIGH is found. The duration of this HIGH value will indicate if this data bit is a 1 or 0. If the HIGH value holds for around 60 to 70 microseconds, then the incoming bit is a 1 and 0 otherwise.
    2. Save the current time since power up in microseconds.
    3. While the value on the DHT11's data pin is 1 or HIGH, do the following:
        1. Read in the value of the DHT11's data pin.
        2. If the value is 0, then do the following:
            1. Find the elapsed time since the start of the while loop.
            2. If the elapsed time is greater than 60 microseconds, then the incoming bit is a 1 so put a 1 in the appropriate bit position in the `Data` array based on the `BitPos` and `BytePos` variables. Remember there are 8 bits in a byte and 5 bytes total in the data we are reading in. The data is coming in most significant bit first (bit 7) for each byte.
            3. If the current `BitPos` is 0 that means the current byte has been read in and we need to go to the next byte. To do this we reset the `BitPos` to 7 which is bit 7 and increase the value of `BytePos` to indicate that we will be reading in data for the next byte.
            4. If the current `BitPos` is not 0, then decrease the `BitPos` by 1.
2. Calculate the checksum as the last 8 bits of the sum of the first 4 bytes of the data that was just read in.
3. If the checksum calculated in step 2 is equal to the checksum byte which was the final byte read in then the data has been validated. If the calculated checksum does not match the checksum byte, then the incoming data was corrupted.
4. Return the checksum validation status to the calling function.

See Listing 6-22.

Listing 6-22 The `ReadDataEntryDHT11()` Function

```
boolean ReadDataEntryDHT11()
{
 boolean result = false;
 byte CheckSum = 0;
 int BytePos = 0;
 int BitPos = 7;
 boolean done = false;
 unsigned long StartTime = 0;
 unsigned long CurrentTime = 0;
 unsigned long ElapsedTime = 0;
 byte temp = 0;

 // Read in all 5 bytes of data 1 bit
 at a time
 // Most significant bit is read in
 first.
 for (int i = 0; i < TotalDataBits; i++)
 {
  // Wait for data bit
  done = false;
  while(!done)
  {
   SensorValue =
  digitalRead(SensorPin);
   if (SensorValue == 1)
   {
    done = true;
   }
  }

  // Read in data bit
  done = false;
```

```
    StartTime = micros();
    while(!done)
    {
     SensorValue =
    digitalRead(SensorPin);
     if (SensorValue == 0)
     {
      CurrentTime = micros();
      ElapsedTime = CurrentTime -
     StartTime;
      if (ElapsedTime > 60)
      {
       // Data is 1
       temp = (1 << BitPos);
       Data[BytePos]= Data[BytePos]|temp;
      }

      // Update Bit and Byte Positions
      if(BitPos == 0)
      {
       BitPos = 7;
       BytePos++;
      }
      else
      {
       BitPos--;
      }
      done = true;
     }
    }
  }

  // Calculate Checksum
  CheckSum = Data[0] + Data[1] + Data[2]
  + Data[3];
  if (CheckSum == Data[4])
  {
   result = true;
  }

  return result;
  }
```

The `MergeFraction()` *Function*

The `MergeFraction()` function merges two numbers which represent the whole part of a number and the fraction part of that number into a single floating point number. See Listing 6-23.

Listing 6-23 The `MergeFraction()` Function

```
float MergeFraction(byte whole, byte
fraction)
{
 float result = 0.0f;
 String WholePart = "";
 String FractionPart = "";
 String FinalNumber = "";

 WholePart = String(whole);
 FractionPart = String(fraction);
 FinalNumber = WholePart + "." +
FractionPart;
 result = FinalNumber.toFloat();

 return result;
}
```

The `ReadDHT11()` *Function*

The `ReadDHT11()` function reads in the temperature and humidity from the DHT11 by doing the following:

1. Resets all the bytes of data in the `Data` array to 0 by calling the `ResetData()` function.
2. Sends the request for a temperature/humidity data reading to the DHT11 sensor from the Arduino by calling the `StartSignalToDHT11()` function.
3. Reads in the response signal from the DHT11 that indicates that the DHT11 has received the Arduino's request and is ready to send the data by calling the `ReadDHT11StartSignal()` function.
4. If the return value from step 3 is false, then print out an error message to the Serial Monitor indicating that the DHT11's response to the Arduino's start signal failed.
5. If the return value from step 3 is true, then read in the 5 byte or 40 bit data entry from the DHT11 that holds the

temperature and humidity data by calling the `ReadDataEntryDHT11()` function.

6. If the return value from step 5 is false, then print out a text message to the Serial Monitor that there was an error in the checksum. Otherwise print out a text message saying that the checksum was ok.
7. Constructs the Celsius temperature and humidity from the incoming data and calculates the Fahrenheit temperature from the Celsius temperature.
8. Prints out the temperature, humidity, and the checksum status to the Serial Monitor.

See Listing 6-24.

Listing 6-24 The `ReadDHT11()` Function

```
boolean ReadDHT11()
{
 boolean result = false;

 // Reset data structures that will hold incoming data from the
 // DHT11 to 0.
 ResetData();

 // Send start signal to the DHT11 sensor
 StartSignalToDHT11();

 // Read in the DHT response to the start signal
 // to indicate that the sensor is ready to send data
 result = ReadDHT11StartSignal();
 if (result == false)
 {
  Serial.println("DHT11 START SIGNAL RESPONSE FAILURE ....");
 }

 // Read in 40 bits or the 5 byte data entry sent by the DHT that represents the
 // relative humidity and the temperature.
 result = ReadDataEntryDHT11();
 if (result == false)
 {
  Serial.print("CheckSum Error .... ");
 }
 else
 {
  Serial.print("CheckSum Ok ... ");
 }

 // Process Data Array for final values of temperature and humidity
 TempC = MergeFraction(Data[2], Data[3]);
 TempF = ConvertCelsius2Fahrenheit(TempC);
 Humidity = MergeFraction(Data[0], Data[1]);

 // Print out Final Temperatures to the Serial Monitor.
 Serial.print(F("TempF: "));
 Serial.print(TempF);
 Serial.print(" , ");
 Serial.print("Humidity: ");
 Serial.println(Humidity);

 return result;
}
```

The `ReadTempF()` Function

The `ReadTempF()` function reads the temperature/humidity value from the DHT11 sensor by calling the `ReadDHT11()` function. If there is no checksum error, then return the Fahrenheit value of the temperature. Otherwise, return a –300 value which indicates that the temperature reading failed. See Listing 6-25.

Listing 6-25 The `ReadTempF()` Function

```
float ReadTempF()
{
 float Temp = -300.0f;
 if (ReadDHT11())
 {
  Temp = TempF;
 }
 return Temp;
}
```

The `ReadTempC()` Function

The `ReadTempC()` function retrieves the temperature and humidity from the DHT11 sensor by calling the `ReadDHT11()` function. If the checksum value is good, then the returned temperature is set to the temperature in Celsius. Otherwise, a –300 value is returned to the calling function which indicates that the temperature reading has failed. See Listing 6-26.

Listing 6-26 The `ReadTempC()` Function

```
float ReadTempC()
{
 float Temp = -300.0f;
 if (ReadDHT11())
 {
  Temp = TempC;
 }
 return Temp;
}
```

The `ReadHumidity()` Function

The `ReadHumidity()` function reads in the temperature and humidity from the DHT11 sensor by calling the `ReadDHT11()` function. If the checksum value from the operation is good, then the returned humidity value is set to the humidity read in from the sensor. Otherwise, a value of 300 is returned to indicate that the operation has failed. See Listing 6-27.

Listing 6-27 The `ReadHumidity()` Function

```
float ReadHumidity()
{
 float H = -300.0f;
 if (ReadDHT11())
 {
  H = Humidity;
 }
 return H;
}
```

The `ProcessSensor()` Function

The `ProcessSensor()` function processes the reading in of the temperature and the issuing of alerts to the Android device and is the same as the code for the TMP36 sensor except for the following changes:

1. The temperature is read in from the DHT11 sensor by calling the `ReadTempF()` function.
2. In addition to the temperature range being checked to determine whether it is within the acceptable temperature range, the temperature value itself is checked for validity. For example, a valid temperature value will be greater than –200 degrees Fahrenheit.

See Listing 6-28. The code modifications have been highlighted in bold print.

The `tempf` Command

The `tempf` command retrieves the current temperature in Fahrenheit from the DHT11 sensor by calling the `ReadTempF()` function.

Listing 6-28 The `ProcessSensor()` Function

```
void ProcessSensor()
{
 // Put Code to process your sensors or other hardware in this function.
 if ((AlarmState == Alarm_ON) || (AlarmState == Alarm_TRIPPED))
 {
  unsigned long ElapsedTime = millis() - MinDelayBetweenReadingsStartTime;
  if (ElapsedTime >= MinDelayBetweenReadings)
  {
   float Temp = ReadTempF();
   if (((Temp < MinTemp) || (Temp > MaxTemp)) && (Temp > -200))
   {
    // Temp Outside Safe Range Just Detected
    NumberHits++;
    Serial.print(F("Current Temp (Farenheit): "));
    Serial.print(Temp);
    Serial.print(F(" , Temperature Exceeded Safe Range ... NumberHits: "));
    Serial.println(NumberHits);

    // Send Alarm Tripped Message
    SendAlarmTrippedMessage();
   }
   MinDelayBetweenReadingsStartTime = millis();
  }
 }
}
```

Listing 6-29 The `tempf` Command

```
if (Data == "tempf\n")
{
 float Temp = ReadTempF();

 String returnstring = "";
 returnstring = returnstring + F("TEMP
(Farenheit): ") + String(Temp) + "\n";
 SendTCPMessage(ClientIndex,
returnstring);
}
```

The temperature is then sent to the Android device by calling the `SendTCPMessage()` function. See Listing 6-29.

The `tempc` Command

The `tempc` command retrieves the temperature in Celsius from the DHT11 sensor by calling the `ReadTempC()` function. The temperature is then sent to the Android device by calling the `SendTCPMessage()` function. See Listing 6-30.

Listing 6-30 The tempc Command

```
if (Data == "tempc\n")
{
 float Temp = ReadTempC();

 String returnstring = "";
 returnstring = returnstring + F("TEMP
(Celsius): ") + String(Temp) + "\n";
 SendTCPMessage(ClientIndex,
returnstring);
}
```

The `humidity` Command

The `humidity` command reads in the humidity from the DHT11 sensor by calling

Listing 6-31 The `humidity` Command

```
if (Data == "humidity\n")
{
 float H = ReadHumidity();

 String returnstring = "";
 returnstring = returnstring +
F("Humidity: ") + String(H) + "\n";
 SendTCPMessage(ClientIndex,
returnstring);
}
```

the `ReadHumidity()` function. The humidity value is then returned to the Android device by calling the `SendTCPMessage()` function. See Listing 6-31.

Operating the Wireless DHT Temperature/Humidity Remote Monitoring and Alarm System

Download and install on your Android cell phone the basic wireless framework version 1.1 if you have not already done so. Note that the Android APK install files for all the chapters are located in the downloads section of Chapter 9. Download and install on your Arduino the program for this chapter related to the DHT11 temperature and humidity sensor. Start up the Serial Monitor and you should see something like the following:

```
****** ESP-01 Arduino Mega 2560 Server
Program ********
***********  DHT11 Sensor Alarm System
*************

AT Command Sent: AT+CIPMUX=1
ESP-01 Response: AT+CIPMUX=1

OK

ESP-01 mode set to allow multiple
connections.

AT Command Sent: AT+CWMODE_CUR=2
ESP-01 Response: AT+CWMODE_CUR=2

OK

ESP-01 mode changed to Access Point
Mode.

AT Command Sent: AT+CIPSERVER=1,80
ESP-01 Response: AT+CIPSERVER=1,80

OK

Server Started on ESP-01.

AT Command Sent: AT+CWSAP_CUR="ESP-
01","esp8266esp01",1,2,4,0
ESP-01 Response: AT+CWSAP_CUR="ESP-
01","esp8266esp01",1,2,4,0

OK

AP Configuration Set Successfullly.

AT Command Sent: AT+CWSAP_CUR?
ESP-01 Response: AT+CWSAP_CUR?

+CWSAP_CUR:"ESP-
01","esp8266esp01",1,2,4,0

OK

This is the curent configuration of the
AP on the ESP-01.

AT Command Sent: AT+CIPSTO?
ESP-01 Response: AT+CIPSTO?

+CIPSTO:180

OK

This is the curent TCP timeout in
seconds on the ESP-01.

******** System Has Started ********
```

Next, connect your Android to the Wi-Fi access point ESP-01 that is running on the ESP-01 module. Start up the Android application. You should automatically be connected to the TCP server running on the ESP-01 module. The Android will send a device name change command to the Arduino. The Arduino should receive the command, change the name of the Android device on the wireless network, and send an "OK" text reply to the Android device.

```
Notification: '0,CONNECT'
Processing New Client Connection ...
NewConnection Name: 0

Notification: '
'Notification: '
```

```
'Notification: '+IPD,0,8:name=lg
'ClientIndex: 0
DataLength: 8
Data: name=lg
Client Name Change Requested,
ClientIndex: 0, NewClientName: lg
Client Name Change Succeeded, OK...
Sending Data ....
TCP Connection ID: 0 , DataLength: 3 ,
Data: OK
```

Next, use the `tempf` command to retrieve the temperature from the DHT11 sensor in Fahrenheit. For me the temperature is 71.60 degrees Fahrenheit. The checksum status, the temperature, and the humidity are also displayed on the Serial Monitor.

```
Notification: '
'Notification: '
'Notification: '+IPD,0,6:tempf
'ClientIndex: 0
DataLength: 6
Data: tempf
CheckSum Ok ... TempF: 71.60 , Humidity:
38.00
Sending Data ....
TCP Connection ID: 0 , DataLength: 24 ,
Data: TEMP (Farenheit): 71.60
```

Next, use the `tempc` command to get the temperature in Celsius from the DHT11 sensor. For me the temperature was 22 degrees Celsius. The checksum status, the temperature in Fahrenheit, and the humidity were also displayed on the Serial Monitor.

```
Notification: '
'Notification: '
'Notification: '+IPD,0,6:tempc
'ClientIndex: 0
DataLength: 6
Data: tempc
CheckSum Ok ... TempF: 71.60 , Humidity:
38.00
Sending Data ....
TCP Connection ID: 0 , DataLength: 22 ,
Data: TEMP (Celsius): 22.00
```

Next, use the `humidity` command to get the amount of humidity. For me the amount was 38 percent.

```
Notification: '
'Notification: '
'Notification: '+IPD,0,9:humidity
'ClientIndex: 0
DataLength: 9
Data: humidity
CheckSum Ok ... TempF: 71.60 , Humidity:
38.00
Sending Data ....
TCP Connection ID: 0 , DataLength: 16 ,
Data: Humidity: 38.00
```

Next, use the `mintemp` command to get the minimum temperature value for the acceptable temperature range. It should return a 40 degrees Fahrenheit value.

```
Notification: '
'Notification: '
'Notification: '+IPD,0,8:mintemp
'ClientIndex: 0
DataLength: 8
Data: mintemp
Sending Data ....
TCP Connection ID: 0 , DataLength: 24 ,
Data: MinTemp (Farenheit): 40
```

Next, use the `maxtemp` command to get the maximum temperature for the acceptable temperature range. The maximum temperature should be 90 degrees Fahrenheit.

```
Notification: '
'Notification: '
'Notification: '+IPD,0,8:maxtemp
'ClientIndex: 0
DataLength: 8
Data: maxtemp
Sending Data ....
TCP Connection ID: 0 , DataLength: 24 ,
Data: MaxTemp (Farenheit): 90
```

Next, turn off the emergency call out and the emergency text messages on the Android application. Activate the alarm using the Android application. The application will send an activate alarm text string to the Arduino. The Arduino receives this, activates the alarm, and sends back an acknowledgment message to the Android device. The alarm should be fully active in 10 seconds.

```
Notification: '
'Notification: '
'Notification: '+IPD,0,15:ACTIVATE_ALARM
'ClientIndex: 0
DataLength: 15
Data: ACTIVATE_ALARM
Sending Data ....
TCP Connection ID: 0 , DataLength: 40
, Data: Alarm Activated...WAIT TIME:10
Seconds.
```

In the about 10 seconds the alarm should be fully active and can be tripped by a temperature reading that is outside the minimum/maximum range of temperatures. An alarm now active text string is sent to the Android device from the Arduino. In addition, the DHT11 continuously reads in the temperature and humidity of the environment every 2.5 seconds. The checksum status, the temperature value in Fahrenheit, and the humidity percentage are also printed out to the Serial Monitor.

```
Notification: '
'Sending Data ....
TCP Connection ID: 0 , DataLength: 20 ,
Data: ALARM NOW ACTIVE...
ALARM WAIT STATE FINISHED ... Alarm Now
ACTIVE ....
CheckSum Ok ... TempF: 71.60 , Humidity:
38.00
Notification: '
'CheckSum Ok ... TempF: 71.60 ,
Humidity: 38.00
CheckSum Ok ... TempF: 71.60 , Humidity:
38.00
CheckSum Ok ... TempF: 71.60 , Humidity:
38.00
CheckSum Ok ... TempF: 71.60 , Humidity:
38.00
CheckSum Ok ... TempF: 71.60 , Humidity:
38.00
CheckSum Ok ... TempF: 71.60 , Humidity:
38.00
CheckSum Ok ... TempF: 71.60 , Humidity:
38.00
CheckSum Ok ... TempF: 71.60 , Humidity:
38.00
CheckSum Ok ... TempF: 71.60 , Humidity:
38.00
CheckSum Ok ... TempF: 71.60 , Humidity:
38.00
CheckSum Ok ... TempF: 71.60 , Humidity:
38.00
CheckSum Ok ... TempF: 71.60 , Humidity:
38.00
CheckSum Ok ... TempF: 71.60 , Humidity:
38.00
CheckSum Ok ... TempF: 71.60 , Humidity:
38.00
```

Next, let's trip the alarm by setting the minimum temperature to a value higher than the one in your current environment. In my case the current temperature is 71.60 degrees Fahrenheit. In order to trip the alarm I will need to set the minimum temperature to a value above 71.60 degrees such as 75 degrees.

```
Notification: '
'Notification: '+IPD,0,11:mintemp=75
'ClientIndex: 0
DataLength: 11
Data: mintemp=75
Sending Data ....
TCP Connection ID: 0 , DataLength: 33 ,
Data: MinTemp set to: 75 Deg Farenheit
```

In a few seconds the alarm should trip and send a text message to your Android device indicating that the alarm has been tripped and the number of times it has been tripped so far. Note also that you may receive an occasional ping command from the Android device that is used to make sure that the TCP connection remains active.

```
Notification: '
'CheckSum Ok ... TempF: 71.60 ,
Humidity: 38.00
Current Temp (Farenheit): 71.60 ,
Temperature Exceeded Safe Range ...
NumberHits: 1
Sending Data ....
TCP Connection ID: 0 , DataLength: 16 ,
Data: ALARM_TRIPPED:1
Notification: '
'CheckSum Ok ... TempF: 71.60 ,
Humidity: 38.00
```

```
Current Temp (Farenheit): 71.60 ,
Temperature Exceeded Safe Range ...
NumberHits: 2
Sending Data ....
TCP Connection ID: 0 , DataLength: 16 ,
Data: ALARM_TRIPPED:2

Notification: '
'Notification: '
'Notification: '+IPD,0,5:ping
'ClientIndex: 0
DataLength: 5
Data: ping

Ping Recieved from Client ...
Sending Data ....
TCP Connection ID: 0 , DataLength: 8 ,
Data: PING_OK

Notification: '
'CheckSum Ok ... TempF: 71.60 ,
Humidity: 38.00
Current Temp (Farenheit): 71.60 ,
Temperature Exceeded Safe Range ...
NumberHits: 3
Sending Data ....
TCP Connection ID: 0 , DataLength: 16 ,
Data: ALARM_TRIPPED:3

Notification: '
'CheckSum Ok ... TempF: 71.60 ,
Humidity: 38.00
Current Temp (Farenheit): 71.60 ,
Temperature Exceeded Safe Range ...
NumberHits: 4
Sending Data ....
TCP Connection ID: 0 , DataLength: 16 ,
Data: ALARM_TRIPPED:4

Notification: '
'CheckSum Ok ... TempF: 71.60 ,
Humidity: 38.00
Current Temp (Farenheit): 71.60 ,
Temperature Exceeded Safe Range ...
NumberHits: 5
Sending Data ....
TCP Connection ID: 0 , DataLength: 16 ,
Data: ALARM_TRIPPED:5

Notification: '
'CheckSum Ok ... TempF: 71.60 ,
Humidity: 38.00
Current Temp (Farenheit): 71.60 ,
Temperature Exceeded Safe Range ...
NumberHits: 6
Sending Data ....
TCP Connection ID: 0 , DataLength: 16 ,
Data: ALARM_TRIPPED:6
```

Next, let's deactivate the alarm system.

```
Notification: '
'Notification: '
'Notification: '+IPD,0,17:DEACTIVATE_
ALARM
'ClientIndex: 0
DataLength: 17
Data: DEACTIVATE_ALARM

Sending Data ....
TCP Connection ID: 0 , DataLength: 32 ,
Data: Alarm Has Been DE-Activated ...
```

Arduino Cameras

The two best cameras for the Arduino that I recommend are the Omnivision OV7670 FIFO camera and the ArduCAM OV2640 Mini camera. The OV7670 comes in two main versions: one with FIFO frame buffer memory and one without any memory. If you are using the camera with the Arduino, then you will need the version with the FIFO memory. The OV7670 is complex to operate and requires an Arduino Mega to use in order to take a picture. The benefit over the ArduCAM mini is that it costs around $9 to $10 each which is less than half the cost of the ArduCAM mini which is around $25.00 as of this writing. You can learn more about how to use this camera in a book I wrote called *Beginning Arduino ov7670 Camera Development* which covers the most popular version of the FIFO version of the OV7670. The ArduCAM OV2640 costs more than the OV7670 but is much easier to use and only requires an Arduino Uno to take a picture and save it on a SD card. However, the OV7670 does appear to capture a sharper image. The maximum resolution of the ArduCAM mini is also higher than the OV7670. I will cover the ArduCAM OV2640 Mini camera in this book.

ArduCAM OV2640 Camera

The specifications for the ArduCAM OV2640 Mini camera are as follows:

- 2MP image sensor
- I2C interface for the sensor configuration
- SPI interface for camera commands and data stream
- All IO ports are 5 V/3.3 V tolerant
- Supports JPEG compression mode, single and multiple shoot mode, one time capture multiple read operation, burst read operation, low power mode, etc.
- Well mated with standard Arduino boards
- Provides open source code library for Arduino, STM32, Chipkit, Raspberry Pi, BeagleBone Black
- Small form factor
- Power supply: 5 V/70 mA
- Low power mode: 5 V/20 mA
- Frame buffer: 384 KB
- Resolution support: UXGA, SVGA, VGA, QVGA, CIF, QCIF
- Image format support: RAW, YUV, RGB, JPEG

See Figure 6-22.

Figure 6-22 The ArduCAM OV2640 Mini camera.

ArduCAM Mini Camera Library Software Installation

In order to use the ArduCAM Mini camera with your Arduino, you will need to download and install the ArduCAM libraries. You will need to go to the ArduCAM website http://www.arducam.com and download the library. Once you have downloaded the zip file, you will need to uncompress it using a program like 7-zip and install the two directories "ArduCAM" and "UTFT4ArduCAM_SPI" under the "libraries" directory for Arduino. For example, on my Windows XP system I have installed the ArduCAM libraries in my "Program Files/Arduino/libraries" directory by copying the two directories to this "libraries" directory. After doing this you should be able to compile source code that includes the ArduCAM library. For this chapter I used version 3.4.7 of the ArduCAM library that was released on August 8, 2015.

The Memory Saver Include File

Change the "`memorysaver.h`" include file located in the ArduCAM library directory so that the following line is uncommented:

```
#define OV2640_CAM
```

This includes the camera register information that is needed for the ArduCAM OV2640 Mini camera.

Hands-on Example: The ArduCAM OV2640 Camera Wireless Remote Surveillance System

In this hands-on example I show you how to build and operate a wireless remote controlled ArduCAM OV2640 Mini-based surveillance system. With this surveillance system you will be able to use your Android device to take photos with the ArduCAM OV2640 Mini camera and store them on your Android device. The photos will be taken with the camera at one location and the photo will be transferred wirelessly to another location and displayed on your Android device. You will have the option of saving this image to local Android storage on your device.

Parts List

- 1 Android cell phone
- 1 Arduino Mega 2560 microcontroller
- 1 5- to 3.3-V step-down voltage regulator
- 1 Logic level converter from 5 to 3.3 V
- 1 ESP-01 Wi-Fi module
- 1 Arduino development station such as a desktop or a notebook
- 1 Breadboard
- 1 ArduCAM OV2640 Mini camera
- 1 Package of wires both male-to-male and female-to-male to connect the components

Setting Up the Hardware

Step-Down Voltage Regulator Connections

1. Connect the Vin pin on the voltage regulator to the 5-V power pin on the Arduino Mega 2560.
2. Connect the Vout pin on the voltage regulator to a node on your breadboard represented by a horizontal row of empty slots. This is the 3.3-V power node that will supply the 3.3-V power to the ESP-01 module as well as provide the reference voltage for the logic level shifter.
3. Connect the GND pin on the voltage regulator to the GND pin on the Arduino or form a GND node similar to the 3.3-V power node in the previous step.

ESP-01 Wi-Fi Module Connections

1. Connect the Tx or TxD pin to the Rx3 pin on the Arduino Mega 2560. This is the receive pin for the serial hardware port 3.
2. Connect the EN or CH_PD pin to the 3.3-V power node from the step-down voltage regulator.
3. Connect the IO16 or RST pin to the 3.3-V power node from the step-down voltage regulator.
4. Connect the 3V3 or VCC pin to the 3.3-V power node from the step-down voltage regulator.
5. Connect the GND pin to the ground pin or node.
6. Connect the Rx or RxD pin to pin B0 on the 3.3-V side of the logic level shifter. The voltage from this pin will be shifted from 5 V from the Arduino to 3.3 V for the ESP-01 module.

Logic Level Shifter

1. Connect the 3.3-V pin to the 3.3-V power node from the step-down voltage regulator.
2. Connect the GND pin on the 3.3-V side to the ground node for the Arduino.
3. Connect the B0 pin on the 3.3-V side to Rx or RxD pin on the ESP-01. This corresponds to the A0 pin on the 5-V side.

4. Connect the 5-V pin to the 5-V power pin on the Arduino.
5. Connect the GND pin on the 5-V side to the ground node for the Arduino.
6. Connect the A0 pin on the 5-V side to Tx3 pin on the Arduino. This pin corresponds to the B0 pin on the 3.3-V side.

ArduCAM OV2640 Mini

1. Connect the CS pin to digital pin 8 on the Arduino.
2. Connect the SDA pin to digital pin 20 on the Arduino.
3. Connect the SCL pin to digital pin 21 on the Arduino.
4. Connect the MISO pin to digital pin 50 on the Arduino.
5. Connect the MOSI pin to digital pin 51 on the Arduino.
6. Connect the SCK pin to digital pin 52 on the Arduino.
7. Connect the VCC pin to the 5 volt node for the circuit.
8. Connect the GND pin to the common ground node for the circuit.

See Figure 6-23.

Setting Up the Software

The code for the Arduino in this hands-on example is based on the code for the infrared motion detection alarm system that was presented in Chapter 5. I will cover only the new code that is specific to the ArduCAM Mini camera in this section.

You must include the `SPI` library into the source code since the ArduCAM Mini camera module uses it.

```
#include <SPI.h>
```

`Include` the `Wire` library since the camera uses the I2C interface which requires this library.

```
#include <Wire.h>
```

The following `include` files are the standard ArduCAM libraries:

```
#include <ArduCAM.h>
#include <UTFT_SPI.h>
```

The memorysaver.h include file is needed to define the specific details for the OV2640 Mini camera.

```
#include "memorysaver.h"
```

Set digital pin 8 on the Arduino as the slave select or chip select pin for the camera. This pin is connected to the CS pin on the camera and is part of the `SPI`.

```
const int SPI_CS = 8;
```

The myCAM variable is used to control the camera. It is initialized with the specific model of camera being used in this example which is the OV2640 camera and the Arduino pin that is used as the slave select or chip select pin. The chip select pin is used to connect or disconnect the camera from the serial peripheral interface (SPI).

```
ArduCAM myCAM(OV2640, SPI_CS);
```

The `ResolutionType` enumeration defines the camera resolutions that are used in this hands-on example which are the VGA, QVGA, and QQVGA image resolutions or None if no resolution has been set yet.

```
enum ResolutionType
{
  VGA,
  QVGA,
  QQVGA,
  None
};
```

The `Resolution` variable holds the current camera resolution and is initialized to None.

```
ResolutionType Resolution = None;
```

The `MaxPacketSize` is the largest number of bytes that can be sent at one time using the `AT+CIPSEND` command on the ESP-01 module.

```
const int MaxPacketSize = 2048;
```

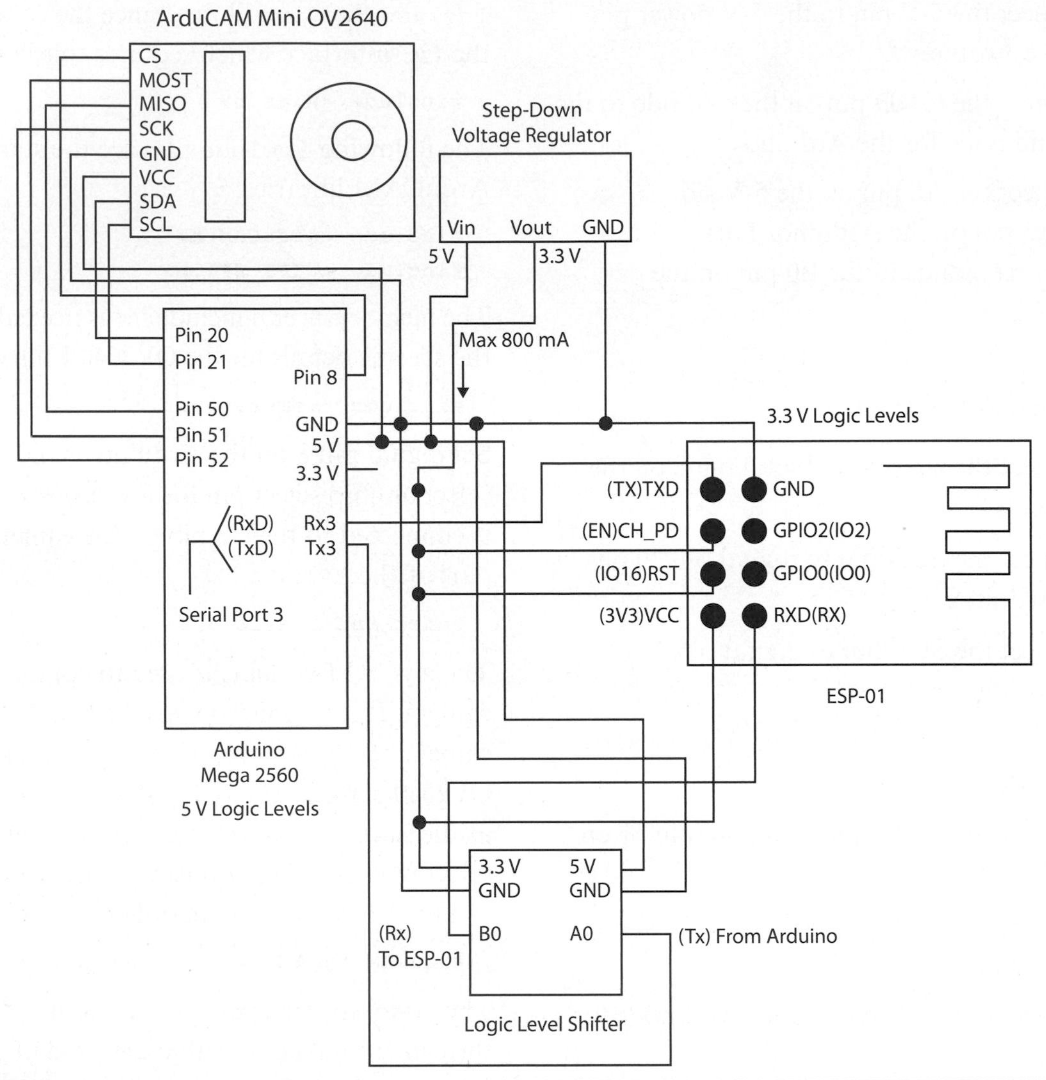

Figure 6-23 The ArduCAM OV2640 Mini camera system.

The `TCPBuffer` variable holds the image data that will be sent to the Android device.

```
byte   TCPBuffer[MaxPacketSize];
```

The `TCPBufferIndex` variable holds the location of the next available empty slot in the `TCPBuffer` array. It is initialized to the first slot which is 0.

```
int    TCPBufferIndex  = 0;
```

The `TCPBufferLength` variable holds the number of valid bytes in the `TCPBuffer` and is initialized to 0.

```
int    TCPBufferLength = 0;
```

The `InitializeCamera()` Function

The `InitializeCamera()` function initializes the ArduCAM Mini camera by:

1. Starting up the I2C interface by calling the `Wire.begin()` or `Wire1.begin()` function.
2. Setting the chip select pin number on the Arduino to be an `OUTPUT` pin by calling the `pinMode(SPI_CS, OUTPUT)` function with the pin number `SPI_CS` and the type of pin which is `OUTPUT`. This outputs a voltage to the CS pin on the camera.
3. Initializing the `SPI` bus on the Arduino by calling the `SPI.begin()` function.

4. Testing to see if the `SPI` bus is working by writing the hex value 0x55 to the register `ARDUCHIP_TEST1` and then reading the value from the `ARDUCHIP_TEST1` register. If the returned value is hex 0x55, then the `SPI` bus is working correctly. Otherwise, print out an error to the Serial Monitor and go into an infinite loop to halt program execution.
5. Testing to see if the ov2640 camera is detected.
6. Setting the image format that the ArduCAM Mini camera generates to JPEG by calling the `myCAM.set_format (JPEG)` function with the input parameter JPEG.
7. Performing further camera initialization by calling the `myCAM.InitCAM()` function.

See Listing 6-32.

Listing 6-32 The `InitializeCamera()` Function

```
void InitializeCamera()
{
  uint8_t vid, pid;
  uint8_t temp;
#if defined (__AVR__)
  Wire.begin();
#endif
#if defined(__arm__)
  Wire1.begin();
#endif

  Serial.println(F("ArduCAM Starting
.........."));
  // set the SPI_CS as an output:
  pinMode(SPI_CS, OUTPUT);

  // initialize SPI:
  SPI.begin();
  //Check if the ArduCAM SPI bus is OK
  myCAM.write_reg(ARDUCHIP_TEST1, 0x55);
  temp = myCAM.read_reg(ARDUCHIP_
TEST1);
  if (temp != 0x55)
  {
    Serial.println(F("SPI interface
    Error!"));
    while (1);
  }

  //Check if the camera module type is
OV2640
  myCAM.rdSensorReg8_8(OV2640_CHIPID_
HIGH, &vid);
  myCAM.rdSensorReg8_8(OV2640_CHIPID_
LOW, &pid);
  if ((vid != 0x26) || (pid != 0x42))
   Serial.println(F("Can't find OV2640
  module!"));
  else
   Serial.println(F("OV2640 detected
  ..........."));

  // Change to JPEG capture mode and
initialize the OV2640 module
  myCAM.set_format(JPEG);
  myCAM.InitCAM();
}
```

The `CaptureImage()` Function

The `CaptureImage()` function captures a camera image to the camera's on board FIFO memory by doing the following:

1. Resets the FIFO read pointer to 0 which is the beginning of the image frame by calling the `myCAM.flush_fifo()` function. This sets the camera to start writing a new image at the beginning of the camera's frame buffer memory or FIFO memory.
2. Resets the FIFO image capture flag by calling the `myCAM.clear_fifo_flag()` function. This flag is set every time an image has been captured and written to the FIFO buffer. We need to clear this flag before capturing another image.
3. Starts the image capture by calling the `myCAM.start_capture()` function.
4. Waits until the image capture is done and then continues execution of the program.

See Listing 6-33.

Listing 6-33 The `CaptureImage()` Function

```
void CaptureImage()
{
  // Capture Image from camera to
camera's FIFO memory buffer
  Serial.println(F("............
Starting Image Capture ........
.......")); 

  //Flush the FIFO
  myCAM.flush_fifo();

  //Clear the capture done flag
  myCAM.clear_fifo_flag();

  //Start capture
  myCAM.start_capture();
  Serial.println(F("Start Capture"));

  // Wait till capture is finished
  while (!myCAM.get_bit(ARDUCHIP_TRIG ,
CAP_DONE_MASK));
  Serial.println(F("Capture Done!"));
}
```

The `SendTCPBinaryData()` Function

The `SendTCPBinaryData()` function is generally the same as the `SendTCPMessage()` function that was shown and discussed in Listing 4-72 in Chapter 4. However a key change was made to the code, in that we are now sending binary image data instead of encoded text characters. So instead of using the `Serial3.print()` function we use the `Serial3.write()` function to send binary data. The key code change is the following:

```
Serial3.write(Buffer, BufferLength);
```

This function sends `BufferLength` bytes from the Buffer array over serial port 3 to the ESP-01 module so that the module can transmit this binary image data over Wi-Fi to the Android device. See Listing 6-34. The modifications from the code from Listing 4-72 are highlighted in bold.

Listing 6-34 The `SendTCPBinaryData()` Function

```
boolean SendTCPBinaryData(String
ClientIndex, byte* Buffer, int
BufferLength)
{
  boolean result = true;
  boolean ReadyToSend = false;
  boolean OKReceived = false;
  char Out;
  int TimeOutValue = 4000;
  boolean TimedOut = false;
  unsigned int TimeStart  = 0;
  unsigned int TimePassed = 0;

  ESP01ResponseString = "";

  // Send Command to ESP-01: AT+CIPSEN
  D=ClientIndex,DataLength
  // Wait for ">" from ESP-01
  // Send Actual Data to ESP-01
  // Wait for OK from ESP-01
  indicating send was successfull

  // Send AT Command to ESP-01
  Serial3.print(F("AT+CIPSEND="));
  Serial3.print(ClientIndex);
  Serial3.print(F(","));
  Serial3.print(String(BufferLength));
  Serial3.print(F("\r\n"));

  // Wait until ESP-01 responds with a
  '>' symbol
  TimeStart = millis();
  while (!ReadyToSend && !TimedOut)
  {
    // Read incoming data
    if (Serial3.available()>0)
    {
      Out = (char)Serial3.read();
      ESP01ResponseString += Out;
      if (Out == '>')
      {
        ReadyToSend = true;
      }
    }

    // Check for time out
    TimePassed = millis() - TimeStart;
```

```
  if (TimePassed > TimeOutValue)
  {
  TimedOut = true;
  }
 }

 if (TimedOut)
 {
  Serial.println(F(" ***********
 ERROR, SendTCPMessage() Timed Out On
 Waiting For > "));
  Serial.print(ESP01ResponseString);
  return false;
 }

 ESP01ResponseString = "";

 // Send Debug Info
 Serial.println(F("Sending Binary
 Data ...."));
 Serial.print(F("TCP Connection ID: "));
 Serial.print(ClientIndex);
 Serial.print(F(" , BufferLength: "));
 Serial.println(BufferLength);

 // Send data to ESP-01
 Serial3.write(Buffer, BufferLength);

 // Wait for OK from ESP-01
 TimeStart = millis();
 while (!OKReceived && !TimedOut)
 {
  if (Serial3.available()>0)
  {
   Out = (char)Serial3.read();
   ESP01ResponseString += Out;
   if (ESP01ResponseString.
  indexOf("OK") >= 0)
   {
    OKReceived = true;
   }
  }

  // Check for time out
  TimePassed = millis() - TimeStart;
  if (TimePassed > TimeOutValue)
  {
   TimedOut = true;
  }
 }
 // Check for Time Out
 if (TimedOut)
 {
  Serial.println(F(" ****** ERROR,
 SendTCPMessage() Timed Out On Wating
 for OK after Send ... "));
  Serial.print(ESP01ResponseString);
  return false;
 }

 return result;
}
```

The ***ReadFifoBurst()*** *Function*

The `ReadFifoBurst()` function reads the image data that is stored in the camera's FIFO memory and sends it to the Android using the TCP connection.

The function does this by:

1. Reading in the length of the image stored in the FIFO memory buffer by calling the `myCAM.read_fifo_length()` function.
2. If the length of the image is greater or equal to 393,216 bytes which is greater than the size of the entire FIFO memory buffer, then the end of the image has not been found and an error condition is returned.
3. The camera is selected to be active on the `SPI` bus by calling the `myCAM.CS_LOW()` function.
4. Setting the camera into FIFO burst mode which speeds up reading from the camera's memory by calling the `myCAM.set_fifo_burst()` function.
5. Reading in and discarding the first byte of the image which is a dummy byte that does not contain any image information by calling the `SPI.transfer(0x00)` function.
6. Resetting TCP buffer by setting the `TCPBufferIndex` and `TCPBufferLength` variables to 0.
7. For each byte in the image buffer read the byte by calling the `SPI.transfer(0x00)`

function with the parameter 0x00 and transmit this byte to the Android device over Wi-Fi by doing the following:

1. Reading in a byte from the camera's FIFO memory buffer.
2. Putting the byte from camera's memory that was just read in into the TCP buffer and increasing the value of the variable that keeps track of the length of the data in the buffer by 1.
3. If the TCP buffer is full, then send the data in the buffer to the Android by calling the `SendTCPBinaryData(ClientIndex, TCPBuffer, TCPBufferLength)` function and then reset the buffer. The `TCPBufferIndex` variable is set to the first element in the array which is 0. The `TCPBufferLength` is also set to 0 which indicates that the TCP buffer is empty.
4. If the TCP buffer is not full, then increase the buffer index variable `TCPBufferIndex` by 1.
5. Increasing the variable that keeps track of the number of bytes read in by 1.
6. Delaying program execution 10 microseconds.

8. If the length of the TCP buffer is not zero which means that there is data that still needs to be sent, then send this data by calling the `SendTCPBinaryData(ClientIndex, TCPBuffer, TCPBufferLength)` function.
9. Disconnecting the camera from the `SPI` bus after the image is read and transmitting to the Android device by calling the `myCAM.CS_HIGH()` function.
10. Printing out the total number of bytes transmitted and the total time it takes for transmission to the Serial Monitor.

See Listing 6-35.

Listing 6-35 The `ReadFifoBurst()` Function

```
uint8_t ReadFifoBurst(String
ClientIndex)
{
 uint32_t length = 0;
 uint8_t temp,temp_last;
 long bytecount = 0;
 int total_time = 0;

 length = myCAM.read_fifo_length();
 if(length >= 393216)  //384 kb
 {
  Serial.println(F("Not found the
 end."));
  return 0;
 }

 Serial.print(F("READ_FIFO_LENGTH()
= "));
 Serial.println(length);
 total_time = millis();
 myCAM.CS_LOW();
 myCAM.set_fifo_burst();
 temp = SPI.transfer(0x00); // Added
in to original source code, read in
dummy byte
 length--;

 // Reset TCP Buffer
 TCPBufferIndex = 0;
 TCPBufferLength = 0;
 while( length-- )
 {
  temp_last = temp;
  temp = SPI.transfer(0x00);

  // Put newly read in byte from
 ArduCAM FIFO into TCP buffer
  TCPBuffer[TCPBufferIndex] = temp;
  TCPBufferLength++;

  // Process TCP Buffer
  if (TCPBufferLength ==
 MaxPacketSize)
  {
   // Buffer Full so Send Data
   SendTCPBinaryData(ClientIndex,
  TCPBuffer, TCPBufferLength);
```

```
      // Reset Buffer
      TCPBufferIndex = 0;
      TCPBufferLength = 0;
    }
    else
    {
      // Buffer Not full so continue
    reading in Data from
      // ArduCAM FIFO Memory
      TCPBufferIndex++;
    }

    // Total Bytes read in from ArduCAM
  fifo
    bytecount++;
    delayMicroseconds(10);
  }

  // Check to see if data is still in
TCP Buffer
  if (TCPBufferLength != 0)
  {
    // Send remaining data in buffer
    SendTCPBinaryData(ClientIndex,
  TCPBuffer, TCPBufferLength);
  }

  myCAM.CS_HIGH();

  Serial.print(F("ByteCount = "));
  Serial.println(bytecount);

  total_time = millis() - total_time;
  Serial.print("Total time used:");
  Serial.print(total_time/1000, DEC);
  Serial.println(" seconds ...");
}
```

The `TransmitImageSize()` *Function*

The `TransmitImageSize()` function sends the size of the image that was captured by the camera to the Android by doing the following:

1. Retrieving the length of the captured image from the camera.
2. Adjusting the length of the image by 1 to get the final length of the captured image.

Listing 6-36 The `TransmitImageSize()` Function

```
void TransmitImageSize(String
ClientIndex)
{
  // Get Image Size
  uint32_t FIFOImageLength =  myCAM.
read_fifo_length();
  long Size = FIFOImageLength - 1; //
Account for 1 dummy byte at beginning
of image

  Serial.print(F("FIFO RAW IMAGE SIZE
= "));
  Serial.println(FIFOImageLength);

  Serial.print(F("Transmitting Image
Length = "));
  Serial.println(Size);

  // Transmit Image Size to Android
  String returnstring = String(Size) +
"\n";
  SendTCPMessage(ClientIndex,
returnstring);
}
```

3. Printing some debug messages to the Serial Monitor indicating the raw image size and the size of the image that will be transmitted to the Android.
4. Sending the size of the image to the Android by calling the `SendTCPMessage (ClientIndex, returnstring)` function.

See Listing 6-36.

The `SetCameraResolution()` *Function*

The `SetCameraResolution()` function changes the resolution of the camera by doing the following:

1. If the resolution of the image to be captured is QQVGA, then set the output JPEG size to 160×120.

Listing 6-37 The `SetCameraResolution()` Function

```
void SetCameraResolution()
{
  // Set Screen Size
  if (Resolution == QQVGA)
  {
   myCAM.OV2640_set_JPEG_
  size(OV2640_160x120);
  }
  else
  if (Resolution == QVGA)
  {
   myCAM.OV2640_set_JPEG_
  size(OV2640_320x240);
  }
  else
  if (Resolution == VGA)
  {
   myCAM.OV2640_set_JPEG_
  size(OV2640_640x480);
  }
  else
  {
   Serial.println(F("ERROR in setting
  Camera Resolution"));
  }
}
```

2. If the resolution of the image to be captured is QVGA, then set the output JPEG size to 320×240.
3. If the resolution of the image to be captured is VGA, then set the output JPEG size to 640×480.
4. If the resolution of the image is none of the above then print out an error to the Serial Monitor.

See Listing 6-37.

The `setup()` Function

The `setup()` function is changed by adding the `InitializeCamera()` function to initialize the camera.

Listing 6-38 The `vga` Command

```
if (Data == "vga\n")
{
 if (Resolution != VGA)
 {
  Resolution = VGA;
  SetCameraResolution();
 }
 CaptureImage();
 TransmitImageSize(ClientIndex);
}
```

The `vga` Command

The `vga` command captures a vga image by doing the following:

1. If the current camera resolution is not vga, then set the resolution to vga.
2. Telling the camera to capture an image by calling the `CaptureImage()` function.
3. Transmitting the captured image size to the Android device by calling the `TransmitImageSize(ClientIndex)` function.

See Listing 6-38.

The `qvga` Command

The `qvga` command captures a qvga image by doing the following:

1. If the current camera resolution is not qvga, then set the resolution to qvga.
2. Telling the camera to capture an image by calling the `CaptureImage()` function.
3. Transmitting the captured image size to the Android device by calling the `TransmitImageSize(ClientIndex)` function.

See Listing 6-39.

The `qqvga` Command

The `qqvga` command captures a qqvga image by doing the following:

1. If the current camera resolution is not qqvga, then set the resolution to qqvga.

Listing 6-39 The qvga Command

```
if (Data == "qvga\n")
{
 if (Resolution != QVGA)
 {
  Resolution = QVGA;
  SetCameraResolution();
 }
 CaptureImage();
 TransmitImageSize(ClientIndex);
}
```

Listing 6-40 The qqvga Command

```
if (Data == "qqvga\n")
{
 if (Resolution != QQVGA)
 {
  Resolution = QQVGA;
  SetCameraResolution();
 }
 CaptureImage();
 TransmitImageSize(ClientIndex);
}
```

2. Telling the camera to capture an image by calling the `CaptureImage()` function.
3. Transmitting the captured image size to the Android device by calling the `TransmitImageSize(ClientIndex)` function.

See Listing 6-40.

The `GetImageData` Command

The `GetImageData` command retrieves the camera's captured image data and transmits it to the Android device by calling the `ReadFifoBurst(ClientIndex)` function. See Listing 6-41.

The code for the Android application's basic wireless framework will also need to be modified in order to work with this hands-on example.

Listing 6-41 The `GetImageData` Command

```
if (Data == "GetImageData\n")
{
 ReadFifoBurst(ClientIndex);
}
```

Modifying the `onCreate()` Function in the `MainActivity` Class

This section discusses the changes and additions that need to be made to the `onCreate()` function in order to have this application work with the camera.

The Android's file storage system needs to be initialized so that pictures can be saved by calling the `InitExternalStorage()` function.

The "Send Data" Button's `onClick()` Function

The `onClick()` function for the "Send Data" button is modified so that:

1. The current time is stored in the `m_TimeLastTCPActivity` variable.
2. If the command that is being sent to the Arduino is vga, qvga, or qqvga, then set up the return processing of the next TCP message received to expect the size of the image that was captured by the camera. Otherwise, set up the return processing to expect a plain text message.
3. Add a newline or "\n" character to the command that is being sent to the Arduino.

See Listing 6-42.

Setting Up the Ping Handler

The `PingHandler()` function execution is set up using the variable r which is a `Runnable` object. The changes to setting up the ping handler are as follows:

1. The `PingHandler()` function now returns the time delay until the next `PingHandler()` function execution.

Listing 6-42 The "Send Data" Button `onClick()` Function

```
m_Button.setOnClickListener(new View.OnClickListener() {
 public void onClick(View v) {
  // Set current time as the tie of most recent TCP Activity
  m_TimeLastTCPActivity = SystemClock.elapsedRealtime();

  // Perform action on click
  m_Alert1SFX.PlaySound(false);

  // Basic Framework
  String Command = m_PhoneEntryView.getText().toString();

  // Set Up Data Handler for all Commands
  m_WifiMessageHandler.ResetData();

  // Command is a take photo command
  if ((Command.equalsIgnoreCase("vga") ||
   (Command.equalsIgnoreCase("qvga")) ||
   Command.equalsIgnoreCase("qqvga")))
   {
    m_TakePhotoState = TakePhotoCallbackState.GetImageSize;
    m_WifiMessageHandler.SetCommand("GetImageSize");
   }
   else
   {
    // Command is not a take photo command
    // Set return data to text
    m_WifiMessageHandler.SetCommand("GetTextData");
   }

   // Add newline character to all Commands
   Command += "\n";

   // Send actual command over TCP connection
   if (m_ClientConnectThread != null) {
    if (m_ClientConnectThread.IsConnected()) {
     m_ButtonActive = false;
     m_Button.setEnabled(false);
     m_ClientConnectThread.write(Command.getBytes());
     m_DebugMsg += "Sending COMMAND = " + Command + "\n";
    } else {
     m_DebugMsg += "ERROR! ClientConnectedThread is not connected!! \n";
     Log.e(TAG, "ERROR! ClientConnectedThread is not connected!!");
     if (m_TTS != null) {
     m_TTS.speak("ERROR
     ClientConnectedThread is not connected", TextToSpeech.QUEUE_ADD, null);
    }
   }
  } else {
   m_DebugMsg += "ClientConnectThread is Null!! \n";
  }
  m_DebugMsgView.setText(m_DebugMsg.toCharArray(), 0, m_DebugMsg.length());
 }
});
```

2. This time delay is set by calling the `m_PingHandler.postDelayed(this, NextTimeInterval)` function.

See Listing 6-43.

The `InitExternalStorage()` Function

The `InitExternalStorage()` function retrieves the directory in which the images from the camera will be stored.

Listing 6-43 Setting Up the Ping Handler

```
Runnable r = new Runnable() {
 public void run(){
  long NextTimeInterval =
 PingHandler();
  m_PingHandler.postDelayed(this,
 NextTimeInterval);
 }
};
```

The function does the following:

1. Tests if the storage is writable by calling the `isExternalStorageWritable()` function. Displays a message in the Debug Window with the result.
2. Retrieves and prints out to the debug window the directory path to the user's "pictures" directory.
3. The final path to the directory is retrieved by calling the `GetAlbumStorageDir(Environment.DIRECTORY_PICTURES, m_AndroidPicsDirectory)` function with the input parameters for the directory type which is the pictures directory and the album name which will be "CAMPics".
4. The final path is printed to the debug window.

See Listing 6-44.

Listing 6-44 The `InitExternalStorage()` Function

```
void InitExternalStorage()
{
 // Test for External Storage and check if writable
 if (isExternalStorageWritable())
 {
  String Message = "External Storage is PRESENT and WRITABLE ...\n";
  AddDebugMessage(Message);
 }
 else
 {
  String Message = "External Storage is GONE or NOT Writable !!!! \n";
  AddDebugMessage(Message);
 }

 // Get the user's Pictures directory
 File PicsDir = GetExternalStorageDirectoryPath(Environment.DIRECTORY_PICTURES);
 String Path = "User's Picture Directory: " + PicsDir.getAbsolutePath() + " ...\n";
 AddDebugMessage(Path);

 // Get/Create Directory for Arduino Camera pictures
 m_CameraDir = GetAlbumStorageDir(Environment.DIRECTORY_PICTURES,
 m_AndroidPicsDirectory);
 Path = "Camera Pics Directory: " + m_CameraDir.getAbsolutePath() + " ...\n";
 AddDebugMessage(Path);
}
```

The `isExternalStorageWritable()` Function

The `isExternalStorageWritable()` function determines if the external storage on the Android file system is writable by:

1. Getting the external storage state of the Android system by calling the `Environment.getExternalStorageState()` function.
2. If the result is equal to "`MEDIA_MOUNTED`", then we can save the images to storage.

See Listing 6-45.

The `GetExternalStorageDirectoryPath()` Function

The `GetExternalStorageDirectoryPath()` function gets the path of the type of directory input by the user.

See Listing 6-46.

The `GetAlbumStorageDir()` Function

The `GetAlbumStorageDir()` function creates and returns a new File object by:

1. Creating a new File class object using the input parameters of `DirectoryType` and `AlbumName`. This File object retrieves the system file path for the `DirectoryType` such as Pictures that is on the Android file system and adds on the `AlbumName` to the end of the path.
2. Attempting to create this new path by calling the `file.mkdirs()` function which makes all the directories in the path contained in the file variable from step 1. Adding a message to the Debug Window if the path could not be created.

See Listing 6-47.

The `PingHandler()` Function

The `PingHandler()` function was modified and now does the following:

1. The elapsed time since the last command was issued is calculated.

Listing 6-45 The `isExternalStorageWritable()` Function

```
public boolean
isExternalStorageWritable()
{
 String state = Environment.
getExternalStorageState();
 if (Environment.MEDIA_MOUNTED.
equals(state))
 {
  return true;
 }
 return false;
}
```

Listing 6-46 The `GetExternalStorageDirectoryPath()` Function

```
public File GetExternalStorageDirectory
Path(String DirectoryType)
{
 File Path = Environment.getExternal
StoragePublicDirectory(DirectoryType);
 return Path;
}
```

Listing 6-47 The `GetAlbumStorageDir()` Function

```
public File GetAlbumStorageDir(String
DirectoryType, String AlbumName)
{
 // Get the directory for the user's
album
 File file = new File(Environment.
getExternalStoragePublicDirectory
(DirectoryType), AlbumName);
 if (!file.mkdirs()) {
  Log.e(TAG, "External Storage Album
 Directory not created !!!!");
  String Error = "External Storage
 Album Directory not created !!!!\n";
  AddDebugMessage(Error);
 }
 return file;
}
```

2. If this elapsed time is equal to or greater than `m_PingInterval`, then the next ping interval value is set to `m_PingInterval`.
3. If this elapsed time is less than `m_PingInterval`, then set the next ping interval value to the difference between the `m_PingInterval` and the elapsed time which is the remaining time in the ping interval. This value is then returned and the function is exited.
4. If the elapsed time was equal to or greater than the `m_PingInterval` variable, then send a ping command to the Arduino in order to keep the TCP connection active.
5. The time delay for the next ping interval is returned.

See Listing 6-48. The additional code is highlighted in bold print.

Listing 6-48 The `PingHandler()` Function

```
long PingHandler()
{
 long NextTimeInterval = 0;
 long ElapsedTime = 0;

 ElapsedTime = SystemClock.elapsedRealtime() - m_TimeLastTCPActivity;
 if (ElapsedTime >= m_PingInterval)
 {
  // Time to send a ping to the server
  // and set the next ping interval to the default ping interval.
  NextTimeInterval = m_PingInterval;
 }
 else
 {
  // Not yet time to send a ping to the server
  // Set the next time interval so that the ping handler will be called when
  // the time between the last tcp activity and the next ping handler call will be
  // the m_PingInterval time.
  NextTimeInterval = m_PingInterval - ElapsedTime;
  return NextTimeInterval;
 }

 // Send Ping if Connected
 String Command = "ping\n";

 // Set Up Data Handler
 m_WifiMessageHandler.ResetData();
 m_WifiMessageHandler.SetCommand("GetTextData");
 if (m_ClientConnectThread != null)
 {
  if (m_ClientConnectThread.IsConnected())
  {
   // Send Ping to Server
   m_ClientConnectThread.write(Command.getBytes());

   // Print out Ping sent debug message
   AddDebugMessageFront("PingSent..." + "\n");
   Log.e(TAG, "PingHandler() Executed ....");
  }
 }
 return NextTimeInterval;
}
```

Listing 6-49 The `TakePhotoCommand Callback()` Function

```
void TakePhotoCommandCallback()
{
 // Photo is now ready for viewing
 m_TakePhotoCallbackDone = true;
 if (m_TakePhotoState == TakePhoto
CallbackState.GetImageData)
 {
  m_PhotoData = m_WifiMessageHandler.
 GetBinaryData();
 }
}
```

The `TakePhotoCommandCallback()` Function

The `TakePhotoCommandCallback()` function is called from the `WifiMessageHandler` class after a request for information on either the image size or the image data regarding a "take photo" command (vga, qvga, qqvga) has been completed and the data is ready to be processed. If it is the image data, then the `m_PhotoData` variable is set to point to the binary image data. The `m_TakePhotoCallbackDone` variable is set to true so that the incoming data will be processed in the `run()` command. See Listing 6-49.

The `run()` Function in the `MainActivity` Class

The `run()` function was modified to include code that handled the processing of the incoming image data. If a command to get the image size or get the image binary data has completed, then:

1. If the command that was just completed retrieved the image size, then call the `ProcessGetImageSizeCallback()` function and reset the `m_TakePhotoCallbackDone` variable to false.
2. If the command that was just completed retrieved the image data, then do the following:
   1. Reset the `m_TakePhotoCallbackDone` to false.
   2. Display the image on the Android screen by calling the `LoadJpegPhotoUI(m_PhotoData)` function.
   3. If the surveillance feature is active, then do the following:
      1. Disable the "Send Data" button.
      2. If surveillance sound effects are on, then play a sound.
      3. Increase the surveillance frame number by 1.
      4. Print out the surveillance frame number to the Android debug message window.
      5. If the storing of incoming images is turned on, then save the jpeg to the Android's storage system by calling `SavePicJPEG()` function and print out the filename that it was saved under to the debug message window.
      6. Continue the surveillance by calling the `SendTakePhotoCommand()` function which takes another photo.
   4. If the surveillance feature is not active, then do the following:
      1. If the storing of incoming images is turned on, then save the jpeg to the Android's storage system by calling `SavePicJPEG()` function and print out the filename that it was saved under to the debug message window.
      2. Enable the "Send Data" button.
      3. Play a sound effect to indicate that a new image has been received and displayed on the Android device.

See Listing 6-50. The additional code is highlighted in bold print.

Listing 6-50 The `run()` Function

```
public void run()
{
 // Receive amd Display Text Data
 if (m_RecieveTextDataCallBackDone)
 {
  ReceiveDisplayTextData();
  m_RecieveTextDataCallBackDone =
 false;
  m_ButtonActive = true;
 }
 else
 if (m_RefreshMessageWindows)
 {
  // Refresh Window Messages
  m_DebugMsgView.setText(m_DebugMsg.
 toCharArray(), 0, m_DebugMsg.
 length());
  m_Alert2SFX.PlaySound(false);
  m_RefreshMessageWindows = false;
 }
 else
 if(m_TakePhotoCallbackDone)
 {
  // Take Photo Button callback
  if (m_TakePhotoState ==
 TakePhotoCallbackState.GetImageSize)
  {
   m_TakePhotoCallbackDone = false;
   ProcessGetImageSizeCallback();
  }
  else
  if (m_TakePhotoState ==
 TakePhotoCallbackState.GetImageData)
  {
   // Display New Photo received from
  the Arduino
   m_TakePhotoCallbackDone = false;
   LoadJpegPhotoUI(m_PhotoData);

   // If surveillance is active
   if (m_SurveillanceActive)
   {
    m_Button.setEnabled(false);
    if (m_SurveillanceSFX)
    {
     m_Alert1SFX.PlaySound(false);
    }
    m_SurveillanceFrameNumber++;
    m_DebugMsg = "";
    m_DebugMsg += "Frame Number: " +
    m_SurveillanceFrameNumber + "\n";
    m_SurveillanceFileName = "";
    if (m_AndroidStorage ==
   AndroidStorage.StoreYes)
    {
     String filename = SavePicJPEG();
     m_DebugMsg+= "File Name: " +
    filename + "\n";
    }
    m_DebugMsg += "\n\n\n\n";
    m_DebugMsgView.setText(m_
   DebugMsg.toCharArray(), 0, m_
   DebugMsg.length());
    // Send another take photo
   command to Arduino
    SendTakePhotoCommand();
   }
   else
   {
    // If surveillance is not active
   and storage is yes
    // then store image on Android
    m_SurveillanceFileName = "";
    if (m_AndroidStorage ==
   AndroidStorage.StoreYes)
    {
     String filename = SavePicJPEG();
     m_DebugMsg+= "File Name: " +
    filename + "\n";
     m_DebugMsgView.setText(
     m_DebugMsg.toCharArray(), 0, m_
    DebugMsg.length());
    }
    m_ButtonActive = true;
    m_Alert2SFX.PlaySound(true);
   }
  }
 }
 else
 {
  // Error
  Log.e(TAG, "RUN:: ERROR IN RUN()
 ... No callback executed !!!!");
 }

 // Set Take Photo Button active
state
 if (m_ButtonActive == true)
 {
  m_Button.setEnabled(true);
 }
 else
 {
  m_Button.setEnabled(false);
 }
}
```

The `ProcessGetImageSizeCallback()` Function

The `ProcessGetImageSizeCallback()` function processes the image size data by doing the following:

1. The incoming text data is retrieved from the `WifiMessageHandler` class.
2. The data is then trimmed of whitespace characters.
3. The trimmed text data is then converted to an integer and assigned to the `m_PhotoSize` variable.
4. Add a message to the debug window displaying the size of the incoming image.
5. Send the command to the Arduino to transmit the actual image data to the Android device by calling the `SendGetImageDataCommand()` function.

See Listing 6-51.

The `SendGetImageDataCommand()` Function

The `SendGetImageDataCommand()` function sends the command to retrieve the image data to the Arduino over Wi-Fi by doing the following:

1. Defining the command to be sent to the Arduino as "`GetImageData\n`".
2. Storing the current time in the `m_TimeLastTCPActivity` variable.
3. Resetting the data structures in the Wi-Fi message handler that is used to process incoming Wi-Fi data.
4. Setting the binary data length that is to be received to the size of the expected binary image data.
5. The status of the take photo command is set to getting the image data. The `m_TakePhotoState` variable is set to `TakePhotoCallbackState.GetImageData`.
6. Setting the command in the Wi-Fi message handler to "`GetImageData`". This command string will be used to process the incoming data.
7. If there is a Wi-Fi connection between the Android device and the Arduino, then disable the "Send Data" button and send that actual command to the Arduino by calling the `m_ClientConnectThread.write(Command.getBytes())` function with the command string converted to an array of bytes. A message is printed to the debug message window notifying the user that the "`GetImageData`" command has been sent.

See Listing 6-52.

Listing 6-51 The `ProcessGetImageSizeCallback()` Function

```
void ProcessGetImageSizeCallback()
{
 String Data = m_WifiMessageHandler.
GetStringData();
 String DataTrimmed = Data.trim();

 m_PhotoSize = Integer.
parseInt(DataTrimmed);

 // Debug Print Outs
 AddDebugMessage("RUN:: Returned
Image Size = " + m_PhotoSize + "\n");
 SendGetImageDataCommand();
}
```

The `LoadJpegPhotoUI()` Function

The `LoadJpegPhotoUI()` function displays a JPEG image on the Android device's user interface based on the input `JpegData`.

The function does the following:

1. The data in the `JpegData` byte array is decoded into a bitmap image. This is

Listing 6-52 The `SendGetImageDataCommand()` Function

```
void SendGetImageDataCommand()
{
 String Command = "GetImageData\n";

 // Set current time for this TCP command
 m_TimeLastTCPActivity = SystemClock.elapsedRealtime();

 // Set Up Data Handler
 m_WifiMessageHandler.ResetData();
 m_WifiMessageHandler.SetDataReceiveLength(m_PhotoSize);
 m_TakePhotoState = TakePhotoCallbackState.GetImageData;
 m_WifiMessageHandler.SetCommand("GetImageData");
 if (m_ClientConnectThread != null)
 {
  if (m_ClientConnectThread.IsConnected())
  {
   m_Button.setEnabled(false);
   m_ClientConnectThread.write(Command.getBytes());
   m_DebugMsg += "Writing GetImageData command!! Command = " + Command;
  }
  else
  {
   m_DebugMsg += "ConnectedThread is Null!! \n";
  }
 }
 else
 {
  m_DebugMsg += "ClientConnectThread is Null!! \n";
 }

 m_DebugMsgView.setText(m_DebugMsg.toCharArray(), 0, m_DebugMsg.length());
}
```

done by calling the `BitmapFactory.decodeByteArray(JpegData, offset, length)` function with the image data, the offset from the beginning of the array to where the image data starts, and the length of the data that forms the image.

2. The `ImageView` window that will hold the image is found by calling the `findViewById(R.id.imageView1)` function with the resource ID of the window we want to put the image in.
3. If the image was successfully decoded, then the image is put into the window by calling the `ImageView.setImageBitmap(BM)` function with the bitmap image generated from the JPEG data from step 1.
4. If the image from step 1 is null which means that the decoding failed then issue an error in the log window by calling the `Log.e()` function.

See Listing 6-53.

Listing 6-53 The `LoadJpegPhotoUI()` Function

```
void LoadJpegPhotoUI(byte[] JpegData)
{
 /*
 public static Bitmap decodeByteArray (byte[] data, int offset, int length)
 Added in API level 1

 Decode an immutable bitmap from the specified byte array.

 Parameters:
 data      byte array of compressed image data
 offset    offset into imageData for where the decoder should begin parsing.
 length    the number of bytes, beginning at offset, to parse

 Returns:
 The decoded bitmap, or null if the image could not be decoded.
 */

 Bitmap BM = null;
 int offset = 0;
 int length = JpegData.length;

 BM = BitmapFactory.decodeByteArray(JpegData, offset, length);
 ImageView ImageView = (ImageView) findViewById(R.id.imageView1);
 if (BM != null)
 {
  ImageView.setImageBitmap(BM);
 }
 else
 {
  Log.e("MainActivity", "ImageView Bitmap Creation Failed!!!");
  AddDebugMessage(TAG + ": ImageView Bitmap Creation Failed !!!\n");
 }
}
```

The `SavePicJPEG()` Function

The `SavePicJPEG()` function saves the photo that was transmitted to the Android from the Arduino by:

1. Creating the beginning of the filename from the image resolution such as VGA, QVGA, or QQVGA.
2. Creating the date by first getting a `Calendar` class object by calling the `Calendar.getInstance()` function. Then using this object to get the month, day of the month, year, hour, minute, and second and adding this to the end of the filename.
3. Adding the ".jpg" extension to the filename.
4. Creating a new file using the filename in the directory specified by `m_CameraDir`.
5. Writing out the image data that is in the `m_PhotoData` array to the file.
6. If surveillance is active, then assign the filename created to the `m_Surveillance FileName String` variable.

See Listing 6-54.

Listing 6-54 The `SavePicJPEG()` Function

```
String SavePicJPEG()
{
 // Main Filename
 String Filename = m_Resolution.toString() + "_";

 // Add Date
 Calendar rightNow = Calendar.getInstance();
 int Month    = rightNow.get(Calendar.MONTH);
 int DayOfMonth  =  rightNow.get(Calendar.DAY_OF_MONTH);
 int Year   =  rightNow.get(Calendar.YEAR);
 int Hour   =  rightNow.get(Calendar.HOUR_OF_DAY);
 int Minute   =  rightNow.get(Calendar.MINUTE);
 int Second   =  rightNow.get(Calendar.SECOND);

 String Date = Month + "_" + DayOfMonth + "_" + Year + "_" + Hour + "_" + Minute +
"_" + Second;

 // Add JPG Extension
 String Extension = ".jpg";

 // Build Final Filename
 Filename += Date;
 Filename += Extension;

 // Open File to write out image data
 File file = new File(m_CameraDir, Filename);

 try
 {
  OutputStream os = new FileOutputStream(file);
  os.write(m_PhotoData, 0, m_PhotoSize);
  os.close();
 }
 catch (IOException e)
 {
  // Unable to create file, likely because external storage is
  // not currently mounted.
  Log.e("ExternalStorage", "Error writing " + file, e);
  AddDebugMessage(TAG + ": Error writing " + file);
 }

 // If surveillance is active then record the filename that the photo was saved to.
 if (m_SurveillanceActive)
 {
  m_SurveillanceFileName = Filename;
 }
 return Filename;
}
```

The SendTakePhotoCommand() Function

The SendTakePhotoCommand() function issues a Take Photo command (vga, qvga, qqvga) by:

1. Setting the command that is sent to the Arduino to the current user selected resolution such as VGA, QVGA, or QQVGA and including a new line character at the end of the command.
2. Recording the current time in the m_TimeLastTCPActivity variable.
3. Resetting the data in the Wi-Fi message handler by calling the m_WifiMessageHandler.ResetData() function.
4. Setting the command in the message handler to one that matches the type of return value expected from the Arduino which is the GetImageSize command. The data expected is the size of the photo in bytes that was just taken by the camera.
5. Setting the state of the Take Photo command to the GetImageSize state.
6. Disabling the "Send Data" button.
7. The command created in step 1 is converted from a String object into an array of bytes and transmitted to the Arduino.

See Listing 6-55.

Listing 6-55 The SendTakePhotoCommand() Function

```
void SendTakePhotoCommand()
{
 String Command = m_Resolution.toString() + "\n";

 m_TimeLastTCPActivity = SystemClock.elapsedRealtime();

 // Set Up Data Handler
 m_WifiMessageHandler.ResetData();
 m_WifiMessageHandler.SetCommand("GetImageSize");
 m_TakePhotoState = TakePhotoCallbackState.GetImageSize;

 if (m_ClientConnectThread != null)
 {
  if (m_ClientConnectThread.IsConnected())
  {
   m_Button.setEnabled(false);
   m_ClientConnectThread.write(Command.getBytes());
   m_DebugMsg += "Writing Take Photo Command Message to Arduino!! \n";
  }
  else
  {
   m_DebugMsg += "ConnectedThread is Null!! \n";
  }
 }
 else
 {
  m_DebugMsg += "ClientConnectThread is Null!! \n";
 }

 m_DebugMsgView.setText(m_DebugMsg.toCharArray(), 0, m_DebugMsg.length());
}
```

Modifying the Android Application's Menu

The menu was modified by doing the following:

1. Adding a "Resolution Settings" menu that allows the user to select between vga, qvga, and qqvga camera resolutions.
2. Adding a "Surveillance Settings" menu that allows the user to turn camera surveillance on and off.
3. Adding an "Android Storage Settings" menu that allows the user to save incoming images.

See Listing 6-56.

The `onOptionsItemSelected()` *Function*

This function is called when the user selects a menu item. There are additional entries for the following items:

1. Code is added for the user's selection of the camera's resolution. For the resolutions of vga, qvga, and qqvga the `m_Resolution` variable is set by calling the `SetCameraResolution()` function with the resolution that the user has selected. The `UpdateCommandTextView()` is then called to update the information window on the Android application.
2. Code is added for the user's selection of the camera's surveillance state which can be either on or off. The `ProcessSurveillanceSelection()` is called to set the state. The `UpdateCommandTextView()` function is called to update the information window and a `Toast.makeText()` function is called to briefly display a banner with this information on the Android application.
3. Code is added for the user's selection of the image storage state which is held in the `m_AndroidStorage` variable and can be either `AndroidStorage.StoreYes` or `AndroidStorage.`

Listing 6-56 Adding in Menu Items

```
<item android:id="@+id/resolution_settings" android:title="Resolution Settings">
 <menu>
  <item android:id="@+id/qqvga" android:title="QQVGA (160 x 120)"/>
  <item android:id="@+id/qvga" android:title="QVGA (320 x 240)" />
  <item android:id="@+id/vga" android:title="VGA (640 x 480)"/>
 </menu>
</item>

<item android:id="@+id/surveillance" android:title="Surveillance Settings">
 <menu>
  <item android:id="@+id/surveillance_on" android:title="Turn On Surveillance"/>
  <item android:id="@+id/surveillance_off" android:title="Turn Off Surveillance"/>
 </menu>
</item>

<item android:id="@+id/androidstoragesettings" android:title="Android Storage Settings">
 <menu>
  <item android:id="@+id/androidstorageyes" android:title="Save Incoming Images"/>
  <item android:id="@+id/androidstorageno" android:title="Do NOT Save Incoming Images"/>
 </menu>
</item>
```

StoreNo. The application's information window is also updated by calling the UpdateCommandTextView() function. The Toast.makeText() function is called to show a banner that appears briefly in the application.

See Listing 6-57. The additional code is shown in bold print.

Listing 6-57 The onOptionsItemSelected() Function

```
public boolean onOptionsItemSelected(MenuItem item)
{
 // Handle item selection
 switch (item.getItemId()) {
 ///////////////////////// Alarm Activation
 case R.id.alarm_activate:
 SendCommand("ACTIVATE_ALARM");
 SetAlarmStatus(true);
 UpdateCommandTextView();
 Toast.makeText(this, "Alarm Activated !!!", Toast.LENGTH_LONG).show();
 return true;

 case R.id.alarm_deactivate:
 SendCommand("DEACTIVATE_ALARM");
 SetAlarmStatus(false);
 UpdateCommandTextView();
 Toast.makeText(this, "Alarm Deactivated !!!", Toast.LENGTH_LONG).show();
 return true;

 ///////////////////////// Alarm Emergency Phone Callout
 case R.id.callout_on:
 SetCallOutStatus(true);
 UpdateCommandTextView();
 Toast.makeText(this, "Call Out Activated !!!", Toast.LENGTH_LONG).show();
 return true;

 case R.id.callout_off:
 SetCallOutStatus(false);
 UpdateCommandTextView();
 Toast.makeText(this, "Call Out Deactivated !!!", Toast.LENGTH_LONG).show();
 return true;

 //////////////////////// Text messages
 case R.id.notext:
 SetTextMessageStatus(TextMessageSetting.TEXT_OFF);
 UpdateCommandTextView();
 Toast.makeText(this, "Text Message Alerts Turned Off !!!", Toast.LENGTH_LONG).show();
 return true;

 case R.id.text_only:
 SetTextMessageStatus(TextMessageSetting.TEXT_ON);
 UpdateCommandTextView();
 Toast.makeText(this, "Text Message Alerts Activated !!!", Toast.LENGTH_LONG).show();
 return true;

 ///////////////////////// Resolution
 case R.id.qqvga:
```

```
SetCameraResolution(Resolution.qqvga);
UpdateCommandTextView();
return true;

case R.id.qvga:
SetCameraResolution(Resolution.qvga);
UpdateCommandTextView();
return true;

case R.id.vga:
SetCameraResolution(Resolution.vga);
UpdateCommandTextView();
return true;

////////////////////////////// Surveillance
case R.id.surveillance_on:
ProcessSurveillanceSelection(true);
UpdateCommandTextView();
Toast.makeText(this, "Surveillance Turned On !!!", Toast.LENGTH_LONG).show();
return true;

case R.id.surveillance_off:
ProcessSurveillanceSelection(false);
UpdateCommandTextView();
Toast.makeText(this, "Surveillance Turned Off !!!", Toast.LENGTH_LONG).show();
return true;

///////////////////////////// Android Storage
case R.id.androidstorageyes:
m_AndroidStorage = AndroidStorage.StoreYes;
UpdateCommandTextView();
Toast.makeText(this, "Images will be saved to Android Storage", Toast.LENGTH_LONG).
show();
return true;

case R.id.androidstorageno:
m_AndroidStorage = AndroidStorage.StoreNo;
UpdateCommandTextView();
Toast.makeText(this, "Images will NOT be saved to Android Storage", Toast.LENGTH_
LONG).show();
return true;
//////////////////////////////////////////////////////
case R.id.savedevicename:
m_AndroidClientName = m_PhoneEntryView.getText().toString();
SaveDeviceName();
Toast.makeText(this, "Saving Device Name: " + m_AndroidClientName, Toast.LENGTH_
SHORT).show();
AddDebugMessageFront("Android Device Name Changed to: " + m_AndroidClientName + "\n");
m_PhoneEntryView.setText(m_EmergencyPhoneNumber);
UpdateCommandTextView();
return true;

default:
return super.onOptionsItemSelected(item);
}
}
```

The `UpdateCommandTextView()` Function

The code for the `UpdateCommandTextView()` function displays the system status in the information window in the lower right-hand corner and is updated for this hands-on example. Code has been added that displays the current user selected camera resolution, the camera's surveillance status, and the status of the saving of incoming images. The additions are highlighted in bold print in Listing 6-58.

Listing 6-58 The `UpdateCommandTextView()` Function

```
void UpdateCommandTextView()
{
 String AlarmStatus = "";
 String CallOutStatus = "";
 String SurveillanceStatus = "";

 // Get Current Camera Resolution and
set text view
 if (m_AlarmSet)
 {
  AlarmStatus = "AlarmON";
 }
 else
 {
  AlarmStatus = "AlarmOFF";
 }

 if (m_CallOutSet)
 {
  CallOutStatus = "CallOutON";
 }
 else
 {
  CallOutStatus = "CallOutOFF";
 }

 if (m_SurveillanceActive)
 {
  SurveillanceStatus = "SurveilOn";
 }
 else
 {
 SurveillanceStatus = "SurveilOff";
 }

 String Info =m_AndroidClientName +
"\n" +
 m_EmergencyPhoneNumber + "\n" + "\n" +
 AlarmStatus + "\n" +
 CallOutStatus + "\n" +
 m_TextMessagesSetting + "\n" +
 SurveillanceStatus + "\n" +
 m_Resolution.toString() + "\n" +
 m_AndroidStorage + "\n";
 int length = Info.length();
 m_InfoTextView.setText(Info.
toCharArray(), 0, length);
}
```

The `ProcessSurveillanceSelection()` Function

The `ProcessSurveillanceSelection()` function activates or deactivates the surveillance system based on the input parameter.

The function does the following:

1. If surveillance is already on and the user is trying to turn on surveillance, then do nothing since surveillance is already on and exit the function.
2. If the user selects to turn surveillance on, then activate the surveillance, reset the surveillance frame number, and execute the `SendTakePhotoCommand()` function in order to perform a "`Take Photo`" command (vga, qvga, qqvga) to begin the surveillance.
3. If the user selects to turn surveillance off, then the surveillance is turned off.

See Listing 6-59.

Operating the Camera Surveillance System

For this hands-on example you will need to update the Android program to the basic wireless framework version 1.2. Note that the Android APK install files for all the chapters are

Listing 6-59 The `ProcessSurveillanceSelection()` Function

```
void ProcessSurveillanceSelection
(boolean Activated)
{
  if (m_SurveillanceActive &&
Activated)
  {
   // surveillance is already
  activated so do nothing
   return;
  }

  // If surveillance set to active
  if (Activated)
  {
   // Surveillance is set to on.
   m_SurveillanceActive = true;

   // Disable "Take Photo" button
  while video surveillance is active
   m_ButtonActive = false;

   // Reset Surveillance Frame Number
   m_SurveillanceFrameNumber = 0;

   // Send TakePhoto Command to Arduino
   SendTakePhotoCommand();
  }
  else
  {
   m_SurveillanceActive = false;
  }
}
```

located in the downloads section of Chapter 9. The information window in the lower right-hand corner of the application now contains the current status of the surveillance feature, the camera resolution that will be used to take a photo, and the current status of the incoming image file save feature.

See Figure 6-24.

The menu system has also been modified. The new menu items "Resolution Settings," "Surveillance Settings," and "Android Storage Settings" have been added. See Figure 6-25.

Figure 6-24 The basic wireless framework version 1.2.

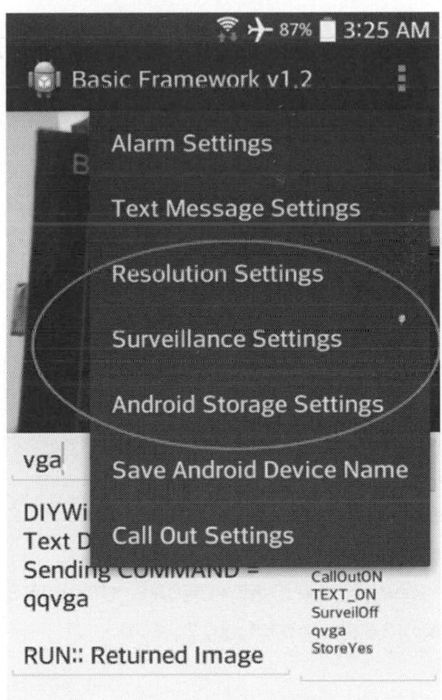

Figure 6-25 The modified menu system.

Next, you will need to download the Arduino ArduCAM project and install it. After you install it press the Serial Monitor button and something like the following should appear on the Serial Monitor:

```
****** ESP-01 Arduino Mega 2560 Server
Program ********
****** ArduCAM Mini OV2640 System
Version 1.1 *************

ArduCAM Starting .........
OV2640 detected ..........

AT Command Sent: AT+CIPMUX=1
ESP-01 Response: AT+CIPMUX=1

OK

ESP-01 mode set to allow multiple
connections.

AT Command Sent: AT+CWMODE_CUR=2
ESP-01 Response: AT+CWMODE_CUR=2

OK

ESP-01 mode changed to Access Point
Mode.

AT Command Sent: AT+CIPSERVER=1,80
ESP-01 Response: AT+CIPSERVER=1,80

no change

OK

Server Started on ESP-01.

AT Command Sent: AT+CWSAP_CUR="ESP-
01","esp8266esp01",1,2,4,0
ESP-01 Response: AT+CWSAP_CUR="ESP-
01","esp8266esp01",1,2,4,0

OK

AP Configuration Set Successfullly.

AT Command Sent: AT+CWSAP_CUR?
ESP-01 Response: AT+CWSAP_CUR?

+CWSAP_CUR:"ESP-
01","esp8266esp01",1,2,4,0

OK

This is the curent configuration of the
AP on the ESP-01.

AT Command Sent: AT+CIPSTO?
ESP-01 Response: AT+CIPSTO?

+CIPSTO:180

OK

This is the curent TCP timeout in
seconds on the ESP-01.

******** System Has Started ********
```

Next, connect your Android's Wi-Fi to the access point running on the ESP-01 module. Start up the Android program. The Android should connect to the TCP server running on the ESP-01 module. The Arduino will receive the name change command from the Android, change the name of the Android to the one specified which in my case is "verizon," and finally send an "OK" text string back to the Android.

```
Notification: '0,CONNECT'
Processing New Client Connection ...
NewConnection Name: 0
Notification: '
'Notification: '
'Notification: '+IPD,0,13:name=verizon
'ClientIndex: 0
DataLength: 13
Data: name=verizon
Client Name Change Requested,
ClientIndex: 0, NewClientName: verizon
Client Name Change Succeeded, OK...
Sending Data ....
TCP Connection ID: 0 , DataLength: 3 ,
Data: OK
```

Next, let's take a photo with the camera using the qqvga command and have it sent back to the Android for display. The Arduino will receive the qqvga command, order the camera to capture an image at qqvga resolution, and send the size of the captured image back to the Android.

The Android will then send a `GetImageData` command to the Arduino. When the Arduino receives this command it will start sending binary image data to the Android in packets of up to 2048 bytes each until the entire image has been transmitted. The image is then displayed on the Android device's screen.

```
Notification: '
'Notification: '
'Notification: '+IPD,0,6:qqvga
'ClientIndex: 0
DataLength: 6
Data: qqvga
............ Starting Image Capture
...............
Start Capture
Capture Done!
FIFO RAW IMAGE SIZE = 3072
Transmitting Image Length = 3071
Sending Data ....
TCP Connection ID: 0 , DataLength: 5 ,
Data: 3071
Notification: '
'Notification: '
'Notification: '+IPD,0,13:GetImageData
'ClientIndex: 0
DataLength: 13
Data: GetImageData
READ_FIFO_LENGTH() = 3072
Sending Binary Data ....
TCP Connection ID: 0 , BufferLength:
2048
Sending Binary Data ....
TCP Connection ID: 0 , BufferLength:
1023
ByteCount = 3071
Total time used:0 seconds ...
```

Next, try to take some pictures at the qvga and vga resolutions. Note that when trying to take a picture at vga resolution the light sensitivity may take a frame or so to adjust. See Figure 6-26 for a screen capture of the Android application after loading in a vga image that was captured by the camera. The image is of a previous book I wrote covering the OV7670 camera called *Beginning Arduino OV7670 Camera Development*.

Figure 6-26 An image captured at VGA resolution.

Figure 6-27 A VGA image that was saved to the local Android file system when surveillance was turned on.

Next, turn on the option to save incoming images and then turn on the surveillance mode. This will start a continuous loop of taking an image, transmitting it to the Android device, displaying the image on the device, and then saving that image to the Android device.
The images should be saved in the CAMPics directory under your Pictures directory. An example of a VGA image that was saved to the Android storage system is shown in Figure 6-27.

Summary

In this chapter I covered many different wireless and remote control hands-on example projects involving a variety of sensors. The Android cell phone was used as a controller which wirelessly connected to an Arduino equipped with an ESP8266-based ESP-01 Wi-Fi module. The ESP-01 provided the Wi-Fi connectivity between the Arduino and the Android device. This chapter covered the reed switch sensor, the flame sensor, the tilt switch sensor, the TMP36 analog temperature sensor, the light detecting photo resistor sensor, the digital DHT11 temperature and humidity sensor, and the ArduCAM OV2640 Mini camera.

CHAPTER 7

Standalone ESP8266 (Model: NodeMCU ESP-12E) and Android Wireless Sensor and Remote Control Projects

THIS CHAPTER WILL PRESENT a wireless communication system using the standalone ESP8266 ESP-12E-based NodeMCU microcontroller and an Android device. This system will be used for a variety of remote sensing and remote control projects.

First, a general overview of the Android and NodeMCU wireless system is given. Next, modifications needed for the basic wireless framework program for the Android are discussed. The basic wireless framework code for the NodeMCU is also covered. This is followed by hands-on examples instructing the reader how to build and operate wireless, remote controlled systems using the Android and NodeMCU with a variety of sensors.

The types of sensors covered in this chapter are:

- The sound detector sensor
- The infrared motion detector sensor
- The SW520D tilt and vibration sensor
- The KeyesIR obstacle detection sensor
- The water level sensor

The Android and NodeMCU (ESP-12E) Wireless System Overview

The system that is discussed in this chapter is similar to the basic wireless framework presented in Chapter 4. This system consists of an Android cell phone and an ESP8266-based NodeMCU microcontroller. The sensor or other hardware that is to be controlled is attached to the NodeMCU microcontroller. The Android cell phone acts as a user interface and a controller for the NodeMCU microcontroller and the attached sensors and other hardware. For example, the Android can send commands such as "`ACTIVATE_ALARM\n`" and "`DEACTIVATE_ALARM\n`" to turn an alarm system located on the NodeMCU on and off. The NodeMCU processes these commands and may send a response such as "`OK\n`" or "`ERROR\n`" to acknowledge that the command was either processed successfully or failed. Alerts can be sent from the NodeMCU to the Android at any time which might trigger an emergency phone call out or text message from the Android cell phone. For example, a motion sensor that detects motion might send out an

"ALARM_TRIPPED\n" or "ALARM_TRIPPED:3\n" message to the Android cell phone. Notifications can also be sent from the NodeMCU microcontroller to the Android that do not require any action to be taken but are displayed in a window on the Basic Wireless Framework application. For example, the messages "Temp is HIGH\n" and "Humidity is HIGH\n" could be notifications that relay the temperature and humidity information to the user. See Figure 7-1.

Modifying the Android Basic Wireless Framework for the NodeMCU (ESP-12E) Platform

In this chapter the basic wireless framework application for the Android will need to be modified. The key change is the addition of code that will allow the user to select the model of server that will be connected to. The two server types are the ESP-01 and the NodeMCU. Each server will require that the Android device set its name on the wireless network in a different way.

The ServerType enumeration specifies the type of server that Android device will connect to. The choices are the ESP-01 module, the NodeMCU microprocessor, and None which indicate that the user has not selected a server type yet.

```
enum ServerType
{
  None,
  ESP01,
  NodeMCU
};
```

The ServerModel variable holds the type of server that is to be connected to and is by default set to None.

```
ServerType ServerModel = ServerType.
None;
```

Modifying the onCreate() Function

The onCreate() function in the MainActivity class needs to be modified by commenting out the CreateClientConnection() function that

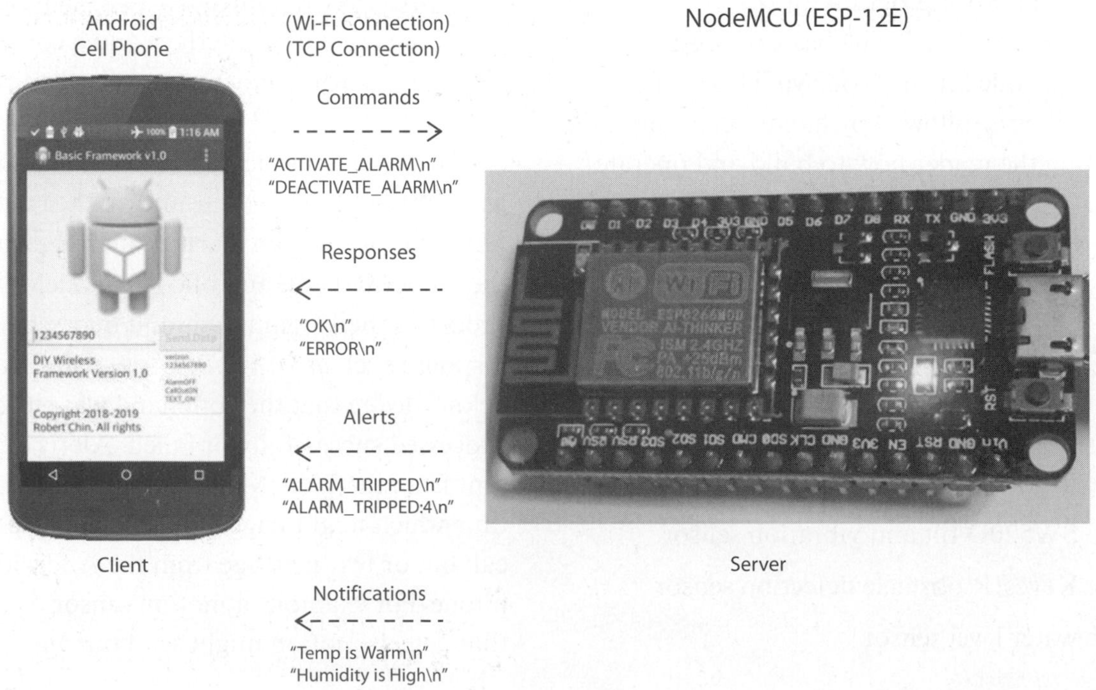

Figure 7-1 The Android and NodeMCU basic wireless framework.

initializes and creates the client connection to the server.

```
// Create the Client Socket
Connection Thread
//CreateClientConnection
(m_ServerStaticIP, m_PortNumber);
```

Also, the `WifiMessageHandler` variable needs to be created here.

```
m_WifiMessageHandler = new
WifiMessageHandler(this);
```

Modifying the Android Basic Framework Menu

A "Connect to Server" menu is added to the basic framework with the sub menu items of "`Connect to ESP-01`" and "`Connect to NodeMCU.`" The new menu code is shown in Listing 7-1.

Modifying the `CreateClientConnection()` Function

The `CreateClientConnection()` function is modified by commenting out the creation of the `m_WifiMessageHandler` variable. See Listing 7-2.

Listing 7-1 Additions to the Basic Wireless Framework Menu System

```
<item android:id="@+id/server_
settings" android:title="Connect to
Server">
  <menu>
    <item android:id="@+id/connect_
  esp01" android:title="Connect to
  ESP-01"/>
    <item android:id="@+id/connect_
  nodemcu" android:title="Connect to
  NodeMCU"/>
  </menu>
<item>
```

Listing 7-2 The `CreateClient Connection()` Function

```
void CreateClientConnection(String
ServerStaticIP, int PortNumber)
{
 //m_WifiMessageHandler = new
WifiMessageHandler(this);
 m_ClientConnectThread = new
Client(ServerStaticIP, PortNumber,
m_WifiMessageHandler, this);
 if (m_ClientConnectThread != null)
 {
  AddDebugMessage(TAG + ": Starting
 ClientConnectThread ...\n");
  m_ClientConnectThread.start();
 }
}
```

Modifying the `onOptionsItemSelected()` Function

New code is added in to handle the user's selection of the "Connect to ESP-01" and "Connect to NodeMCU" menu item selections.

If the "Connect to ESP-01" menu item is selected then the case statement for the R.id.connect_esp01 value is executed which does the following:

1. If there is currently no TCP connection to the server then do the following:
    1. Set the `ServerModel` variable to `ServerType.ESP01`.
    2. Create the TCP connection between the Android device and the server by calling the `CreateClientConnection(m_ServerStaticIP, m_PortNumber)` function with the IP address of the server and the port number that the server is listening to.
    3. Display a notice on the Android screen that a TCP connection attempt is in progress by calling the `Toast.makeText()` function.

2. If there is currently an active TCP connection with the server then display a notice on the Android screen indicating that the Android is already connected to the server by calling the `Toast.makeText()` function.

If the "Connect to NodeMCU" menu item was selected then the code for the case `R.id.connect_nodemcu` is executed which does the following:

1. If there is no TCP connection to the server then do the following:
    1. Set the `ServerModel` variable to `ServerType.NodeMCU`
    2. Connect the Android device to the server by calling the `CreateClientConnection(m_ServerStaticIP, m_PortNumber)` function with the IP address of the server and the port number that the server is listening to for an incoming connection.
    3. Display a notice on the Android screen indicating the Android is connecting to the server by calling the `Toast.makeText()` function.
2. If there is already a TCP connection to the server then display a notice on the Android screen indicating that a TCP connection already exists by calling the `Toast.makeText()` function.

See Listing 7-3.

Listing 7-3 Handling the Server Connection Menu Items

```
case R.id.connect_esp01:
if (m_ClientConnectThread == null)
 {
  // Create the Client Socket Connection Thread
  ServerModel = ServerType.ESP01;
  CreateClientConnection(m_ServerStaticIP, m_PortNumber);
  Toast.makeText(this, "Connecting Android to ESP-01 Server...", Toast.LENGTH_LONG).
 show();
 }
 else
 {
  // Client already connected
  Toast.makeText(this, "Android Already Connected to Server!!!", Toast.LENGTH_LONG).
 show();
 }
 return true;

case R.id.connect_nodemcu:
 if (m_ClientConnectThread == null)
 {
  // Create the Client Socket Connection Thread
  ServerModel = ServerType.NodeMCU;
  CreateClientConnection(m_ServerStaticIP, m_PortNumber);
  Toast.makeText(this, "Connecting Android to NodeMCU Server...", Toast.LENGTH_
 LONG).show();
 }
 else
```

```
{
 // Client already connected
 Toast.makeText(this, "Android Already Connected to Server!!!", Toast.LENGTH_LONG).
show();
}
return true;
```

Modifying the `SetAndroidControllerName()` Function

The `SetAndroidControllerName()` function is modified so that the command that changes the name of the Android device on the wireless network is based on the type of server that is being connected to.

If the server type is the ESP-01 module then the command to change the name of the Android is of the form:

```
"name=AndroidClientName\n"
```

If the server type is the NodeMCU microcontroller then the command to change the name of the Android is of the form:

```
"AndroidClientName\n"
```

If the server type is neither the ESP-01 nor the NodeMCU then an error is generated.

See Listing 7-4. The changes to the code are highlighted in bold print.

Listing 7-4 The `SetAndroidControllerName()` Function

```
void SetAndroidControllerName()
{
 String Command = "";

 if (ServerModel == ServerType.ESP01)
 {
  Command = "name=" + m_
 AndroidClientName +"\n";
 }
 else
 if (ServerModel == ServerType.
NodeMCU)
 {
  Command = m_AndroidClientName
 +"\n";
 }
 else
 {
  // ERROR
  Log.e(TAG, "Android Server Model
 ERROR ....");
  return;
 }
 // Prepare to send data to server
 m_WifiMessageHandler.ResetData();
 m_WifiMessageHandler.SetCommand
("GetTextData");
 if (m_ClientConnectThread != null)
 {
  if (m_ClientConnectThread.
 IsConnected()) {
   // Send Ping to Server
 m_ClientConnectThread.write(Command.
getBytes());

  // Print out Ping sent debug
 message
  Log.e(TAG, "Android Client Name Set
 ....");
  }
 }
}
```

The NodeMCU (ESP-12E) Basic Wireless Server Framework

In this section I cover the code for the NodeMCU-based server that will be connecting with the Android device.

The "`ESP8266WiFi.h`" include file is needed in order to use Wi-Fi-related classes and functions in the program.

```
#include <ESP8266WiFi.h>
```

The `ssid` variable holds the name of the access point that will be created on the NodeMCU device.

```
const char* ssid = "ESP8266-12E-AP";
```

The password variable holds the password needed to connect to the access point that will be created on the NodeMCU device.

```
const char* password = "robtestESP12E";
```

The `server` variable represents the server that is running on the NodeMCU and is listening to port 80 for incoming connection requests.

```
WiFiServer server(80);
```

The `myIP` variable holds the Internet protocol address of the access point that was created on the NodeMCU.

```
IPAddress myIP;
```

The `ClientInfo` structure holds the information about each client that is connected to the server running on the NodeMCU and holds the ID for the client and a `WiFiClient` object that is used to communicate with the client.

```
struct ClientInfo
{
 String ID;
 WiFiClient ClientObject;
};
```

The `MaxClients` variable holds the maximum number of clients that can be connected at one time to the TCP server running on the NodeMCU.

```
const int MaxClients = 4;
```

The `Clients` variable holds all the clients that are connected to the TCP server running on the NodeMCU.

```
ClientInfo Clients[MaxClients];
```

The `Client1` variable is used to check for new client connections and is nonzero if a new client has just connected to the server.

```
WiFiClient Client1;
```

The `StartupInfoDisplayed` variable is false if the start-up information has not yet been displayed after a new client has connected. Once the information is displayed the variable changes to true.

```
boolean StartupInfoDisplayed = false;
```

The `setup()` Function

The `setup()` function initializes the basic wireless framework version 1.3 by doing the following:

1. Initializing the Arduino's Serial Monitor to 9600 baud.
2. Printing some debug text to the Serial Monitor.
3. Initializing the pin connected to the built-in LED on the NodeMCU to be an output pin.
4. Turning off the built-in LED by setting the pin connected to the LED to HIGH. The LED is low enabled so that a HIGH value will shut it off.
5. Setting the Wi-Fi mode of the NodeMCU to the Access Point (AP) mode by calling the `WiFi.mode(WIFI_AP)` function with `WIFI_AP` as the parameter.
6. Creating the access point on the NodeMCU by calling the `WiFi.softAP(ssid, password)` function with the ssid of the access point and the password for the access point as parameters.
7. Retrieving the IP address of the access point by calling the `WiFi.softAPIP()` function.
8. Starting up the TCP server on the NodeMCU by calling the `server.begin()` function.
9. Printing out the server address to the Serial Monitor.
10. Initializing the array that contains the clients that are connected to the server by calling the `InitializeClients()` function.
11. Printing out the Wi-Fi mode of the NodeMCU to the Serial Monitor which is retrieved by calling the `(WiFi.getMode())` function.

12. Turning on the LED light on the NodeMCU device to indicate that the server and access point are up and running by calling the `digitalWrite(LED_BUILTIN, LOW)` function to turn on the LED. A LOW value is written to the pin attached to the built-in LED which turns on the LED. Remember that the LED is active LOW which means that a LOW voltage value will turn the LED on.

See Listing 7-5.

The `loop()` Function

The `loop()` function is executed after the `setup()` function and loops continuously. The `loop()` function does the following:

1. Checks for new clients that have just connected to the server by calling the `CheckForNewConnections()` function.
2. Processes the clients that are already connected to this server by calling the `ProcessConnectedClients()` function.

Listing 7-5 The `setup()` Function

```
void setup()
{
 Serial.begin(9600);
 Serial.println();

 Serial.println(F("*************************************************************"));
 Serial.println(F("***** NodeMCU ESP-12E Access Point Server Program *******"));
 Serial.println(F("************* Basic Framework v1.0 **********************"));

 // Light Initialization
 pinMode(LED_BUILTIN, OUTPUT);      // Initialize the LED_BUILTIN pin as an output
 digitalWrite(LED_BUILTIN, HIGH);   // turn off

 // Set Wifi Mode to Access Point (AP) mode
 WiFi.mode(WIFI_AP);

 // Create Access Point on this module
 WiFi.softAP(ssid, password);
 myIP = WiFi.softAPIP();

 server.begin();

 Serial.print(F("TCP Server started, Server IP Address: "));
 Serial.println(myIP.toString());

 // Initialize Active Clients Array
 InitializeClients();

 // Display Wifi mode of the ESP-12E module
 Serial.print(F("WIFI Mode: "));
 Serial.println(WiFi.getMode());

 // System Active
 Serial.println(F("************* System is Now Active *****************"));
 digitalWrite(LED_BUILTIN, LOW);  // turn on
}
```

Listing 7-6 The `loop()` Function

```
void loop()
{
  // Check for new clients
  CheckForNewConnections();

  // Process the clients that are
already connected to this Access Point
  ProcessConnectedClients();

  // Process Sensor
  ProcessSensor();
}
```

3. Processes the sensors by calling the `ProcessSensor()` function. This function is currently a placeholder function with no code since we are covering just the basic framework now without any sensors.

See Listing 7-6.

The `CheckForNewConnections()` Function

The `CheckForNewConnections()` function checks for and processes new connections to the TCP server running on the NodeMCU by doing the following:

1. Checking to see if there is a new client trying to connect to the server.
2. If there is a new client trying to connect to the server then do the following:
   1. If the program connection information has not been displayed yet then display that information by calling the `ProgramInfo()` function.
   2. Search through the array that holds the clients that are connected to the server. If an available slot is found and the new client is still connected to the server then do the following:
      1. Print out a debug message to the Serial Monitor indicating that the NodeMCU is waiting for the Android device to send its network name to the server.
      2. Get the current time and save it in the `StartTime` variable.
      3. While the client name has not been received from the Android device and a timeout has not occurred do the following:
         1. If data from the new client is available then do the following:
            1. Read in the data from the client until a newline or "`\n`" character is found. This is the client's name.
            2. Print out a debug message to the Serial Monitor displaying the new client's name.
            3. Kill any other clients that use this name by calling the `KillCLient(ClientID)` function with the new name as a parameter.
            4. Save the new client's name and the `WiFiClient` object that represents the new client to the available slot in the array that holds the clients that are connected to the server.
            5. Send a text string to the Android device indicating that the name has been received.
         2. Calculate the elapsed time since the start of waiting to receive the new client's name. If this time has exceeded the timeout value then stop waiting for the new client

    name and continue execution of the program.

  4. If a timeout occurred while waiting for the new client's name then disconnect the new client from the server.

3. If an available slot in the array that holds the clients that are connected to the server was not found this means that all the available client connections to the server are being used. If an available slot was not found then disconnect the new client from the server and print out an error message to the Serial Monitor.

See Listing 7-7.

Listing 7-7 The `CheckForNewConnections()` Function

```
void CheckForNewConnections()
{
 boolean EmptySlotFound   = false;
 String  ClientID = "NA";
 boolean ClientIDReceived = false;

 unsigned long Timeout = 1000 * 60; // 60 seconds
 unsigned long StartTime = 0;
 unsigned long DelayTimeCount = 0;
 boolean ClientIDTimeout = false;

 // Check if there is a new client that has connected to the server
 Client1 = server.available();

 // If a new client has connected then process it
 if (Client1)
 {
  // Display start up info once
  if (StartupInfoDisplayed == false)
  {
   ProgramInfo();
   StartupInfoDisplayed = true;
  }

  Serial.println(F("\n[Client connected]"));
  for (int i = 0; (i < MaxClients) && (EmptySlotFound == false); i++)
  {
   if (Clients[i].ID == "")
   {
    // Empty slot found
    if (Client1.connected())
    {
     Serial.println(F("Waiting to Recieve New CLient's ID"));
     StartTime = millis();
     while((ClientIDReceived == false) && (ClientIDTimeout == false))
     {
      if (Client1.available())
      {
       ClientID = Client1.readStringUntil('\n');
       ClientIDReceived = true;
```

```
        Serial.print(F("Incoming Client's ClientID: "));
        Serial.println(ClientID);

        // Kill all other Clients with same name as incoming client name
        if (KillCLient(ClientID) > 0)
        {
         // Client(s) with same name found
         Serial.println(F("Existing Client(s) with same name DISCONNECTED ...."));
        }

        // Copy New Client Info to Active Clients
        Clients[i].ClientObject = Client1;
        Clients[i].ID = ClientID;

        // Send reply to client
        SendStringClient(Clients[i].ID, Clients[i].ClientObject, "Client ID Recieved
       ...\n");
       }
       DelayTimeCount = millis() - StartTime;
       if (DelayTimeCount > Timeout)
       {
        ClientIDTimeout = true;
       }
      }
       // Process Timeout case for client ID
       if (ClientIDTimeout == true)
       {
        // Close Connection
        Serial.println(F("ClientID Recieve Timed Out ............."));
        Serial.println(F("Disconnecting Client ...................."));
        Serial.println(F("Try To Connect Again ...................."));
        Client1.stop();
       }
      }

      EmptySlotFound = true;
     }
    }

    if (EmptySlotFound == false)
    {
     // Too many clients are trying to connect
     Serial.println(F("Too many clients are trying to connect ..."));
     Serial.print(F("Maximum Number is "));
     Serial.println(MaxClients);
     Serial.println(F("Closing Client Connection ...."));
     Client1.stop();
    }
  }
}
```

The `ProgramInfo()` Function

The `ProgramInfo()` function displays the name of the program being executed to the Arduino's Serial Monitor, the TCP server's IP address, and the current Wi-Fi mode.

See Listing 7-8.

The `KillCLient()` Function

The `KillCLient()` function disconnects a client from the server by doing the following:

1. For each of the clients that are connected to the server do the following:
   1. If the ID of the client matches the target ID of the client to be disconnected then do the following:
      1. Disconnect the client from the server.
      2. Set the ID of the client that was just disconnected to the null string or "".
      3. Increase the value of the variable that keeps track of the number of clients disconnected by 1.
2. Print out the total number of clients that were disconnected to the Serial Monitor.
3. Return the total number of clients that were disconnected.

See Listing 7-9.

The `SendStringClient()` Function

The `SendStringClient()` function sends a text string to a client that is currently connected to the server.

See Listing 7-10.

Listing 7-8 The `ProgramInfo()` Function

```
void ProgramInfo()
{
 Serial.println();
 Serial.println(F("*************
**********************************
****"));
 Serial.println(F("****** NodeMCU
ESP-12E Access Point Server Program
******"));
 Serial.println(F("**************
Basic Framework v1.0
*********************"));

 Serial.println();
 Serial.print(F("TCP Server started,
Server IP Address: "));
 Serial.println(myIP.toString());

 // Display Wifi mode of the ESP-12E
module
 Serial.print(F("WIFI Mode: "));
 Serial.println(WiFi.getMode());
}
```

Listing 7-9 The `KillClient()` Function

```
int KillCLient(String ID)
{
 int result = 0;

 for (int i = 0; i < MaxClients; i++)
 {
  if (Clients[i].ID == ID)
  {
   // Found, Disconnect Client
   Clients[i].ClientObject.stop();
   Clients[i].ID = "";
   result++;
  }
 }

 Serial.print(F("Number of Clients
Killed: "));
 Serial.println(String(result));
 return result;
}
```

Listing 7-10 The `SendStringClient()` Function

```
void SendStringClient(String ClientID,
WiFiClient c, String text)
{
 Serial.print(F("Sending Text to
Client, ClientID: "));
 Serial.print(ClientID);
 Serial.print(F(", Text: "));
 Serial.println(text);

 if (c.connected())
 {
  c.print(text);
 }
 else
 {
  Serial.println(F("ERROR - Client is
 not connected!!"));
 }
}
```

The `ProcessConnectedClients()` Function

The `ProcessConnectedClients()` function processes incoming commands and data from clients by doing the following:

1. For each of the clients that is currently connected to the server do the following:
    1. If data is available from the client then do the following:
        1. Read in the incoming data until there is a new line character.
        2. Process the incoming data by calling the `ProcessIncomingClientData()` function.
    2. If the client is not connected to the server then print out a debug message to the Serial Monitor indicating the name of the client that was disconnected.

See Listing 7-11.

Listing 7-11 The `ProcessConnectedClients()` Function

```
void ProcessConnectedClients()
{
 String line = "NA";

 for (int i = 0; i < MaxClients; i++)
 {
  if (Clients[i].ID != "")
  {
   // Client is valid so check to see
  if there is data to read in
   if (Clients[i].ClientObject.
  connected())
   {
    if (Clients[i].ClientObject.
   available())
    {
     line = Clients[i].ClientObject.
    readStringUntil('\n');
     ProcessIncomingClientData(Client
    s[i].ID, line);
    }
   }
   else
   {
    // Client has disconnected
    // Free up previously used slot
   and free up resources
    Serial.print(F("Client has
   disconnected, Client ID: "));
    Serial.println(Clients[i].ID);

    Clients[i].ID = "";
    Clients[i].ClientObject.stop();
   }
  }
 }
}
```

The `ProcessIncomingClient-Data()` Function

The `ProcessIncomingClientData()` function processes the incoming commands from the Android cell phone. The following commands are processed.

The `list` Command

The `list` command is processed by:

1. Calling the `GetListClients()` function which retrieves the list of clients that are currently connected to the server.
2. Sending the list of clients to the Android device by calling the `SendDataToClient()` function.
3. Printing the list of clients to the Serial Monitor.

The `numberclients` Command

The `numberclients` command is processed by:

1. Getting the number of clients connected to the server by calling the `WiFi.softAPgetStationNum()` function.
2. Sending the data back to the Android device by calling the `SendDataToClient()` function.
3. Printing the number of clients to the Serial Monitor.

The `ledon` and `ledoff` Commands

The `ledon` and `ledoff` commands are processed by turning the LED on the NodeMCU on or off by calling the `ProcessBuiltInLED()` function and then sending an acknowledgment text string back to the Android device by calling the `SendDataToClient()` function.

The `:` Command

The `:` command sends a message to the client that is specified. The exact format of the command is "`ClientName:MessageToClient`". This command is processed by doing the following:

1. The position of the colon is determined.
2. The client name or ID is determined.
3. The message that will be sent to the client is determined.
4. The message is then sent to the designated client by calling the `SendDataToClient()` function.
5. If the message was sent successfully then send an "OK" back to the Android device by calling the `SendDataToClient(ClientID, "OK\n")` function.
6. If the message was not sent successfully then send an error message back to the Android device by calling the `SendDataToClient(ClientID, "ERROR\n")` function.

The `ping` Command

The `ping` command is processed by:

1. Printing out on the Serial Monitor that a ping command was received from a client.
2. Sending a "`PING_OK`" text string back to the Android device by calling the `SendDataToClient(ClientID, "PING_OK\n")` function with the client ID that requested the ping and the string.

Unknown Command

If the incoming data cannot be determined to be one of the commands then an unknown command text string is sent back to the Android device by calling the `SendDataToClient()` function.

See Listing 7-12.

The `SendDataToClient()` Function

The `SendDataToClient()` function sends a text string to a client that is connected to the server. The function searches through the clients that are connected to the server and if the client with a matching name or client ID is found then the text data is sent to that client. The client name and the data sent to the client are also printed to the Serial Monitor. If the client is found then a 1 value is returned otherwise a 0 value is returned.

See Listing 7-13.

Listing 7-12 The `ProcessIncomingClientData()` Function

```
void ProcessIncomingClientData(String ClientID, String Data)
{
 Serial.print(F("ClientID: "));
 Serial.print(ClientID);
 Serial.print(F(" Incoming Data: "));
 Serial.println(Data);

 if (Data == "list")
 {
  String list = GetListClients();
  SendDataToClient(ClientID, "ActiveClients: " + list + "\n");
  Serial.println(String(F("ActiveClients: ")) + list + "\n");
 }
 else
 if (Data == "numberclients")
 {
  int number = WiFi.softAPgetStationNum();
  SendDataToClient(ClientID,"NumberClients: " + String(number) + "\n");
  Serial.println(String(F("NumberClients: ")) + String(number) + "\n");
 }
 else
 if ((Data == "ledon") || (Data == "ledoff"))
 {
  String response = ProcessBuiltInLED(Data);
  if (!SendDataToClient(ClientID, response))
  {
   Serial.println(F("*************** ERROR ... Client Not Found"));
  }
 }
 else
 if (Data.indexOf(":") >= 0)
 {
  // Send data to specific client by name
  // ClientName:MessageToClient
  // Send message to destination client
  // Send confirmation or failure message to sending client
  boolean MessageSent = true;
  int destseparator = Data.indexOf(":");
  String  clientid = Data.substring(0,destseparator);
  String  message = Data.substring(destseparator+1);

  if (!SendDataToClient(clientid, message + "\n"))
  {
   Serial.println(F("*************** ERROR ... Client Not Found"));
   MessageSent = false;
  }
  // Check if message sent successfully and send back appropriate
  // response to the sending client
  if (MessageSent)
```

```
    {
     // Send OK response message to requesting client
     SendDataToClient(ClientID, "OK\n");
    }
    else
    {
     // Send ERROR response message to requesting client
     SendDataToClient(ClientID, "ERROR\n");
    }
  }
  else
  if (Data == "ping")
  {
   // Ping from client
   Serial.print(F("Received PING from Client: "));
   Serial.println(ClientID);
   SendDataToClient(ClientID, "PING_OK\n");
  }
  else
  {
   SendDataToClient(ClientID, "Unknown Command: " + Data + "\n");
  }
}
```

Listing 7-13 The `SendDataToClient()` Function

```
int SendDataToClient(String ClientID, String Data)
{
 boolean found = false;
 int result = 0;

 // send Data to client with ClientID
 for (int i = 0; (i < MaxClients) && !found; i++)
 {
  if (Clients[i].ID == ClientID)
  {
   Clients[i].ClientObject.print(Data);
   found = true;
  }
 }
 if (found)
 {
  result = 1;
 }

 Serial.print(F("Sending Data to Client ID: "));
 Serial.print(ClientID);
 Serial.print(F(" , Data: "));
 Serial.println(Data);

 return result;
}
```

Listing 7-14 The `InitializeClients()` Function

```
void InitializeClients()
{
 for (int i = 0; i < MaxClients; i++)
  {
   Clients[i].ID = "";
  }
}
```

The `InitializeClients()` Function

The `InitializeClients()` function initializes the `Clients` array that holds the clients that are connected to the server by setting the client ID for each slot in the array to the null string. This indicates that these slots are available to be used for new clients.

See Listing 7-14.

The `GetListClients()` Function

The `GetListClients()` function returns the list of clients that are connected to the server by searching through the `Clients` array, finding all the client entries that have a client ID that is not the null string, and then returning these clients to the caller.

See Listing 7-15.

The `ProcessBuiltInLED()` Function

The `ProcessBuiltInLED()` function processes an LED-related command by doing the following:

1. If the command is `ledon` then turn the LED on.
2. If the command is `ledoff` then turn the LED off.
3. Returning a string to the caller of the function indicating what command was executed.

See Listing 7-16.

Listing 7-15 The `GetListClients()` Function

```
String GetListClients()
{
 String result = "";

 for (int i = 0; i < MaxClients; i++)
 {
  if (Clients[i].ID != "")
  {
   result += Clients[i].ID;
   result += ", ";
  }
 }

 return result;
}
```

Listing 7-16 The `ProcessBuiltInLED()` Function

```
String ProcessBuiltInLED(String Data)
{
 String result = "Server Response:";
 if (Data == "ledon")
 {
  // Turn LED on - active low
  digitalWrite(LED_BUILTIN, LOW);
  result += "LED_ON";
 }
 else if(Data == "ledoff")
 {
  // Turn LED off - active low
  digitalWrite(LED_BUILTIN, HIGH);
  result += "LED_OFF";
 }

 result += "\n";

 return result;
}
```

The `ProcessSensor()` Function

The `ProcessSensor()` function contains no code but serves as a placeholder function where you will be able to add in sensor and other hardware-related code in the future.

See Listing 7-17.

Listing 7-17 The `ProcessSensor()` Function

```
void ProcessSensor()
{
  // Put your sensor and other
hardware related code here.
}
```

Hands-on Example: The Basic Wireless Framework Version 1.3 for the Android and NodeMCU Microprocessor

In this hands-on example I demonstrate how to build and use version 1.3 of the Android basic wireless framework with the NodeMCU-based Wi-Fi TCP server. This basic framework will be the basis for the rest of the hands-on examples for this chapter.

Parts List

- 1 Android cell phone
- 1 NodeMCU microprocessor

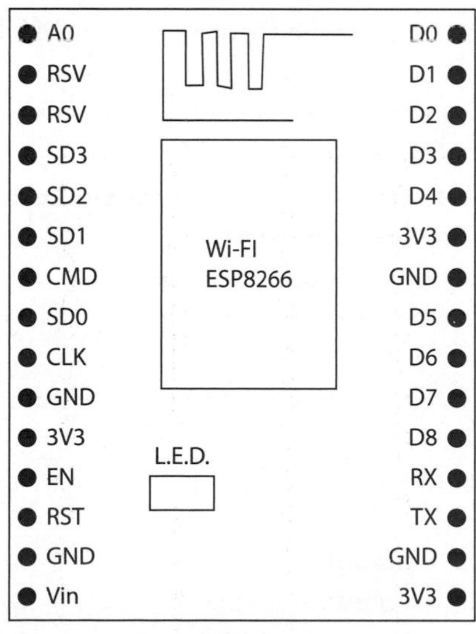

Figure 7-2 The ESP8266-based NodeMCU microprocessor.

Setting Up the Hardware

All the hardware you need for this hands-on example is a NodeMCU device. See Figure 7-2.

Setting Up the Software

The modified software for the Android basic wireless framework and the server code for the NodeMCU were discussed earlier in this chapter.

Operating the System

Download the Android basic framework version 1.3 APK or the Android Studio project for this chapter and install the application on your Android cell phone. Also, download the basic framework for the NodeMCU for this chapter and install that on your NodeMCU device.

Start up the Android basic framework application. You should see something similar to Figure 7-3.

A new item called "Connect to Server" has been added to the main menu. See Figure 7-4.

Figure 7-3 The Android basic wireless framework version 1.3.

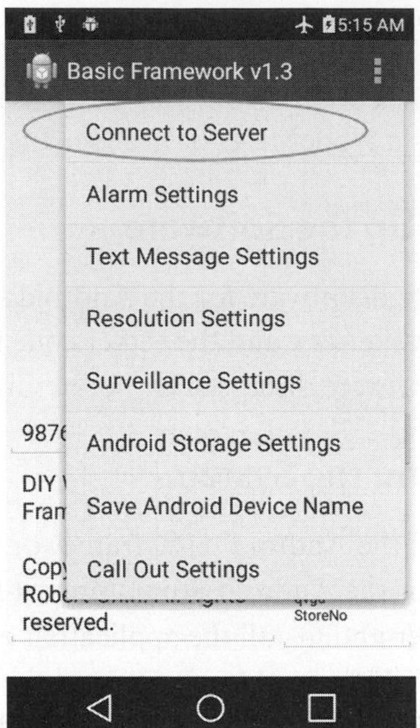

Figure 7-4 The new Connect to Server menu selection.

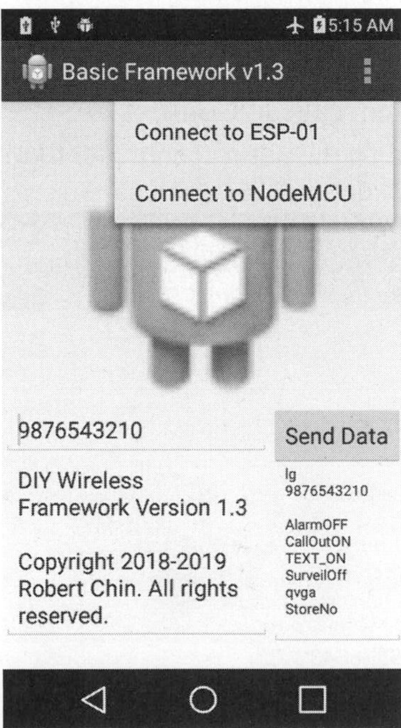

Figure 7-5 Connecting to the ESP-01 or the NodeMCU.

If you select the "Connect to Server" menu selection, the exact type of server to connect to will appear. The menu selections for this sub menu are "Connect to ESP-01" and "Connect to NodeMCU". See Figure 7-5.

After you finish viewing the changes made to the Android application exit out of the program by pressing the back button.

Next, you need to start up the Serial Monitor on the Arduino IDE. For more information on how to use the Arduino IDE with the NodeMCU, please refer to Chapter 2 in this book which is on the ESP8266. You may see some start-up debug information or you may see nothing at startup. On my XP system I don't see the start-up screen indicating the program name. However, on my Windows 10 system I do see the start-up screen. Connect your Android to the Wi-Fi access point running on the NodeMCU. Next, start up the Android basic framework version 1.3. Connect to the TCP server running on the NodeMCU using the new menu selections that were just discussed. You should see the name of the program that is running on the NodeMCU which is the "Basic Framework v1.0," the TCP server address, and the Wi-Fi mode (which should be the number 2 that indicates the NodeMCU is in access point mode) show up on the Serial Monitor. The Android application should automatically send the device's name to the server and a "`Client ID Received`..." text string should be sent back to the Android device from the NodeMCU.

```
****************************************
*****************
****** NodeMCU ESP-12E Access Point
Server Program ******
************** Basic Framework v1.0
*********************

TCP Server started, Server IP Address:
192.168.4.1
WIFI Mode: 2

[Client connected]
Waiting to Recieve New CLient's ID
Incoming Client's ClientID: lg
Number of Clients Killed: 0
Sending Text to Client, ClientID: lg,
Text: Client ID Recieved ...
```

Next, send a `list` command to the NodeMCU from the Android application. The NodeMCU should receive the command from the client and send a list of clients that are connected to the server back to the Android device. In my case there is one client which is my phone which has the name `lg` on the wireless network. The list of active clients should also be printed to the Serial Monitor.

```
ClientID: lg Incoming Data: list
Sending Data to Client ID: lg , Data:
ActiveClients: lg,
ActiveClients: lg,
```

Next, send a `ping` command to the NodeMCU from the Android application. The NodeMCU should receive the `ping` command, and then send a "`PING_OK`" text string back to the Android device.

```
ClientID: lg Incoming Data: pingRecieved
PING from Client: lgSending Data to
Client ID: lg , Data: PING_OK
```

Next, send a `numberclients` command to the NodeMCU. The NodeMCU will receive the command, and send the number of clients that are connected to the server back to the Android device. The number of clients is also printed to the Serial Monitor.

```
ClientID: lg Incoming Data:
numberclients
Sending Data to Client ID: lg , Data:
NumberClients: 1

NumberClients: 1
```

Next, send a `ledoff` command to the NodeMCU. The NodeMCU will receive this command, turn off the LED on the device, and send a "`Server Response: LED_OFF`" text string back to the Android.

```
ClientID: lg Incoming Data: ledoff
Sending Data to Client ID: lg , Data:
Server Response:LED_OFF
```

Next, send a `ledon` command to the NodeMCU. The NodeMCU will turn on the LED on the device and send back a "`Server Response:LED_ON`" string to the Android.

```
ClientID: lg Incoming Data: ledon
Sending Data to Client ID: lg , Data:
Server Response:LED_ON
```

Next, send a message using the `:` or colon operator which sends a message to a specific client by name. For example, for me the command `lg:test1234567890` will send the string "`test1234567890`" back to my Android device which is named `lg`. An "OK" string will also be sent to my device indicating that the text string was sent.

```
ClientID: lg Incoming Data:
lg:test1234567890
Sending Data to Client ID: lg , Data:
test1234567890

Sending Data to Client ID: lg , Data: OK
```

Next, send a command to the NodeMCU that will not be recognized as a valid command such as `test1234567890`. The NodeMCU will send back an "`Unknown Command: test1234567890`" to the Android.

```
ClientID: lg Incoming Data:
test1234567890
Sending Data to Client ID: lg , Data:
Unknown Command: test1234567890
```

Finally, disconnect your Android device from the NodeMCU server by hitting the back button on the Android device. You should see a client has disconnected notice on the Arduino's Serial Monitor.

```
Client has disconnected, Client ID: lg
```

Hands-on Example: The Wireless Glass Break/Sound Detector Alarm System

In this hands-on example I demonstrate how to build and operate a wireless sound detector alarm system that can detect loud noises such as the sound of breaking glass from a forced entry by a burglar through a closed and locked window. It can also detect other loud noises such as from attempts to knock down or kick in a front or back door.

Parts List

- 1 Android cell phone
- 1 NodeMCU microprocessor
- 1 FC-04 sound sensor (this sensor was discussed in a previous chapter)
- 1 Breadboard (optional)
- 1 Package of wires male-to-male or female-to-female (depending on the exact connection method)

Setting Up the Hardware

1. Connect the Vcc pin on the sound sensor to the 3V3 pin on the NodeMCU.
2. Connect the GND pin on the sound sensor to the GND pin on the NodeMCU.
3. Connect the Out pin on the Sound Sensor to the D1 pin on the NodeMCU.

See Figure 7-6.

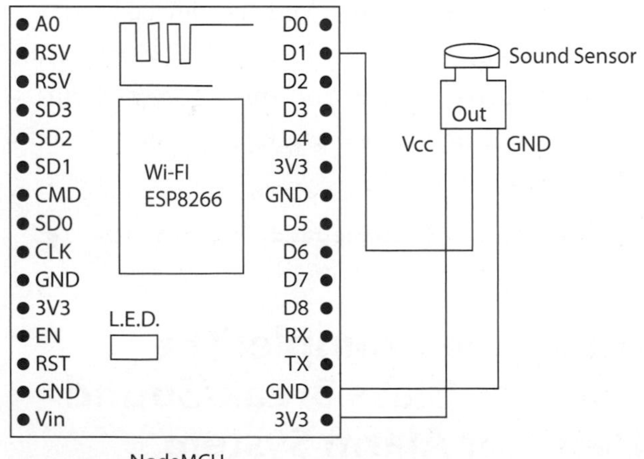

Figure 7-6 The wireless sound detector alarm system.

Setting Up the Software

No change to the Android code is needed for this hands-on example and you will be able to use the same Android application you used in the previous hands-on example.

The NodeMCU code for this hands-on example is based on the basic server framework that was covered in the previous hands-on example. It is also based on the Arduino Mega-based sound detection alarm system presented in Chapter 5. In this section I will cover only the changes made to the basic server framework code and Arduino Mega-based sound detection system code.

The `SensorPin` variable holds the NodeMCU pin that is attached to the sensor and is set to pin D1.

```
int SensorPin = D1;
```

The `SendAlarmNowActiveMessage()` Function

The `SendAlarmNowActiveMessage()` function sends an alarm is now active text string to the Android device by calling the `SendDataToClient(ConnectionIDForAlarm, returnstring)` function. See Listing 7-18.

The `SendAlarmTrippedMessage()` Function

The `SendAlarmTrippedMessage()` function sends an alarm tripped string and the total number of times that the alarm has been tripped since it was activated to the Android cell phone that activated the alarm. See Listing 7-19.

Listing 7-18 The `SendAlarmNowActiveMessage()` Function

```
void SendAlarmNowActiveMessage()
{
  // Send the Alarm tripped message
  String returnstring = "ALARM NOW
ACTIVE...\n";
  SendDataToClient(ConnectionIDFor
Alarm, returnstring);
}
```

Listing 7-19 The `SendAlarmTripped Message()` Function

```
void SendAlarmTrippedMessage()
{
 // Send the Alarm tripped message
 String returnstring = "ALARM_
TRIPPED:" + String(NumberHits) + "\n";
 SendDataToClient(ConnectionIDFor
Alarm, returnstring);
}
```

Listing 7-20 The `ProgramInfo()` Function

```
void ProgramInfo()
{
 Serial.println();
 Serial.println(F("***********
***********************
*********************"));
 Serial.println(F("****** NodeMCU
ESP-12E Access Point Server Program
******"));
 Serial.println(F("************
Sound Detection Sensor System
**************"));

 Serial.println();
 Serial.print(F("TCP Server started,
Server IP Address: "));
 Serial.println(myIP.toString());

 // Display Wifi mode of the ESP-12E
module
 Serial.print(F("WIFI Mode: "));
 Serial.println(WiFi.getMode());
}
```

The `ProgramInfo()` Function

The `ProgramInfo()` function is modified to indicate that the program uses a sound detector as the sensor. See Listing 7-20.

Modifying the `setup()` Function

The `setup()` function is modified by adding in a line of code that initializes the sensor pin to be an INPUT pin that can read voltage levels.

```
pinMode(SensorPin, INPUT);
```

Modifying the `Activate Alarm` Command

The `activate alarm` command is modified by using the `SendDataToClient(ClientID, returnstring)` function to send a text string over TCP to the Android device instead of the `SendTCPMessage(ClientIndex, returnstring)` function. See Listing 7-21.

Listing 7-21 The `Activate Alarm` Command

```
if (Data == "ACTIVATE_ALARM")
{
 if (AlarmState == Alarm_OFF)
 {
  NumberHits = 0;
  AlarmState = Alarm_WAIT;
  String returnstring = "Alarm
 Activated...WAIT TIME:" +
 String(WaitTime/1000) + "
 Seconds.\n";
  SendDataToClient(ClientID,
 returnstring);
  WaitTimeStart = millis();
  ConnectionIDForAlarm = ClientID;
 }
 else
 {
  String returnstring = "Alarm
 Already Active...\n";
  SendDataToClient(ClientID,
 returnstring);
 }
}
```

The `Deactivate Alarm` Command

The `deactivate alarm` command is modified by using the `SendDataToClient(ClientID, returnstring)` function to send a text string to the Android cell phone instead of

Listing 7-22 The `Deactivate Alarm` Command

```
if (Data == "DEACTIVATE_ALARM")
{
 AlarmState = Alarm_OFF;
 String returnstring = "Alarm Has
Been DE-Activated ...\n";
 SendDataToClient(ClientID,
returnstring);
}
```

Listing 7-23 The `Hits` Command

```
if (Data == "hits")
{
 // Return number of motions detected
to requesting client
 String returnstring = "Number Motions
Detected: " + String(NumberHits) + "\n";
 SendDataToClient(ClientID,
returnstring);
}
```

using the `SendTCPMessage(ClientIndex, returnstring)` function. See Listing 7-22.

The `Hits` Command

The `hits` command is modified by using the `SendDataToClient(ClientID, returnstring)` function to send a text string to the Android cell phone instead of using the `SendTCPMessage(ClientIndex, returnstring)` function. See Listing 7-23.

The `Reset` Command

The `reset` command is modified by using the `SendDataToClient(ClientID, returnstring)` function to send a text string to the Android cell phone instead of using the `SendTCPMessage(ClientIndex, returnstring)` function. See Listing 7-24.

The `Wait` Command

The `wait` command is modified by using the `SendDataToClient(ClientID, returnstring)` function to send a text string to the Android cell phone instead of using the `SendTCPMessage(ClientIndex, returnstring)` function. See Listing 7-25.

Listing 7-24 The `Reset` Command

```
if (Data == "reset")
{
 // Reset number hits to 0
 NumberHits = 0;
 String returnstring = "Number Hits
Reset to 0.\n";
 SendDataToClient(ClientID,
returnstring);
}
```

Listing 7-25 The `Wait` Command

```
if (Data.indexOf("wait=") >= 0)
{
 int indexwait = Data.indexOf("=");
 String waitlength = Data.
substring(indexwait+1);
 WaitTime = waitlength.toInt();

 String returnstring = "";
 returnstring = returnstring +
F("WaitTime set to: ") + WaitTime/1000
+ F(" Seconds\n");
 SendDataToClient(ClientID,
returnstring);
}
```

The `loop()` Function

The `loop()` function is modified by adding in an `UpdateAlarm()` function to update the status of the alarm system and by adding code to the formerly empty `ProcessSensor()` function which now processes the sound detector sensor. See Listing 7-26.

Operating the System

Download and install the Android basic wireless framework version 1.3 program if you have not done so already. Download the NodeMCU

Listing 7-26 The loop() Function

```
void loop()
{
 // Check for new clients
 CheckForNewConnections();

 // Process the clients that are
already connected to this Access Point
 ProcessConnectedClients();

 // Update the Alarm Status
 UpdateAlarm();

 // Process Sensor
 ProcessSensor();
}
```

sound detector project for this chapter and install it on your NodeMCU device using the Arduino IDE. Start up the Arduino IDE's Serial Monitor. Connect the Android device to the Wi-Fi access point running on the NodeMCU. Start up the Android basic wireless framework version 1.3 program. Using the Android program connect to the TCP server running on the NodeMCU. The Android should connect to the TCP server, send the device's name or client ID to the NodeMCU, and then receive a text string back from the NodeMCU that indicates that the client ID was received. You should see something similar to the following on the Serial Monitor. Note that the header portion that displays the program name, server IP address, and Wi-Fi mode may be duplicated depending on your operating system. The name of my Android cell phone was `lg`.

```
****************************************
****************
****** NodeMCU ESP-12E Access Point
Server Program ******
************ Sound Detection Sensor
System **************

TCP Server started, Server IP Address:
192.168.4.1
WIFI Mode: 2
```

```
[Client connected]
Waiting to Recieve New CLient's ID
Incoming Client's ClientID: lg
Number of Clients Killed: 0
Sending Text to Client, ClientID: lg,
Text: Client ID Recieved ...
```

Next, using the Android application turn off the emergency call out and the emergency text messages. Activate the alarm. The NodeMCU should receive the activate alarm text string, set the alarm into waiting mode, and then after approximately 10 seconds send an alarm now active text string back to the Android device. A debug message should also be printed out to the Serial Monitor indicating that the alarm is now fully activated and ready to be tripped.

```
ClientID: lg Incoming Data: ACTIVATE_
ALARM
Sending Data to Client ID: lg , Data:
Alarm Activated...WAIT TIME:10 Seconds.
Sending Data to Client ID: lg , Data:
ALARM NOW ACTIVE...
ALARM WAIT STATE FINISHED ... Alarm Now
ACTIVE ....
```

Next, trip the alarm several times by making loud noises. You should see the total number of sound detections or number of hits displayed along with name of the Android device that the alarm tripped text string was sent to which in my case was `lg`. The following output from the Serial Monitor indicates that the alarm has been tripped 5 times and that 5 alarm tripped text strings have been sent to the Android device over the TCP connection.

```
SensorValue: 0 , Motion Detected ...
NumberHits: 1
Sending Data to Client ID: lg , Data:
ALARM_TRIPPED:1

SensorValue: 1
SensorValue: 0 , Motion Detected ...
NumberHits: 2
Sending Data to Client ID: lg , Data:
ALARM_TRIPPED:2
```

```
SensorValue: 1
SensorValue: 0 , Motion Detected ...
NumberHits: 3
Sending Data to Client ID: lg , Data:
ALARM_TRIPPED:3

SensorValue: 1
SensorValue: 0 , Motion Detected ...
NumberHits: 4
Sending Data to Client ID: lg , Data:
ALARM_TRIPPED:4

SensorValue: 1
SensorValue: 0 , Motion Detected ...
NumberHits: 5
Sending Data to Client ID: lg , Data:
ALARM_TRIPPED:5
```

Next, deactivate the alarm using the Android application. The NodeMCU should receive the deactivate alarm text string and then should send the Android device a text string indicating that the alarm has been deactivated.

```
SensorValue: 1
ClientID: lg Incoming Data: DEACTIVATE_
ALARM
Sending Data to Client ID: lg , Data:
Alarm Has Been DE-Activated ...
```

Finally, press the back button on the Android device to exit the program. You should see on the Serial Monitor that the Android client has disconnected from the TCP server running on the NodeMCU device.

```
Client has disconnected, Client ID: lg
```

Hands-on Example: The Wireless HC-SR501 Infrared Motion Detector Alarm System

In this hands-on example I will show you how to build and operate a wireless motion detector alarm system using the HC-SR501 sensor.

Parts List

- 1 Android cell phone
- 1 NodeMCU microprocessor
- 1 HC-SR501 infrared motion detector
- 1 Package of wires to connect the motion detector to the NodeMCU

Setting Up the Hardware

1. Connect the Vcc pin on the motion detector to the Vin pin on the NodeMCU. Note that when connected to the USB port the Vin pin on the NodeMCU provides a 5-V output on most brands.
2. Connect the GND pin on the motion detector to a GND pin on the NodeMCU.
3. Connect the Out pin on the motion detector to pin D1 on the NodeMCU.

See Figure 7-7.

Setting Up the Software

The NodeMCU code for this hands-on example is based on the code for the sound detector alarm system covered in the last hands-on example. I will only cover the changes to the

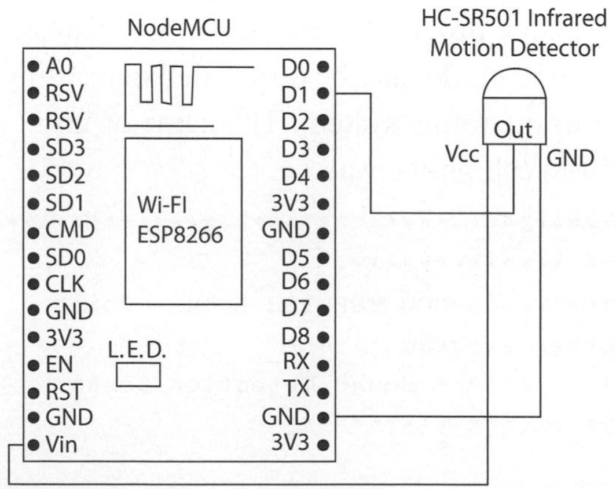

Figure 7-7 The NodeMCU wireless infrared motion detector alarm system.

code made specifically for the HC-SR501 sensor in this section.

The `MotionDetected` variable indicates the value the sensor will output when a motion has been detected. The variable is set by default to 1 or HIGH voltage.

```
int MotionDetected = 1;
```

The `MotionNotDetected` variable indicates the value the sensor will output when a motion is not detected. The variable is set by default to 0 or LOW voltage.

```
int MotionNotDetected  = 0;
```

The `SensorValue` holds the current motion detection status of the sensor and is initialized to a value that indicates no motion is detected.

```
int SensorValue = MotionNotDetected;
```

The `PrevSensorValue` variable holds the previously read in state of the motion detector and is initialized to a value that indicates that no motion was detected.

```
int PrevSensorValue = MotionNotDetected;
```

The `ProcessSensor()` Function

The `ProcessSensor()` function handles the reading in of the sensor data and the text data sent to the Android device as the result of a tripped alarm. The code is the same as the sound detector alarm system code presented in the previous hands-on example but the `SoundNotDetected` and `SoundDetected` variables are replaced with the `MotionNotDetected` and `MotionDetected` variables. See Listing 7-27. The modified code is shown in bold print.

Operating the System

Download and install on your Android cell phone the basic framework version 1.3 APK or Android Studio Project if you have not done so already. Download and install on your NodeMCU the motion detector project file for this chapter. Bring up the Serial Monitor. Next, connect the Android to the Wi-Fi access point running on the NodeMCU.

Run the Android basic framework program and turn off the emergency call out and the emergency text messages. Next, connect the Android to the TCP server running on the NodeMCU. The program name, the TCP server address, and the Wi-Fi mode of the NodeMCU are displayed. The name of the Android is sent to the NodeMCU and a client ID received text string is sent back to the Android from the NodeMCU. You should see something like the following on the Serial Monitor:

```
*****************************************
*****************
****** NodeMCU ESP-12E Access Point
Server Program ******
************ Motion Detection Sensor
System **************

TCP Server started, Server IP Address:
192.168.4.1
WIFI Mode: 2

[Client connected]
Waiting to Recieve New CLient's ID
Incoming Client's ClientID: lg
Number of Clients Killed: 0
Sending Text to Client, ClientID: lg,
Text: Client ID Recieved ...
```

Next, turn on the alarm using the Android application. The NodeMCU will receive the activate alarm text string, activate the alarm, and put it into waiting mode. The NodeMCU will send a text string back to the Android indicating that the alarm has been activated and put into the wait state. After around 10 seconds the alarm will become fully active. Another text string will be sent to the Android indicating that the alarm can be tripped. A debug message is also printed to the Serial Monitor.

```
ClientID: lg Incoming Data: ACTIVATE_
ALARM
```

Listing 7-27 The `ProcessSensor()` Function

```
void ProcessSensor()
{
 // Put Code to process your sensors or other hardware in this function.
 if ((AlarmState == Alarm_ON) || (AlarmState == Alarm_TRIPPED))
 {
  // Process Sensor Input
  SensorValue = digitalRead(SensorPin);
  if ((PrevSensorValue == MotionNotDetected) && (SensorValue == MotionDetected))
  {
   unsigned long ElapsedTime = millis() - MinDelayBetweenHitsStartTime;
   if (ElapsedTime >= MinDelayBetweenHits)
   {
    // Motion Just Detected
    NumberHits++;
    Serial.print(F("SensorValue: "));
    Serial.print(SensorValue);
    Serial.print(F(" , Motion Detected ... NumberHits: "));
    Serial.println(NumberHits);
    PrevSensorValue = MotionDetected;

    // Send Alarm Tripped Message
    SendAlarmTrippedMessage();

    // Time this hit occured
    MinDelayBetweenHitsStartTime = millis();
   }
  }
  else
  if ((PrevSensorValue == MotionDetected) && (SensorValue == MotionNotDetected))
  {
   Serial.print(F("SensorValue: "));
   Serial.println(SensorValue);
   PrevSensorValue = MotionNotDetected;
  }
 }
}
```

```
Sending Data to Client ID: lg , Data:
Alarm Activated...WAIT TIME:10 Seconds.

Sending Data to Client ID: lg , Data:
ALARM NOW ACTIVE...

ALARM WAIT STATE FINISHED ... Alarm Now
ACTIVE ....
```

Next, trip the alarm several times by putting your hand in front of the motion sensor. This should trip the alarm and send an alarm tripped text string to the Android device. The Android device should receive and display the alarm tripped text string and produce a sound to indicate that it received the alert. After around 2.5 seconds the alarm will be able to be tripped again.

```
SensorValue: 1 , Motion Detected ...
NumberHits: 1
```

```
Sending Data to Client ID: lg , Data:
ALARM_TRIPPED:1

SensorValue: 0
SensorValue: 1 , Motion Detected ...
NumberHits: 2
Sending Data to Client ID: lg , Data:
ALARM_TRIPPED:2

SensorValue: 0
SensorValue: 1 , Motion Detected ...
NumberHits: 3
Sending Data to Client ID: lg , Data:
ALARM_TRIPPED:3

SensorValue: 0
SensorValue: 1 , Motion Detected ...
NumberHits: 4
Sending Data to Client ID: lg , Data:
ALARM_TRIPPED:4
```

Next, deactivate the alarm using the Android application. The NodeMCU will receive a deactivate alarm text string from the Android, deactivate the alarm, and then send a text string back to the Android to indicate that the alarm has been deactivated.

```
ClientID: lg Incoming Data: DEACTIVATE_
ALARM
Sending Data to Client ID: lg , Data:
Alarm Has Been DE-Activated ...
```

Finally, exit the Android program by pressing the back button. You will also be disconnected from the NodeMCU.

```
Client has disconnected, Client ID: lg
```

The SW520D Tilt and Vibration Sensor

The SW520D tilt and vibration sensor is a switch that conducts current when placed upright and is open and does not conduct current when laying on its side or when there is a vibration. Inside the switch are loose metal balls that conduct current between the switch's two terminals depending on the orientation of the switch. See Figure 7-8.

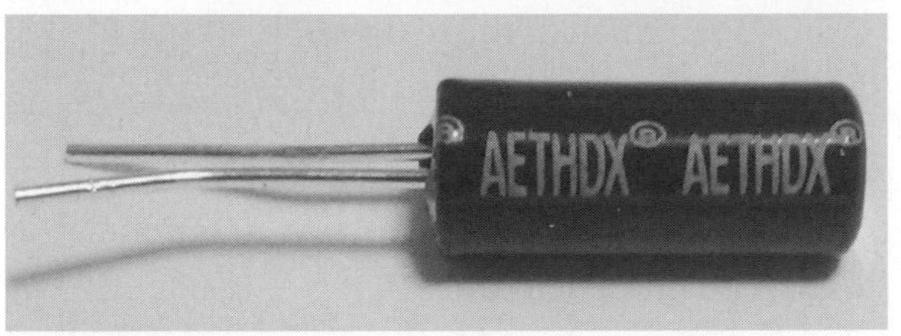

Figure 7-8 The SW520D tilt and vibration switch.

Hands-on Example: The Wireless Tilt and Vibration Sensor System

In this hands-on example I discuss how to use a tilt and vibration switch to build a wireless alarm system that will detect when the switch vibrates or is tilted on its side.

Parts List

- 1 Android cell phone
- 1 NodeMCU microprocessor
- 1 SW520D tilt and vibration switch
- 1 Resistor (10,000 Ohm recommended)
- 1 Breadboard
- 1 Package of wires to connect the components together

Setting Up the Hardware

1. Connect one end of the tilt switch to the 3.3-V pin on the NodeMCU.
2. Connect the other end of the tilt switch to one end of a 10,000-ohm resistor.
3. Connect the other end of the resistor to a GND pin on the NodeMCU.
4. Connect the D1 pin on the NodeMCU to the node that contains one end of the tilt switch and one end of the resistor.

See Figure 7-9.

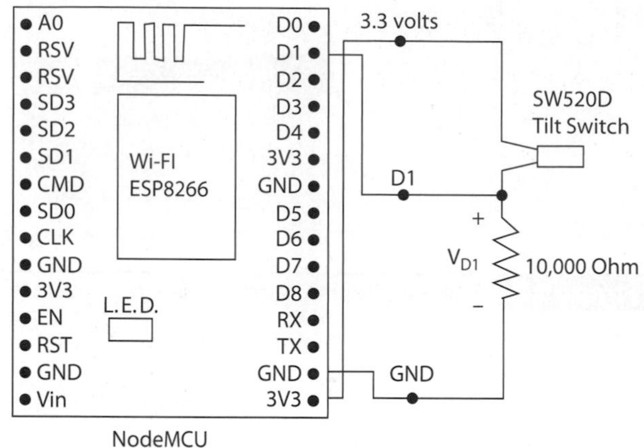

Figure 7-9 The wireless tilt and vibration switch alarm system.

Setting Up the Software

The Arduino software for this example is almost identical as the last hands-on example for the motion detector.

The `MotionDetected` variable holds the value that is read in from the tilt sensor when motion has been detected. This means that the tilt switch is on its side or has experienced a vibration. The value is initialized to 0 or LOW voltage.

```
int MotionDetected = 0; // Tilt Switch
on Side
```

The `MotionNotDetected` variable holds the value that is read in from the tilt sensor when a motion is not detected. This means that the tilt switch is standing upright. The value is initialized to 1 or HIGH voltage.

```
int MotionNotDetected  = 1; // Tilt
Switch Upright
```

Operating the System

Install the tilt motion sensor project for this chapter on your NodeMCU using the Arduino IDE. Start up the Serial Monitor. Connect your Android device to the Wi-Fi access point running on the NodeMCU using the password shown in the code. Next, run the Android basic framework application version 1.3 and connect to the NodeMCU. The program name, the TCP server address, and Wi-Fi mode should be displayed. The NodeMCU should receive the Android device name and reply with a text string that indicates that the name was received. You should see something like the following on the Serial Monitor:

```
****************************************
*****************
****** NodeMCU ESP-12E Access Point
Server Program ******
********** Tilt Motion Detection Sensor
System **********

TCP Server started, Server IP Address:
192.168.4.1
WIFI Mode: 2

[Client connected]
Waiting to Recieve New CLient's ID
Incoming Client's ClientID: lg
Number of Clients Killed: 0
Sending Text to Client, ClientID: lg,
Text: Client ID Recieved ...
```

Next, deactivate the emergency call out and emergency text messages using the Android application. Then activate the alarm system from the Android application. The NodeMCU will receive the activate alarm text string from the Android device and then reply with a text string indicating that the alarm has been put into the wait state.

```
ClientID: lg Incoming Data: ACTIVATE_
ALARM
Sending Data to Client ID: lg , Data:
Alarm Activated...WAIT TIME:10 Seconds.
```

Wait approximately 10 seconds and the NodeMCU will send a text string to the Android indicating that the alarm is now active. An alarm now active text debug message will also be printed to the Serial Monitor.

```
Sending Data to Client ID: lg , Data:
ALARM NOW ACTIVE...
```

```
ALARM WAIT STATE FINISHED ... Alarm Now
ACTIVE ....
```

Next, trip the alarm several times by tilting or shaking the tilt switch. Each time the alarm is tripped it sends a text string to the Android indicating that the alarm has been tripped and the number of times it has been tripped since it was activated.

```
SensorValue: 0 , Motion Detected ...
NumberHits: 1
Sending Data to Client ID: 1g , Data:
ALARM_TRIPPED:1

SensorValue: 1
SensorValue: 0 , Motion Detected ...
NumberHits: 2
Sending Data to Client ID: 1g , Data:
ALARM_TRIPPED:2

SensorValue: 1
SensorValue: 0 , Motion Detected ...
NumberHits: 3
Sending Data to Client ID: 1g , Data:
ALARM_TRIPPED:3

SensorValue: 1
SensorValue: 0 , Motion Detected ...
NumberHits: 4
Sending Data to Client ID: 1g , Data:
ALARM_TRIPPED:4
```

Next, deactivate the alarm using the Android application. The NodeMCU should receive the deactivate alarm text string, deactivate the alarm, and then send back a text string to the Android indicating that the alarm has been deactivated.

```
ClientID: 1g Incoming Data: DEACTIVATE_
ALARM
Sending Data to Client ID: 1g , Data:
Alarm Has Been DE-Activated ...
```

Finally, push the back button on the Android to exit the application and disconnect from the NodeMCU.

```
Client has disconnected, Client ID: 1g
```

The KeyesIR Obstacle Avoidance Sensor

The KeysIR obstacle avoidance sensor detects objects. The specifications for the sensor are as follows:

- Working voltage: 3.3- to 5-V DC
- Working current: Greater or equal to 20 mA
- Operating temperature: Between –10 and 50 degrees Centigrade
- Effective object detection angle: 35 degrees
- Outputs LOW voltage if object is detected
- Outputs HIGH voltage if no object is detected

See Figure 7-10.

Hands-on Example: The Wireless KeyesIR Obstacle Avoidance Alarm System

In this hands-on example I show you how to build a KeyesIR-based alarm system that will detect the presence of an object. This system is useful for such things as detecting the opening of a door or window.

Parts List

- 1 Android cell phone
- 1 NodeMCU microprocessor
- 1 KeyesIR obstacle avoidance sensor

Figure 7-10 The KeyesIR obstacle avoidance sensor.

- 1 Breadboard
- 1 Package of wires to connect the components together

Setting Up the Hardware

1. Connect the + pin on the sensor to the 3.3-V pin on the NodeMCU.
2. Connect the GND pin on the sensor to a GND pin on the NodeMCU.
3. Connect the out pin on the sensor to the D1 pin on the NodeMCU.

See Figure 7-11.

Setting Up the Software

The code for the NodeMCU is almost identical to the code for the previous example. I will only cover the changes made to the code for this hands-on example. The changes are only made so that the code is more relevant to the KeyesIR sensor. The code functions exactly the same with the variable `MotionDetected` changed to `ObjectDetected` and `MotionNotDetected` changed to `ObjectNotDetected`. Listing 7-28 shows the code changes.

Operating the System

Download and install the KeyesIR program for this chapter on your NodeMCU. Start up

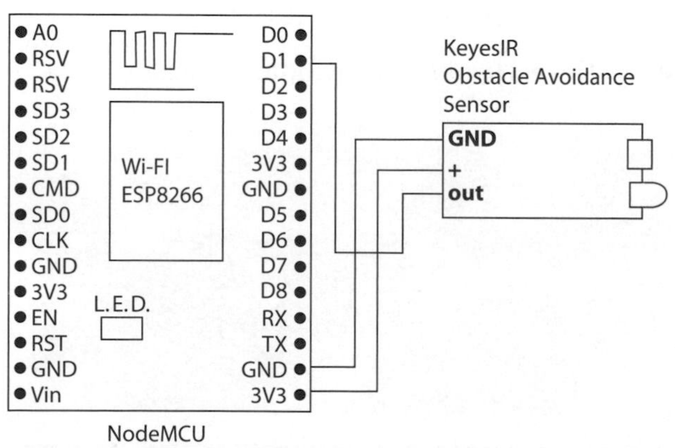

Figure 7-11 The KeyesIR obstacle avoidance alarm system.

the Serial Monitor. Connect your Android cell phone to the Wi-Fi access point running on the NodeMCU. Start up the Android basic framework application and connect to the NodeMCU. After you are connected you should see something like the following on your Serial Monitor:

```
****************************************
*****************
****** NodeMCU ESP-12E Access Point
Server Program ******
********** KeyesIR Object Detection
Sensor System *******

TCP Server started, Server IP Address:
192.168.4.1
WIFI Mode: 2

[Client connected]
Waiting to Recieve New CLient's ID
Incoming Client's ClientID: lg
Number of Clients Killed: 0
Sending Text to Client, ClientID: lg,
Text: Client ID Recieved ...
```

Next, turn off the emergency call out and the emergency text messages from the Android application. Activate the alarm. After 10 seconds the alarm will be fully active and can now be tripped.

```
ClientID: lg Incoming Data: ACTIVATE_
ALARM
Sending Data to Client ID: lg , Data:
Alarm Activated...WAIT TIME:10 Seconds.

Sending Data to Client ID: lg , Data:
ALARM NOW ACTIVE...

ALARM WAIT STATE FINISHED ... Alarm Now
ACTIVE ....
```

Next, trip the alarm by putting your hand very close to the sensor. The alarm will trip and send a text string to the Android device that will display it and produce a sound effect until you tap the debug window on the application. Try to trip the alarm several times.

```
SensorValue: 0 , Motion Detected ...
NumberHits: 1
```

Listing 7-28 Code Changes for the KeyesIR Sensor

```
int ObjectDetected   = 0;
int ObjectNotDetected  = 1;
int SensorValue      = ObjectNotDetected;
int PrevSensorValue = ObjectNotDetected;
void ProcessSensor()
{
 // Put Code to process your sensors or other hardware in this function.
 if ((AlarmState == Alarm_ON) || (AlarmState == Alarm_TRIPPED))
 {
  // Process Sensor Input
  SensorValue = digitalRead(SensorPin);
  if ((PrevSensorValue == ObjectNotDetected) && (SensorValue == ObjectDetected))
  {
   unsigned long ElapsedTime = millis() - MinDelayBetweenHitsStartTime;
   if (ElapsedTime >= MinDelayBetweenHits)
   {
    // Motion Just Detected
    NumberHits++;
    Serial.print(F("SensorValue: "));
    Serial.print(SensorValue);
    Serial.print(F(" , Motion Detected ... NumberHits: "));
    Serial.println(NumberHits);
    PrevSensorValue = ObjectDetected;

    // Send Alarm Tripped Message
    SendAlarmTrippedMessage();

    // Time this hit occured
    MinDelayBetweenHitsStartTime = millis();
   }
  }
  else
  if ((PrevSensorValue == ObjectDetected) && (SensorValue == ObjectNotDetected))
  {
   Serial.print(F("Object Not Detected, SensorValue: "));
   Serial.println(SensorValue);
   PrevSensorValue = ObjectNotDetected;
  }
 }
}
```

```
Sending Data to Client ID: 1g , Data:
ALARM_TRIPPED:1

Object Not Detected, SensorValue: 1
SensorValue: 0 , Motion Detected ...
NumberHits: 2

Sending Data to Client ID: 1g , Data:
ALARM_TRIPPED:2

Object Not Detected, SensorValue: 1
SensorValue: 0 , Motion Detected ...
NumberHits: 3
```

```
Sending Data to Client ID: lg , Data:
ALARM_TRIPPED:3

Object Not Detected, SensorValue: 1
SensorValue: 0 , Motion Detected ...
NumberHits: 4
Sending Data to Client ID: lg , Data:
ALARM_TRIPPED:4
```

Next, deactivate the alarm.

```
Object Not Detected, SensorValue: 1
ClientID: lg Incoming Data: DEACTIVATE_
ALARM
Sending Data to Client ID: lg , Data:
Alarm Has Been DE-Activated ...
```

Finally, exit the Android application by pressing the back button.

```
Client has disconnected, Client ID: lg
```

The Funduino Water Level Sensor

The Funduino water level sensor is an analog sensor that measures the level of water. It can operate using 3.3 and 5 V. See Figure 7-12.

Analog Input on the NodeMCU

The NodeMCU has an analog A0 input pin that is used to measure voltage. The maximum safe voltage that can be measured is 3.3 V. Reading a 5-V value using the A0 pin might damage the NodeMCU. On a bare bones ESP8266 you can only measure a maximum of 1 V using the analog to digital converter (ADC). A greater value might damage the ESP8266. However, the NodeMCU uses a voltage divider to reduce the input voltage so that a value of 3.3 V would be reduced to approximately 1 V which is then read by the NodeMCU's ADC. The actual data is read by calling the `analogRead(A0)` function with A0 as the parameter.

Figure 7-12 The Funduino water level sensor.

Hands-on Example: The Wireless Water Level Detector Alarm System

In this hands-on example I show you how to build a wireless water level detector alarm system that will detect the level of water in an environment and issue an alarm if the water level is outside a certain range of values. This system can be used to monitor for flooding in a basement or an application that requires that the level of water be maintained within a certain range.

Parts List

- 1 Android cell phone
- 1 NodeMCU microprocessor
- 1 Funduino water level sensor
- 1 Package of wires to connect the components
- 1 Glass of water for testing the sensor

Setting Up the Hardware

1. Connect the + pin on the sensor to a 3.3-V pin on the NodeMCU.
2. Connect the – pin on the sensor to a GND pin on the NodeMCU.
3. Connect the S pin on the sensor to pin A0 on the NodeMCU.

See Figure 7-13.

Setting Up the Software

Most of the code for the NodeMCU is the same as for other hands-on examples presented

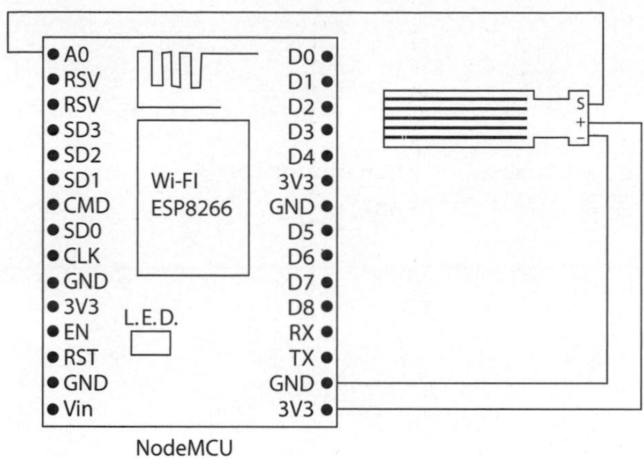

Figure 7-13 The wireless water level alarm system.

previously in this book. The unique code for this example is discussed in this section.

The `MinimumWaterLevel` variable holds the minimum value the sensor must read for the alarm not to trip. If the sensor reads a lower value then the alarm will trip if it is set. This variable is initialized to 0.

```
int MinimumWaterLevel = 0;
```

The `MaximumWaterLevel` variable holds the maximum value the sensor can read for the alarm not to trip. If the value read is greater than this maximum value then the alarm will trip if it is set. The default value is set to 100.

```
int MaximumWaterLevel = 100;
```

The `SensorPin` variable is the NodeMCU pin that the sensor is connected to and is set by default to the analog pin on the NodeMCU.

```
int SensorPin = A0;
```

The `ProcessSensor()` Function

The `ProcessSensor()` function is modified from previous code by doing the following:

1. A `delay(100)` function is added to have a minimum delay between executions of reading in the analog value from the sensor. This is needed because without it the NodeMCU malfunctions. Specifically the Wi-Fi access point is turned off.
2. The value from the sensor is read in by calling the `analogRead(SensorPin)` function which converts the voltage at the A0 pin to a number and stores it in the `SensorValue` variable.
3. If the sensor value is less than the `MinimumWaterLevel` or the sensor value is greater than the `MaximumWaterLevel` value then the sensor value is out of the acceptable range. This case might trigger an alarm tripped status and a text string to be sent to the Android device.
4. If an alarm tripped status has occurred then print out a text string to the Serial Monitor indicating that a water level outside the allowable range has been detected and the number of times this has occurred since the alarm was activated.

See Listing 7-29. The key new code is shown in bold print.

The `wl` Command

The `wl` command retrieves a water level reading from the sensor and sends it back to the Android device.

See Listing 7-30.

The `wlmin` Command

The `wlmin` command returns the minimum water level for the allowable range of water level values. See Listing 7-31.

The `wlmax` Command

The `wlmax` command retrieves the maximum allowable water level for the alarm system. See Listing 7-32.

Listing 7-29 The `ProcessSensor()` Function

```
void ProcessSensor()
{
  // Put Code to process your sensors or other hardware in this function.
  if ((AlarmState == Alarm_ON) || (AlarmState == Alarm_TRIPPED))
  {
    // Process Sensor Input
    delay(100);
    SensorValue = analogRead(SensorPin);
    if ((SensorValue < MinimumWaterLevel) ||
       (SensorValue > MaximumWaterLevel))
    {
     unsigned long ElapsedTime = millis() - MinDelayBetweenHitsStartTime;
     if (ElapsedTime >= MinDelayBetweenHits)
     {
       // Motion Just Detected
       NumberHits++;
       Serial.print(F("SensorValue: "));
       Serial.print(SensorValue);
       Serial.print(F(" , Water Level Event Detected ... NumberHits: "));
       Serial.println(NumberHits);

       // Send Alarm Tripped Message
       SendAlarmTrippedMessage();

       // Time this hit occured
       MinDelayBetweenHitsStartTime = millis();
     }
    }
  }
}
```

Listing 7-30 The `wl` Command

```
if (Data == "wl")
{
  // Return water level to requesting
client
  int wl = analogRead(SensorPin);

  String returnstring = "Water Level:
" + String(wl) + "\n";
  SendDataToClient(ClientID,
returnstring);
}
```

The `wlmin` Set Command

The `wlmin=` command sets the minimum allowable water level for the alarm system. See Listing 7-33.

Listing 7-31 The `wlmin` Command

```
if (Data == "wlmin")
{
  String returnstring = "MinimumWater
Level: " + String(MinimumWaterLevel)
+ "\n";
  SendDataToClient(ClientID,
returnstring);
}
```

The `wlmax` Set Command

The `wlmax=` command sets the maximum allowable water level for the alarm system. See Listing 7-34.

Listing 7-32 The `wlmax` Command

```
if (Data == "wlmax")
{
 String returnstring = "MaximumWater
Level: " + String(MaximumWaterLevel)
+ "\n";
 SendDataToClient(ClientID,
returnstring);
}
```

Listing 7-33 The `wlmin` Command

```
if (Data.indexOf("wlmin=") >= 0)
{
 int index = Data.indexOf("=");
 String value = Data.substring
(index+1);
 MinimumWaterLevel = value.toInt();

 String returnstring = "";
 returnstring = returnstring +
F("MinimumWaterLevel set to: ") +
  MinimumWaterLevel + F("\n");
 SendDataToClient(ClientID,
returnstring);
}
```

Operating the System

Download and install on your NodeMCU the water level project for this chapter. Start up the Serial Monitor. Connect your Android to the Wi-Fi access point running on the NodeMCU. Start up the Android basic framework version 1.3 application. Connect to the NodeMCU using the application. You should see something like the following:

```
****************************************
*****************
****** NodeMCU ESP-12E Access Point
Server Program ******
********** Water Level Sensor Alarm
System **************
```

Listing 7-34 The `wlmax` Command

```
if (Data.indexOf("wlmax=") >= 0)
{
 int index = Data.indexOf("=");
 String value = Data.substring
(index+1);
 MaximumWaterLevel = value.toInt();

 String returnstring = "";
 returnstring = returnstring +
F("MaximumWaterLevel set to: ") +
  MaximumWaterLevel + F("\n");
 SendDataToClient(ClientID,
returnstring);
}
```

```
TCP Server started, Server IP Address:
192.168.4.1
WIFI Mode: 2

[Client connected]
Waiting to Recieve New CLient's ID
Incoming Client's ClientID: lg
Number of Clients Killed: 0
Sending Text to Client, ClientID: lg,
Text: Client ID Recieved ...
```

Next, while the water level sensor is dry take a water level reading by issuing a `wl` command. You should see the water level appear on your Android phone. In my case the water level is 3.

```
ClientID: lg Incoming Data: wl
Sending Data to Client ID: lg , Data:
Water Level: 3
```

Next, retrieve the current minimum water level by issuing a `wlmin` command. The return value should be 0.

```
ClientID: lg Incoming Data: wlmin
Sending Data to Client ID: lg , Data:
MinimumWaterLevel: 0
```

Next, retrieve the maximum allowable water level by issuing the `wlmax` command. The return value should be 100.

```
ClientID: lg Incoming Data: wlmax
Sending Data to Client ID: lg , Data:
MaximumWaterLevel: 100
```

Next, activate the alarm using the Android application. After waiting 10 seconds the alarm should be fully active and ready to be tripped.

```
ClientID: lg Incoming Data: ACTIVATE_
ALARM
Sending Data to Client ID: lg , Data:
Alarm Activated...WAIT TIME:10 Seconds.

Sending Data to Client ID: lg , Data:
ALARM NOW ACTIVE...

ALARM WAIT STATE FINISHED ... Alarm Now
ACTIVE ....
```

Next, trip the alarm by putting the water level sensor in a cup of water until the alarm is tripped. In each of the following cases the alarm has tripped because the water level is greater than the maximum allowable level which is 100.

```
SensorValue: 168 , Water Level Event
Detected ... NumberHits: 1
Sending Data to Client ID: lg , Data:
ALARM_TRIPPED:1

SensorValue: 265 , Water Level Event
Detected ... NumberHits: 2
Sending Data to Client ID: lg , Data:
ALARM_TRIPPED:2

SensorValue: 261 , Water Level Event
Detected ... NumberHits: 3
Sending Data to Client ID: lg , Data:
ALARM_TRIPPED:3

SensorValue: 254 , Water Level Event
Detected ... NumberHits: 4
Sending Data to Client ID: lg , Data:
ALARM_TRIPPED:4
```

Next, deactivate the alarm system using the Android application.

```
ClientID: lg Incoming Data: DEACTIVATE_
ALARM
Sending Data to Client ID: lg , Data:
Alarm Has Been DE-Activated ...
```

Next, take a water level reading with the sensor submerged in the water. Then set the maximum water level to a value that is well above that value. In my case it is 500. Use the `wlmax=500` to set the maximum water level to 500.

```
ClientID: lg Incoming Data: wlmax=500
Sending Data to Client ID: lg , Data:
MaximumWaterLevel set to: 500
```

Next, set the minimum water level to a level that is below your current water level. In my case it is 100. When the minimum level of water goes below this value then the alarm will trip.

```
ClientID: lg Incoming Data: wlmin=100
Sending Data to Client ID: lg , Data:
MinimumWaterLevel set to: 100
```

Next, activate the alarm using the Android application and wait till it is fully active and can be tripped.

```
ClientID: lg Incoming Data: ACTIVATE_
ALARM
Sending Data to Client ID: lg , Data:
Alarm Activated...WAIT TIME:10 Seconds.

Sending Data to Client ID: lg , Data:
ALARM NOW ACTIVE...

ALARM WAIT STATE FINISHED ... Alarm Now
ACTIVE ....
```

Next, start pulling the water sensor out of the water until the alarm trips. The alarm should trip when the measured water level value is below 100 in my case.

```
SensorValue: 45 , Water Level Event
Detected ... NumberHits: 1
Sending Data to Client ID: lg , Data:
ALARM_TRIPPED:1

SensorValue: 49 , Water Level Event
Detected ... NumberHits: 2
Sending Data to Client ID: lg , Data:
ALARM_TRIPPED:2

SensorValue: 99 , Water Level Event
Detected ... NumberHits: 3
Sending Data to Client ID: lg , Data:
ALARM_TRIPPED:3
```

```
SensorValue: 96 , Water Level Event
Detected ... NumberHits: 4
Sending Data to Client ID: lg , Data:
ALARM_TRIPPED:4

SensorValue: 68 , Water Level Event
Detected ... NumberHits: 5
Sending Data to Client ID: lg , Data:
ALARM_TRIPPED:5
```

Thus, in this hands-on example we have tested the cases where the water level has exceeded the maximum value as well as the water level being below the minimum value.

Next, deactivate the alarm using the Android application.

```
ClientID: lg Incoming Data: DEACTIVATE_
ALARM
Sending Data to Client ID: lg , Data:
Alarm Has Been DE-Activated ...
```

Finally, exit the Android application by pressing the back button. You should also be disconnected from the NodeMCU server.

```
Client has disconnected, Client ID: lg
```

Summary

In this chapter I have presented a wireless remote control system based on an Android device and a NodeMCU. I started by giving a general overview of the basic features of the system. I then covered the Android and NodeMCU code that was required to operate the basic wireless framework that was used for the rest of the examples in this chapter. Next, I covered many different hands-on examples using various sensors which were:

- The sound detector sensor
- The infrared motion detector sensor
- The SW520D tilt and vibration sensor
- The KeyesIR obstacle detection sensor
- The water level sensor

CHAPTER 8

Android, Arduino, ESP-01, and NodeMCU ESP-12E Wireless Sensor and Remote Control Projects

In this chapter I cover wireless projects that involve the Android, Arduino with ESP-01, and the NodeMCU. I start off covering the ArduCAM Mini OV2640 2MP Plus camera which is a newer version of the ArduCAM camera that I covered earlier in the book. I then use this updated camera version in a hands-on example that involves a wireless Arduino-based video surveillance system. Next, I expand on this Arduino server-based framework in another hands-on example where I add an infrared motion sensor and an alarm system. This new system is able to take a photo whenever the motion sensor is tripped and to send this image to the Android controller for display. Next, I discuss a multi-client basic framework that is based on the NodeMCU server and Arduino clients. I then give a general overview of version 2.0 of the basic wireless framework program for Android. Next, I present the additional code needed for the Android basic wireless framework version 2.0. Then I cover the new basic framework version 2.0 for the NodeMCU server. Next, I show an example of setting up the Arduino for station/client mode. The rest of the chapter is then dedicated to using hands-on examples to illustrate the NodeMCU server and the multi-client system.

The hands-on examples relating to the NodeMCU server-based systems are as follows:

- Hands-on example: The ArduCAM OV2640 2MP Mini Plus camera Arduino Mega 2560 client surveillance system with NodeMCU server
- Hands-on example: The ArduCAM OV2640 2MP Mini Plus infrared motion detection Arduino Mega 2560 client surveillance and alarm system with NodeMCU server
- Hands-on example: The infrared motion detection alarm system using the NodeMCU server
- Hands-on example: The ArduCAM OV2640 2MP Mini Plus and infrared motion detection Arduino Mega 2560 client surveillance and alarm system with NodeMCU server with an infrared motion detection alarm system

The ArduCAM Mini OV2640 2MP Plus

The ArduCAM Mini OV2640 2MP Plus is a more recent version of the Mini OV2640. The important thing to take note of is that there appears to be a hardware change that requires a software change in order to correctly read in an image from the camera's FIFO memory. The front of this camera is shown in Figure 8-1.

Figure 8-1 The ArduCAM Mini OV2640 2 MP Plus digital camera.

Hands-on Example: The Wireless ArduCAM Mini 2MP Plus Camera Surveillance System

This hands-on example covers the building and the operation of an ArduCAM Mini OV2640 2MP Plus digital camera-based surveillance system.

Parts List

- 1 Android cell phone
- 1 Arduino Mega 2560 microcontroller
- 1 5- to 3.3-V step-down voltage regulator
- 1 Logic level converter from 5 to 3.3 V
- 1 ESP-01 Wi-Fi module
- 1 Arduino development station such as a desktop or a notebook
- 1 Breadboard
- 1 ArduCAM OV2640 Mini Plus camera
- 1 Package of wires both male-to-male and female-to-male to connect the components

Setting Up the Hardware

Step-Down Voltage Regulator Connections

1. Connect the Vin pin on the voltage regulator to the 5-V power pin on the Arduino Mega 2560.
2. Connect the Vout pin on the voltage regulator to a node on your breadboard represented by a horizontal row of empty slots. This is the 3.3-V power node that will supply the 3.3-V power to the ESP-01 module as well as provide the reference voltage for the logic level shifter.
3. Connect the GND pin on the voltage regulator to the GND pin on the Arduino or form a GND node similar to the 3.3-V power node in the previous step.

ESP-01 Wi-Fi Module Connections

1. Connect the Tx or TxD pin to the Rx3 pin on the Arduino Mega 2560. This is the receive pin for the serial hardware port 3.
2. Connect the EN or CH_PD pin to the 3.3-V power node from the step-down voltage regulator.
3. Connect the IO16 or RST pin to the 3.3-V power node from the step-down voltage regulator.
4. Connect the 3V3 or VCC pin to the 3.3-V power node from the step-down voltage regulator.
5. Connect the GND pin to the ground pin or node.
6. Connect the Rx or RxD pin to pin B0 on the 3.3-V side of the logic level shifter. The voltage from this pin will be shifted from 5 V from the Arduino to 3.3 V for the ESP-01 module.

Logic Level Shifter

1. Connect the 3.3-V pin to the 3.3-V power node from the step-down voltage regulator.
2. Connect the GND pin on the 3.3-V side to the ground node for the Arduino.
3. Connect the B0 pin on the 3.3-V side to Rx or RxD pin on the ESP-01. This corresponds to the A0 pin on the 5-V side.
4. Connect the 5-V pin to the 5-V power pin on the Arduino.
5. Connect the GND pin on the 5-V side to the ground node for the Arduino.
6. Connect the A0 pin on the 5-V side to Tx3 pin on the Arduino. This pin corresponds to the B0 pin on the 3.3-V side.

ArduCAM OV2640 Mini Plus

1. Connect the CS pin to digital pin 8 on the Arduino.
2. Connect the SDA pin to digital pin 20 on the Arduino.
3. Connect the SCL pin to digital pin 21 on the Arduino.
4. Connect the MISO pin to digital pin 50 on the Arduino.
5. Connect the MOSI pin to digital pin 51 on the Arduino.
6. Connect the SCK pin to digital pin 52 on the Arduino.
7. Connect the VCC pin to the 5-V node for the circuit.
8. Connect the GND pin to the common ground node for the circuit.

See Figure 8-2.

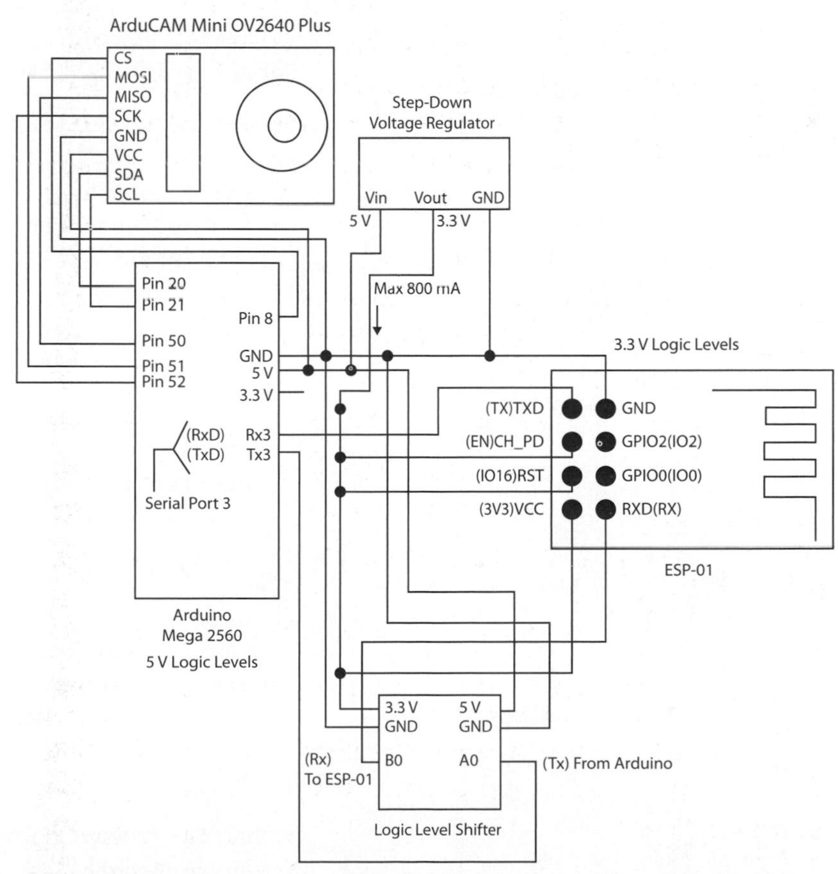

Figure 8-2 The ArduCAM OV2640 Mini Plus surveillance system.

Setting Up the Software

The code for the ArduCAM Mini Plus camera surveillance system is the same as for the older version of the Mini camera presented earlier in the book except for some of the code in the `ReadFifoBurst()` function that controls how the image from the camera's frame buffer is read in.

The `ReadFifoBurst()` Function

The `ReadFifoBurst()` function is modified to accommodate the ArduCAM Mini Plus camera. The key change is that the image header is now checked for. In the older version of the camera the first byte read in was a dummy byte that was not part of the image data. In the new version of the camera the first byte may be part of the image header. Thus, the old code would need to be modified.

See Listing 8-1. The code changes are highlighted in bold print.

Listing 8-1 The `ReadFifoBurst()` Function

```
uint8_t ReadFifoBurst(String
ClientIndex)
{
 uint32_t length = 0;
 uint8_t temp,temp_last;

 long bytecount = 0;
 int total_time = 0;

 boolean is_header = false;

 length = myCAM.read_fifo_length();
 if(length >= 393216)  //384 kb
 {
  Serial.println(F("Not found the
 end."));
  return 0;
 }
 Serial.print(F("READ_FIFO_LENGTH() =
"));
 Serial.println(length);

 total_time = millis();

 myCAM.CS_LOW();
 myCAM.set_fifo_burst();
 temp = SPI.transfer(0x00); // Added
in to original source code, read in
dummy byte
 length--;

 // Reset TCP Buffer
 TCPBufferIndex = 0;
 TCPBufferLength = 0;
 while( length-- )
 {
  temp_last = temp;
  temp = SPI.transfer(0x00);

  // Total Bytes read in from ArduCAM
 fifo
  bytecount++;

  // New Code for Arducam Mini Plus
  if (is_header == true)
  {
   TCPBuffer[TCPBufferIndex] = temp;
   TCPBufferLength++;
  }
  else if ((temp == 0xD8) & (temp_
 last == 0xFF))
  {
   is_header = true;
   Serial.println(F("ACK IMG"));

   TCPBuffer[TCPBufferIndex] = temp_
  last;
   TCPBufferLength++;
   if (TCPBufferLength ==
  MaxPacketSize)
   {
    // Buffer Full so Send Data
    SendTCPBinaryData(ClientIndex,
   TCPBuffer, TCPBufferLength);

    // Reset Buffer
    TCPBufferIndex = 0;
    TCPBufferLength = 0;
   }
   else
   {
    // Buffer Not full so continue
   reading in Data from
    // ArduCAM FIFO Memory
    TCPBufferIndex++;
   }

   TCPBuffer[TCPBufferIndex] = temp;
   TCPBufferLength++;
  }
```

```
 // Process TCP Buffer
 if (TCPBufferLength ==
MaxPacketSize)
 {
  // Buffer Full so Send Data
  SendTCPBinaryData(ClientIndex,
 TCPBuffer, TCPBufferLength);

  // Reset Buffer
  TCPBufferIndex = 0;
  TCPBufferLength = 0;
 }
 else
 {
  // Buffer Not full so continue
 reading in Data from
  // ArduCAM FIFO Memory
  TCPBufferIndex++;
 }

 // Commented out because it causes
loss of data in transfer
 // so that data does not match
amount of data read from read fifo
length command
 //if( (temp == 0xD9) && (temp_last
== 0xFF) )
 //  break;
 delayMicroseconds(10);
 }

 // Check to see if data is still in
TCP Buffer
 if (TCPBufferLength != 0)
 {
  // Send remaining data in buffer
  SendTCPBinaryData( ClientIndex,
 TCPBuffer, TCPBufferLength);
 }

 myCAM.CS_HIGH();

 Serial.print(F("ByteCount = "));
 Serial.println(bytecount);

 total_time = millis() - total_time;
 Serial.print("Total time used:");
 Serial.print(total_time/1000, DEC);
 Serial.println(" seconds ...");
}
```

Operating the System

Download the ArduCAM Mini Plus server version project for this chapter and install it on your Arduino. This camera system operates the same way as the camera surveillance system using the older ArduCAM mini version presented earlier in this book.

Hands-on Example: The Wireless ArduCAM Mini OV2640 2MP Plus Camera and HC-SR501 Infrared Motion Detector Surveillance and Alarm System for the Arduino Mega 2560 Server

In this hands-on example I will show you how to build and operate a wireless surveillance and alarm system for an Arduino Mega 2560 server using an ArduCAM OV2640 Mini 2MP Plus camera and HC-SR501 infrared motion detector. An Android cell phone will be used as the wireless controller for the system.

Parts List

- 1 Android cell phone
- 1 Arduino Mega 2560 microcontroller
- 1 5- to 3.3-V step-down voltage regulator
- 1 Logic level converter from 5 to 3.3 V
- 1 ESP-01 Wi-Fi module
- 1 Arduino development station such as a desktop or a notebook
- 1 Breadboard
- 1 ArduCAM OV2640 Mini Plus camera
- 1 HC-SR501 infrared motion detector
- 1 Package of wires to connect the components

Setting Up the Hardware

Step-Down Voltage Regulator Connections

1. Connect the Vin pin on the voltage regulator to the 5-V power pin on the Arduino Mega 2560.
2. Connect the Vout pin on the voltage regulator to a node on your breadboard represented by a horizontal row of empty slots. This is the 3.3-V power node that will supply the 3.3-V power to the ESP-01 module as well as provide the reference voltage for the logic level shifter.
3. Connect the GND pin on the voltage regulator to the GND pin on the Arduino or form a GND node similar to the 3.3-V power node in the previous step.

ESP-01 Wi-Fi Module Connections

1. Connect the Tx or TxD pin to the Rx3 pin on the Arduino Mega 2560. This is the receive pin for the serial hardware port 3.
2. Connect the EN or CH_PD pin to the 3.3-V power node from the step-down voltage regulator.
3. Connect the IO16 or RST pin to the 3.3-V power node from the step-down voltage regulator.
4. Connect the 3V3 or VCC pin to the 3.3-V power node from the step-down voltage regulator.
5. Connect the GND pin to the ground pin or node.
6. Connect the Rx or RxD pin to pin B0 on the 3.3-V side of the logic level shifter. The voltage from this pin will be shifted from 5 V from the Arduino to 3.3 V for the ESP-01 module.

Logic Level Shifter

1. Connect the 3.3-V pin to the 3.3-V power node from the step-down voltage regulator.
2. Connect the GND pin on the 3.3-V side to the ground node for the Arduino.
3. Connect the B0 pin on the 3.3-V side to Rx or RxD pin on the ESP-01. This corresponds to the A0 pin on the 5-V side.
4. Connect the 5-V pin to the 5-V power pin on the Arduino.
5. Connect the GND pin on the 5-V side to the ground node for the Arduino.
6. Connect the A0 pin on the 5-V side to Tx3 pin on the Arduino. This pin corresponds to the B0 pin on the 3.3-V side.

ArduCAM OV2640 Mini Plus

1. Connect the CS pin to digital pin 8 on the Arduino.
2. Connect the SDA pin to digital pin 20 on the Arduino.
3. Connect the SCL pin to digital pin 21 on the Arduino.
4. Connect the MISO pin to digital pin 50 on the Arduino.
5. Connect the MOSI pin to digital pin 51 on the Arduino.
6. Connect the SCK pin to digital pin 52 on the Arduino.
7. Connect the VCC pin to the 5-V node for the circuit.
8. Connect the GND pin to the common ground node for the circuit.

The HC-SR501 Infrared Motion Detector

1. Connect the Vcc pin on the motion detector to the 5-V power node on the Arduino Mega 2560.
2. Connect the GND pin on the motion detector to the GND node on the Arduino.
3. Connect the Out pin on the motion detector to digital pin 7 on the Arduino.

See Figure 8-3.

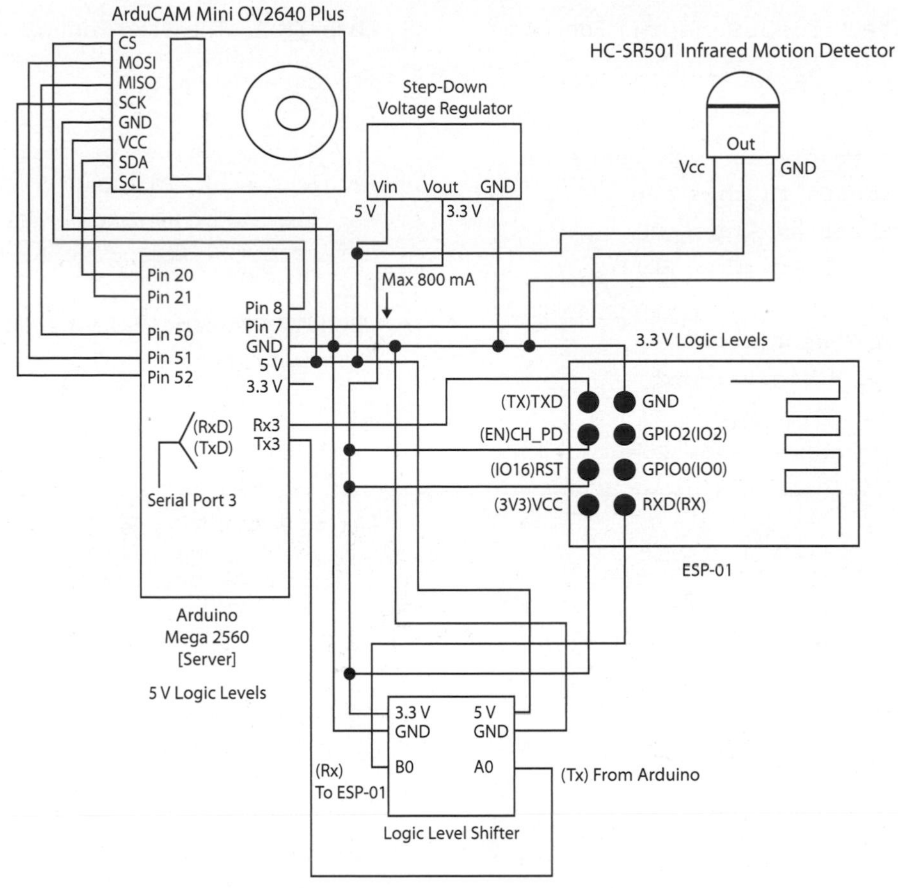

Figure 8-3 The wireless surveillance and infrared motion detection system.

Setting Up the Software

The key code change for this example is that when the alarm is tripped the number of hits will be updated but no text string will be sent to the Android device. The Android device must now poll the alarm system to determine if the alarm has been tripped. The `ProcessSensor()` function is changed to reflect this. The code that handles the tripping of the alarm system is shown in bold print in Listing 8-2. The code that sends a text string to the Android indicating that the alarm has been tripped has been removed.

The hits command is changed slightly so that the beginning of the return value of a hits command starts with `"HITS DETECTED:"` followed by the number of times that alarm system has been tripped. See Listing 8-3.

The `reset` command is modified by removing the code that returns an acknowledgment string to the Android. See Listing 8-4.

Operating the System

Download and install on your Android device the basic framework application version 2.0. There should be an Android Studio version and a directly installable Android APK file. Next, download and install on your Arduino device the "ArducamMiniPlus-MotionDetector-ESP01-Server.ino" project file for this chapter. Bring up the Arduino's Serial Monitor. You should see something similar to the following:

```
****** ESP-01 Arduino Mega 2560 Server
Program ********
**** ArduCAM Mini Plus OV2640 System
Version 2.0 ******
```

Listing 8-2 The `ProcessSensor()` Function

```
void ProcessSensor()
{
  // Put Code to process your sensors
or other hardware in this function.
  if ((AlarmState == Alarm_ON) ||
(AlarmState == Alarm_TRIPPED))
  {
    // Process Sensor Input
    SensorValue = digitalRead
   (SensorPin);
    if ((PrevSensorValue == EventNot
  Detected) && (SensorValue == Event
  Detected))
    {
      unsigned long ElapsedTime =
    millis() - MinDelayBetweenHits
    StartTime;
      if (ElapsedTime >= MinDelayBetween
    Hits)
      {
        // Motion Just Detected
        NumberHits++;
        Serial.print(F("SensorValue: "));
        Serial.print(SensorValue);
        Serial.print(F(" , Event Detected
      ... NumberHits: "));
        Serial.println(NumberHits);
        PrevSensorValue = EventDetected;

        // Time this hit occured
        MinDelayBetweenHitsStartTime =
      millis();
      }
    }
    else
    if ((PrevSensorValue ==
  EventDetected) && (SensorValue ==
  EventNotDetected))
    {
      Serial.print(F("SensorValue: "));
      Serial.println(SensorValue);
      PrevSensorValue = EventNotDetected;
    }
  }
}
```

Listing 8-3 The `Hits` Command

```
if (Data == "hits\n")
{
  // Return number of motions detected
to requesting client
  String returnstring = "HITS
DETECTED:" + String(NumberHits) +
"\n";
  SendTCPMessage(ClientIndex,
returnstring);
}
```

Listing 8-4 The `Reset` Command

```
if (Data == "reset\n")
{
  // Reset number hits to 0
  NumberHits = 0;
  Serial.println(F("Reset NumberHits
to 0."));
}
```

```
********** Infrared Motion Detection
******************
**** Surveillance and Motion Detection
Alarm System ***

ArduCAM Starting ..........
OV2640 detected ...........

AT Command Sent: AT+CIPMUX=1
ESP-01 Response: AT+CIPMUX=1

OK

ESP-01 mode set to allow multiple
connections.

AT Command Sent: AT+CWMODE_CUR=2
ESP-01 Response: AT+CWMODE_CUR=2

OK

ESP-01 mode changed to Access Point
Mode.

AT Command Sent: AT+CIPSERVER=1,80
ESP-01 Response: AT+CIPSERVER=1,80

OK

Server Started on ESP-01.
```

```
AT Command Sent: AT+CWSAP_CUR="ESP-
01","esp8266esp01",1,2,4,0
ESP-01 Response: AT+CWSAP_CUR="ESP-
01","esp8266esp01",1,2,4,0
OK
AP Configuration Set Successfullly.
AT Command Sent: AT+CWSAP_CUR?
ESP-01 Response: AT+CWSAP_CUR?
+CWSAP_CUR:"ESP-01","esp8266esp01",1,
2,4,0
OK
This is the curent configuration of the
AP on the ESP-01.
AT Command Sent: AT+CIPSTO?
ESP-01 Response: AT+CIPSTO?
+CIPSTO:180
OK
This is the curent TCP timeout in
seconds on the ESP-01.
************** System Has Started
********************
```

Connect your Android cell phone to the access point located on the ESP-01 module. Start up the Android basic framework application version 2.0. Connect to the TCP server running on the ESP-01 module using the application. On the Serial Monitor you should see the Android connecting to the server. The name of the Android device should be changed to the name that you have specified for the wireless network. In my case the name of my Android device is "`verizon`". Finally, an "OK" string should be sent back to the Android device indicating that the name change was successful.

```
Notification: '0,CONNECT'
Processing New Client Connection ...
NewConnection Name: 0
Notification: '
'Notification: '
'Notification: '+IPD,0,13:name=verizon
'ClientIndex: 0
DataLength: 13
Data: name=verizon
Client Name Change Requested,
ClientIndex: 0, NewClientName: verizon
Client Name Change Succeeded, OK...
Sending Data ....
TCP Connection ID: 0 , DataLength: 3 ,
Data: OK
```

With the new version of the Android basic wireless framework you will be able to select the network device or target that you will send commands to. The default target for commands is the server which you just connected to. Select "Monitor Target Device" to select the server as one of the devices that will be monitored. Select the "Camera Status Target Device" menu item and select "Set to Camera Available" to have the camera on the server take a photo whenever the motion detector is tripped. Turn off the call out and the emergency text message options.

Next, activate the alarm system on the server using the Android application. After the default wait time of 10 seconds the alarm is activated.

```
Notification: '
'Notification: '
'Notification: '+IPD,0,15:ACTIVATE_ALARM
'ClientIndex: 0
DataLength: 15
Data: ACTIVATE_ALARM
Sending Data ....
TCP Connection ID: 0 , DataLength: 40 ,
Data: Alarm Activated...WAIT TIME:10
Seconds.
Notification: '
'Sending Data ....
TCP Connection ID: 0 , DataLength: 20 ,
Data: ALARM NOW ACTIVE...
ALARM WAIT STATE FINISHED ... Alarm Now
ACTIVE ....
```

Next, under the "Network Monitoring" menu select the "Activate Network Monitoring" option to start the monitoring of the alarm status of the monitored devices that were selected previously.

In this case it would be just the server that has the motion detector and the camera. A hits command is sent to the server to determine if the alarm has been tripped. If the return value is 0 then no motion has been detected.

```
Notification: '
'Notification: '
'Notification: '+IPD,0,5:hits
'ClientIndex: 0
DataLength: 5
Data: hits
Sending Data ....
TCP Connection ID: 0 , DataLength: 16 ,
Data: HITS DETECTED:0
Notification: '
'Notification: '
'Notification: '+IPD,0,5:hits
'ClientIndex: 0
DataLength: 5
Data: hits
Sending Data ....
TCP Connection ID: 0 , DataLength:
16 , Data: HITS DETECTED:0
```

Next, trip the alarm by putting your hand in front of the motion sensor and camera and moving it around. Once the alarm has been tripped then the next hits command should return a nonzero number of hits that have been detected.

```
Notification: '
'SensorValue: 1 , Event Detected ...
NumberHits: 1
Notification: '
'Notification: '+IPD,0,5:hits
'ClientIndex: 0
DataLength: 5
Data: hits
Sending Data ....
TCP Connection ID: 0 , DataLength: 16 ,
Data: HITS DETECTED:1
```

Next, after processing the "HITS DETECTED" value a reset command is issued to the server to set the number of alarm trips back to zero.

```
Notification: '
'Notification: '
'Notification: '+IPD,0,6:reset
'ClientIndex: 0
DataLength: 6
Data: reset
Reset NumberHits to 0.
```

Next, since the server has a camera a take photo command is issued to the server. In my case the camera resolution was set to qvga so a `qvga` command was sent to the server. The Arduino receives this command, takes the photo, and sends back the size in bytes of the image that was taken. In my case the number of bytes was 6152.

```
Notification: '
'Notification: '+IPD,0,5:qvga
'ClientIndex: 0
DataLength: 5
Data: qvga
............. Starting Image Capture
...............
Start Capture
Capture Done!
FIFO RAW IMAGE SIZE = 6152
Transmitting Image Length = 6152
Sending Data ....
TCP Connection ID: 0 , DataLength: 5 ,
Data: 6152
```

Next, a `GetImageData` command is issued by the Android to the server in order to retrieve the actual image data. The Arduino receives this command and sends the data in chunks up to the maximum allowable packet size which is 2048.

```
Notification: '
'Notification: '
'Notification: '+IPD,0,13:GetImageData
'ClientIndex: 0
DataLength: 13
```

```
Data: GetImageData
READ_FIFO_LENGTH() = 6152
ACK IMG
Sending Binary Data ....
TCP Connection ID: 0 , BufferLength:
2048
Sending Binary Data ....
TCP Connection ID: 0 , BufferLength:
2048
Sending Binary Data ....
TCP Connection ID: 0 , BufferLength:
2048
Sending Binary Data ....
TCP Connection ID: 0 , BufferLength: 8
ByteCount = 6151
Total time used:0 seconds ...
```

Test out the alarm a few more times to verify that the motion detector and image capture features of the alarm system work correctly.

Finally, turn off the network monitoring and then deactivate the alarm using the Android application.

```
Notification: '
'Notification: '
'Notification: '+IPD,0,17:DEACTIVATE_
ALARM
'ClientIndex: 0
DataLength: 17
Data: DEACTIVATE_ALARM
Sending Data ....
TCP Connection ID: 0 , DataLength: 32 ,
Data: Alarm Has Been DE-Activated ...
```

The Basic Android, Arduino with ESP-01, and NodeMCU Wireless Multi-Client Framework

In this section I will give an overview of the multi-client framework where one Android cell phone can serve as the controller for multiple devices. These devices can have sensors and alarm systems on them and can be controlled and monitored from the Android device. The server for our multi-client wireless framework will be the NodeMCU. The NodeMCU will be able to connect up to 4 clients at one time. One of the clients will be the Android controller. The alarm system on each device can be polled by the Android controller. If the alarm on the device has been tripped then an emergency text message can be sent out and if the device has a camera then a picture can be taken of the object that caused the alarm to be tripped. See Figure 8-4.

Overview of the Android Basic Wireless Framework Version 2.0 Application

For the 2.0 version of the Android basic wireless framework you can get a list of clients that are connected to the server. You are also able to switch the destination device for the commands you send using the application to one of these clients or to the server. The destination device that will receive the commands is called the "Target". In Figure 8-5 the target device is circled and is by default the server.

In order to find the clients that are currently connected to the server you would select the "Refresh Clients" menu item. To select one of the clients that are connected to the server or the server itself as the target device you would select the "Switch Target Device" menu item and then select the device you want to send commands to.

Another key change is that the alarm systems located on devices will not automatically send a text string to the Android controller when an alarm is tripped. Instead they will have the status of their alarms polled so that they will send the Android controller the number of alarm trips that have occurred since the last time the device was polled. To select a device for monitoring you must first choose the device as the target and then select

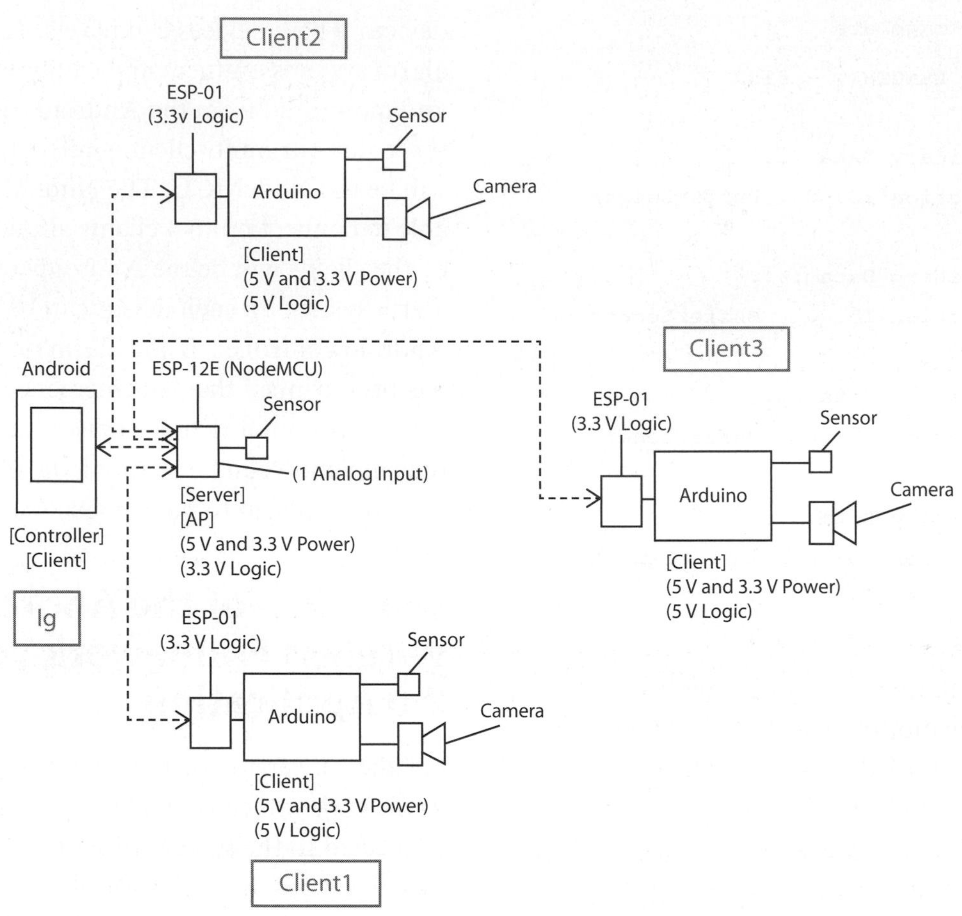

Figure 8-4 An example of the Android, Arduino with ESP-01, and NodeMCU basic wireless multi-client framework.

Figure 8-5 The targeted device on Android basic wireless framework 2.0.

"Monitor Target Device" to put this in a list of devices that will have the status of their alarms monitored. To remove a device from the list of monitored devices select the device as the target and select "Unmonitor Target Device" from the menu. If the monitored device has a camera then you can select the "Set To Camera Available" menu option under the "Camera Status Target Device" menu to take a picture whenever the alarm is tripped on the device. Figure 8-6 shows that both the device called `client1` which has a camera and the device that is the server which does not have a camera will be monitored for changes in their alarm tripped status.

To activate the monitoring of the wireless network select the "Activate Network Monitoring" under the "Network Monitoring"

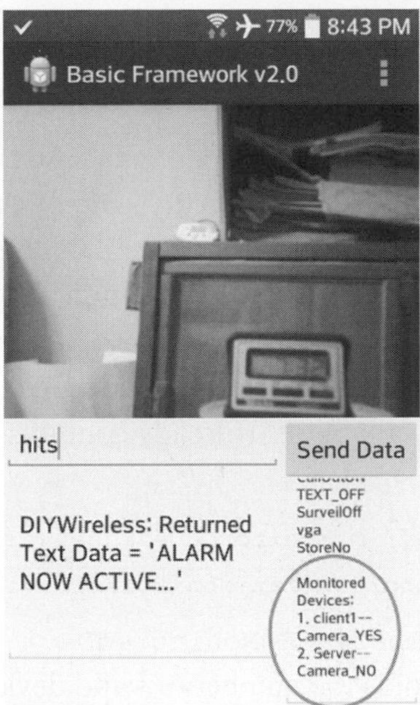

Figure 8-6 The devices that will be monitored for alarm tripped status.

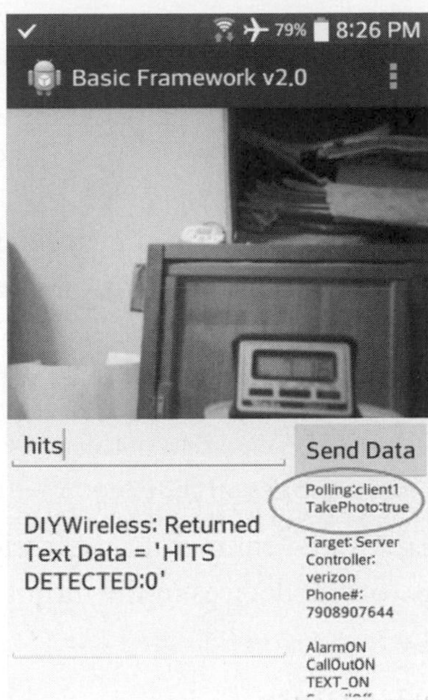

Figure 8-7 Polling a device when network monitoring is on.

menu item. This will determine if the alarm in each device in the monitored devices list has been tripped by sending a hits command. If the hits command returns a 1 or greater value then the alarm has been tripped. If the alarm was tripped then send a reset command to the device to set the total number of trips back to 0. If the device has a camera then take a photo with the camera and send this photo back to the android for display. While the wireless network of devices is being monitored the lower right window displays the device being polled and a picture will be taken with a camera located on this device when an intruder is detected. See Figure 8-7.

Code Additions for the Android Basic Wireless Framework Version 2.0

This section covers the addition code for version 2.0 of the Android basic wireless framework.

The `MaxClients` variable holds the maximum number of clients that can be connected to the server which is by default 4.

```
int MaxClients = 4;
```

The `m_NumberClients` variable holds the current number of clients that are connected to the server and is set by default to 0. Refreshing the clients will update this variable to the current status.

```
int m_NumberClients = 0;
```

The `m_Clients` variable is an array of strings that represents the names of the clients that are currently connected to the server.

```
String[] m_Clients = new
String[MaxClients];
```

The `m_TargetDevice` variable holds the name of the network device being sent commands from the Android basic wireless framework application and is set by default to the server.

```
String m_TargetDevice = "Server";
```

The `m_NetworkGroupID` variable holds the group ID of a client that is connected to the server and is used in selecting a target device by the user.

```
int m_NetworkGroupID = 11;
```

The m_NetworkItemID variable holds the ID of a client that is connected to the server. This is set to the same value for all clients.

```
int m_NetworkItemID = 22;
```

The m_NetworkMonitoringOn variable is true when network monitoring is on and the devices in the monitored devices list are being polled for their alarm status. The default value is set to false.

```
boolean m_NetworkMonitoringOn = false;
```

The PollingState enumeration describes the polling state of the devices in the monitored devices list.

```
enum PollingState
{
 NotStarted,
 InProgress,
 Finished
}
```

The Max_Monitored_Devices variable holds the maximum number of devices that can be monitored which is the maximum number of clients that are able to connect to the server plus the server itself.

```
int Max_Monitored_Devices = MaxClients
+ 1;
```

The m_MonitoredDevices variable holds an array of strings that represents the names of the devices that will have their alarm systems monitored for alarm trips.

```
String[] m_MonitoredDevices = new
String[Max_Monitored_Devices];
```

The m_CameraAvailability variable holds the availability status for the camera that may be attached to a monitored device.

```
boolean[] m_CameraAvailability = new
boolean[Max_Monitored_Devices];
```

The m_MonitoredDevicesPollingState variable holds the current polling state of the monitored devices.

```
PollingState[]
m_MonitoredDevicesPollingState = new
PollingState[Max_Monitored_Devices];
```

The m_MonitoredDevicesAvailability variable holds the availability state of the corresponding slot in the m_MonitoredDevices variable array. A value of true indicates that the slot in the m_MonitoredDevices array is available and a value of false indicates that the slot is unavailable and already contains a valid device.

```
boolean[] m_MonitoredDevicesAvailability
= new boolean[Max_Monitored_Devices];
```

The m_NumberValidMonitoredDevices variable holds the number of valid devices held in the m_MonitoredDevices array.

```
int m_NumberValidMonitoredDevices = 0;
```

The m_ValidMonitoredDevices variable holds the indices from the m_MonitoredDevices array that represents valid devices.

```
int[] m_ValidMonitoredDevices = new
int[Max_Monitored_Devices];
```

The m_CurrentPolledDevice variable holds a number that represents the current device being polled and is used as an index into the m_ValidMonitoredDevices array that returns a number that is used again as an index into the m_MonitoredDevices array to determine the name of the device currently being polled.

```
int m_CurrentPolledDevice = 0;
```

The m_PollingTimeOut variable holds the time in milliseconds that polling will time out for a device if no response is received. The polling time out value is set by default to 30 seconds.

```
int m_PollingTimeOut = 1000 * 30;  //30
seconds timeout for polling
```

The m_TimeBeginPoll variable holds the time just before a device starts to be polled.

```
long m_TimeBeginPoll = 0;
```

The m_PollingTakePhoto variable is true if the current device that is being polled has been set by the user to have a camera activated for taking a photo after the alarm has been tripped.

```
boolean m_PollingTakePhoto = false;
```

The `m_MonitorDevicesHandler` variable is the alarm monitoring handler for the devices in the monitored devices' list. It is called repeatedly at fixed time intervals if network monitoring is activated.

```
Handler m_MonitorDevicesHandler = new
Handler();
```

The `m_MonitoredDevicesRunnable` variable holds the actual code that is executed during the network monitoring of devices in the monitored devices list.

```
Runnable m_MonitoredDevicesRunnable =
null;
```

The `m_MonitoredDevicesInterval` variable holds the time between processing of the devices in the monitored devices list and is set by default to 1000 milliseconds or 1 second.

```
long m_MonitoredDevicesInterval = 1000;
// Number milliseconds
```

Modifying the `onCreate()` Function

The `onCreate()` function in the `MainActivity` class is modified.

The `m_MonitoredDevicesRunnable` variable is defined here and the function that is defined is the `run()` function which does the following:

1. Handles the polling of the devices in the monitored devices' list by calling the `NetworkAlarmPollingHandler()` function.
2. If network monitoring is active then the monitor devices' handler executes this `run()` function again after `m_MonitoredDevicesInterval number of` microseconds.

See Listing 8-5.

Listing 8-5 The `Network Monitoring` Code

```
m_MonitoredDevicesRunnable = new
Runnable() {
  public void run(){
    NetworkAlarmPollingHandler();
    if (m_NetworkMonitoringOn)
    {
     m_MonitorDevicesHandler.
    postDelayed(this, m_
    MonitoredDevicesInterval);
    }
  }
};
```

The data structures that are used to keep track of the devices that need to be tracked for network monitoring are initialized by calling the `InitMonitoredDevicesDataStructure()` function.

The `NetworkAlarmPollingHandler()` Function

The `NetworkAlarmPollingHandler()` function monitors the alarm status of the devices in the monitored devices' list by doing the following:

1. If network monitoring is not on then exit the function.
2. If the number of valid monitored devices is zero then exit the function.
3. Gets the information for the device that is currently being polled or is set to be polled.
4. If the device is currently not being polled then do the following:
    1. Save the current time as the time that the polling for this device has started.
    2. Update the polling state of this device to polling in progress.
    3. Set the target device to this current device that is being polled.

    4. Send a hits command to the microcontroller to retrieve the current state of the alarm system on the microcontroller.
5. If the device is currently being polled then do the following:
    1. Calculate the elapsed time since the start of polling the device.
    2. If the elapsed time is greater than the timeout time then set the polling status of the device to not started and set the currently polled device to the next device in the list of monitored devices. If the next device index is greater or equal to the number of valid monitored devices then reset the next device index back to 0.
6. If the polling of the device has finished then do the following:
    1. Set the currently polled device to the next device in the list of monitored devices. If the next device index is greater or equal to the number of valid monitored devices then reset the next device index back to 0.
    2. Set the poll status of the device that just finished polling to not started.
7. Update the information window on the Android application by calling the `UpdateCommandTextView()` function.

See Listing 8-6.

Listing 8-6 The `NetworkAlarmPollingHandler()` Function

```
void NetworkAlarmPollingHandler()
{
  // Poll each of the devices in the monitored devices list to determine if the alarm has
  // tripped which means that hits > 0
  // If the alarm has tripped then if the device has a camera then send a take photo
comamand
  // to the device.
  if (m_NetworkMonitoringOn == false)
  {
    // exit function if network monitoring is not on.
    return;
  }
  // If no valid monitored devices then exit function.
  if (m_NumberValidMonitoredDevices == 0)
  {
    return;
  }

  int ValidDeviceIndex = m_ValidMonitoredDevices[m_CurrentPolledDevice];
  String MonitoredDevice = m_MonitoredDevices[ValidDeviceIndex];
  PollingState pollstate = m_MonitoredDevicesPollingState[ValidDeviceIndex];
  if (pollstate == PollingState.NotStarted)
  {
    // Polling for this device has not been started yet so begin polling
    // Save starting time of polling
    m_TimeBeginPoll = SystemClock.elapsedRealtime();

    // Update Polling Progress
    m_MonitoredDevicesPollingState[ValidDeviceIndex] = PollingState.InProgress;
```

```
    // Change Targetdevice to device being polled
    m_TargetDevice = MonitoredDevice;

    // Send hits command to the device that is currently being polled
    SendCommand("hits");
   }
   else if (pollstate == PollingState.InProgress)
   {
    // Polling for this device is in progress
    // Check for elapsed time
    long ElapsedTime = SystemClock.elapsedRealtime() - m_TimeBeginPoll;
    if (ElapsedTime > m_PollingTimeOut)
    {
     AddDebugMessageFront("Polling Timeout: " + MonitoredDevice + "\n");
     Toast.makeText(this, "Polling Timeout: " + MonitoredDevice, Toast.LENGTH_SHORT).
    show();

     // Stop Polling and move on to next device
     m_MonitoredDevicesPollingState[ValidDeviceIndex] = PollingState.NotStarted;
     m_CurrentPolledDevice++;
     if (m_CurrentPolledDevice >= m_NumberValidMonitoredDevices)
     {
      // reset to beginning of list
      m_CurrentPolledDevice = 0;
     }
    }
   }
   else if (pollstate == PollingState.Finished)
   {
    // Polling for current device is finished
    // Set up for next device to be polled
    m_CurrentPolledDevice++;
    if (m_CurrentPolledDevice >= m_NumberValidMonitoredDevices)
    {
     // reset to beginning of list
     m_CurrentPolledDevice = 0;
    }
    m_MonitoredDevicesPollingState[ValidDeviceIndex] = PollingState.NotStarted;
   }
   // Update GUI
   UpdateCommandTextView();
  }
```

The `UpdateCommandTextView()` Function

The `UpdateCommandTextView()` function is modified by doing the following:

1. The list of monitored devices is retrieved by calling the `GetMonitoredDevices()` function.
2. If network monitoring is on then display the current device that is being polled and if a photo is to be taken on the alarm tripping.
3. Additions are made to the information window adding in network monitoring-related and monitored devices' information.

See Listing 8-7. The additional code is shown in bold print.

The `GetMonitoredDevices()` Function

The `GetMonitoredDevices()` function returns a string that contains the list of devices that will have their alarm systems monitored when the network monitoring feature is turned on by the user.

Listing 8-7 The `UpdateCommandTextView()` Function

```
void UpdateCommandTextView()
{
 String AlarmStatus = "";
 String CallOutStatus = "";
 String SurveillanceStatus = "";
 String ListMonitoredDevices = "";
 String NetworkMonitoring = "";

 ListMonitoredDevices =
GetMonitoredDevices();
 if (ListMonitoredDevices.equals(""))
 {
  ListMonitoredDevices = "None\n";
 }

 // Get Current Camera Resolution and
set text view
 if (m_AlarmSet)
 {
  AlarmStatus = "AlarmON";
 }
 else
 {
  AlarmStatus = "AlarmOFF";
 }

 if (m_CallOutSet)
 {
  CallOutStatus = "CallOutON";
 }
 else
 {
  CallOutStatus = "CallOutOFF";
 }
 if (m_SurveillanceActive)
 {
  SurveillanceStatus = "SurveilOn";
 }
 else
 {
  SurveillanceStatus = "SurveilOff";
 }

 if (m_NetworkMonitoringOn)
 {
  NetworkMonitoring = "Polling:" +
 GetCurrentPolledDevice() + "\n";
  NetworkMonitoring += "TakePhoto:" +
 m_PollingTakePhoto + "\n";
  NetworkMonitoring += "\n";
 }
 else
 {
  NetworkMonitoring = "";
 }

 String Info =    NetworkMonitoring +
  "Target: " + m_TargetDevice + "\n" +
  "Controller: " + m_
 AndroidClientName + "\n" +
  "Phone#: " + m_EmergencyPhoneNumber
 + "\n" + "\n" +
  AlarmStatus + "\n" +
  CallOutStatus + "\n" +
  m_TextMessagesSetting + "\n" +
  SurveillanceStatus + "\n" +
  m_Resolution.toString() + "\n" +
  m_AndroidStorage + "\n" + "\n" +
  "Monitored Devices:\n" +
  ListMonitoredDevices;
 int length = Info.length();
 m_InfoTextView.setText(Info.
toCharArray(), 0, length);
}
```

See Listing 8-8.

The `GetCurrentPolledDevice()` Function

The `GetCurrentPolledDevice()` function returns the name of the device on the wireless network that is currently being polled by the Android controller. See Listing 8-9.

Listing 8-8 The `GetMonitoredDevices()` Function

```
String GetMonitoredDevices()
{
 String result = "";
 int counter = 1;
 String CameraStatus = "";

 // Return monitored devices in form of
 // 1. DeviceName \n
 for (int i = 0; i < Max_Monitored_
Devices; i++)
 {
  if (!m_MonitoredDevicesAvailability
  [i])
  {
   // slot not available so it
  conatains a valid device name
   if (m_CameraAvailability[i])
   {
    CameraStatus = "Camera_YES";
   }
   else
   {
    CameraStatus = "Camera_NO";
   }
   result += counter + ". " + m_
  MonitoredDevices[i] + "--" +
  CameraStatus + "\n";
   counter++;
  }
 }
 return result;
}
```

Listing 8-9 The `GetCurrentPolledDevice()` Function

```
String GetCurrentPolledDevice()
{
 String result = "";

 int ValidDeviceIndex =
m_ValidMonitoredDevices[m_Current
PolledDevice];
 result = m_MonitoredDevices[ValidDev
iceIndex];
 return result;
}
```

Listing 8-10 The `InitMonitoredDevicesDataStructure()` Function

```
void
InitMonitoredDevicesDataStructure()
{
 for (int i=0; i < Max_Monitored_
Devices; i++)
 {
  m_MonitoredDevices[i] = "";
  m_MonitoredDevices Availability[i]
 = true;
  m_CameraAvailability[i] = false;
  m_MonitoredDevicesPollingState[i] =
 PollingState.NotStarted;
 }
}
```

The `InitMonitoredDevicesDataStructure()` Function

The `InitMonitoredDevicesDataStructure()` function initializes the data structures needed for the list of devices that will have their alarm systems monitored during network monitoring. See Listing 8-10.

Menu Additions

For version 2.0 of the Android basic wireless framework we add in the following new main menu items:

1. A `"Refresh Clients"` menu item.
2. A `"Switch Target Device"` menu item.
3. A `"Monitor Target Device"` menu item.
4. An `"Unmonitor Target Device"` menu item.
5. A `"Camera Status Target Device"` menu item.
6. A `"Network Monitoring"` menu item.

See Listing 8-11.

Listing 8-11 Menu Additions for Version 2.0

```
<item android:id="@+id/refresh_
clients" android:title="Refresh
Clients">
</item>
<item android:id="@+id/switch_to_
device" android:title="Switch Target
Device">
  <menu>
  </menu>
</item>

<item android:id="@+id/monitor_
current_client" android:title="Monitor
Target Device">
</item>

<item android:id="@+id/
unmonitor_current_client"
android:title="Unmonitor Target
Device">
</item>

<item android:id="@+id/Link_Camera"
android:title="Camera Status Target
Device">
  <menu>
    <item android:id="@+id/link_camera_
  activate" android:title="Set to
  Camera Available"/>
    <item android:id="@+id/link_camera_
  deactivate" android:title="Set to
  Camera Unavailable"/>
  </menu>
</item>

<item android:id="@+id/network_
monitoring" android:title="Network
Monitoring">
  <menu>
    <item android:id="@+id/
  network_monitoring_activate"
  android:title="Activate Network
  Monitoring"/>
    <item android:id="@+id/
  network_monitoring_deactivate"
  android:title="Deactivate Network
  Monitoring"/>
  </menu>
</item>
```

The `onOptionsItemSelected()` Function

New menu items and code were added to the `onOptionsItemSelected()` function which include the following:

1. Code was added to detect the user's selection of a network device such as a client or server and to set the target device to this new selection.
2. If the user selects the refresh clients menu item then retrieve the list of current clients connected to the server by calling the `SendCommandToServer("list")` function with the list command.
3. If the user selects the switch target device menu item then a list of network devices is shown by calling the `DisplayNetwork(item)` function with the currently selected menu item as a parameter.
4. If the user selects the monitor target device menu item then the current target device is added to the list of devices that will be monitored when network monitoring is turned on. A call to the `AddDeviceToMonitoredDevices(m_TargetDevice)` function with the target device as the parameter actually adds the device. The information window is also updated.
5. If the user selects the unmonitor target device menu item then the current target device is removed from the list of devices that will be monitored when network monitoring is activated. The `RemoveMonitoredDevice(m_TargetDevice)` function with the target device set as the parameter is called to actually implement the removal. The information window is also updated.
6. If the user selects the camera as being available to the target device and the target device is already being monitored then the

camera status is changed by calling the `SetCameraStatus(m_TargetDevice,true)` function with the target device and the value true as parameters. The information window is also updated.

7. If the user selects the camera as being unavailable to the target device and the target device is already being monitored then the camera status is changed by calling the `SetCameraStatus(m_TargetDevice,false)` function with the target device and false as parameters. The information window is also updated.

8. If the user activates network monitoring then the network monitoring is turned on by doing the following:

    1. The number of valid monitored devices is determined by calling the `GetValidMonitoredDevices()` function.

    2. If the number of valid monitored devices is greater than 0 then set the global network monitoring variable to true, set the current polled device to 0, reset the polling state on all the monitored devices, and initialize the running of the monitored devices' handler.

    3. Updating the information window.

9. If the user deactivates network monitoring then the global network monitoring variable is set to false and the information window is updated.

See Listing 8-12. The new code is shown in bold print.

The `DisplayNetwork` Function

The `DisplayNetwork` function displays all the devices on the wireless network by doing the following:

1. Retrieving the submenu from the current `menuitem` object.
2. Deleting all the items in the submenu.
3. Adding in the server as the first selection in the submenu.
4. Adding in the clients that are connected to the server as the remaining submenu item selections.

See Listing 8-13.

The `AddDeviceToMonitoredDevices()` Function

The `AddDeviceToMonitoredDevices()` function adds a device to the list of devices that will be monitored when network monitoring is active by doing the following:

1. Finding if the device is already in the list of monitored devices. If it is already in the list then exit the function.
2. Finding an empty slot in the list of monitored devices' array by calling `FindMonitoredDeviceEmptySlot()` function.
3. If an empty slot exists then put the device name in the list of monitored devices and mark the slot as being not available.

See Listing 8-14.

The `FindMonitoredDevice()` Function

The `FindMonitoredDevice()` function attempts to find the input string that represents the name of the device. It returns a value of 0 or greater if the device is found which is the index where the device is located in the monitored devices array and –1 if it is not found.

See Listing 8-15.

The `FindMonitored DeviceEmpty Slot()` Function

The `FindMonitoredDeviceEmptySlot()` function returns the index of the first

Listing 8-12 The `onOptionsItemSelected()` Function

```
@Override
public boolean onOptionsItemSelected(MenuItem item)
{
 // Item selected is a device on the wireless network
 if (item.getGroupId() == m_NetworkGroupID)
 {
  m_TargetDevice = item.getTitle().toString();
  UpdateCommandTextView();
  Toast.makeText(this, "Switching to "+ item.getTitle() + " on Wireless Network",
 Toast.LENGTH_LONG).show();
  return true;
 }

 // Handle item selection
 switch (item.getItemId()) {

  //////////////////// Server Connection
  case R.id.connect_esp01:
   if (m_ClientConnectThread == null)
   {
    // Create the Client Socket Connection Thread
    ServerModel = ServerType.ESP01;
    CreateClientConnection(m_ServerStaticIP, m_PortNumber);
    Toast.makeText(this, "Connecting Android to ESP-01 Server...", Toast.LENGTH_
   LONG).show();
   }
   else
   {
    // Client already connected
    Toast.makeText(this, "Android Already Connected to Server!!!", Toast.LENGTH_
   LONG).show();
   }
   return true;

  case R.id.connect_nodemcu:
   if (m_ClientConnectThread == null)
   {
    // Create the Client Socket Connection Thread
    ServerModel = ServerType.NodeMCU;
    CreateClientConnection(m_ServerStaticIP, m_PortNumber);
    Toast.makeText(this, "Connecting Android to NodeMCU Server...", Toast.LENGTH_
   LONG).show();
   }
   else
   {
    // Client already connected
    Toast.makeText(this, "Android Already Connected to Server!!!", Toast.LENGTH_
   LONG).show();
   }
```

```
 return true;

////////////////// Refresh Clients
case R.id.refresh_clients:
 SendCommandToServer("list");
 Toast.makeText(this, "Refreshing Clients", Toast.LENGTH_LONG).show();
 return true;

//////////////////// Switch to Device
case R.id.switch_to_device:
 DisplayNetwork(item);
 return true;

//////////////////// Monitor Current Client
case R.id.monitor_current_client:
 if (AddDeviceToMonitoredDevices(m_TargetDevice))
 {
  Toast.makeText(this, "Monitoring Set For " + m_TargetDevice, Toast.LENGTH_LONG).
 show();
 }
 else
 {
  Toast.makeText(this, "ERROR! Monitoring NOT Set For " + m_TargetDevice, Toast.
 LENGTH_LONG).show();
 }
 UpdateCommandTextView();
 return true;

///////////////////// Unmonitor Current Client
case R.id.unmonitor_current_client:
 if (RemoveMonitoredDevice(m_TargetDevice))
 {
  Toast.makeText(this, "Monitoring Removed For " + m_TargetDevice, Toast.LENGTH_
 LONG).show();
 }
 else
 {
  Toast.makeText(this, "ERROR! " + m_TargetDevice +  " Not In Monitored Devices
 List", Toast.LENGTH_LONG).show();
 }
 UpdateCommandTextView();
 return true;

////////////////////// Set Availability of Camera
case R.id.link_camera_activate:
 if (SetCameraStatus(m_TargetDevice,true))
 {
  Toast.makeText(this, "Camera is set to available for " + m_TargetDevice, Toast.
 LENGTH_LONG).show();
 }
 else
```

```
 {
  Toast.makeText(this, "ERROR! Camera could NOT be set to available for " + m_
 TargetDevice, Toast.LENGTH_LONG).show();
 }
 UpdateCommandTextView();
 return true;

case R.id.link_camera_deactivate:
 if (SetCameraStatus(m_TargetDevice,false))
 {
  Toast.makeText(this, "Camera is set to unavailable for " + m_TargetDevice,
 Toast.LENGTH_LONG).show();
 }
 else
 {
  Toast.makeText(this, "ERROR! Camera could NOT be set to unavailable for " + m_
 TargetDevice, Toast.LENGTH_LONG).show();
 }
 UpdateCommandTextView();
 return true;

////////////////// Start Monitoring Network Devices
case R.id.network_monitoring_activate:
 m_NumberValidMonitoredDevices = GetValidMonitoredDevices();
 if (m_NumberValidMonitoredDevices > 0)
 {
  m_NetworkMonitoringOn = true;
  m_CurrentPolledDevice = 0;
  ResetMonitoredDevicesPollingState();
  m_MonitorDevicesHandler.postDelayed(
  m_MonitoredDevicesRunnable,
   m_MonitoredDevicesInterval);
  Toast.makeText(this, "Starting Network Monitoring, Number Monitored Devices: " +
 m_NumberValidMonitoredDevices, Toast.LENGTH_LONG).show();
 }
 else
 {
  Toast.makeText(this, "ERROR! No Devices Set to Monitor ", Toast.LENGTH_LONG).
 show();
 }
 UpdateCommandTextView();
 return true;

case R.id.network_monitoring_deactivate:
 m_NetworkMonitoringOn = false;
 Toast.makeText(this, "Stopping Network Monitoring", Toast.LENGTH_LONG).show();
 UpdateCommandTextView();
 return true;

///////////////////////// Alarm Activation
case R.id.alarm_activate:
```

```
 SendCommand("ACTIVATE_ALARM");
 SetAlarmStatus(true);
 UpdateCommandTextView();
 Toast.makeText(this, "Alarm Activated !!!", Toast.LENGTH_LONG).show();
 return true;

case R.id.alarm_deactivate:
 SendCommand("DEACTIVATE_ALARM");
 SetAlarmStatus(false);
 UpdateCommandTextView();
 Toast.makeText(this, "Alarm Deactivated !!!", Toast.LENGTH_LONG).show();
 return true;

//////////////////////// Alarm Emergency Phone Callout
case R.id.callout_on:
 SetCallOutStatus(true);
 UpdateCommandTextView();
 Toast.makeText(this, "Call Out Activated !!!", Toast.LENGTH_LONG).show();
 return true;

case R.id.callout_off:
 SetCallOutStatus(false);
 UpdateCommandTextView();
  Toast.makeText(this, "Call Out Deactivated !!!", Toast.LENGTH_LONG).show();
 return true;

////////////////////// Text messages
case R.id.notext:
 SetTextMessageStatus(TextMessageSetting.TEXT_OFF);
 UpdateCommandTextView();
 Toast.makeText(this, "Text Message Alerts Turned Off !!!", Toast.LENGTH_LONG).
show();
 return true;

case R.id.text_only:
 SetTextMessageStatus(TextMessageSetting.TEXT_ON);
 UpdateCommandTextView();
 Toast.makeText(this, "Text Message Alerts Activated !!!", Toast.LENGTH_LONG).
show();
 return true;

//////////////////////// Resolution
case R.id.qqvga:
 SetCameraResolution(Resolution.qqvga);
 UpdateCommandTextView();
 return true;

case R.id.qvga:
 SetCameraResolution(Resolution.qvga);
 UpdateCommandTextView();
return true;
```

```
    case R.id.vga:
     SetCameraResolution(Resolution.vga);
     UpdateCommandTextView();
     return true;

    //////////////////////////// Surveillance
    case R.id.surveillance_on:
     ProcessSurveillanceSelection(true);
     UpdateCommandTextView();
     Toast.makeText(this, "Surveillance Turned On !!!", Toast.LENGTH_LONG).show();
     return true;

    case R.id.surveillance_off:
     ProcessSurveillanceSelection(false);
     UpdateCommandTextView();
     Toast.makeText(this, "Surveillance Turned Off !!!", Toast.LENGTH_LONG).show();
     return true;

    //////////////////////////// Android Storage
    case R.id.androidstorageyes:
     m_AndroidStorage = AndroidStorage.StoreYes;
     UpdateCommandTextView();
     Toast.makeText(this, "Images will be saved to Android Storage", Toast.LENGTH_
    LONG).show();
     return true;

    case R.id.androidstorageno:
     m_AndroidStorage = AndroidStorage.StoreNo;
     UpdateCommandTextView();
     Toast.makeText(this, "Images will NOT be saved to Android Storage", Toast.LENGTH_
    LONG).show();
     return true;

    //////////////////////////////////////////////////
    case R.id.savedevicename:
     m_AndroidClientName = m_PhoneEntryView.getText().toString();
     SaveDeviceName();
     Toast.makeText(this, "Saving Device Name: " + m_AndroidClientName, Toast.LENGTH_
    SHORT).show();
     AddDebugMessageFront("Android Device Name Changed to: " + m_AndroidClientName +
    "\n");

  m_PhoneEntryView.setText(m_EmergencyPhoneNumber);
     UpdateCommandTextView();
     return true;

     default:
      return super.onOptionsItemSelected(item);
  }
}
```

Listing 8-13 The `DisplayNetwork()` Function

```
void DisplayNetwork(MenuItem menuitem)
{
 Log.e("MainActivity", "In
DisplayNetwork");
 SubMenu submenu1 = menuitem.
getSubMenu();
 if (submenu1 != null)
 {
  submenu1.clear();
  submenu1.add(m_NetworkGroupID, m_
 NetworkItemID, 0, "Server");
  for (int i = 0; i < m_
 NumberClients; i++)
   {
    submenu1.add(m_NetworkGroupID, m_
  NetworkItemID, i+1, m_Clients[i]);
   }
 }
 else
 {
  Log.e("MainActivity", "ERROR, Can
 not get Network List SubMenu");
 }
}
```

empty slot in the list of monitored devices `m_MonitoredDevices` or –1 if all the slots are full and contain valid devices. See Listing 8-16.

The `RemoveMonitoredDevice()` Function

The `RemoveMonitoredDevice()` function removes a device from the list of monitored devices and resets the slot that the device previously occupied to its default values. See Listing 8-17.

The `SetCameraStatus()` Function

The `SetCameraStatus()` function sets the availability of the camera on a device that is in the list of devices to be monitored by doing the following:

1. Finding the index that the device is located in by calling the

Listing 8-14 The `AddDeviceToMonitoredDevices()` Function

```
boolean AddDeviceToMonitoredDevices
(String device)
{
 boolean result = false;

 // Check for duplicates
 if (FindMonitoredDevice(device) >= 0)
 {
  // device already here
  return false;
 }

 // Find Empty Slot
 int emptyslot =
FindMonitoredDeviceEmptySlot();
 if (emptyslot >= 0)
 {
  // slot available
  m_MonitoredDevices[emptyslot] =
 device;
  m_MonitoredDevicesAvailability
 [emptyslot] = false;
  result = true;
 }
 else
 {
  result =  false;
 }
 return result;
}
```

`FindMonitoredDevice(device)` function with the device name as the parameter.

2. If the index is equal to or greater than 0 then the device is found, so set the corresponding value of the `m_CameraAvailability` array to the function's input parameter variable status.

See Listing 8-18.

The `GetValidMonitoredDevices()` Function

The `GetValidMonitoredDevices()` function returns the number of valid devices that are in

Listing 8-15 The `FindMonitoredDevice()` Function

```
int FindMonitoredDevice(String device)
{
 int result = -1;

 for (int i = 0; i < Max_Monitored_
Devices; i++)
 {
  if (m_MonitoredDevices[i].
 equals(device))
  {
   return i;
  }
 }
 return result;
}
```

Listing 8-16 The `FindMonitoredDeviceEmptySlot()` Function

```
int FindMonitoredDeviceEmptySlot()
{
 int result = -1;
 for (int i = 0; i < Max_Monitored_
Devices; i++)
 {
  if (m_MonitoredDevicesAvailability
 [i] == true)
  {
   // Return first available slot
   return i;
  }
 }
 return result;
}
```

the list of monitored devices. It also fills the `m_ValidMonitoredDevices` array with the indices of the valid devices that are in the `m_MonitoredDevices` array. See Listing 8-19.

The `ResetMonitoredDevicesPollingState()` Function

The `ResetMonitoredDevicesPollingState()` function resets the polling state of all the

Listing 8-17 The `RemoveMonitoredDevice()` Function

```
boolean RemoveMonitoredDevice(String
device)
{
 boolean result = false;

 // Find index of device if exists
 int index = FindMonitored
Device(device);
 if (index >= 0)
 {
  // device found
  m_MonitoredDevices[index] = "";
  m_MonitoredDevicesAvailability
 [index] = true;
  m_CameraAvailability[index] =
 false;
  m_MonitoredDevicesPollingState
 [index] = PollingState.NotStarted;
  result = true;
 }
 else
 {
  // device not found
  result = false;
 }
 return result;
}
```

Listing 8-18 The `SetCameraStatus()` Function

```
boolean SetCameraStatus(String device,
boolean status)
{
 boolean result = false;
 int index =
FindMonitoredDevice(device);
 if (index >= 0) {
  // Device found
  m_CameraAvailability[index] =
 status;
  result = true;
 }
 return result;
}
```

Listing 8-19 The GetValidMonitoredDevices() Function

```
int GetValidMonitoredDevices()
{
 int j = 0;
 for (int i = 0; i < Max_Monitored_
Devices; i++)
 {
  if (!m_MonitoredDevicesAvailability
  [i])
  {
   // slot not available so it
  conatains a valid device name
  m_ValidMonitoredDevices[j] = i;
   j++;
  }
 }
 return j;
}
```

Listing 8-20 The ResetMonitoredDevicesPollingState() Function

```
void ResetMonitoredDevicesPolling
State()
{
 for (int i = 0; i < Max_Monitored_
Devices; i++)
 {
  // Reset each monitored devices
 polling to NotStarted
  m_MonitoredDevicesPollingState[i] =
 PollingState.NotStarted;
 }
}
```

monitored devices to the not started state. See Listing 8-20.

The `ReceiveDisplayTextData()` Function

The `ReceiveDisplayTextData()` function receives incoming text data from the server and is modified for version 2.0 by doing the following:

1. If the incoming data from the server contains the sub string `"ActiveClients:"` then call the `SetClients(DataTrimmed)` function with the incoming data to process the list of clients that are connected to the server.
2. If the incoming data from the server contains the sub string `"HITS DETECTED:"` then this is the response to the hits command issued by the network monitoring handler from the device that is being polled. To process this data do the following:
   1. If network monitoring is not on then exit the function.
   2. Find the number of hits detected from the number after the colon in the string.
   3. If the number of hits is greater than 0 then do the following:
      1. Set the alarm to the tripped state.
      2. Play a sound effect to indicate that the alarm has been tripped.
      3. Process the alarm event by calling the `ProcessAlarmTripped()` function.
      4. Switch the target device to the device that is being polled.
      5. Send the reset command to set the number of hits to 0 on the target device by calling the `SendCommand("reset")` function with the "reset" string as the parameter.
      6. If a camera for the target device is available for use then do the following:
         1. Send a command to take a photo to the target device by calling the `SendTakePhotoCommand()` function.
         2. Set the `PollingFinished` variable to false since we are not done receiving data from the currently polled device. Specifically we still need to

receive the image from the camera.

3. Set the `m_PollingTakePhoto` to true to indicate the device that is currently being polled will take a photo.

4. If polling is finished then set the polling status for the device currently being polled to finished.

See Listing 8-21. The new code is highlighted in bold print.

The `SetClients()` Function

The `SetClients()` function parses the input string which is the result of the list command. The function determines the number of clients as well as the name of each client which is put in the `m_Clients` array. The list of clients is also

Listing 8-21 The `ReceiveDisplayTextData()` Function

```
void ReceiveDisplayTextData()
{
 String Data = m_WifiMessageHandler.GetStringData();
 String DataTrimmed = Data.trim();

 // Debug Print Outs
 Log.e(TAG, "Returned Text Data = " + DataTrimmed);
 m_DebugMsg = "";
 AddDebugMessage(TAG + ": Returned Text Data = " + "'" + DataTrimmed + "'" + "\n");

 // Process ping response from server
 if (DataTrimmed.equals("PING_OK"))
 {
  m_Alert1SFX.PlaySound(false);
  Log.e(TAG,"PING_OK RECEIVED !!!!!!!!!!!!!!!!!!!!!!!!!!!!!!!!!!");
 }
 else if (DataTrimmed.equals("ALARM_TRIPPED"))
 {
  m_AlarmTripped = true;
  ProcessAlarmTripped();
 }
 else if (DataTrimmed.contains("ALARM_TRIPPED:"))
 {
  // Process Alarm tripped message that contains the total number of hits
  int indexstart = DataTrimmed.indexOf(":");
  String NumberHits = DataTrimmed.substring(indexstart+1);

  m_TotalNumberEventsDetected = Integer.valueOf(NumberHits);
  m_AlarmTripped = true;
  m_TextMessageSent = false;

  m_Alert2SFX.PlaySound(true);
  ProcessAlarmTripped();
 }
 else if (DataTrimmed.contains("ActiveClients:"))
 {
  SetClients(DataTrimmed);
```

```
  }
  else if (DataTrimmed.contains("HITS DETECTED:"))
  {
   if (!m_NetworkMonitoringOn)
   {
    return;
   }

   boolean PollingFinished = true;
   int ValidDeviceIndex = m_ValidMonitoredDevices[m_CurrentPolledDevice];

   // Process Alarm tripped message that contains the total number of hits
   int indexstart = DataTrimmed.indexOf(":");
   String NumberHits = DataTrimmed.substring(indexstart+1);

   m_TotalNumberEventsDetected = Integer.valueOf(NumberHits);
   if (m_TotalNumberEventsDetected > 0)
   {
    m_AlarmTripped = true;
    m_TextMessageSent = false;

    m_Alert2SFX.PlaySound(true);
    ProcessAlarmTripped();

    // Switch TargetDevice to device being polled
    m_TargetDevice = m_MonitoredDevices[ValidDeviceIndex];

    // Send rest command to device being polled
    SendCommand("reset");

    // Camera Code
    // If camera for this device is available
    if (m_CameraAvailability[ValidDeviceIndex])
    {
     // Take Photo with Camera
     SendTakePhotoCommand();

     // Set Polling Finished to false
     PollingFinished = false;

     // Set taking photo to true
     m_PollingTakePhoto = true;

     // Display Message
     Toast.makeText(this, "Taking Photo", Toast.LENGTH_SHORT).show();
    }
   }

   // Update Polling Status
   if (PollingFinished)
   {
    m_MonitoredDevicesPollingState[ValidDeviceIndex] = PollingState.Finished;
   }
  }
}
```

Listing 8-22 The `SetClients()` Function

```
void SetClients(String cl)
{
 // Process list of clients from
"ActiveClients: " return string
 int indexcolon = cl.indexOf(":");
 String list = cl.substring(indexcolon
+ 1);
 String[] temp = list.split(",");
 int length = temp.length;
 String tempdebug = "m_Clients:\n";

 m_NumberClients = 0;
 for (int i = 0; i < length; i++)
 {
  if (!temp[i].equals(" "))
  {
   m_Clients[i] = temp[i].trim();
   tempdebug += i + ". " + "'" + m_
  Clients[i] + "'" + "\n";
   m_NumberClients++;
  }
 }
 AddDebugMessageFront(tempdebug);
}
```

shown in the debug message window. See Listing 8-22.

Modifying the `run()` Function

The `run()` function is modified for version 2.0 of the basic wireless framework. The key code addition is in the code that handles the processing of a completed image data transfer. After the photo has finished being displayed then the polling status of the device that is currently being polled is changed to finished. See Listing 8-23. The code changes are highlighted in bold print.

Basic Framework 2.0 for the NodeMCU Server

This section covers the basic framework 2.0 code for the NodeMCU server which

Listing 8-23 `run()` Function Additions for Version 2.0

```
if (m_TakePhotoState == TakePhoto
CallbackState.GetImageData)
{
 // Display New Photo received from
the Arduino
 m_TakePhotoCallbackDone = false;
 LoadJpegPhotoUI(m_PhotoData);

 // If surveillance is active
 if (m_SurveillanceActive)
 {
  m_Button.setEnabled(false);
  if (m_SurveillanceSFX)
  {
   m_Alert1SFX.PlaySound(false);
  }
  m_SurveillanceFrameNumber++;

  m_DebugMsg = "";
  m_DebugMsg += "Frame Number: " + m_
 SurveillanceFrameNumber + "\n";

  m_SurveillanceFileName = "";
  if (m_AndroidStorage ==
 AndroidStorage.StoreYes)
  {
   String filename = SavePicJPEG();
   m_DebugMsg+= "File Name: " +
  filename + "\n";
  }
  m_DebugMsg += "\n\n\n\n";
  m_DebugMsgView.setText(m_DebugMsg.
 toCharArray(), 0, m_DebugMsg.
 length());

  // Send another take photo command
 to Arduino
  SendTakePhotoCommand();
 }
 else
 {
  // If surveillance is not active
 and storage is yes
  // then store image on Android
  m_SurveillanceFileName = "";
  if (m_AndroidStorage ==
 AndroidStorage.StoreYes)
```

```
    {
     String filename = SavePicJPEG();
     m_DebugMsg+= "File Name: " +
    filename + "\n";
     m_DebugMsgView.setText(m_DebugMsg.
    toCharArray(), 0, m_DebugMsg.
    length());
    }
    m_ButtonActive = true;
    m_Alert2SFX.PlaySound(true);

    // Check for Network Monitoring
    if (m_NetworkMonitoringOn)
    {
     // If network monitoring is on then
     // update Polling Status of
    currently polled device
     int ValidDeviceIndex = m_
    ValidMonitoredDevices[m_
    CurrentPolledDevice];
     m_MonitoredDevicesPollingState
    [ValidDeviceIndex] = PollingState.
    Finished;
    }
  }
}
```

is designed to work with the Android basic wireless framework code version 2.0 in a multi-client environment that can transmit both text and binary data. This section will cover only the changes made from the previous version of the NodeMCU server.

The `BinaryDataSource` variable holds the name of client that will be the source of the binary data that will be transferred.

```
String BinaryDataSource = "";
```

The `BinaryDataDest` variable holds the name of the client that will receive the binary data that will be sent by the source.

```
String BinaryDataDest = "";
```

The `BinaryDataSize` variable is the size of the binary data to be transferred in bytes.

```
int BinaryDataSize = 0;
```

The `TotalBinaryDataSent` variable holds the total number of bytes that have been transmitted so far from the source client to the destination client.

```
int TotalBinaryDataSent = 0;
```

The `BinaryMode` variable is true if binary transfer mode is active and false otherwise.

```
boolean BinaryMode = false;
```

Modifying the `ProcessIncoming ClientData()` Function

The `ProcessIncomingClientData()` function processes incoming commands from clients.

The addition to this function involves recognizing and processing the command to specify a binary transfer mode. The format for the binary command is as follows:

```
BIN:DestClientName:SizeOfBinaryData
```

If an incoming command contains "BIN:" then start the processing for a binary command by doing the following:

1. Set binary mode to true.
2. Set the binary data source to the client that issued the `BIN` command.
3. Process the binary command by calling the `ProcessBinaryCommand(message)` function by sending the function the portion of the command that is after the "`BIN:`" part of the incoming command string.
4. Send a text string back to the calling client indicating the binary transfer mode has been enabled.
5. Exit the function.

See Listing 8-24. The changes are highlighted in bold print.

The `ProcessBinaryCommand()` Function

The `ProcessBinaryCommand()` function parses the binary transfer command sent by the

Listing 8-24 Modifying the `ProcessIncomingClientData()` Function

```
if (Data.indexOf(":") >= 0)
{
 // Send data to specific client by
name
 // ClientName:MessageToClient
 // Send message to destination
client
 // Send confirmation or failure
message to sending client
 boolean MessageSent   = true;
 int     destseparator = Data.
indexOf(":");
 String  clientid      = Data.
substring(0,destseparator);
 String  message       = Data.
substring(destseparator+1);

 // Process BIN command
 if (clientid == "BIN")
 {
  BinaryMode = true;
  BinaryDataSource = ClientID;
  ProcessBinaryCommand(message);
  SendDataToClient(ClientID, "Binary
 Transfer Enabled.\n");
  return;
 }

 if (!SendDataToClient(clientid,
message + "\n"))
 {
 Serial.println(F("***************
ERROR ... Client Not Found"));
 MessageSent = false;
 }

 // Check if message sent
successfully and send back appropriate
 // response to the sending client
 if (MessageSent)
 {
  // Send OK response message to
 requesting client
  //SendDataToClient(ClientID,
 "OK\n");
 }
 else
 {
  // Send ERROR response message to
 requesting client
  SendDataToClient(ClientID,
 "ERROR\n");
 }
}
```

source client. The string that is parsed is of the form:

`DestClientName:SizeOfBinaryData`

The destination client name for the binary transfer and the size of the binary data in bytes are retrieved. Debug information is also printed out to the Serial Monitor. See Listing 8-25.

The `ProcessConnectedClients()` Function

The `ProcessConnectedClients()` function processes incoming data if it is available for all the clients that are connected to the server. Code is added to the function that processes binary data by doing the following:

1. When data is available for reading from a connected client and if the client is the binary data source and binary mode is on then process the incoming binary data from the client by calling the `ProcessBinaryData()` function.
2. Otherwise process the incoming data from the client as text data.

See Listing 8-26. The modifications are shown in bold print.

The `ProcessBinaryData()` Function

The `ProcessBinaryData()` function processes the incoming binary data from a client by doing the following:

1. Find the indices in the `Clients` array that correspond to the client that is the binary

Listing 8-25 The `ProcessBinaryCommand()` Function

```
void ProcessBinaryCommand(String cmd)
{
 // Process the binary command
 // BIN:DestClientName:SizeOfBinary
Data

 // cmd is in the form:
 // DestClientName:SizeOfBinaryData

 int sizeseparator = cmd.indexOf(":");
 String  clientname  = cmd.
substring(0,sizeseparator);
 String  sizedata    = cmd.
substring(sizeseparator+1);

 BinaryDataDest = clientname;
 BinaryDataSize = sizedata.toInt();

 Serial.print(F("Set Binary
Mode Command Recieved from
BinaryDataSource: "));
 Serial.print(BinaryDataSource);
 Serial.print(F(" , to BinaryData
Dest: "));
 Serial.print(BinaryDataDest);
 Serial.print(F(" , for
BinaryDataSize: "));
 Serial.println(BinaryDataSize);
}
```

data source and the client that is the binary data destination.

2. While the source client has data that is available for reading then do the following:
   1. Read in a byte of the data from the source client.
   2. Write out a byte of data to the destination client.
   3. Update the total number of bytes of binary data transferred.

Listing 8-26 The `ProcessConnectedClients()` Function

```
void ProcessConnectedClients()
{
 String line = "NA";
 for (int i = 0; i < MaxClients; i++)
 {
  if (Clients[i].ID != "")
  {
   // Client is valid so check to see
  if there is data to read in
   if (Clients[i].ClientObject.
  connected())
   {
    if (Clients[i].ClientObject.
   available())
    {
     // Process Binary Data
     if ((Clients[i].ID ==
    BinaryDataSource) &&
        (BinaryMode == true))
     {
      // Process incoming data
      ProcessBinaryData();
     }
     else
     {
      // Process Text Data
      line = Clients[i].ClientObject.
     readStringUntil('\n');

      ProcessIncomingClientData
     (Clients[i].ID, line);
     }
    }
   }
   else
   {
    // Client has disconnected
    // Free up previously used slot
   and free up resources
    Serial.print(F("Client has
   disconnected, Client ID: "));
    Serial.println(Clients[i].ID);

    Clients[i].ID = "";
    Clients[i].ClientObject.stop();
   }
  }
 }
}
```

3. If the total number of bytes transferred is greater than or equal to the size of the binary data that is to be transferred then do the following:
   1. Print out the total number of bytes transferred to the Serial Monitor.
   2. Send a text string to the source client that indicates that binary transfer has completed.
   3. Reset the variables that control the binary data transfer to their initial states.

See Listing 8-27.

Listing 8-27 The `ProcessBinaryData()` Function

```
void ProcessBinaryData()
{
  int sourceindex = 0;
  int destindex = 0;

  // Find Source and dest indexes
  sourceindex = FindClientIndex(BinaryDataSource);
  destindex = FindClientIndex(BinaryDataDest);

  // While more data from source process it
  if (Clients[sourceindex].ClientObject.available())
  {
    //  Read data
    byte b = Clients[sourceindex].ClientObject.read();

    // Send data
    Clients[destindex].ClientObject.write(b);
    TotalBinaryDataSent++;
  }

  // Check to see if all the binary data has been sent
  if (TotalBinaryDataSent >= BinaryDataSize)
  {
    // Binary recieving/sending is done
    // Debug output to Serial Monitor
    Serial.print(F("Binary Transfer Finished. Bytes Transfered: "));
    Serial.println(TotalBinaryDataSent);

    // Send Finished message to source
    String returnstring = "";
    returnstring = returnstring + F("Binary Transfer Finished. Bytes Transfered: ") +
                   TotalBinaryDataSent + "\n";
    SendDataToClient(BinaryDataSource, returnstring);

    // Reset Variables
    BinaryDataSource = "";
    BinaryDataDest = "";
    TotalBinaryDataSent = 0;
    BinaryDataSize = 0;
    BinaryMode = false;
  }
}
```

Example of Setting Up the Arduino with ESP-01 for Station/Client Mode

The `setup()` function sets the ESP-01 module for station mode by doing the following:

1. Initializing debugging with the Serial Monitor at 9600 baud.
2. Initializing serial communications with the ESP-01 module on serial port 3 at 115,200 baud.
3. Delaying program execution for 1000 milliseconds or 1 second.
4. Initializing sensors and other hardware. In this specific case a camera is initialized by calling the `InitializeCamera()` function.
5. Turning off the Arduino's built-in LED to indicate that the system has not yet become active.
6. Enabling single connection mode by calling the `SendATCommand(F("AT+CIPMUX=0\r\n"))` function with the parameter as shown.
7. Setting the mode to station mode by calling the `SendATCommand(F("AT+CWMODE_CUR=1\r\n"))` function with the parameter shown.
8. Joining the access point by sending the join access point command which is `AT+CWJAP_CUR="ESP8266-12E-AP","robtestESP12E"` to the ESP-01 module.
9. Starting the connection to the TCP server by sending the command which is `AT+CIPSTART="TCP","192.168.4.1",80` to the ESP-01 module.
10. Setting the name of the client by calling the `SendTCPMessage(ClientName)` function.
11. Turning on the built-in LED on the Arduino if there are no errors and printing the error status to the Serial Monitor.

See Listing 8-28.

Hands-on Example: The ArduCAM OV2640 2MP Mini Plus Camera Arduino Mega 2560 Client Surveillance System with NodeMCU Server

In this hands-on example I demonstrate how to build and operate a wireless surveillance system based on an Android controller, a NodeMCU server, and an Arduino client that contains an ArduCAM OV2640 Mini Plus camera. See Figure 8-8.

Parts List

- 1 Android cell phone
- 1 NodeMCU microcontroller
- 1 Arduino Mega 2560 microcontroller
- 1 5- to 3.3-V step-down voltage regulator
- 1 Logic level converter from 5 to 3.3 V
- 1 ESP-01 Wi-Fi module
- 1 Arduino development station such as a desktop or a notebook
- 1 Breadboard (minimum)
- 1 ArduCAM OV2640 Mini Plus camera
- 1 Package of wires both male-to-male and female-to-male to connect the components

Setting Up the Hardware

Step-Down Voltage Regulator Connections

1. Connect the Vin pin on the voltage regulator to the 5-V power pin on the Arduino Mega 2560.
2. Connect the Vout pin on the voltage regulator to a node on your breadboard represented by a horizontal row of empty slots. This is the 3.3-V power node that will supply the 3.3-V power to the ESP-01 module as well as provide the reference voltage for the logic level shifter.

Listing 8-28 The `setup()` Function for the Arduino with ESP-01 Station or Client Configuration

```
void setup()
{
  // Initialize Serial
  Serial.begin(9600);
  Serial.println(F("*****************************************************************
**"));
  Serial.println(F("****** ESP-01 Arduino Mega 2560 Station/Client Program
********"));
  Serial.println(F("*********** ArduCAM Mini 2MP OV2640 Plus
**********************"));
  Serial.println(F("************** Surveillance System ***************************"));
  Serial.println();
  Serial.println();
  Serial.println();
  Serial.println();

  // Initialize ESP-01 on Serial Port 3 on Arduino Mega 2560
  Serial3.begin(115200);
  delay(1000);

  // Set up ESP-01 as a Station and Client using AT commands
  Serial.println();
  Serial.println();

  // Initialize ArduCAM Mini Camera
  InitializeCamera();

  // Initialize digital pin LED_BUILTIN as an output and turn off.
  pinMode(LED_BUILTIN, OUTPUT);
  digitalWrite(LED_BUILTIN, LOW);

  // Enable Single Connection
  if (SendATCommand(F("AT+CIPMUX=0\r\n")))
  {
   Serial.println(F("ESP-01 mode set for single connection."));
  }
  else
  {
   Serial.println(F("ERROR - ESP-01 FAILED to set single connection mode for TCP."));
   StartUpOk = false;
  }
   Serial.println();
   Serial.println();

   // Set Station Mode
   if (SendATCommand(F("AT+CWMODE_CUR=1\r\n")))
   {
    Serial.println(F("ESP-01 mode changed to Station Mode."));
   }
   else
```

```
  {
   Serial.println(F("ERROR - ESP-01 FAILED to change to Station Mode."));
   StartUpOk = false;
  }
  Serial.println();
  Serial.println();
  // Join AP
  // AT+CWJAP_CUR="ESP8266-12E-AP","robtestESP12E"
  String JoinAP = String(F("AT+CWJAP_CUR=\"ESP8266-12E-AP\",\"robtestESP12E\"\
r\n"));
  if (SendATCommand(JoinAP))
  {
   Serial.println(F("ESP-01 station has joined AP."));
  }
  else
  {
   Serial.println(F("ERROR - ESP-01 station FAILED to join AP. RESTART SYSTEM TO
  ATTEMPT AGAIN ..."));
   while(1){}
  }
  Serial.println();
  Serial.println();
  // Start TCP server connection
  // AT+CIPSTART="TCP","192.168.4.1",80
  String TCPString = String(F("AT+CIPSTART=\"TCP\",\"192.168.4.1\",80\r\n"));
  if (SendATCommand(TCPString))
  {
   Serial.println(F("ESP-01 Connected to TCP Server."));
  }
  else
  {
   Serial.println(F("ERROR - ESP-01 FAILED to Connect to TCP Server. RESTART SYSTEM
  TO TRY AND RECONNECT."));
   while(1);
  }
  Serial.println();
  Serial.println();

  // Send Name To Server using TCP connection
  int lengthname = ClientName.length();
  if (SendTCPMessage(ClientName))
  {
   Serial.print(F("Client Name Set TO: "));
   Serial.print(ClientName);
  }
  else
  {
   Serial.println(F("ERROR. Failed to Set Client Name.. RESTART SYSTEM TO TRY
  AGAIN."));
```

```
    StartUpOk = false;
  }
  Serial.println();
  Serial.println();

  // Start Up Diagnotic Messages
  if (StartUpOk)
  {
    Serial.println(F("************ Start Up Ok **************"));
    digitalWrite(LED_BUILTIN, HIGH);
  }
  else
  {
    Serial.println(F("****** ERROR ****** ERROR(S) on Start Up **************"));
    digitalWrite(LED_BUILTIN, LOW);
  }
}
```

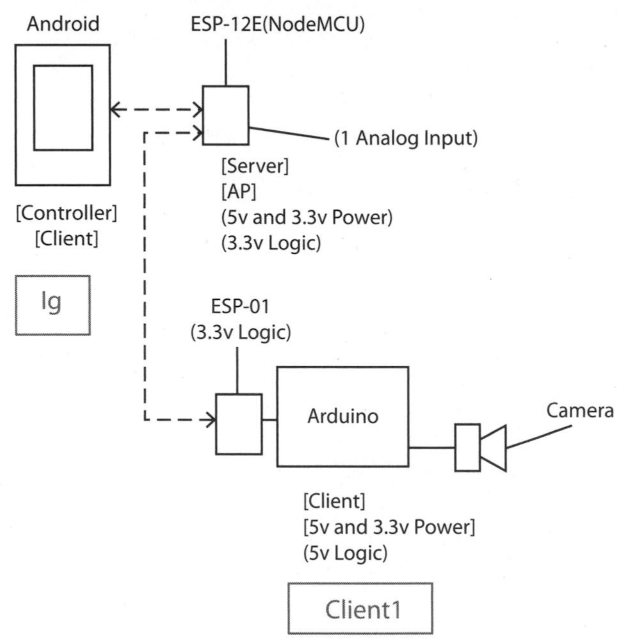

Figure 8-8 The Android, NodeMCU server, and Arduino ArduCAM camera client system.

3. Connect the GND pin on the voltage regulator to the GND pin on the Arduino or form a GND node similar to the 3.3-V power node in the previous step.

ESP-01 Wi-Fi Module Connections

1. Connect the Tx or TxD pin to the Rx3 pin on the Arduino Mega 2560. This is the receive pin for the serial hardware port 3.
2. Connect the EN or CH_PD pin to the 3.3-V power node from the step-down voltage regulator.
3. Connect the IO16 or RST pin to the 3.3-V power node from the step-down voltage regulator.
4. Connect the 3V3 or VCC pin to the 3.3-V power node from the step-down voltage regulator.
5. Connect the GND pin to the ground pin or node.
6. Connect the Rx or RxD pin to pin B0 on the 3.3-V side of the logic level shifter. The voltage from this pin will be shifted from 5 V from the Arduino to 3.3 V for the ESP-01 module.

Logic Level Shifter

1. Connect the 3.3-V pin to the 3.3-V power node from the step-down voltage regulator.
2. Connect the GND pin on the 3.3-V side to the ground node for the Arduino.
3. Connect the B0 pin on the 3.3-V side to Rx or RxD pin on the ESP-01. This corresponds to the A0 pin on the 5-V side.

4. Connect the 5-V pin to the 5-V power pin on the Arduino.
5. Connect the GND pin on the 5-V side to the ground node for the Arduino.
6. Connect the A0 pin on the 5-V side to Tx3 pin on the Arduino. This pin corresponds to the B0 pin on the 3.3-V side.

ArduCAM OV2640 Mini Plus

1. Connect the CS pin to digital pin 8 on the Arduino.
2. Connect the SDA pin to digital pin 20 on the Arduino.
3. Connect the SCL pin to digital pin 21 on the Arduino.
4. Connect the MISO pin to digital pin 50 on the Arduino.
5. Connect the MOSI pin to digital pin 51 on the Arduino.
6. Connect the SCK pin to digital pin 52 on the Arduino.
7. Connect the VCC pin to the 5-V node for the circuit.
8. Connect the GND pin to the common ground node for the circuit.

See Figure 8-9.

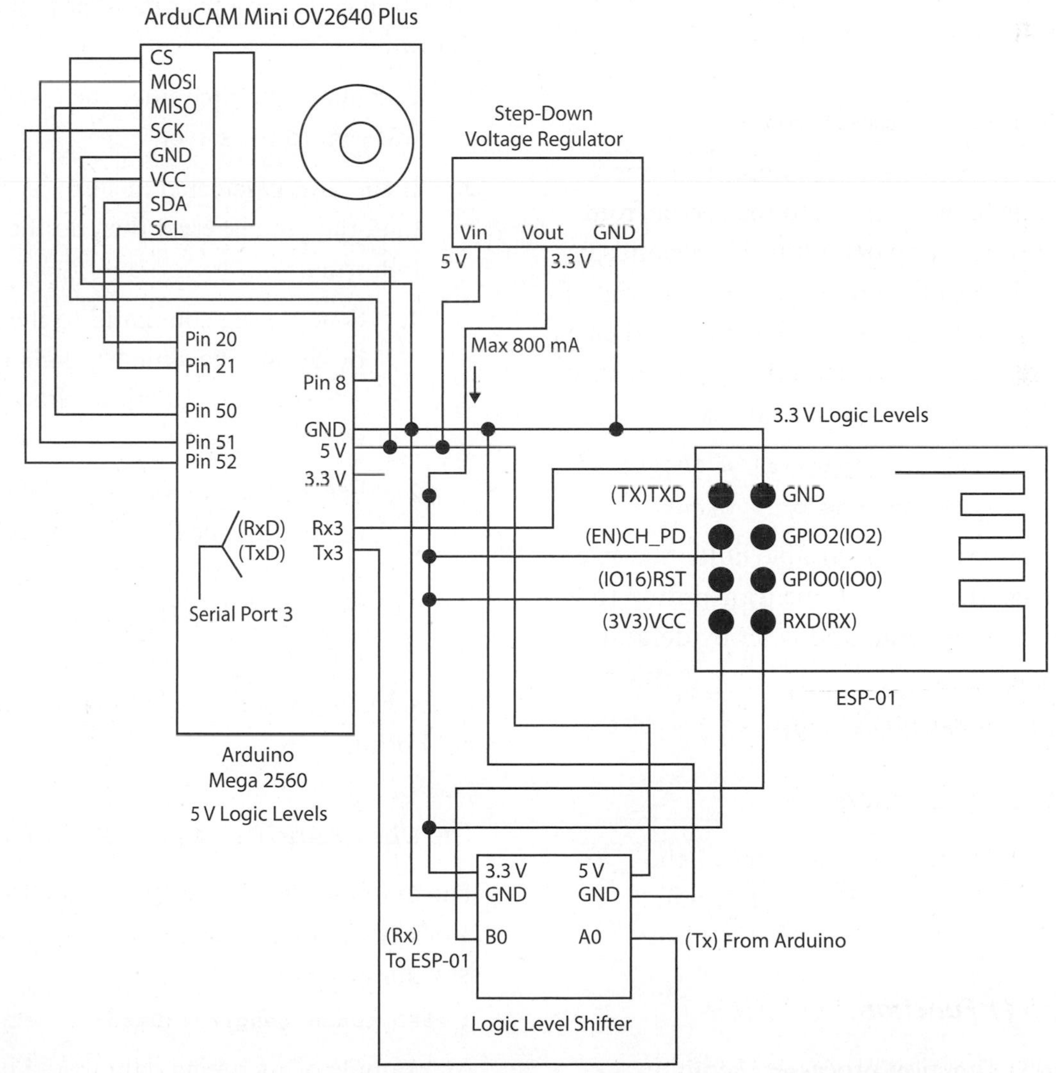

Figure 8-9 The ArduCAM OV2640 Mini Plus camera Arduino client surveillance system with NodeMCU server.

Setting Up the Software

This section will cover the new code specific for the client version of this camera surveillance system. Code that remains the same from the server version was already discussed and will not be covered here.

The `ClientName` is the name of this client which is by default `client1`.

```
String ClientName = "client1";
```

The `TimeStarted` variable holds the time that a command was last sent by the client to the server.

```
unsigned long TimeStarted = 0;
```

The `TimeDelayed` variable holds the time in milliseconds since the last command was sent to the server by this client.

```
unsigned long TimeDelayed = 0;
```

The `PingInterval` variable holds the time in milliseconds between pings to the server from this client. It is set by default to 120 seconds or 2 minutes. If the client does not send any data to the server for a time period of 2 minutes then a ping will be sent to the server. Pings are needed to keep the client connected to the server.

```
unsigned long PingInterval = 1000UL *
120UL; // 120 seconds or 2 minutes
```

The `BinaryDataSize` variable holds the size of the image that needs to be transmitted to the Android controller and is set by default to 0.

```
long BinaryDataSize = 0;
```

The `setup()` Function

The `setup()` function is the same as the example client set-up function shown in Listing 8-28.

The `loop()` Function

The `loop()` function processes incoming commands from the Android controller by calling the `ProcessESP01Notifications()` function. See Listing 8-29.

Listing 8-29 The `loop()` Function

```
void loop()
{
  // Process Wifi Notifications
  ProcessESP01Notifications();
}
```

The `ProcessESP01Notifications()` Function

The `ProcessESP01Notifications()` function is modified from previous versions by adding in code for a ping handler that does the following:

1. Calculates the time since the last command was sent to the server.
2. If the time calculated in step 1 is greater than the ping interval then do the following:
    1. Send a ping command to the server by calling the `SendTCPMessage("ping\n")` function with "ping\n" as the parameter.
    2. Set the `TimeStarted` variable which holds the time that the last command was sent to the server to the current time by calling the `millis()` function.

See Listing 8-30. Code additions are shown in bold print.

The `CheckAndProcessIPD()` Function

The `CheckAndProcessIPD()` function processes incoming data to this client. The data is of the format

```
+IPD,<Data Length>:<Data>
```

An example of incoming data would be the camera command `qvga` which tells the camera to

Listing 8-30 The `ProcessESP01 Notifications()` Function

```
void ProcessESP01Notifications()
{
 char Out;

 if (Serial3.available()>0)
 {
  Out = (char)Serial3.read();
  Notification += Out;

  if (Out == '\n')
  {
   if (Notification.indexOf("+IPD")>=0)
   {
    // Client Data notification found
    Serial.print(F("Notification:'"));
    Serial.print(Notification);
    Serial.print(F("'"));
    CheckAndProcessIPD(Notification);
    Notification = "";
   }
   else
   {
    // Print out to Serial Monitor
   and ignore new line
    // Reset for next input line of
   incoming data
    Serial.print(F("Notification:'"));
    Serial.print(Notification);
    Serial.print(F("'"));
    Notification = "";
   }
  }
 }

 // Update Time since last
notifcation and check if need to
 // send a ping message to server
 TimeDelayed = millis() - TimeStarted;
 if (TimeDelayed > PingInterval)
 {
  // Send Ping to server
  Serial.println(F("\n.....Sending
 ping to server ..."));
  Serial.print(F("PingInterval
 (seconds): "));
  Serial.println(PingInterval/1000);
  SendTCPMessage("ping\n");
  TimeStarted = millis();
 }
}
```

take a photo of qvga resolution. The number 5 is the length of the data which includes a newline character at the end.

```
+IPD,5:QVGA
```

The function processes incoming data by doing the following:

1. If the incoming data contains the substring "+IPD" then continue with the function otherwise exit the function.
2. Parse the incoming data to find the data length and the data portions of the string.
3. Print out the data length and the data that were found from step 2 to the Serial Monitor.
4. If a ">" symbol is found then assume that we need to process a command of the form `Command>ReturnClientName` where a command is given to this client and the return data must be sent to the `ReturnClientName` client by doing the following:
    1. The `ReturnClientname` string value is parsed and saved based on the position of the ">" character.
    2. The `Command` value is parsed and saved based on the position of the ">" character.
    3. The command is processed by calling the `ProcessCommand()` function and the return value is stored in the `ResponseString` string variable.
    4. If the `ResponseString` is not empty then do the following:
        1. If the `ResponseString` is "BIN" then send the binary transfer command to the server. The command is of the form **`BIN:DestClientName:SizeOfBinaryData`**.
        2. If the `ResponseString` is not "BIN" then send a response string to the server of the form `ReturnClient:ResponseString`.

This sends the `ResponseString` to the client specified after the ">" character.

3. Send the actual response to the server by calling the `SendTCPMessage()` function.
4. Set the `TimeStarted` variable to the current time to indicate that the client has just sent data to the server.

5. If the incoming data is `"Client ID Recieved ...\n"` then send a ping command to the server and reset the `TimeStarted` variable to the current time to indicate that outgoing TCP activity just occurred.

5. If the incoming data is `"PING_OK\n"` then print out a debug message to the Serial Monitor indicating that the server has received the client's ping and has responded.
6. If the incoming data contains the substring `"Binary Transfer Enabled"` then the binary mode was successfully set at the server, so start the image binary data transfer by calling the `ReadFifoBurst()` function and printing out a debug message to the Serial Monitor.
7. If the incoming data contains the substring `"Binary Transfer Finished."` then the binary transfer to the server has completed. Print out the full message to the Serial Monitor.
8. If the incoming data is not recognized as a case of one of the above then the command is unknown. Print out an error message to the Serial Monitor and the actual incoming data.

See Listing 8-31.

The `ProcessCommand()` Function

The **`ProcessCommand()`** function processes commands that are sent to the client by doing the following:

1. If the command is "ping" then set the string to be returned to the caller to `"Client_PING_OK\n"`.
2. If the command is "vga" then do the following:
   1. If the current camera resolution is not set to vga then set the current camera resolution to vga by calling the `SetCameraResolution()` function.
   2. Capture the image using the camera by calling the `CaptureImage()` function.
   3. Get the string that will be used to transmit the image size of the photo that was just taken by calling the `TransmitImageSize()` function.
3. If the command is "qvga" then do the following:
   1. If the current camera resolution is not set to qvga then set the current camera resolution to qvga by calling the `SetCameraResolution()` function.
   2. Capture the image using the camera by calling the `CaptureImage()` function.
   3. Get the string that will be used to transmit the image size of the photo that was just taken by calling the `TransmitImageSize()` function.
4. If the command is "qqvga" then do the following:
   1. If the current camera resolution is not set to qqvga then set the current camera resolution to qqvga by calling the `SetCameraResolution()` function.
   2. Capture the image using the camera by calling the `CaptureImage()` function.

Listing 8-31 The `CheckAndProcessIPD()` Function

```
void CheckAndProcessIPD(String line)
{
 int index1 = 0;
 int index2 = 0;

 String DataLength = "";
 String Data = "";

 // Process +IPD incoming data notification
 // +IPD,5:QVGA
 // +IPD,<Data Length>:<Data>

 // Check for +IPD keyword
 if (line.indexOf("+IPD") >= 0)
 {
  // Incoming Data Dectected
  index1 = line.indexOf(',');
  index2 = line.indexOf(':');

  DataLength  = line.substring(index1+1, index2);
  Data        = line.substring(index2+1);

  Serial.print(F("DataLength: "));
  Serial.println(DataLength);
  Serial.print(F("Data: "));
  Serial.println(Data);

  // Check for Return Message Format
  // Message>ReturnClientName
  int indexseparator = Data.indexOf(">");
  if (indexseparator >= 0)
  {
   String ReturnClient = Data.substring(indexseparator+1);
   ReturnClient.trim();

   // Process Command
   String cmd = Data.substring(0,indexseparator);
   String ResponseString = ProcessCommand(cmd);
   String ReturnString = "";

   if (ResponseString != "")
   {
    if (ResponseString == "BIN")
    {
     // Send Binary Transfer Command to Server
     // BIN:DestClientName:SizeOfBinaryData
     ReturnString = "BIN:" + ReturnClient + ":" + BinaryDataSize + "\n";
    }
```

```
        else
        {
         // Create string to be sent to server
         ReturnString = ReturnClient + ":" + ResponseString;
        }
        // Send text back to reciever
        SendTCPMessage(ReturnString);
        TimeStarted = millis();
      }
    }
    else
    if (Data == "Client ID Recieved ...\n")
    {
     // Client ID has been recieved so send ping to server
     if (SendTCPMessage("ping\n"))
     {
      Serial.println(F("PING Sent to Server ...."));
      TimeStarted = millis();
     }
     else
     {
      Serial.println(F("PING to Server FAILED ...."));
     }
    }
    else
    if (Data == "PING_OK\n")
    {
     Serial.print(F("Server has confirmed our PING ..."));
    }
    else
    if (Data.indexOf("Binary Transfer Enabled") >= 0)
    {
     // Binary Mode Set at Server so start Image Binary Data Transfer
     //"Binary Transfer Enabled.\n" sent from Server
     Serial.println(F("Binary Transfer Mode Enabled on Server ..."));
     ReadFifoBurst();
    }
    else
    if (Data.indexOf("Binary Transfer Finished.") >= 0)
    {
     // Binary data transfer has finished.
     Serial.println(Data);
    }
    else
    {
     Serial.print(F("Unknown Command or Response: "));
     Serial.print(Data);
    }
  }
}
```

3. Get the string that will be used to transmit the image size of the photo that was just taken by calling the `TransmitImageSize()` function.

5. If the command is `"GetImageData"` then request that the server prepare for a binary data transfer by setting the function's return string to "`BIN`".

6 If the command is not recognized as one of the previous commands then return a string indicating that the command is unknown.

See Listing 8-32.

Listing 8-32 The `ProcessCommand()` Function

```
String ProcessCommand(String cmd)
{
 String result = "";

 if (cmd == "ping")
 {
  result = "Client_PING_OK\n";
 }
 else
 if (cmd == "vga")
 {
  if (Resolution != VGA)
  {
   Resolution = VGA;
   SetCameraResolution();
  }
  CaptureImage();
  result = TransmitImageSize();
 }
 else
 if (cmd == "qvga")
 {
  if (Resolution != QVGA)
  {
   Resolution = QVGA;
   SetCameraResolution();
  }
  CaptureImage();
  result = TransmitImageSize();
 }
 else
 if (cmd == "qqvga")
 {
  if (Resolution != QQVGA)
  {
   Resolution = QQVGA;
   SetCameraResolution();
  }
  CaptureImage();
  result = TransmitImageSize();
 }
 else
 if (cmd == "GetImageData")
 {
  // Issue binary command to Server
  // BIN:DestClientName:SizeOf
BinaryData
  Serial.println(F("Binary Data
Transfer Requested ..."));
  result = "BIN";
 }
 else
 {
  // Unknown Command
  result = "Unknown Command: " + cmd
+ "\n";
 }

 return result;
}
```

The `TransmitImageSize()` Function

The `TransmitImageSize()` function is modified from previous versions by doing the following:

1. The function now returns a string.
2. The size of the image captured by the camera is now the same as the value retrieved from the `myCAM.read_fifo_length()` function and not 1 byte less like in the previous version.
3. The size of the captured image is saved in the `BinaryDataSize` variable.
4. A string is returned that holds the size of the image and is in a form designed to

Listing 8-33 The `TransmitImageSize()` Function

```
String TransmitImageSize()
{
 String result = "";

 // Get Image Size
 uint32_t FIFOImageLength =  myCAM.
read_fifo_length();
 //long Size = FIFOImageLength - 1;
 long Size = FIFOImageLength; //
Extra Dummy Byte removed with Mini
Plus Camera Version.

 Serial.print(F("FIFO RAW IMAGE SIZE
= "));
 Serial.println(FIFOImageLength);

 Serial.print(F("Image Length = "));
 Serial.println(Size);

 // Image Size
 BinaryDataSize = Size;
 result = String(Size) + "\n";

 return result;
}
```

be transmitted and read by the Android controller.

See Listing 8-33. The code modifications are highlighted in bold print.

Operating the System

Download and install version 2.0 of the Android basic wireless framework to your Android cell phone if you have not done so already. Download and install the basic framework 2.0 server project for the NodeMCU on your NodeMCU device. Finally, download and install the client version of the ArduCAM mini surveillance system on your Arduino Mega 2560. Connect your Android device to the access point running on the NodeMCU microprocessor. Run the Android basic wireless framework and connect to the TCP server running on the NodeMCU. Turn on the Arduino client and start up the Serial Monitor. The Arduino client should connect to the Wi-Fi access point running on the NodeMCU and then connect to the TCP server running on the NodeMCU. Upon connecting to the TCP server the Arduino client should set its name to client1. The NodeMCU should send an acknowledgment text string to the Arduino that the client ID was received. After receiving the acknowledgment the Arduino should send a ping command to the server. The server should reply with a "PING_OK" text string. You should see something similar to the following on the Serial Monitor for the Arduino client:

```
****************************************
***********************
****** ESP-01 Arduino Mega 2560 Station/
Client Program ********
*********** ArduCAM Mini 2MP OV2640 Plus
**********************
************** Surveillance System
*****************************


ArduCAM Starting ..........
OV2640 detected ...........
AT Command Sent: AT+CIPMUX=0
ESP-01 Response: WIFI CONNECTED
AT+CIPMUX=0

busy p...
WIFI GOT IP

OK
ESP-01 mode set for single connection.

AT Command Sent: AT+CWMODE_CUR=1
ESP-01 Response:
AT+CWMODE_CUR=1

OK
ESP-01 mode changed to Station Mode.

AT Command Sent: AT+CWJAP_CUR="ESP8266-
12E-AP","robtestESP12E"
ESP-01 Response:
AT+CWJAP_CUR="ESP8266-12E-
AP","robtestESP12E"
```

```
WIFI DISCONNEWIFI CONNECTED
WIFI GOT IP

OK
ESP-01 station has joined AP.


AT Command Sent: AT+CIPSTART="T
CP","192.168.4.1",80
ESP-01 Response:
AT+CIPSTART="TCP","192.168.4.1",80

CONNECT

OK
ESP-01 Connected to TCP Server.
Sending Data ....
DataLength: 7 , Data: client1
Client Name Set TO: client1

************ Start Up Ok **************

Notification: '
'Notification: '
'Notification: '+IPD,23:Client ID
Recieved ...
'DataLength: 23
Data: Client ID Recieved ...
Sending Data ....
DataLength: 5 , Data: ping
PING Sent to Server ....
Notification: '
'Notification: '
'Notification: '+IPD,8:PING_OK
'DataLength: 8
Data: PING_OK
Server has confirmed our PING ...
```

Using the Android application refresh the clients. Switch the target device to the client that has the camera which should be `client1`. Type in `qvga` into the phone entry textbox and hit the "Send Data" button to send this command to the target device which should send the command `qvga` and redirect the output back to the Android cell phone which is the controller. The name of my Android device is verizon, so the command that will be received by the Arduino client should be `qvga>verizon`. Next, a photo is taken by the camera and the size of the image in bytes is transferred to the Android controller using the command verizon:6152 which sends the size of the image to my verizon device. After the Android controller receives the image size it issues a command which tells the client to send the binary data. The command is `GetImageData>verizon` in my case where the output of the `GetImageData` command is redirected to my verizon Android controller. In order to transmit binary data the client must first request binary mode, define the destination of the binary data and the size of the binary data which in my case the command is `BIN:verizon:6152`. The server then notifies the client that the binary transfer mode has been enabled. Once the client receives this notification it sends the binary data to the server which redirects it to the destination client. Once all the binary data has been transferred the server sends the client a text string that acknowledges that the binary transfer has completed and the total number of bytes that were transferred. The image that was taken by the ArduCAM camera should appear on your Android device. You should also see something similar to the following on your Serial Monitor:

```
Notification: '
'Notification: '+IPD,13:qvga>verizon
'DataLength: 13
Data: qvga>verizon
............. Starting Image Capture
...............
Start Capture
Capture Done!
FIFO RAW IMAGE SIZE = 6152
Image Length = 6152
Sending Data ....
DataLength: 13 , Data: verizon:6152
Notification: '
'Notification: '
'Notification:
'+IPD,21:GetImageData>verizon
'DataLength: 21
Data: GetImageData>verizon
```

```
Binary Data Transfer Requested ...
Sending Data ....
DataLength: 17 , Data: BIN:verizon:6152
Notification: '
'Notification: '
'Notification: '+IPD,25:Binary Transfer
Enabled.
'DataLength: 25
Data: Binary Transfer Enabled.
Binary Transfer Mode Enabled on Server
...
In ReadFifoBurst, READ_FIFO_LENGTH() =
6152
ACK IMG
Sending Binary Data ....
BufferLength: 2048
Sending Binary Data ....
BufferLength: 2048
Sending Binary Data ....
BufferLength: 2048
Sending Binary Data ....
BufferLength: 8
ByteCount = 6152
Total time used:1 seconds ...
Notification: '
'Notification: '
'Notification: '+IPD,49:Binary Transfer
Finished. Bytes Transfered: 6152
'DataLength: 49
Data: Binary Transfer Finished. Bytes
Transfered: 6152
Binary Transfer Finished. Bytes
Transfered: 6152
```

Next, take a picture using qqvga mode by sending the `qqvga` command to `client1`. Enter `qqvga` in the phone text entry box and press "Send Data" to send the command to the target device which should be set to `client1`. You should see something similar to the following show up on the Serial Monitor. Once the image data has finished being transferred it should be displayed on your Android's screen.

```
Notification: '
'Notification: '+IPD,14:qqvga>verizon
'DataLength: 14
Data: qqvga>verizon
............. Starting Image Capture
..............
Start Capture
Capture Done!
FIFO RAW IMAGE SIZE = 3080
Image Length = 3080
Sending Data ....
DataLength: 13 , Data: verizon:3080
Notification: '
'Notification: '
'Notification:
'+IPD,21:GetImageData>verizon
'DataLength: 21
Data: GetImageData>verizon
Binary Data Transfer Requested ...
Sending Data ....
DataLength: 17 , Data: BIN:verizon:3080
Notification: '
'Notification: '
'Notification: '+IPD,25:Binary Transfer
Enabled.
'DataLength: 25
Data: Binary Transfer Enabled.
Binary Transfer Mode Enabled on Server
...
In ReadFifoBurst, READ_FIFO_LENGTH()=3080
ACK IMG
Sending Binary Data ....
BufferLength: 2048
Sending Binary Data ....
BufferLength: 1032
ByteCount = 3080
Total time used:0 seconds ...
Notification: '
'Notification: '
'Notification: '+IPD,49:Binary Transfer
Finished. Bytes Transfered: 3080
'DataLength: 49
Data: Binary Transfer Finished. Bytes
Transfered: 3080
Binary Transfer Finished. Bytes
Transfered: 3080
```

Next, take a photo at the vga camera resolution by sending a `vga` command to `client1`. The lightning may be low at first so take a photo several times to get a clear fully lit image.

```
Notification: '
'Notification: '+IPD,12:vga>verizon
```

```
'DataLength: 12
Data: vga>verizon
............ Starting Image Capture
..............
Start Capture
Capture Done!
FIFO RAW IMAGE SIZE = 8200
Image Length = 8200
Sending Data ....
DataLength: 13 , Data: verizon:8200
Notification: '
'Notification: '
'Notification:
'+IPD,21:GetImageData>verizon
'DataLength: 21
Data: GetImageData>verizon
Binary Data Transfer Requested ...
Sending Data ....
DataLength: 17 , Data: BIN:verizon:8200
Notification: '
'Notification: '
'Notification: '+IPD,25:Binary Transfer
Enabled.
'DataLength: 25
Data: Binary Transfer Enabled.
Binary Transfer Mode Enabled on Server
...
In ReadFifoBurst, READ_FIFO_LENGTH()
= 8200
ACK IMG
Sending Binary Data ....
BufferLength: 2048
Sending Binary Data ....
BufferLength: 2048
Sending Binary Data ....
BufferLength: 2048
Sending Binary Data ....
BufferLength: 2048
Sending Binary Data ....
BufferLength: 8
ByteCount = 8200
Total time used:1 seconds ...
Notification: '
'Notification: '
'Notification: '+IPD,49:Binary Transfer
Finished. Bytes Transfered: 8200
'DataLength: 49
Data: Binary Transfer Finished. Bytes
Transfered: 8200
Binary Transfer Finished. Bytes
Transfered: 8200
```

Next, turn on the surveillance feature from the Android application. You should get a continuous stream of images from the camera. You can select different resolutions from the "Resolution Settings" menu.

Hands-on Example: The ArduCAM OV2640 2MP Mini Plus Infrared Motion Detection Arduino Mega 2560 Client Surveillance and Alarm System with NodeMCU Server

In this hands-on example I show you how to build a wireless surveillance and motion detection system using an Android cell phone, a NodeMCU server, and an Arduino Mega 2560 client that controls an ArduCAM OV2640 2MP Mini Plus camera and an HC-SR501 infrared motion detector. Note: If you have previously constructed the Arduino server-based ArduCAM camera and infrared motion sensor project then you can use that same Arduino, ArduCAM, and motion sensor hardware setup and just change the software that is installed. See Figure 8-10.

Parts List

- 1 Android cell phone
- 1 NodeMCU microcontroller
- 1 Arduino Mega 2560 microcontroller
- 1 5- to 3.3-V step-down voltage regulator
- 1 Logic level converter from 5 to 3.3 V
- 1 ESP-01 Wi-Fi module
- 1 Arduino development station such as a desktop or a notebook
- 1 Breadboard
- 1 ArduCAM OV2640 Mini Plus camera

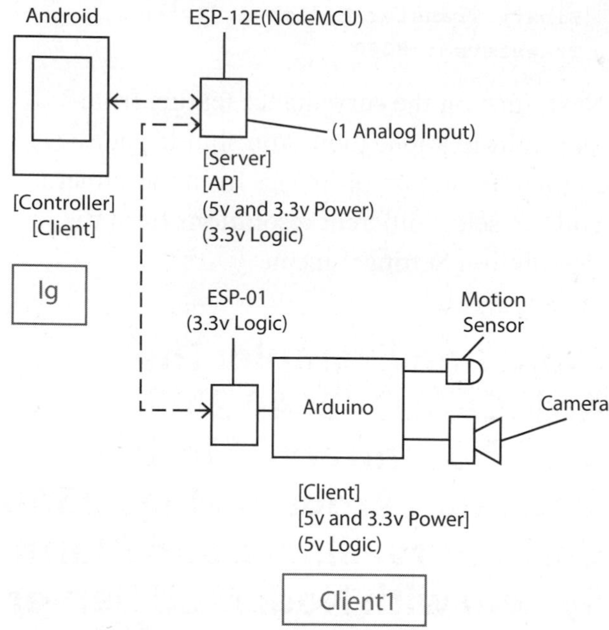

Figure 8-10 The Android, NodeMCU server, and Arduino client ArduCAM Mini Plus camera and infrared motion detection surveillance and alarm system.

- 1 HC-SR501 infrared motion detector
- 1 Package of wires to connect the components with both male-to-male and male-to-female wire connections

Setting Up the Hardware

Step-Down Voltage Regulator Connections

1. Connect the Vin pin on the voltage regulator to the 5-V power pin on the Arduino Mega 2560.
2. Connect the Vout pin on the voltage regulator to a node on your breadboard represented by a horizontal row of empty slots. This is the 3.3-V power node that will supply the 3.3-V power to the ESP-01 module as well as provide the reference voltage for the logic level shifter.
3. Connect the GND pin on the voltage regulator to the GND pin on the Arduino or form a GND node similar to the 3.3-V power node in the previous step.

ESP-01 Wi-Fi Module Connections

1. Connect the Tx or TxD pin to the Rx3 pin on the Arduino Mega 2560. This is the receive pin for the serial hardware port 3.
2. Connect the EN or CH_PD pin to the 3.3-V power node from the step-down voltage regulator.
3. Connect the IO16 or RST pin to the 3.3-V power node from the step-down voltage regulator.
4. Connect the 3V3 or VCC pin to the 3.3-V power node from the step-down voltage regulator.
5. Connect the GND pin to the ground pin or node.
6. Connect the Rx or RxD pin to pin B0 on the 3.3-V side of the logic level shifter. The voltage from this pin will be shifted from 5 V from the Arduino to 3.3 V for the ESP-01 module.

Logic Level Shifter

1. Connect the 3.3-V pin to the 3.3-V power node from the step-down voltage regulator.
2. Connect the GND pin on the 3.3-V side to the ground node for the Arduino.
3. Connect the B0 pin on the 3.3-V side to Rx or RxD pin on the ESP-01. This corresponds to the A0 pin on the 5-V side.
4. Connect the 5-V pin to the 5-V power pin on the Arduino.
5. Connect the GND pin on the 5-V side to the ground node for the Arduino.
6. Connect the A0 pin on the 5-V side to Tx3 pin on the Arduino. This pin corresponds to the B0 pin on the 3.3-V side.

ArduCAM OV2640 Mini Plus

1. Connect the CS pin to digital pin 8 on the Arduino.

2. Connect the SDA pin to digital pin 20 on the Arduino.
3. Connect the SCL pin to digital pin 21 on the Arduino.
4. Connect the MISO pin to digital pin 50 on the Arduino.
5. Connect the MOSI pin to digital pin 51 on the Arduino.
6. Connect the SCK pin to digital pin 52 on the Arduino.
7. Connect the VCC pin to the 5-V node for the circuit.
8. Connect the GND pin to the common ground node for the circuit.

The HC-SR501 Infrared Motion Detector

1. Connect the Vcc pin on the motion detector to the 5-V power node on the Arduino Mega 2560.
2. Connect the GND pin on the motion detector to the GND node on the Arduino.
3. Connect the Out pin on the motion detector to digital pin 7 on the Arduino.

See Figure 8-11.

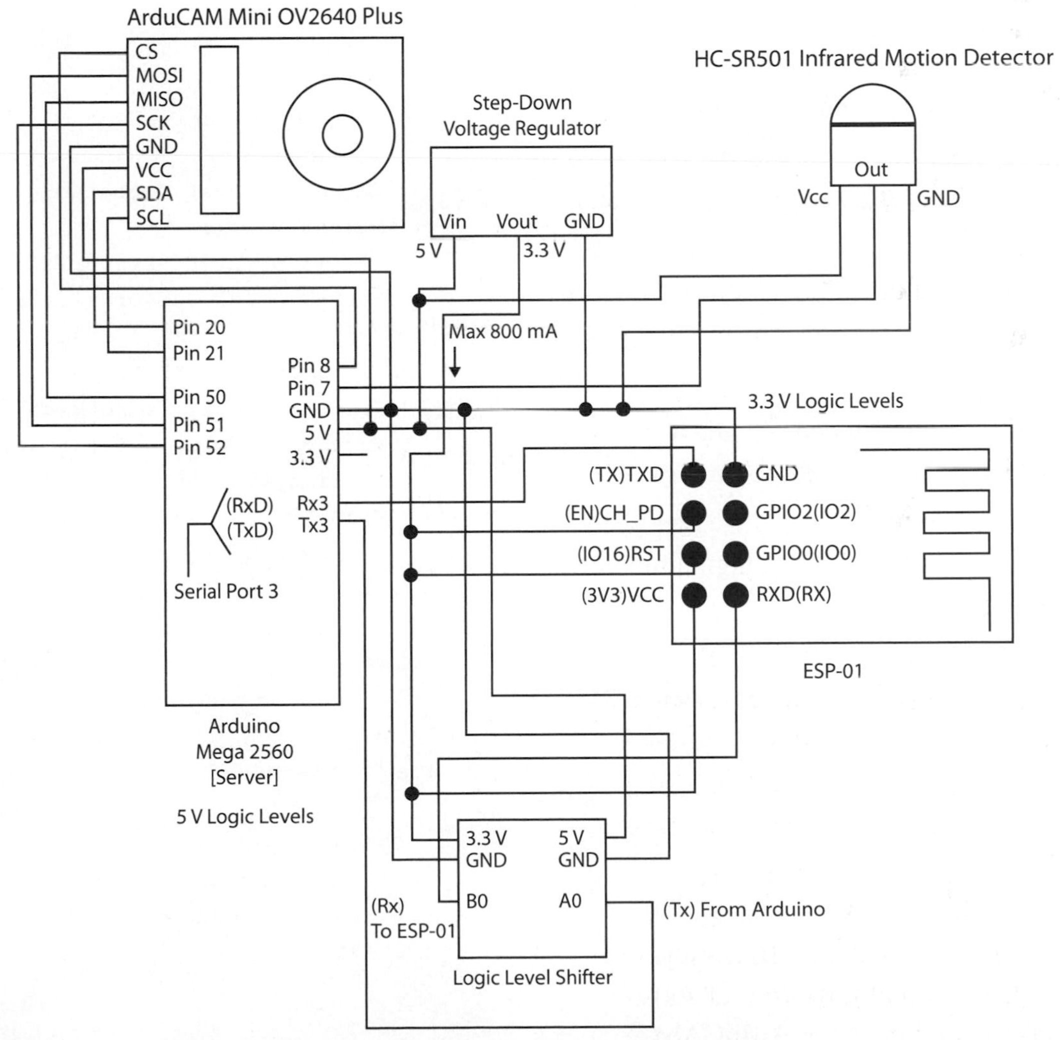

Figure 8-11 The Arduino client with ArduCAM camera and motion detector.

Setting Up the Software

The key new code change for this hands-on example is that when the alarm is tripped the number of hits will be updated but a text string is NOT sent to the Android controller. The alarm system will now be polled instead by the Android controller. The key function that is modified is the `ProcessSensor()` function. The code block highlighted in Listing 8-34 has been modified so that only the number of hits is updated.

The `ProcessCommand()` Function

The `ProcessCommand()` function is modified by adding in recognition of motion sensor-related commands which are:

1. The string "`ACTIVATE_ALARM`" which activates the motion sensor alarm system.
2. The string "`DEACTIVATE_ALARM`" which deactivates the motion sensor alarm system.
3. The string "`hits`" which returns the number of times the alarm has been tripped.
4. The string "`reset`" which sets the number of hits to 0.
5. The string "`wait=`" which sets the initial waiting time in milliseconds before the alarm is activated.

See Listing 8-35. The new code is highlighted in bold print.

Operating the Alarm and Surveillance System

Download and install version 2.0 of the Android basic wireless framework to your Android cell phone if you have not done so already. Download and install the basic framework 2.0 server project for the NodeMCU on your NodeMCU device. Finally, download and install the client version of the ArduCAM mini surveillance and motion detection alarm

Listing 8-34 The `ProcessSensor()` Function

```
void ProcessSensor()
{
 // Put Code to process your sensors
or other hardware in this function.
 if ((AlarmState == Alarm_ON) ||
(AlarmState == Alarm_TRIPPED))
  {
   // Process Sensor Input
   SensorValue =
  digitalRead(SensorPin);
   if ((PrevSensorValue ==
  EventNotDetected) && (SensorValue ==
  EventDetected))
    {
     unsigned long
    ElapsedTime = millis() -
    MinDelayBetweenHitsStartTime;
     if (ElapsedTime >=
    MinDelayBetweenHits)
      {
       // Motion Just Detected
       NumberHits++;
       Serial.print(F("SensorValue: "));
       Serial.print(SensorValue);
       Serial.print(F(" , Event Detected
     ... NumberHits: "));
       Serial.println(NumberHits);
       PrevSensorValue = EventDetected;
       // Time this hit occured
       MinDelayBetweenHitsStartTime =
     millis();
      }
    }
   else
   if ((PrevSensorValue ==
  EventDetected) && (SensorValue ==
  EventNotDetected))
    {
     Serial.print(F("SensorValue: "));
     Serial.println(SensorValue);
     PrevSensorValue = EventNot
   Detected;
    }
  }
}
```

Listing 8-35 The `ProcessCommand()` Function

```
String ProcessCommand(String cmd,
String ReturnClient)
{
 String result = "";

 if (cmd == "ping")
 {
  result = "Client_PING_OK\n";
 }
 else
 if (cmd == "vga")
 {
  if (Resolution != VGA)
  {
   Resolution = VGA;
   SetCameraResolution();
  }
  CaptureImage();
  result = TransmitImageSize();
 }
 else
 if (cmd == "qvga")
 {
  if (Resolution != QVGA)
  {
   Resolution = QVGA;
   SetCameraResolution();
  }
  CaptureImage();
  result = TransmitImageSize();
 }
 else
 if (cmd == "qqvga")
 {
  if (Resolution != QQVGA)
  {
   Resolution = QQVGA;
   SetCameraResolution();
  }
  CaptureImage();
  result = TransmitImageSize();
 }
 else
 if (cmd == "GetImageData")
 {
  // Issue binary command to Server
  // BIN:DestClientName:SizeOfBinary
 Data
  Serial.println(F("Binary Data
  Transfer Requested ..."));
  result = "BIN";
 }
 else
 if (cmd == "ACTIVATE_ALARM")
 {
  if (AlarmState == Alarm_OFF)
  {
   NumberHits = 0;
   AlarmState = Alarm_WAIT;
   result = "Alarm Activated...WAIT
  TIME:" + String(WaitTime/1000) + "
  Seconds.\n";
   WaitTimeStart = millis();
   DeviceRequestingAlarm =
  ReturnClient;
  }
  else
  {
   result = "Alarm Already
  Active...\n";
  }
 }
 else
 if (cmd == "DEACTIVATE_ALARM")
 {
  AlarmState = Alarm_OFF;
  result = "Alarm Has Been DE-
 Activated ...\n";
 }
 else
 if (cmd == "hits")
 {
  // Return number of motions
 detected to requesting client
  result = "HITS DETECTED:" +
 String(NumberHits) + "\n";
 }
 else
 if (cmd == "reset")
 {
  // Reset number hits to 0
  NumberHits = 0;
  Serial.println(F("Reset NumberHits
 to 0."));
 }
 else
 if (cmd == "reset2")
```

```
    {
     // Reset number hits to 0
     NumberHits = 0;
     result = "Number Hits Reset to
    0.\n";
     Serial.println(F("Reset NumberHits
    to 0."));
    }
    else
    if (cmd.indexOf("wait=") >= 0)
    {
     int indexwait = cmd.indexOf("=");
     String waitlength = cmd.
    substring(indexwait+1);
     WaitTime = waitlength.toInt();
     result = result + F("WaitTime
    set to: ") + WaitTime/1000 + F("
    Seconds\n");
    }
    else
    {
     // Unknown Command
     result = "Unknown Command: " + cmd
    + "\n";
    }
  return result;
}
```

system on your Arduino Mega 2560. Connect your Android device to the access point running on the NodeMCU microprocessor. Run the Android basic wireless framework and connect to the TCP server running on the NodeMCU. Turn on the Arduino client and start up the Serial Monitor. The Arduino client should connect to the Wi-Fi access point running on the NodeMCU and then connect to the TCP server running on the NodeMCU. Upon connecting to the TCP server the Arduino client should set its name to `client1`. The NodeMCU should send an acknowledgment text string to the Arduino that the client ID was received. After receiving the acknowledgment the Arduino should send a ping command to the server. The server should reply with a "PING_OK" text string. You should see something similar to the following on the Serial Monitor for the Arduino client:

```
*******************************************
**********************
****** ESP-01 Arduino Mega 2560 Station/
Client Program ********
*********** ArduCAM Mini 2MP OV2640 Plus
**********************
******* Motion Detection Alarm and
Surveillance System ********
ArduCAM Starting ..........
OV2640 detected ..........
AT Command Sent: AT+CIPMUX=0
ESP-01 Response: AT+CIPMUX=0
busy p...
WIFI CONNECTED
WIFI GOT IP
OK
ESP-01 mode set for single connection.
AT Command Sent: AT+CWMODE_CUR=1
ESP-01 Response:
AT+CWMODE_CUR=1
OK
ESP-01 mode changed to Station Mode.
AT Command Sent: AT+CWJAP_CUR="ESP8266-
12E-AP","robtestESP12E"
ESP-01 Response:
AT+CWJAP_CUR="ESP8266-12E-
AP","robtestESP12E"
WIFI DISCONNEWIFI CONNECTED
WIFI GOT IP
OK
ESP-01 station has joined AP.
AT Command Sent: AT+CIPSTART="T
CP","192.168.4.1",80
ESP-01 Response:
AT+CIPSTART="TCP","192.168.4.1",80
CONNECT
OK
ESP-01 Connected to TCP Server.
Sending Data ....
DataLength: 7 , Data: client1
Client Name Set TO: client1
************ Start Up Ok **************
Notification: '
```

```
'Notification: '
'Notification: '+IPD,23:Client ID
Recieved ...
'DataLength: 23
Data: Client ID Recieved ...
Sending Data ....
DataLength: 5 , Data: ping
PING Sent to Server ....
Notification: '
'Notification: '
'Notification: '+IPD,8:PING_OK
'DataLength: 8
Data: PING_OK
Server has confirmed our PING ...
```

Refresh the clients using the Android application and then switch the target device for the commands to the `client1` device that operates the camera and the motion detector. Next, test the ability of the camera to capture an image by sending a `qvga` command to `client1`. You should see something like the following on your Serial Monitor and then you should see the actual captured image appear on your Android device:

```
Notification: '
'Notification: '+IPD,13:qvga>verizon
'DataLength: 13
Data: qvga>verizon

............. Starting Image Capture
...............
Start Capture
Capture Done!
FIFO RAW IMAGE SIZE = 6152
Image Length = 6152
Sending Data ....
DataLength: 13 , Data: verizon:6152

Notification: '
'Notification: '
'Notification:
'+IPD,21:GetImageData>verizon
'DataLength: 21
Data: GetImageData>verizon

Binary Data Transfer Requested ...
Sending Data ....
DataLength: 17 , Data: BIN:verizon:6152

Notification: '
'Notification: '
'Notification: '+IPD,25:Binary Transfer
Enabled.
'DataLength: 25
Data: Binary Transfer Enabled.
Binary Transfer Mode Enabled on Server
...
In ReadFifoBurst, READ_FIFO_LENGTH() =
6152
ACK IMG
Sending Binary Data ....
BufferLength: 2048
Sending Binary Data ....
BufferLength: 2048
Sending Binary Data ....
BufferLength: 2048
Sending Binary Data ....
BufferLength: 8
ByteCount = 6152
Total time used:1 seconds ...
Notification: '
'Notification: '
'Notification: '+IPD,49:Binary Transfer
Finished. Bytes Transfered: 6152
'DataLength: 49
Data: Binary Transfer Finished. Bytes
Transfered: 6152

Binary Transfer Finished. Bytes
Transfered: 6152
```

Next, set up `client1` for network monitoring by adding it to the list of monitored devices by selecting the "Monitor Target Device" menu item. Have `client1` take a photo whenever the alarm is tripped by selecting the "Set to Camera Available" option under the "Camera Status Target Device" menu. You should see the `client1` device listed under the "Monitored Devices:" text located in the information window. Scroll all the way down the information window to view the list of monitored devices. You should also see a "CAMERA_YES" value to indicate that a photo will be taken whenever the motion sensor alarm is tripped.

Next, activate the alarm using the android application. You should see an activate alarm command coming into the client with the result being redirected to the Android controller which is in my case named verizon. In 10 seconds the alarm should be fully active. The output of your Serial Monitor should be similar to the following:

```
Notification: '
'Notification: '+IPD,23:ACTIVATE_
ALARM>verizon
'DataLength: 23
Data: ACTIVATE_ALARM>verizon
Sending Data ....
DataLength: 48 , Data: verizon:Alarm
Activated...WAIT TIME:10 Seconds.
Notification: '
'Sending Data ....
DataLength: 28 , Data: verizon:ALARM NOW
ACTIVE...
ALARM WAIT STATE FINISHED ... Alarm Now
ACTIVE ....
```

Next, start the monitoring of the alarm system on `client1` by selecting the "Activate Network Monitoring" option under the "Network Monitoring" menu item. The Android controller will now continuously poll `client1` by sending the "`hits`" command. The `client1` device should start receiving the "`hits`" command and should reply with the number of hits or alarm trips that have occurred.

```
Notification: '
'Notification: '
'Notification: '+IPD,13:hits>verizon
'DataLength: 13
Data: hits>verizon
Sending Data ....
DataLength: 24 , Data: verizon:HITS
DETECTED:0
```

Next, trip the infrared motion sensor alarm by putting your hand in front of the sensor and holding it there for a few seconds. This will allow time for the polling to recognize that the alarm has been tripped and to send a take photo command to `client1`. Once the motion sensor detects motion it updates the number of hits.

The next time `client1` is polled it returns the updated number of hits which will be 1. The Android controller then sends the command to `client1` to reset the number of hits to 0. The Android controller then sends a take photo command to `client1`. In my case the specific take photo command was "`vga`". The image was then transferred to my Android device. The following is similar to what you should see on your own Serial Monitor:

```
Notification: '
'SensorValue: 1 , Event Detected ...
NumberHits: 1
Notification: '
'Notification: '+IPD,13:hits>verizon
'DataLength: 13
Data: hits>verizon


Sending Data ....
DataLength: 24 , Data: verizon:HITS
DETECTED:1
Notification: '
'Notification: '
'Notification: '+IPD,14:reset>verizon
'DataLength: 14
Data: reset>verizon
Reset NumberHits to 0.
Notification: '
'Notification: '+IPD,12:vga>verizon
'DataLength: 12
Data: vga>verizon
............ Starting Image Capture
...............
Start Capture
Capture Done!
FIFO RAW IMAGE SIZE = 19464
Image Length = 19464
Sending Data ....
DataLength: 14 , Data: verizon:19464
Notification: '
'Notification: '
'Notification:
'+IPD,21:GetImageData>verizon
'DataLength: 21
Data: GetImageData>verizon

Binary Data Transfer Requested ...
```

```
Sending Data ....
DataLength: 18 , Data: BIN:verizon:
19464

Notification: '
'Notification: '
'Notification: '+IPD,25:Binary Transfer
Enabled.
'DataLength: 25
Data: Binary Transfer Enabled.

Binary Transfer Mode Enabled on Server
...
In ReadFifoBurst, READ_FIFO_LENGTH() =
19464
ACK IMG
Sending Binary Data ....
BufferLength: 2048
Sending Binary Data ....
BufferLength: 2048
Sending Binary Data ....
BufferLength: 2048
Sending Binary Data ....
BufferLength: 2048
Sending Binary Data ....
BufferLength: 2048
Sending Binary Data ....
BufferLength: 2048
Sending Binary Data ....
BufferLength: 2048
Sending Binary Data ....
BufferLength: 2048
Sending Binary Data ....
BufferLength: 2048
Sending Binary Data ....
BufferLength: 1032
ByteCount = 19464
Total time used:3 seconds ...
SensorValue: 0
Notification: '
'Notification: '
'Notification: '+IPD,50:Binary
Transfer Finished. Bytes Transfered:
19464
'DataLength: 50
Data: Binary Transfer Finished. Bytes
Transfered: 19464

Binary Transfer Finished. Bytes
Transfered: 19464
```

Hands-on Example: The Infrared Motion Detection Alarm System Using the NodeMCU Server

In this hands-on example I will show you how to build a wireless motion detector alarm system using an Android, a NodeMCU server, and an infrared motion sensor. See Figure 8-12.

Parts List

- 1 Android cell phone
- 1 NodeMCU microprocessor
- 1 HC-SR501 infrared motion sensor
- 1 Package of wires to connect the components
- 1 Breadboard (optional)

Setting Up the Hardware

1. Connect the Vcc pin on the motion detector to the Vin pin on the NodeMCU. Note that when connected to the USB port the Vin pin on the NodeMCU provides a 5-V output on most brands.
2. Connect the GND pin on the motion detector to a GND pin on the NodeMCU.
3. Connect the Out pin on the motion detector to pin D1 on the NodeMCU.

See Figure 8-13.

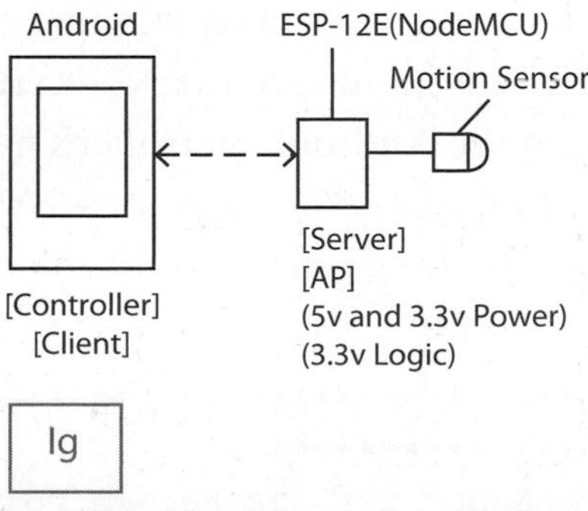

Figure 8-12 The NodeMCU server infrared motion detection alarm system.

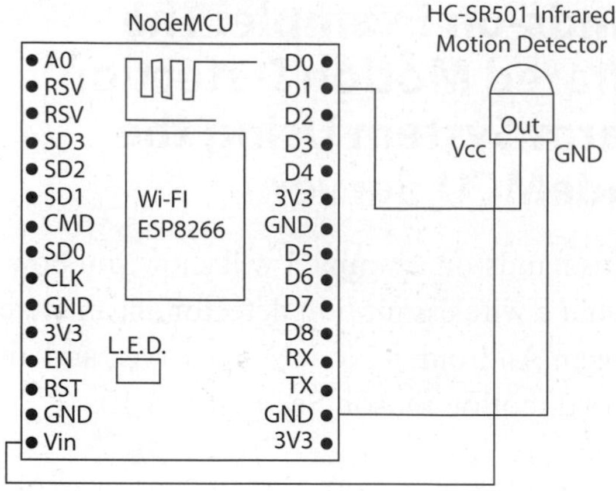

Figure 8-13 The wireless NodeMCU server motion detection alarm system.

Setting Up the Software

The key code change for this hands-on example is that the alarm system is now polled. So instead of sending a text string to the Android controller when the alarm is tripped the code will just update the number of hits. See the highlighted code block in Listing 8-36 to see the changes.

Operating the Alarm System

Download and install on your Android cell phone the basic framework version 2.0 if you have not done so already. Download and install the basic framework 2.0 with motion detection for the NodeMCU. Turn on the NodeMCU and turn on the Serial Monitor. Connect the Android to the Wi-Fi access point running on the NodeMCU. Run the Android application and connect to the TCP server running on the NodeMCU. You should see the NodeMCU server receive the Android controller's name or `clientID`. In my case the name was "`lg`". You should see something similar to the following on your Serial Monitor:

```
*****************************************
*********************
****** NodeMCU ESP-12E Access Point
Server Program **********
************** Basic Framework v2.0
*************************
```

Listing 8-36 The `ProcessSensor()` Function

```
void ProcessSensor()
{
 // Put Code to process your sensors
or other hardware in this function.
 if ((AlarmState == Alarm_ON) ||
(AlarmState == Alarm_TRIPPED))
  {
   // Process Sensor Input
   SensorValue = digitalRead(SensorPin);
   if ((PrevSensorValue ==
  EventNotDetected) && (SensorValue ==
  EventDetected))
    {
     unsigned long
    ElapsedTime = millis() -
    MinDelayBetweenHitsStartTime;
     if (ElapsedTime >=
    MinDelayBetweenHits)
      {
       // Motion Just Detected
       NumberHits++;
       Serial.print(F("SensorValue: "));
       Serial.print(SensorValue);
       Serial.print(F(" , Event Detected
      ... NumberHits: "));
       Serial.println(NumberHits);
       PrevSensorValue = EventDetected;
       // Time this hit occured
       MinDelayBetweenHitsStartTime =
      millis();
      }
    }
    else
   if ((PrevSensorValue ==
  EventDetected) && (SensorValue ==
  EventNotDetected))
    {
     Serial.print(F("SensorValue: "));
     Serial.println(SensorValue);
     PrevSensorValue = EventNotDetected;
    }
  }
}
```

```
* For Use With Android Basic Wireless
Framework Version 2.0 *
********* Infrared Motion Detector Alarm
System *************

TCP Server started, Server IP Address:
192.168.4.1
WIFI Mode: 2

[Client connected]
Waiting to Recieve New CLient's ID
Incoming Client's ClientID: lg
Number of Clients Killed: 0
Sending Text to Client, ClientID: lg,
Text: Client ID Recieved ...
```

Next, refresh the clients using the Android application. This refreshing actually issues a list command to the NodeMCU server which returns the clients that are currently connected to the server.

```
ClientID: lg Incoming Data: list
Sending Data to Client ID: lg , Data:
ActiveClients: lg,

ActiveClients: lg,
```

Next, make sure that the server is selected as the target device. Activate the alarm on the server using the Android application. After waiting for 10 seconds the alarm system on the NodeMCU server becomes active.

```
ClientID: lg Incoming Data: ACTIVATE_
ALARM
Sending Data to Client ID: lg , Data:
Alarm Activated...WAIT TIME:10 Seconds.

Sending Data to Client ID: lg , Data:
ALARM NOW ACTIVE...

ALARM WAIT STATE FINISHED ... Alarm Now
ACTIVE ....
```

Next, turn off the emergency call outs and emergency text messages and add the server to the list of monitored devices. Activate network monitoring. This will start to continuously poll the alarm system on the server by issuing "`hits`" commands. The hits command should return a 0 when no motion has been detected. Next, trip the alarm by waving your hand in front of the motion sensor. A debug message should be printed to the Serial Monitor. The next time a "hits" command is received it will return a 1 as the number of hits detected and trigger a sound effect being played on the Android device. After the Android processes the return value from the hits commands it will send a "`reset`" command that will set the number of hits back to 0.

```
ClientID: lg Incoming Data: hits
Sending Data to Client ID: lg , Data:
HITS DETECTED:0

ClientID: lg Incoming Data: hits
Sending Data to Client ID: lg , Data:
HITS DETECTED:0

ClientID: lg Incoming Data: hits
Sending Data to Client ID: lg , Data:
HITS DETECTED:0

ClientID: lg Incoming Data: hits
Sending Data to Client ID: lg , Data:
HITS DETECTED:0

ClientID: lg Incoming Data: hits
Sending Data to Client ID: lg , Data:
HITS DETECTED:0

ClientID: lg Incoming Data: hits
Sending Data to Client ID: lg , Data:
HITS DETECTED:0

SensorValue: 1 , Event Detected ...
NumberHits: 1
ClientID: lg Incoming Data: hits
Sending Data to Client ID: lg , Data:
HITS DETECTED:1

ClientID: lg Incoming Data: reset
Reset NumberHits to 0.
```

Finally, deactivate the alarm using the Android application and exit the application by pressing the back key.

```
ClientID: lg Incoming Data: DEACTIVATE_
ALARM
Sending Data to Client ID: lg , Data:
Alarm Has Been DE-Activated ...
Client has disconnected, Client ID: lg
```

Hands-on Example: The ArduCAM OV2640 2MP Mini Plus and Infrared Motion Detection Arduino Mega 2560 Client Surveillance and Alarm System with NodeMCU Server with an Infrared Motion Detection Alarm System

In this hands-on example I show you how to use the NodeMCU server-based infrared motion sensor alarm system with an Arduino client-based surveillance and infrared motion-based alarm system. An Android cell phone with version 2.0 of the basic wireless framework is used as the controller for this wireless system. See Figure 8-14.

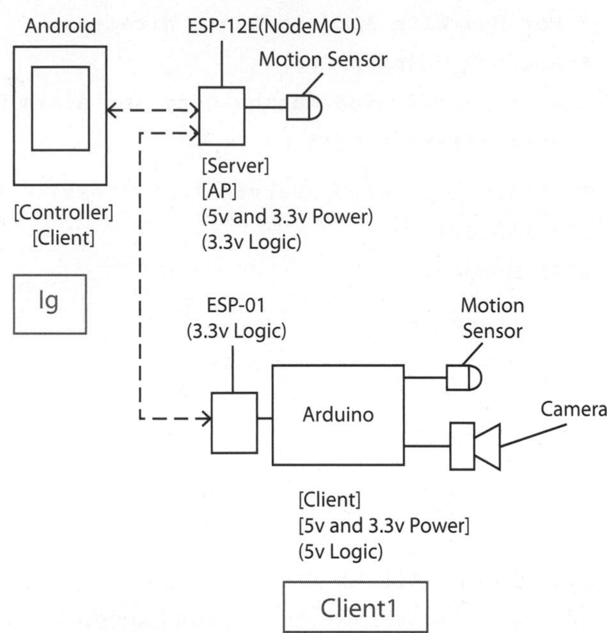

Figure 8-14 The NodeMCU alarm system and Arduino surveillance and alarm system.

Parts List

- 1 Android cell phone
- 1 NodeMCU microcontroller
- 1 Arduino Mega 2560 microcontroller
- 1 5- to 3.3-V step-down voltage regulator
- 1 Logic level converter from 5 to 3.3 V
- 1 ESP-01 Wi-Fi module
- 2 Arduino development stations with the Arduino IDE installed
- 1 Breadboard
- 1 ArduCAM OV2640 Mini Plus camera
- 2 HC-SR501 infrared motion detectors
- 1 Package of wires to connect the components with both male-to-male and male-to-female wire connections

Setting Up the Hardware

1. To build the NodeMCU server-based HC-SR501 infrared motion detection system refer to the previous hands-on example in this chapter that covers this system.
2. To build the Arduino Mega 2560 client-based HC-SR501 and ArduCAM Mini Plus infrared motion detection and surveillance system refer to the previous hands-on example in this chapter that covers this system.

Setting Up the Software

The software for this hands-on example is the same as was used earlier in this chapter for the relevant hands-on examples involving the NodeMCU with motion detector and Arduino client with motion detector and camera.

Operating the System

Connect the Android to the Wi-Fi access point running on the NodeMCU. Start up the Android's basic wireless framework version 2.0 and connect to the TCP server running on the NodeMCU. Refresh the clients using the application and select the switch target device menu item and you should see the server, and the Android device that you just used to connect to the TCP server as options. Start up the Arduino client that has the motion sensor and camera attached to it. The Arduino client should join the Wi-Fi access point on the NodeMCU and then connect to the TCP server running on the NodeMCU. After the Arduino client is connected to the NodeMCU refresh the clients using the Android application. You should now be able to select between the server, the Android device, and the Arduino `client1` as target devices. Select `client1` as the target device. Add `client1` to the list of monitored devices. Also set the camera to be available on `client1` so that when the alarm is tripped on `client1` a photo will also be taken. Send a take photo command such as "`qvga`" to `client1` to make sure that the camera is working correctly and an image is returned to the Android controller. Activate the alarm on `client1` using the Android basic wireless framework application. Next, change the target device to the server. Add the server to the list of monitored devices. Activate the alarm on the server. Finally, turn on network monitoring using the Android application. The application will now continuously retrieve and monitor the alarm status on the server and `client1`.

Summary

In this chapter I have covered wireless sensing and remote control systems involving the Android, Arduino, ESP-01 module, and the NodeMCU microprocessor. I started by discussing the ArduCAM Mini OV2640 2MP Plus camera which is the newest version of the Mini OV2640 2MP model of the camera. I used this new camera in a hands-on example that demonstrated a surveillance system. Next, I extended this surveillance system and added an infrared motion detection and an alarm system. Next, I presented a new basic Android, Arduino with ESP-01, and NodeMCU server-based wireless multi-client framework. I then discussed an overview of the Android basic wireless framework version 2.0 application. Next, I covered the new code additions for the Android basic wireless framework version 2.0. I then discussed the new basic framework 2.0 for the NodeMCU server. Finally, I showed an example of setting up the Arduino with an ESP-01 module for station/client mode. The rest of the chapter consists of hands-on examples that demonstrate the new NodeMCU server-based multi-client system.

The hands-on example are as follows:

- Hands-on example: The ArduCAM OV2640 2MP Mini Plus camera Arduino Mega 2560 client surveillance system with NodeMCU server
- Hands-on example: The ArduCAM OV2640 2MP Mini Plus infrared motion detection Arduino Mega 2560 client surveillance and alarm system with NodeMCU server
- Hands-on Example: The infrared motion detection alarm system using the NodeMCU server
- Hands-on example: The ArduCAM OV2640 2MP Mini Plus and infrared motion detection Arduino Mega 2560 client surveillance and alarm system with NodeMCU server with an infrared motion detection alarm system

CHAPTER 9

The Bonus Chapter: The Emergency Backup Battery Power System, Power Intensive Related Projects, Using the NodeMCU with an ArduCAM Mini Camera, and Some Important Downloads

In this chapter I cover some topics that I believe are important but were not covered previously in the book. Most of this chapter focuses on power intensive related Arduino projects that require a separate power source for the device being controlled. These projects involve a servo, a DC motor with fan attachment, a stepper motor, and a smoke sensor. An emergency backup battery power system for the Arduino is also presented. For the downloadable content for this chapter I have included all the Android APK install files for each version of the Android basic wireless frameworks. I have also included an Android Studio project of the final version of the Android basic wireless framework that was created with a more recent version of Android Studio. I also give some troubleshooting tips for readers that are having trouble getting the examples to work correctly.

Circuit Troubleshooting Tips

The two most common problems that I have encountered when building the projects in this book are poor electrical connections between the jumper wires and the circuit board and USB power cables that are too tight fitting and provide poor electrical contacts for power transmission. I especially found that many USB cables I tried with the official Arduino Mega 2560 were too tight fitting and thus had problems providing adequate power to the Arduino.

Some signs of poor electrical contacts between jumper wires and the breadboard or between the USB cable and the Arduino are:

- Garbage characters appearing on the Serial Monitor.
- Fluctuating or dim power LED lights on components such as the ESP-01.
- No power to the Arduino or components attached to the Arduino such as the ESP-01 or step-down voltage regulator.
- A component in the circuit works intermittently, sometimes working perfectly and other times failing completely.

One solution to wires that are making poor electrical contact would be to shut off the power, wiggle the wires slightly, and then turn the power back on and see if this has solved the problem. For USB cables that are too tight fitting the best solution would be to try a different one. Many clone Arduino boards come with short USB cables and you can buy USB cables online from Amazon and ebay.

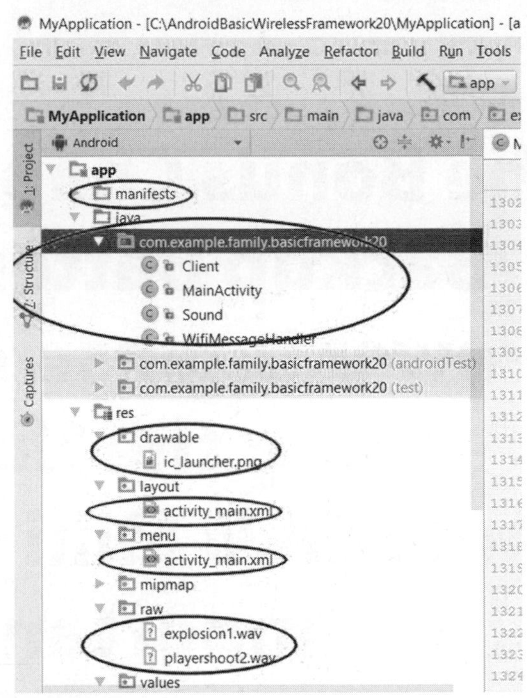

Figure 9-1 Converting to Android Studio 2.3.1.

Android Basic Wireless Framework APKs

For the convenience of the reader I have collected all the executable Android APKs for the Android basic wireless framework versions in this book which are 1.0, 1.1, 1.2, 1.3, 2.0 (Android Studio 1.5), and 2.0 (Android Studio 2.3.1). All these files should be located on the download site provided by McGraw-Hill for this chapter.

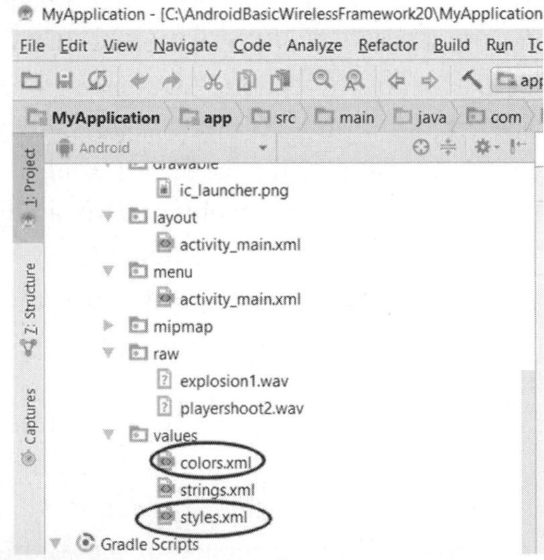

Figure 9-2 Converting to Android Studio 2.3.1.

Converting the Android Basic Wireless Framework Version 2.0 Project from Android Studio 1.5 to Android Studio 2.3.1

The Android projects in this book were mostly created using Android Studio version 1.5. For this chapter I converted the Android basic wireless framework version 2.0 application project from Android Studio 1.5 to Android Studio 2.3.1. What I did was using the newer Android Studio version 2.3.1 I first created a new blank application that had no activity. I copied the source code, resource files, sound effect files, and manifest from the Android Studio version 1.5 project folders to the corresponding Android Studio 2.3.1 project folders. I also had to change the package name in the files to match the new package name for the project. Figure 9-1 shows some of the files that I needed to modify or copy over identified by a red circle.

For this new version of Android Studio there are two files, called colors.xml and styles.xml, that are either new or have been changed from Android Studio 1.5. Both files are located under the "values" directory in the project tree. See Figure 9-2. The files have been circled in red.

Finally, I changed the `MainActivity` class to derive from the `AppCompatActivity` class instead of the `Activity` class.

```
public class MainActivity extends
AppCompatActivity implements Runnable
```

The Automatic Battery Backup Power Supply System

The purpose of this automatic battery backup power supply system is to provide a reliable electrical power source from a 120 VAC receptacle through a 120 VAC to 9 VDC adapter to an Arduino as a primary power source. A reliable backup power source of a battery pack of 9 VDC is provided in case of a power failure or failures. If a power failure or multiple power failures or power interruptions should take place, this superfast solid-state automatic power backup system will connect the backup battery power source to the Arduino unit until the primary power has been restored to normal. When the AC power has been restored the Arduino will be returned to the normal primary power source instantaneously and without any lost operational time.

To understand how this automatic battery backup power supply system functions please refer to Figure 9-3. On this diagram it has two power sources. One power source is from the 120 VAC to 9 VDC adapter output. This 9 VDC output from the adapter is always greater than 9 VDC (usually about 12 VDC with open circuit and not connected to an Arduino). There are two superfast switching Schottky diodes. When the Arduino is connected as shown in Figure 9-3 and energized, the sequence of events is as follows:

1. The Schottky diode D1 will be conducting as the 9 VDC from the 120 VAC adapter is higher than 9 VDC from the backup battery.
2. As the Schottky diode D1 conducts, this provides the +9 VDC as the power source for the Arduino and also provides the

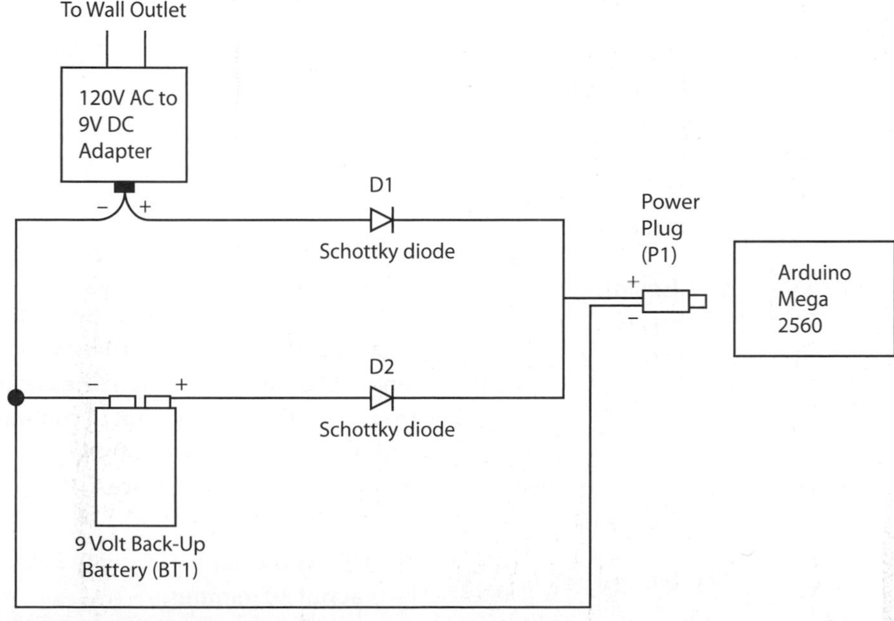

Figure 9-3 The battery backup power supply system schematic.

isolation between the backup battery power source and the +9 VDC adapter power source. A voltage of +9 VDC is placed at the cathode of Schottky diode D2 to ensure that D2 is in cut-off status.

3. If and when the primary power source fails, the superfast switching Schottky diode D1 ceases to conduct instantaneously and the Schottky diode D2 simultaneously conducts to provide a backup power source to the Arduino.
4. At the instant the primary power has been restored, the Schottky diode D1 starts conducting again and providing power to the Arduino unit at the same time cutting Schottky diode D2 off completely because the output voltage from the 9 VDC adapter is higher than the output voltage from the 9 VDC backup battery.

See Figure 9-3.

Parts List

Item #	Part #	Quantity	Product Description
1.	416A4136	1	120 VAC to 9 VDC adapter; tip = – (negative); sleeve = + (positive); MFR: Tamura, Temecula, CA
2.	N/A	1	9-V Energizer Max battery, backup battery power source; also see note #1
3.	N/A	1	9-V battery clip with bare end leads
4.	N/A	1	9-V battery holder
5.	NTE 585	2	Schottky barrier diode, fast switching
6.	PB 83E	1	Externally powered breadboard; Global Specialties
7.	N/A	1	SIM&NAT mslr FV 2.1 mm × 5.5 mm wire power pigtails adapter barrel plug
8.	ADA 373	1	Adafruit jack breadboard-friendly 2.1 mm DC barrel
9.		6	1.5-V “C” cell Energizer Max battery, 6 each
10.	BH261D	1	Velleman BH261D battery holder for six C-cell with solder tag grade to 12
11.	N/A	6	1.5-V “D” cell Energizer battery or Duracell, Quantum Alkaline battery, 6 each
12.	N/A	1	Battery holder for six “D” cell batteries
13.	N/A	1	Pack of jumper wires, as needed assorted wires from a small pack
14.	N/A	1	Roll of vinyl electrical tape

Note 1: **Please choose one or two of the following backup battery source or sources primarily for the length of time that you want the backup battery to support the Arduino unit to continue operating without any interruption after a power failure or power interruption has taken place.**

1a: **A 9-V Energizer battery with 565 mAh capacity that lasts about 30 minutes.**

1b: **A 9-V Energizer battery pack with six "C" cells of 1.5 V each that have a capacity of 5000 mAh and last greater than 16 hours of operation.**

1c: **A 9-V Energizer battery pack with six "D" cells of 1.5 V each that have a capacity of 10,000 mAh and will last about 28 hours.**

Note 2: **Item 1 and item 5, part #416A4136 and part #NTE 585, above may be purchased through the Internet from https://www.ebay.com. Items 2, 3, 4, 6, 7, 8, 9, 10, 11, 12, 13, and 14 may be purchased through the Internet from https://www.amazon.com.**

Building the System

1. Begin with item #6 from Parts List, have the Global Specialties breadboard in front of you with the black terminal post on your right-hand side.
2. Select item #4 and place the 9-V battery holder on the left end of the breadboard. Next, using a 3/16-inch long small screw mount the battery holder on the end of the breadboard.
3. Take one Schottky diode, item #5, with the silver band end, insert its pig tail into the breadboard hole located at "+ (positive), 20" (row +, column 20) just above the red line of row A, column 25. The opposite end of the pig tail is to be inserted into "i, 25" hole (row i, column 25).
4. Take another Schottky diode, item #5, with the silver band end, insert its pig tail into the breadboard hole located at "+, 32" (row +, column 32) just above the red line of row A, column 40. The opposite end of the pig tail is to be inserted into "i, 40" hole (row i, column 40).
5. Select item #8, the Adafruit jack. The rear center pin on the jack must be in line with the hole "C, 58" (row C, column 58). At the same time the middle center pin on the jack must be in line with hole "C, 60" (row C, column 60). Gently push the jack body on to the breadboard ensuring that the anchoring pin on the right side of the jack is secured in place. However, if needed, apply a drop of super glue to the corners of the jack that are in contact with the breadboard.
6. Select item #13, a jumper wire, insert one end of the wire into hole "A, 58" and the other end into the hole "– (negative), 45" (row – (negative), column 45).
7. Take another jumper wire and insert one end of the wire into hole "– (negative), 46" (row – "negative," column 46) and the other end to the black terminal post and secure it.
8. Take another jumper wire and insert one end of the wire into hole "+ (positive), 36" (row +, column 36) and the other end of the wire to the red terminal post and secure it.
9. Take another jumper wire and insert one end of the wire into hole "A, 60" (row A, column 60) and the other end into hole "J, 40" (row J, column 40).
10. Select item #7, adapter barrel plug with pig tail black and red leads, insert the black lead into the black terminal post along with the existing wire already attached, and then secure the black terminal. And insert the red lead into the red terminal post along with the existing wire already attached, and then secure the red terminal post.
11. Select items #2, #3, #4, take one 9-V battery clip, and line it up with the 9-V battery. Next, push the battery clip on to the battery and install the battery into the 9-V battery holder. Then take the black wire from the 9-V battery clip and insert the end of the wire to hole located "– (negative), 20" (row –, column 20), and take red wire from the 9-V battery clip and insert the end of the wire to

hole located "j, 25" (row j, column 25) on the breadboard. This rectangular 9-V battery has a capacity of 565 mAh and lasts for about 30 minutes of Arduino operation. See Figure 9-4.

A. To choose a 9-V backup battery with a larger capacity like a "C" cell battery pack with 5000 mAh that will last about 16 hours, you simply remove the two wires (the black and the red wire) from the present existing hole locations and install the two new wires from the 9-V "C" cell battery pack in place. Hole locations are "–, 20" (row –, column 20) for black wire and "j, 25" (row j, column 25) for the red wire.

B. A 9-V backup battery with a higher capacity like a "D" cell battery pack with 10,000 mAh will last about 28 hours. For the proper output wiring hook up from the "D" cell battery pack, take the black wire and insert it in hole location "–`, 20" (row – , column 20) and take the red wire and insert it in hole location " J, 25" (row J, column 25).

12. To make the C cell battery pack select items #9, #10, and #13 and insert each of the 6 "C" cell 1.5-V batteries into the "C" cell battery holder. Be sure to observe the proper polarity of the batteries accordingly with the battery holder. Take a jumper wire (preferably black color) and attach one end of the wire to the "–" (negative) output terminal of the battery holder. Secure it with a strip of vinyl electrical tape. The other end of the jumper wire remains free (no connection). Take another jumper wire (preferably red color) and attach one end of the wire to the "+" (positive) output terminal of the battery holder and secure it with a strip of vinyl electrical tape. The other end of the jumper wire remains free (no connection). The 9-V "C" battery pack has about 5000 mAh and may last for about 16 hours. See Figure 9-5.

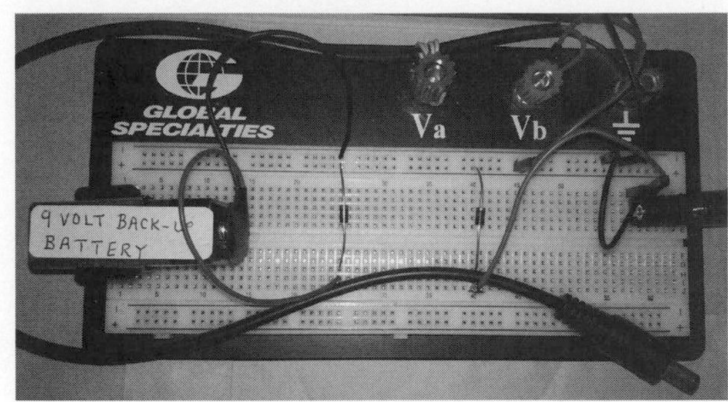

Figure 9-4 Building the automatic battery backup power supply system.

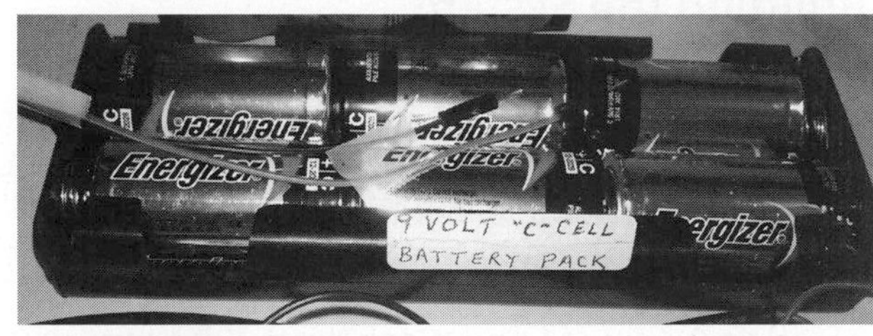

Figure 9-5 The C cell battery pack.

13. To make the D cell battery pack select items #11, #12, and #13 and insert each of the 6 "D" cell 1.5-V batteries into the "D" cell battery holder. Be sure to observe the proper polarity of the batteries accordingly with the battery holder. Take a jumper wire (preferably black color) and attach one end of the wire to the "–" (negative) output terminal of the battery holder and secure it with a strip of vinyl electrical tape. The other end of the jumper wire remains free (no connection). Take another jumper wire (preferably red color) and attach one end of the wire to the "+" (positive) output terminal of the battery holder and secure it with a strip of vinyl electrical tape. The other end of the jumper wire remains free (no connection). The 9-V "D" battery pack has about 10,000 mAh and may last about 28 hours. See Figure 9-6.

Figure 9-6 The D battery pack.

Figure 9-7 The 120-V AC to 9-V DC power adapter.

The AC adapter, which converts power from 120 V AC into 9 V DC power, is shown in Figure 9-7.

The entire backup battery power system with the C cell and the D cell battery packs along with the AC adapter is shown in Figure 9-8.

The SG90 9g Micro Servo

The SG90 9g micro servo is a powerful servo that can be used in applications such as robotics,

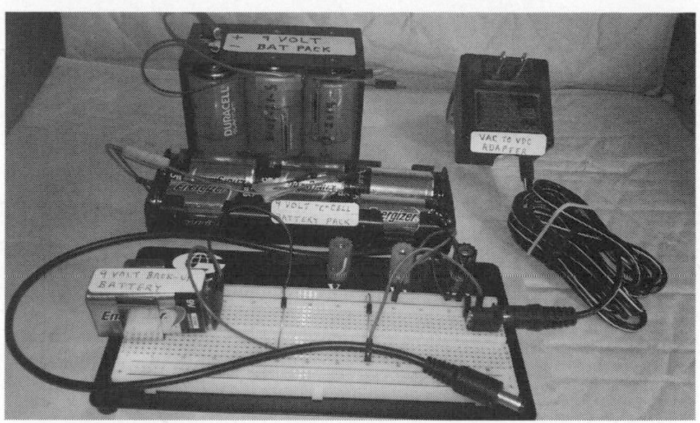

Figure 9-8 The entire backup battery power system with battery packs.

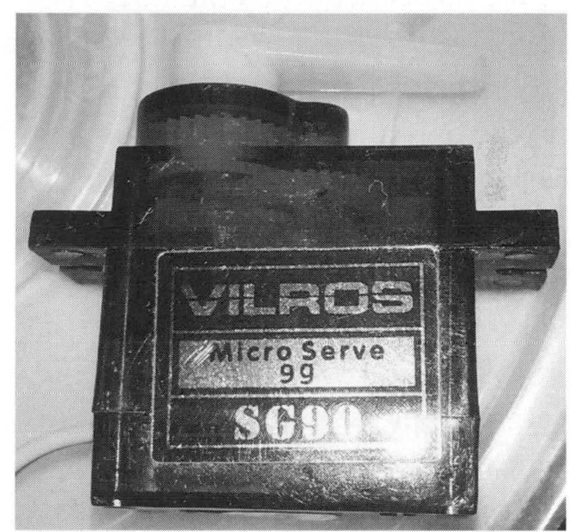

Figure 9-9 The SG90 9g micro servo.

model airplanes, and home security. The specifications of this servo are as follows:

- Operating voltage: 4.8 to 6 V
- Operating temperature: 0 to 55 degrees Centigrade
- Idle current: 6 mA (+ or −10 mA)
- Operating current (no load): 220 mA (+ or −50 mA)
- Operating current (maximum load): 650 mA (+ or −80 mA)
- Vcc connector color: Red
- Gnd connector color: Brown
- Servo input connector color: Orange/Yellow

See Figure 9-9 for a picture of the servo.

The YwRobot 545043 Power Supply Module for a Breadboard

The SG90 micro servo draws so much power when used in an actual situation containing a load that you need a separate power supply for the servo. The Arduino Mega 2560 can use at most 800 mA of current to power itself and the devices that are attached to it. This by itself is near the maximum current usage of the micro servo when under load. If you add in the power drawn by the ESP-01 you can easily see there is a need for an additional separate power source for the servo. One such power source is a power supply module that can be used with a breadboard and is made by YwRobot. This module is shown in Figure 9-10.

This power module has a DC barrel jack so that you can use a 9-V battery or a 9-V AC adapter to power the module. There is also a switch next to the DC barrel jack that is used to turn the power module on or off. This power module from YwRobot also has a selection jumper for the output of 5 or 3.3 V. To select the 5-V option set the jumpers to the configuration shown in Figure 9-11.

Figure 9-10 The YwRobot 545043 power supply module for a breadboard.

Figure 9-11 Selecting 5 V as an output from the power supply.

Hands-on Example: The Wireless Remote Control Servo System Using the Arduino Mega 2560 and ESP-01 Server

In this hands-on example I show you how to build a wireless remote control system that controls a micro servo using an Android, and Arduino Mega 2560 with an ESP-01 Wi-Fi module configured as the server.

Parts List

- 1 Android cell phone
- 1 Arduino Mega 2560 microcontroller
- 1 5- to 3.3-V step-down voltage regulator
- 1 Logic level converter from 5 to 3.3 V
- 1 ESP-01 Wi-Fi module
- 1 Arduino development station such as a desktop or a notebook
- 1 Breadboard
- 1 SG90 Servo
- 1 Power supply module for a breadboard (such as the YwRobot 545043 Power Supply Module)

- 1 Package of wires to connect the components

Setting Up the Hardware

Step-Down Voltage Regulator Connections

1. Connect the Vin pin on the voltage regulator to the 5-V power pin on the Arduino Mega 2560.
2. Connect the Vout pin on the voltage regulator to a node on your breadboard represented by a horizontal row of empty slots. This is the 3.3-V power node that will supply the 3.3-V power to the ESP-01 module as well as provide the reference voltage for the logic level shifter.
3. Connect the GND pin on the voltage regulator to the GND pin on the Arduino or form a GND node similar to the 3.3-V power node in the previous step.

ESP-01 Wi-Fi Module Connections

1. Connect the Tx or TxD pin to the Rx3 pin on the Arduino Mega 2560. This is the receive pin for the serial hardware port 3.
2. Connect the EN or CH_PD pin to the 3.3-V power node from the step-down voltage regulator.
3. Connect the IO16 or RST pin to the 3.3-V power node from the step-down voltage regulator.
4. Connect the 3V3 or VCC pin to the 3.3-V power node from the step-down voltage regulator.
5. Connect the GND pin to the ground pin or node.
6. Connect the Rx or RxD pin to pin B0 on the 3.3-V side of the logic level shifter. The voltage from this pin will be shifted from 5 V from the Arduino to 3.3 V for the ESP-01 module.

Logic Level Shifter

1. Connect the 3.3-V pin to the 3.3-V power node from the step-down voltage regulator.
2. Connect the GND pin on the 3.3-V side to the ground node for the Arduino.
3. Connect the B0 pin on the 3.3-V side to Rx or RxD pin on the ESP-01. This corresponds to the A0 pin on the 5-V side.
4. Connect the 5-V pin to the 5-V power pin on the Arduino.
5. Connect the GND pin on the 5-V side to the ground node for the Arduino.
6. Connect the A0 pin on the 5-V side to Tx3 pin on the Arduino. This pin corresponds to the B0 pin on the 3.3-V side.

SG90 Micro Servo

1. Connect the GND pin to the GND node for the Arduino.
2. Connect the Vcc pin to the 5-V output on the power supply module on the breadboard.
3. Connect the input pin to digital pin 7 on the Arduino.

YwRobot 545043 Power Supply Module for Breadboard

1. Put the power supply module on the breadboard so that the power and ground pins line up with the power and ground columns on the breadboard.
2. Make sure that the jumper selection for the output power pins you will be using are set to a 5-V output.
3. Connect the 5-V output of the power supply module to the Vcc pin on the servo.
4. Connect the GND on the power supply to the ground node for the Arduino.

See Figure 9-12.

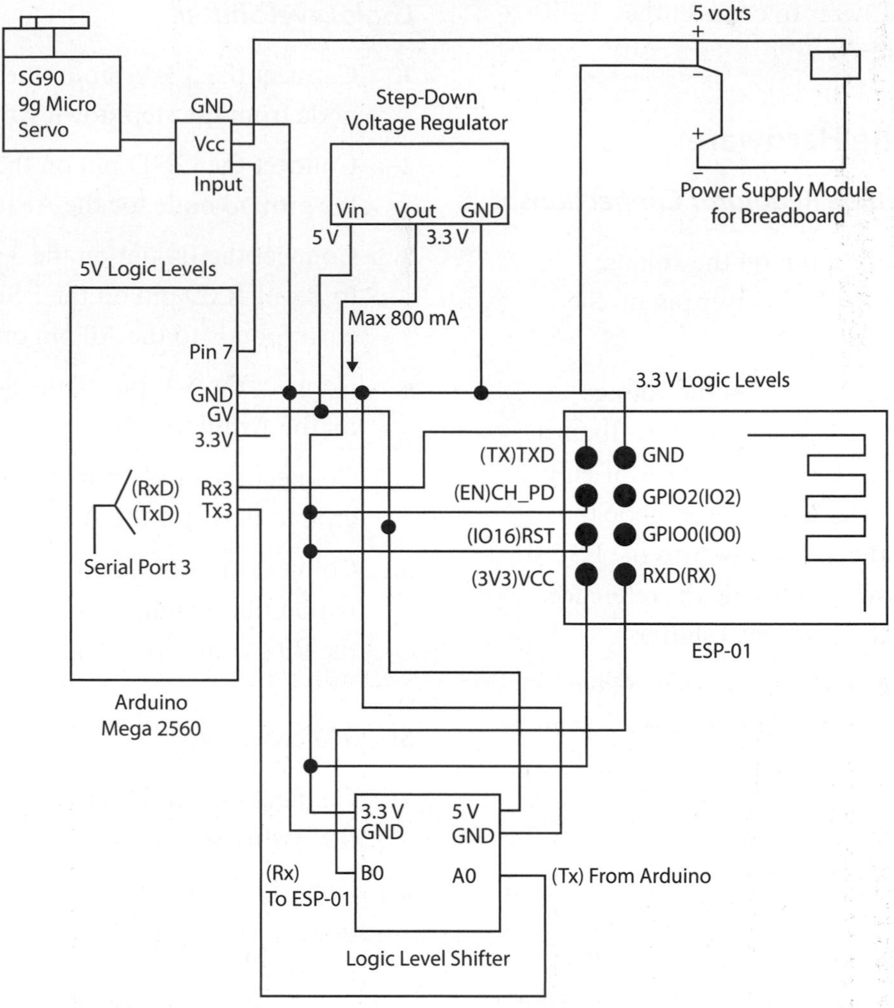

Figure 9-12 The wireless SG90 micro servo system.

Setting Up the Software

The Arduino code for this project is based on the basic wireless framework for the Arduino with the ESP-01 configured as the server. In this section I will cover only the changes to the basic wireless framework needed for this project.

In order to use the servo with the Arduino we need to first include the `Servo` library in the Arduino sketch.

```
#include <Servo.h>
```

The servo itself is declared as a variable called `Servo1` and the type is of the `Servo` class.

```
Servo Servo1;
```

The `Serovo1Pin` variable holds the Arduino digital pin that is connected to the servo's input pin and will be used to control the servo. The default setting is digital pin 7.

```
int Servo1Pin = 7;
```

The `CurrentAngle` variable holds the current angle of the servo.

```
int CurrentAngle = 0;
```

The `InitializeServo()` Function

The `InitializeServo()` function initializes the servo by telling the `Servo` class which pin the servo is connected to by executing the `attach()` command. After the servo is attached to the Arduino's digital pin the angle is then set to the

CurrentAngle value which is 0 by issuing a write() command. Program execution is then halted for 1 second or 1000 milliseconds. See Listing 9-1.

Modifying the setup() Function

The setup() function is modified by adding the call to the function InitializeServo() which initializes the servo.

Modifying the CheckAndProcessIPD() Function

The code for the angleset command is added to the CheckAndProcessIPD() function. This command sets the angle of the servo to the value after the equals sign by doing the following:

1. The angle value is parsed based on the position of the equals sign.
2. The servo's angle value is actually set by executing the Servo1.write (CurrentAngle) function with the value of the angle found in step 1.
3. A return text string is sent to the Android controller indicating the value the servo was set to. The return text was sent by calling the SendTCPMessage() function.

See Listing 9-2.

The code for the angledelta command is added to the CheckAndProcessIPD() function.

Listing 9-1 The InitializeServo() Function

```
void InitializeServo()
{
 Serial.println(F("Initializing Servo
..."));
 Servo1.attach(Servo1Pin);
 Servo1.write(CurrentAngle);
 delay(1000);
}
```

The angledelta command increases or decreases the servo angle by doing the following:

1. Parsing the command and determining the amount the current angle should be increased or decreased.
2. Adding the delta amount to the current angle.
3. Clamping the final angle to a range between 0 and 180 inclusive.
4. Setting the servo to the angle determined from step 3.
5. Sending a text string back to the Android controller indicating that the angle was changed to its current value.

See Listing 9-3.

Listing 9-2 The angleset Command

```
if (Data.indexOf("angleset=") >= 0)
{
 int index = Data.indexOf("=");
 String angle = Data.
substring(index+1);
 CurrentAngle = angle.toInt();
 Servo1.write(CurrentAngle);

 String returnstring = "";
 returnstring = returnstring +
F("Angle set to: ") + CurrentAngle +
"\n";
 SendTCPMessage(ClientIndex,
returnstring);
}
```

Listing 9-3 The angledelta Command

```
if (Data.indexOf("angledelta=") >= 0)
{
 int index = Data.indexOf("=");
 String delta = Data.
substring(index+1);
 CurrentAngle = CurrentAngle + delta.
toInt();
 if (CurrentAngle < 0)
 {
  CurrentAngle = 0;
 }
```

```
  else
  if (CurrentAngle > 180)
  {
   CurrentAngle = 180;
  }
  Servo1.write(CurrentAngle);

  String returnstring = "";
  returnstring = returnstring +
F("Angle set to: ") + CurrentAngle +
"\n";
  SendTCPMessage(ClientIndex,
returnstring);
}
```

Listing 9-4 The `angleread` command

```
if (Data == "angleread\n")
{
  String returnstring = "";
  returnstring = returnstring +
F("CurrentAngle is: ") + CurrentAngle
+ "\n";
  SendTCPMessage(ClientIndex,
returnstring);
}
```

The `angleread` command returns the current servo angle to the Android controller. See Listing 9-4.

Operating the Servo System

Download and install the Arduino servo project for this chapter on your Arduino Mega 2560. Connect a power source to the external power supply module located on the breadboard and turn on the power supply module by pressing down the on/off switch located next to the DC barrel jack. You should see a green light activate on the power supply. Next, start up the Arduino's Serial Monitor. You should see something like the following initialization sequence:

```
****** ESP-01 Arduino Mega 2560 Server
Program ********
***********    Servo Control Program
**************
Initializing Servo ...

AT Command Sent: AT+CIPMUX=1
ESP-01 Response: AT+CIPMUX=1

OK
ESP-01 mode set to allow multiple
connections.

AT Command Sent: AT+CWMODE_CUR=2
ESP-01 Response: AT+CWMODE_CUR=2

OK

ESP-01 mode changed to Access Point
Mode.

AT Command Sent: AT+CIPSERVER=1,80
ESP-01 Response: AT+CIPSERVER=1,80

OK

Server Started on ESP-01.

AT Command Sent: AT+CWSAP_CUR="ESP-
01","esp8266esp01",1,2,4,0
ESP-01 Response: AT+CWSAP_CUR="ESP-
01","esp8266esp01",1,2,4,0

OK

AP Configuration Set Successfullly.

AT Command Sent: AT+CWSAP_CUR?
ESP-01 Response: AT+CWSAP_CUR?

+CWSAP_CUR:"ESP-
01","esp8266esp01",1,2,4,0

OK

This is the curent configuration of the
AP on the ESP-01.

AT Command Sent: AT+CIPSTO?
ESP-01 Response: AT+CIPSTO?

+CIPSTO:180
```

```
OK

This is the curent TCP timeout in
seconds on the ESP-01.

************** System Has Started
********************
```

Connect your Android phone to the Wi-Fi access point located on the ESP-01 module. Start up the Android basic wireless framework application and connect to the TCP server running on the ESP-01 module. The Android should connect to the TCP server and change the name of the device. In my case the name of my Android device is set to "`lg`".

```
Notification: '0,CONNECT'
Processing New Client Connection ...
NewConnection Name: 0

Notification: '
'Notification: '
'Notification: '+IPD,0,8:name=lg
'ClientIndex: 0
DataLength: 8
Data: name=lg

Client Name Change Requested,
ClientIndex: 0, NewClientName: lg
Client Name Change Succeeded, OK...
Sending Data ....
TCP Connection ID: 0 , DataLength: 3 ,
Data: OK
```

Next, read the current angle of the servo by sending an "`angleread`" command from the Android application. The current angle should be returned to the Android application and should be 0 degrees.

```
Notification: '
'Notification: '
'Notification: '+IPD,0,10:angleread
'ClientIndex: 0
DataLength: 10
Data: angleread

Sending Data ....
TCP Connection ID: 0 , DataLength: 19 ,
Data: CurrentAngle is: 0
```

Put one of the plastic arms that should have come with the servo on the servo's rotating knob. Next, send an "`angleset=180`" command to the Arduino through the Android application to set the servo angle to 180 degrees. The Arduino will receive the command, set the servo angle to 180 degrees, and then send a text string back to the Android indicating that the angle was set to 180 degrees. Test the servo's movement again by setting the angle to 90 degrees. You should see the following on your Arduino's Serial Monitor:

```
Notification: '
'Notification: '
'Notification: '+IPD,0,13:angleset=180
'ClientIndex: 0
DataLength: 13
Data: angleset=180

Sending Data ....
TCP Connection ID: 0 , DataLength: 18 ,
Data: Angle set to: 180

Notification: '
'Notification: '
'Notification: '+IPD,0,12:angleset=90
'ClientIndex: 0
DataLength: 12
Data: angleset=90

Sending Data ....
TCP Connection ID: 0 , DataLength: 17 ,
Data: Angle set to: 90
```

Next, increase the servo's angle by 10 degrees by sending an "`angledelta=10`" command to the Arduino using the Android basic wireless framework. The servo's angle should now be set to 100 degrees from 90 degrees.

```
Notification: '
'Notification: '
'Notification: '+IPD,0,14:angledelta=10
'ClientIndex: 0
DataLength: 14
Data: angledelta=10

Sending Data ....
```

```
TCP Connection ID: 0 , DataLength: 18 ,
Data: Angle set to: 100
```

Keep increasing the servo's angle until it reaches 180 degrees and can go no farther.

```
Notification: '
'Notification: '
'Notification: '+IPD,0,14:angledelta=10
'ClientIndex: 0
DataLength: 14
Data: angledelta=10

Sending Data ....
TCP Connection ID: 0 , DataLength: 18 ,
Data: Angle set to: 180
```

Next, subtract 10 degrees from the current servo angle by sending the Arduino an "`angledelta=-10`" command from the Android application. The new servo angle should be 170 degrees. Keep sending this command until the angle reaches 0. Confirm that the servo moves each time this command is sent.

```
Notification: '
'Notification: '
'Notification: '+IPD,0,15:angledelta=-10
'ClientIndex: 0
DataLength: 15
Data: angledelta=-10

Sending Data ....
TCP Connection ID: 0 , DataLength: 18 ,
Data: Angle set to: 170
```

The DC Motor, the L9110 DC Motor Driver IC, and Fan Attachment

The DC motor that I use in this chapter operates within the range of 3 through 6 V, has two leads that connect to a power source, and a shaft that rotates at high speed. See Figure 9-13.

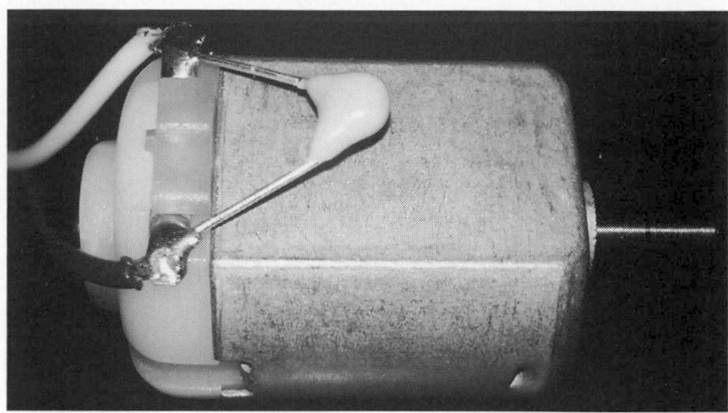

Figure 9-13 The DC motor.

In order to use the DC motor with the Arduino I use the L9110 DC motor driver chip. This chip has an operating voltage of 2.5 V up to 12 V. It can also provide a maximum of 800 mA of operating current to a motor. The operating temperature is between 0 and 80 degrees Centigrade. The chip itself has 4 pins on each side and has a small notch to indicate the top portion of the chip. See Figure 9-14.

Figure 9-14 The L9110 DC motor driver chip.

In order to test the DC motor with a load a small plastic fan was put on the tip of the DC motor's shaft. See Figure 9-15.

Figure 9-15 Fan attachment for the DC motor.

Hands-on Example: The Remote Control Wireless DC Motor Control System Using the Arduino Mega 2560 with an ESP-01 Server

In this hands-on example I show you how to build and operate a remote control wireless system that will control a DC motor using an Arduino Mega 2560 and an ESP-01 server.

Parts List

- 1 Android cell phone
- 1 Arduino Mega 2560 microcontroller
- 1 5- to 3.3-V step-down voltage regulator
- 1 Logic level converter from 5 to 3.3 V
- 1 ESP-01 Wi-Fi module
- 1 Arduino development station such as a desktop or a notebook
- 1 Breadboard
- 1 DC motor that can operate on 5 V and up to 800 mA of current
- 1 L9110 DC motor driver chip
- 1 Plastic fan that can be attached to the shaft of the DC motor
- 1 Power supply module for a breadboard (such as the YwRobot 545043 power supply module)
- 1 Package of wires to connect the components

Setting Up the Hardware

Step-Down Voltage Regulator Connections

1. Connect the Vin pin on the voltage regulator to the 5-V power pin on the Arduino Mega 2560.
2. Connect the Vout pin on the voltage regulator to a node on your breadboard represented by a horizontal row of empty slots. This is the 3.3-V power node that will supply the 3.3-V power to the ESP-01 module as well as provide the reference voltage for the logic level shifter.
3. Connect the GND pin on the voltage regulator to the GND pin on the Arduino or form a GND node similar to the 3.3-V power node in the previous step.

ESP-01 Wi-Fi Module Connections

1. Connect the Tx or TxD pin to the Rx3 pin on the Arduino Mega 2560. This is the receive pin for the serial hardware port 3.
2. Connect the EN or CH_PD pin to the 3.3-V power node from the step-down voltage regulator.
3. Connect the IO16 or RST pin to the 3.3-V power node from the step-down voltage regulator.
4. Connect the 3V3 or VCC pin to the 3.3-V power node from the step-down voltage regulator.
5. Connect the GND pin to the ground pin or node.
6. Connect the Rx or RxD pin to pin B0 on the 3.3-V side of the logic level shifter. The voltage from this pin will be shifted from 5 V from the Arduino to 3.3 V for the ESP-01 module.

Logic Level Shifter

1. Connect the 3.3-V pin to the 3.3-V power node from the step-down voltage regulator.
2. Connect the GND pin on the 3.3-V side to the ground node for the Arduino.
3. Connect the B0 pin on the 3.3-V side to Rx or RxD pin on the ESP-01. This corresponds to the A0 pin on the 5-V side.
4. Connect the 5-V pin to the 5-V power pin on the Arduino.

5. Connect the GND pin on the 5-V side to the ground node for the Arduino.
6. Connect the A0 pin on the 5-V side to Tx3 pin on the Arduino. This pin corresponds to the B0 pin on the 3.3-V side.

DC Motor

1. Connect the GND terminal to the OB pin on the L9110 DC motor driver chip.
2. Connect the Vcc terminal to the OA pin on the L9110 DC motor driver chip.
3. Attach the plastic fan to the shaft of the DC motor.

L9110 DC Motor Driver Chip

1. Connect the OA pin to the positive terminal of the DC motor.
2. Connect the OB pin to the negative terminal of the DC motor.
3. Connect the Vcc pins to the 5-V output from the power supply module located on the breadboard.
4. Connect the GND pins to the ground node for the Arduino.
5. Connect the forward signal or IB pin to digital pin 7 on the Arduino.
6. Connect the backward signal or IA pin to digital pin 8 on the Arduino.

YwRobot 545043 Power Supply Module for Breadboard

1. Put the power supply module on the breadboard so that the power and ground pins line up with the power and ground columns on the breadboard.
2. Make sure that the jumper selection for the output power pins you will be using are set to a 5-V output.
3. Connect the 5-V output of the power supply module to the Vcc pins on the L9110 chip.
4. Connect the GND on the power supply to the ground node for the Arduino.

See Figure 9-16.

Setting Up the Software

The `IBForwardSignal` variable holds the Arduino pin number that is connected to the IB forward signal pin on the L9110 DC motor driver chip. The default pin is set to digital pin 7.

```
int IBForwardSignal = 7;
```

The `IABackwardSignal` variable holds the Arduino pin that is connected to the IA or backward signal pin on the L9110 DC motor driver chip. The default value is digital pin 8 on the Arduino.

```
int IABackwardSignal = 8;
```

The `InitializeDCMotor()` Function

The `InitializeDCMotor()` function initializes the DC motor by setting the pins on the Arduino that control the spin direction of the motor to `OUTPUT` pins that will output voltages to the L9110 DC motor control chip. See Listing 9-5.

Modifying the `setup()` function

Elements related to the DC motor are initialized in the `setup()` function by calling the `InitializeDCMotor()` function.

Modifying the `CheckAndProcessIPD()` Function

The "`forward`" command spins the DC motor's shaft in the forward direction by sending a HIGH voltage signal to the IB or forward signal pin on the L9110 DC motor control chip. A text string is also sent to the Android controller indicating that the DC motor has been activated to spin in the forward direction. See Listing 9-6.

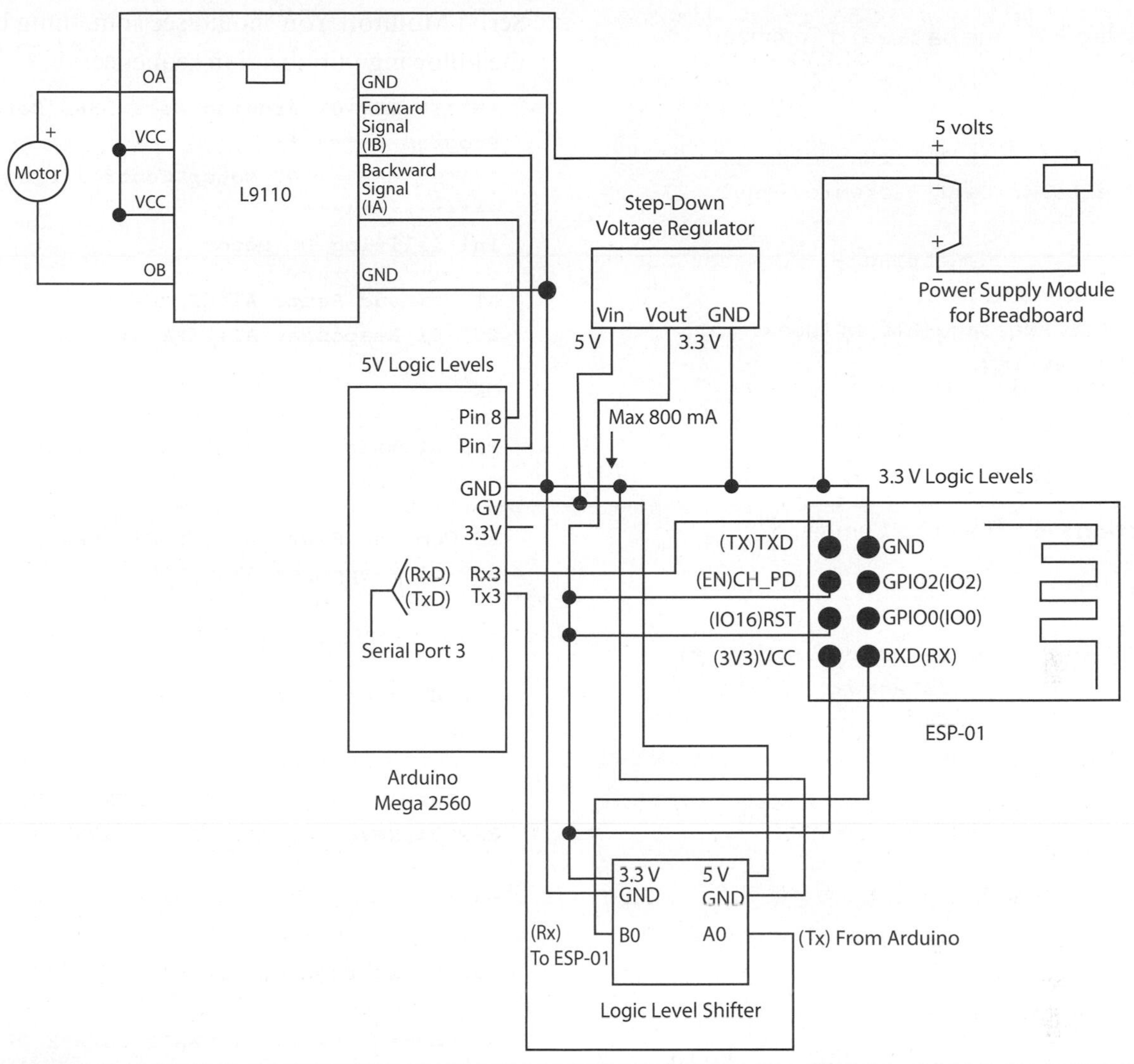

Figure 9-16 The wireless remote controlled DC motor system.

Listing 9-5 The `InitializeDCMotor()` Function

```
void InitializeDCMotor()
{
 Serial.println(F("Initializing DC
Motor ..."));

 pinMode(IBForwardSignal,OUTPUT);
 pinMode(IABackwardSignal,OUTPUT);
}
```

Listing 9-6 The `forward` Command

```
if (Data == "forward\n")
{
 digitalWrite(IBForwardSignal, 1);
 digitalWrite(IABackwardSignal, 0);

 String returnstring = "DC Motor Set
To Forward.\n";
 SendTCPMessage(ClientIndex,
returnstring);
}
```

The `backward` command activates the DC motor and spins the shaft in the backward direction by sending a HIGH voltage signal to the IA or backward signal pin on the L9110 DC motor chip. A text string is also sent to the Android controller indicating that the DC motor has been activated and is spinning in the backward direction. See Listing 9-7.

Listing 9-7 The `backward` Command

```
if (Data == "backward\n")
{
 digitalWrite(IBForwardSignal, 0);
 digitalWrite(IABackwardSignal, 1);

 String returnstring = "DC Motor Set
To Backward.\n";
 SendTCPMessage(ClientIndex,
returnstring);
}
```

Listing 9-8 The `stop` Command

```
if (Data == "stop\n")
{
 digitalWrite(IBForwardSignal, 0);
 digitalWrite(IABackwardSignal, 0);

 String returnstring = "DC Motor Set
To Stop.\n";
 SendTCPMessage(ClientIndex,
returnstring);
}
```

The `stop` command deactivates the DC motor by sending a LOW voltage signal to both the IB and IA pins located on the L9110 DC motor control chip. A text string is also sent to the Android controller indicating that the DC motor has been stopped. See Listing 9-8.

Operating the DC Motor Control System

Download and install the Arduino DC motor project for this chapter on your Arduino Mega 2560. Connect a power source to the external power supply module located on the breadboard and turn on the power supply module by pressing down the on/off switch located next to the DC barrel jack. You should see a green light activate on the power supply. Next, start up the Arduino's Serial Monitor. You should see something like the following initialization sequence:

```
****** ESP-01 Arduino Mega 2560 Server
Program ********
***********   DC Motor Control Program
**************
Initializing DC Motor ...

AT Command Sent: AT+CIPMUX=1
ESP-01 Response: AT+CIPMUX=1

OK

ESP-01 mode set to allow multiple
connections.

AT Command Sent: AT+CWMODE_CUR=2
ESP-01 Response: AT+CWMODE_CUR=2

OK


ESP-01 mode changed to Access Point
Mode.


AT Command Sent: AT+CIPSERVER=1,80
ESP-01 Response: AT+CIPSERVER=1,80


OK


Server Started on ESP-01.


AT Command Sent: AT+CWSAP_CUR="ESP-
01","esp8266esp01",1,2,4,0
ESP-01 Response: AT+CWSAP_CUR="ESP-
01","esp8266esp01",1,2,4,0


OK


AP Configuration Set Successfullly.


AT Command Sent: AT+CWSAP_CUR?
ESP-01 Response: AT+CWSAP_CUR?


+CWSAP_CUR:"ESP-
01","esp8266esp01",1,2,4,0


OK


This is the curent configuration of the
AP on the ESP-01.


AT Command Sent: AT+CIPSTO?
```

```
ESP-01 Response: AT+CIPSTO?

+CIPSTO:180

OK

This is the curent TCP timeout in
seconds on the ESP-01.
```

Connect your Android phone to the Wi-Fi access point located on the ESP-01 module. Start up the Android basic wireless framework application and connect to the TCP server running on the ESP-01 module. The Android should connect to the TCP server and change the name of the device. In my case the name of my Android device is set to "verizon".

```
************ System Has Started
************
Notification: '0,CONNECT'
Processing New Client Connection ...
NewConnection Name: 0

Notification: '
'Notification: '
'Notification: '+IPD,0,13:name=verizon
'ClientIndex: 0
DataLength: 13
Data: name=verizon

Client Name Change Requested,
ClientIndex: 0, NewClientName: verizon
Client Name Change Succeeded, OK...
Sending Data ....
TCP Connection ID: 0 , DataLength: 3 ,
Data: OK
```

Issue a "forward" command to the Arduino using the Android application. The DC motor's fan should start rotating and a text string is sent to the Android controller indicating that the motor is set to forward rotation.

```
Notification: '
'Notification: '
'Notification: '+IPD,0,8:forward
'ClientIndex: 0
DataLength: 8
Data: forward
Sending Data ....
TCP Connection ID: 0 , DataLength: 25 ,
Data: DC Motor Set To Forward.
```

Next, issue a "stop" command to the Arduino. This should stop the rotation of the fan and send a text string to the Android controller indicating that the DC motor has been stopped.

```
Notification: '
'Notification: '
'Notification: '+IPD,0,5:stop
'ClientIndex: 0
DataLength: 5
Data: stop

Sending Data ....
TCP Connection ID: 0 , DataLength: 22 ,
Data: DC Motor Set To Stop.
```

Next, issue a "backward" command to the Arduino. This starts the rotation of the fan in the backward direction and also sends a text string to the Android controller indicating that the DC motor is rotating in the backward direction. You should also feel a breeze coming from the rotating fan now.

```
Notification: '
'Notification: '
'Notification: '+IPD,0,9:backward
'ClientIndex: 0
DataLength: 9
Data: backward

Sending Data ....
TCP Connection ID: 0 , DataLength: 26 ,
Data: DC Motor Set To Backward.
```

Next, issue a "forward" command to the Arduino. You should see the fan blades slow down and change direction.

```
Notification: '
'Notification: '
'Notification: '+IPD,0,8:forward
'ClientIndex: 0
DataLength: 8
Data: forward

Sending Data ....
```

```
TCP Connection ID: 0 , DataLength: 25 ,
Data: DC Motor Set To Forward.
```

Figure 9-17 shows the DC motor with an attached fan, the L9110 DC motor driver chip, the external breadboard power supply module, and an external battery power source while the DC motor is active and is turning the fan.

A front view of the rotating fan is shown in Figure 9-18.

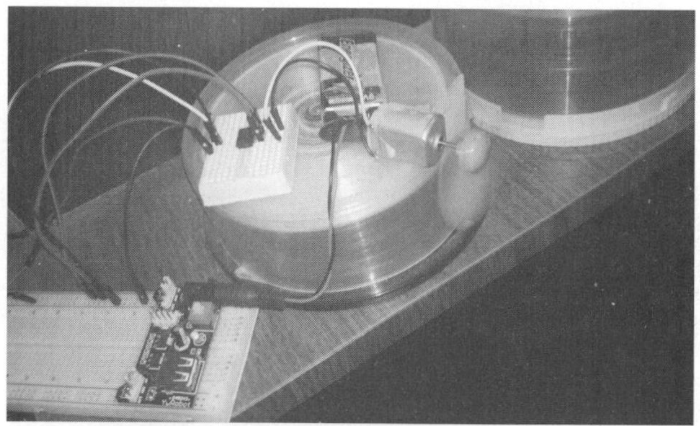

Figure 9-17 The DC motor with fan setup.

Figure 9-18 Front view of an active DC motor with attached fan.

The Stepper Motor and the ULN2003 Stepper Motor Driver Board

The stepper motor is designed to provide a full 360 degree continuous rotation ability (in the form of a shaft that moves in small discrete steps) combined with the strength to hold the shaft's position. The 28byj-48 stepper motor is a common inexpensive stepper motor that is found on websites such as Amazon. See Figure 9-19.

In order to use this stepper motor with the Arduino we need the ULN2003 stepper motor driver board. See Figure 9-20.

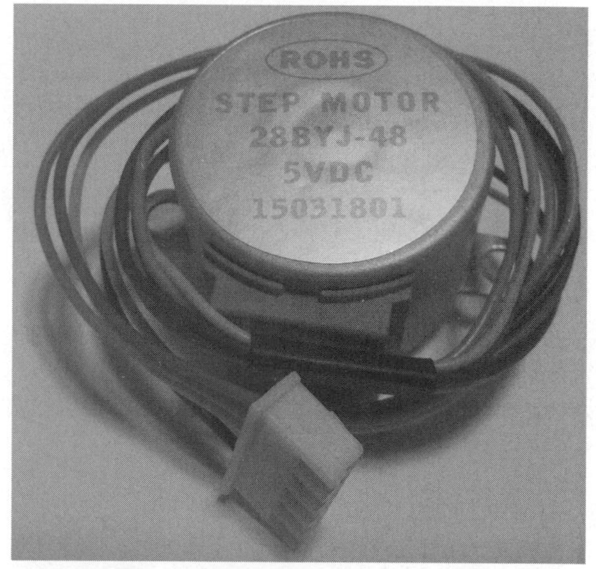

Figure 9-19 The 28byj-48 stepper motor.

Figure 9-20 The ULN2003 stepper motor driver board.

Hands-on Example: The Wireless Remote Controlled Stepper Motor System Using the Arduino Mega 2560 and the ESP-01 Server

In this hands-on example I show you how to build and operate a wireless remote control system for a stepper motor using an Arduino Mega 2560 and an ESP-01 module that functions as a server.

Parts List

- 1 Android cell phone
- 1 Arduino Mega 2560 microcontroller
- 1 5- to 3.3-V step-down voltage regulator
- 1 Logic level converter from 5 to 3.3 V
- 1 ESP-01 Wi-Fi module
- 1 Arduino development station such as a desktop or a notebook
- 1 Breadboard
- 1 28byj-48 stepper motor
- 1 ULN2003 stepper motor driver board
- 1 Power supply module for a breadboard (such as the YwRobot 545043 power supply module)
- 1 Package of wires to connect the components

Setting Up the Hardware

Step-Down Voltage Regulator Connections

1. Connect the Vin pin on the voltage regulator to the 5-V power pin on the Arduino Mega 2560.
2. Connect the Vout pin on the voltage regulator to a node on your breadboard represented by a horizontal row of empty slots. This is the 3.3-V power node that will supply the 3.3-V power to the ESP-01 module as well as provide the reference voltage for the logic level shifter.
3. Connect the GND pin on the voltage regulator to the GND pin on the Arduino or form a GND node similar to the 3.3-V power node in the previous step.

ESP-01 Wi-Fi Module Connections

1. Connect the Tx or TxD pin to the Rx3 pin on the Arduino Mega 2560. This is the receive pin for the serial hardware port 3.
2. Connect the EN or CH_PD pin to the 3.3-V power node from the step-down voltage regulator.
3. Connect the IO16 or RST pin to the 3.3-V power node from the step-down voltage regulator.
4. Connect the 3V3 or VCC pin to the 3.3-V power node from the step-down voltage regulator.
5. Connect the GND pin to the ground pin or node.
6. Connect the Rx or RxD pin to pin B0 on the 3.3-V side of the logic level shifter. The voltage from this pin will be shifted from 5 V from the Arduino to 3.3 V for the ESP-01 module.

Logic Level Shifter

1. Connect the 3.3-V pin to the 3.3-V power node from the step-down voltage regulator.
2. Connect the GND pin on the 3.3-V side to the ground node for the Arduino.
3. Connect the B0 pin on the 3.3-V side to Rx or RxD pin on the ESP-01. This corresponds to the A0 pin on the 5-V side.
4. Connect the 5-V pin to the 5-V power pin on the Arduino.
5. Connect the GND pin on the 5-V side to the ground node for the Arduino.

6. Connect the A0 pin on the 5-V side to Tx3 pin on the Arduino. This pin corresponds to the B0 pin on the 3.3-V side.

28byj-48 Stepper Motor

1. Connect the cable from the stepper motor to the ULN2003 stepper motor driver board.

ULN2003 Stepper Motor Driver Board

1. Connect the negative or GND pin to the Arduino's common ground node.
2. Connect the positive or Vcc pin to the external power supply's 5-V pin.
3. Connect the board to the cable from the stepper motor.
4. Connect pin IN1 to digital pin 7 on the Arduino.
5. Connect pin IN2 to digital pin 8 on the Arduino.
6. Connect pin IN3 to digital pin 9 on the Arduino.
7. Connect pin IN4 to digital pin 10 on the Arduino.

YwRobot 545043 Power Supply Module for Breadboard

1. Put the power supply module on the breadboard so that the power and ground pins line up with the power and ground columns on the breadboard.
2. Make sure that the jumper selection for the output power pins you will be using is set to a 5-V output.
3. Connect the 5-V output of the power supply module to the + or Vcc pin on the ULN2003 board.
4. Connect the GND on the power supply to the ground node for the Arduino.

See Figure 9-21.

Setting Up the Software

This section covers the new code specifically related to the stepper motor. The Arduino has a built-in stepper motor library that is included with the standard Arduino IDE. The library is called "`Stepper.h`" and must be included in the Arduino sketch.

```
#include <Stepper.h>
```

The `MotorPinX` variables define the Arduino pins that will be connected to the ULN2003 board. The pins by default range from digital pin 7 through digital pin 10.

```
int MotorPin1 = 7;
int MotorPin2 = 8;
int MotorPin3 = 9;
int MotorPin4 = 10;
```

The `StepsPerRevolution` variable holds the number of steps that the stepper motor's shaft must take in order for 1 revolution or 360 degrees to occur. For 28byj-48 stepper motor the number of steps it takes for 1 revolution of the shaft is 2038.

```
const int StepsPerRevolution = 2038;
```

The `AngleToSteps` variable holds the conversion value that is used to convert an angle in degrees to the corresponding number of steps the motor must take in order to move the shaft that angle amount in degrees.

```
const float AngleToSteps = 2038/360.0f;
```

The `Speed` variable holds the speed in rotations per minute that the stepper motor will move. The default value is set to 6.

```
int Speed = 6;
```

The `StepperMotor1` variable represents the stepper motor and is initialized by setting the steps the motor needs to take for 1 complete rotation, and the Arduino pins that are connected to the ULN2003 motor driver board.

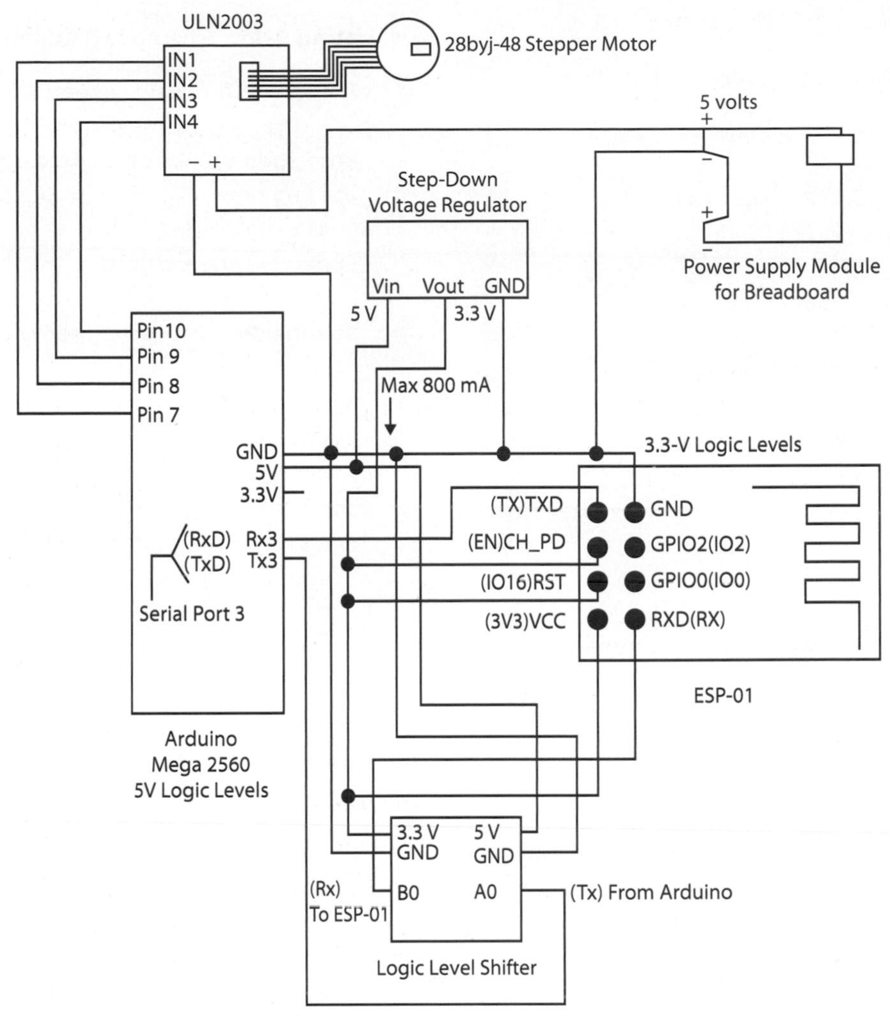

Figure 9-21 The wireless remote controlled stepper motor control system.

```
Stepper StepperMotor1(StepsPerRevolution,
MotorPin1, MotorPin3, MotorPin2, MotorPin4);
```

The `InitializeStepperMotor()` Function

The `InitializeStepperMotor()` function initializes the stepper motor by setting the stepper motor's default rotations per minute and setting the Arduino pins that are connected to the ULN2003 motor driver board to be OUTPUT pins that can provide output voltage signals to the board. See Listing 9-9.

Modifying the `setup()` Function

The stepper motor is initialized in the `setup()` function by calling the `InitializeStepperMotor()` function.

Modifying the `CheckAndProcessIPD (String line)` Function

The angledelta command directs the stepper motor to rotate the shaft by doing the following:

1. The rotation angle is found by parsing the value after the equals sign and converting this string into an integer value.
2. The rotation angle is converted to the number of steps the motor must execute in order to achieve this angle.
3. The number of steps is then printed out to the Arduino Serial Monitor.
4. The stepper motor moves the shaft the required number of steps by calling the `step()` function.

Listing 9-9 The `InitializeStepperMotor()` Function

```
void InitializeStepperMotor()
{
 Serial.println(F("Initializing
Stepper Motor ..."));

 StepperMotor1.setSpeed(Speed);

 pinMode(MotorPin1, OUTPUT);
 pinMode(MotorPin2, OUTPUT);
 pinMode(MotorPin3, OUTPUT);
 pinMode(MotorPin4, OUTPUT);
}
```

Listing 9-10 The `angledelta` Command

```
if (Data.indexOf("angledelta=") >= 0)
{
 int index = Data.indexOf("=");
 String deltastring = Data.
substring(index+1);
 int AngleDelta = deltastring.
toInt();

 int Steps = AngleDelta *
AngleToSteps;

 Serial.print(F("Steps: "));
 Serial.println(Steps);
 StepperMotor1.step(Steps);

 String returnstring = "";
 returnstring = returnstring
+ F("Stepper Motor Moved: ") +
AngleDelta + " Degrees\n";
 SendTCPMessage(ClientIndex,
returnstring);
}
```

5. A text string is sent back to the Android controller indicating that the stepper motor has finished moving the shaft the rotation angle amount specified by the user.

See Listing 9-10.

Listing 9-11 The `speed` Command

```
if (Data.indexOf("speed=") >= 0)
{
 int index = Data.indexOf("=");
 String speedstring = Data.
substring(index+1);
 Speed = speedstring.toInt();

 StepperMotor1.setSpeed(Speed);

 String returnstring = "";
 returnstring = returnstring +
F("Stepper Motor Speed Set To: ") +
Speed + "\n";
 SendTCPMessage(ClientIndex,
returnstring);
}
```

The `speed` command changes the rotations per minute of the stepper motor's shaft by doing the following:

1. The command is parsed and the speed which is located after the equals sign is converted from a string to an integer.
2. The speed of the stepper motor is then set by calling the `setSpeed(Speed)` function with the new speed as the parameter.
3. A text string is sent back to the Android controller indicating that the speed has been set for the stepper motor.

See Listing 9-11.

Operating the Wireless Remotely Controlled Stepper Motor System

Download and install the Arduino stepper motor project for this chapter on your Arduino Mega 2560. Connect a power source to the external power supply module located on the breadboard and turn on the power supply module by pressing down the on/off switch located next to the DC barrel jack. You should see a green light activate on the power supply. Next, start up the Arduino's Serial Monitor.

You should see something like the following initialization sequence:

```
****** ESP-01 Arduino Mega 2560 Server
Program ********
***********   Stepper Motor Control
Program   **************

Initializing Stepper Motor ...

AT Command Sent: AT+CIPMUX=1
ESP-01 Response: AT+CIPMUX=1

OK

ESP-01 mode set to allow multiple
connections.

AT Command Sent: AT+CWMODE_CUR=2
ESP-01 Response: AT+CWMODE_CUR=2

OK

ESP-01 mode changed to Access Point
Mode.

AT Command Sent: AT+CIPSERVER=1,80
ESP-01 Response: AT+CIPSERVER=1,80

OK

Server Started on ESP-01.

AT Command Sent: AT+CWSAP_CUR="ESP-
01","esp8266esp01",1,2,4,0
ESP-01 Response: AT+CWSAP_CUR="ESP-
01","esp8266esp01",1,2,4,0

OK

AP Configuration Set Successfullly.

AT Command Sent: AT+CWSAP_CUR?
ESP-01 Response: AT+CWSAP_CUR?

+CWSAP_CUR:"ESP-
01","esp8266esp01",1,2,4,0

OK

This is the curent configuration of the
AP on the ESP-01.
AT Command Sent: AT+CIPSTO?
ESP-01 Response: AT+CIPSTO?
+CIPSTO:180

OK

This is the curent TCP timeout in
seconds on the ESP-01.

************ System Has Started
************
```

Connect your Android phone to the Wi-Fi access point located on the ESP-01 module. Start up the Android basic wireless framework application and connect to the TCP server running on the ESP-01 module. The Android should connect to the TCP server and change the name of the device. In my case the name of my Android device is set to "`verizon`".

```
Notification: '0,CONNECT'
Processing New Client Connection ...
NewConnection Name: 0

Notification: '
'Notification: '
'Notification: '+IPD,0,13:name=verizon
'ClientIndex: 0
DataLength: 13
Data: name=verizon

Client Name Change Requested,
ClientIndex: 0, NewClientName: verizon
Client Name Change Succeeded, OK...
Sending Data ....
TCP Connection ID: 0 , DataLength: 3 ,
Data: OK
```

Next, send an "`angledelta=90`" command to the Arduino from the Android application. This should turn the stepper motor's shaft by 90 degrees. After the stepper motor finishes the angle change a text string is sent back to the Android controller indicating that the rotation has been completed.

```
Notification: '
'Notification: '
'Notification: '+IPD,0,14:angledelta=90
'ClientIndex: 0
DataLength: 14
Data: angledelta=90
```

```
Steps: 509
Sending Data ....
TCP Connection ID: 0 , DataLength: 32 ,
Data: Stepper Motor Moved: 90 Degrees
```

Next, send an "`angledelta=-90`" command to the Arduino. This should move the stepper motor's shaft 90 degrees in the opposite direction. Note that the number of steps should be –509 where earlier the number of steps was 509. A text string acknowledging the rotation is sent back to the Android controller once the rotation has been completed.

```
Notification: '
'Notification: '
'Notification: '+IPD,0,15:angledelta=-90
'ClientIndex: 0
DataLength: 15
Data: angledelta=-90

Steps: -509
Sending Data ....
TCP Connection ID: 0 , DataLength: 33 ,
Data: Stepper Motor Moved: -90 Degrees
```

Next, issue the command "`angledelta=360`" to move the shaft in 1 full rotation. Note the speed of the rotation.

```
Notification: '
'Notification: '
'Notification: '+IPD,0,15:angledelta=360
'ClientIndex: 0
DataLength: 15
Data: angledelta=360

Steps: 2037
Sending Data ....
TCP Connection ID: 0 , DataLength: 33 ,
Data: Stepper Motor Moved: 360 Degrees
```

Next, send a "`speed=12`" command to the Arduino to increase the speed of the rotations to 12 rotations per minute from the default 6 rotations per minute.

```
Notification: '
'Notification: '
'Notification: '+IPD,0,9:speed=12
'ClientIndex: 0
DataLength: 9
Data: speed=12

Sending Data ....
TCP Connection ID: 0 , DataLength: 31 ,
Data: Stepper Motor Speed Set To: 12
```

Next, send an "`angledelta=360`" command to the Arduino to make the shaft rotate 1 full rotation. The shaft should be rotating about twice as fast as before.

```
Notification: '
'Notification: '
'Notification: '+IPD,0,15:angledelta=360
'ClientIndex: 0
DataLength: 15
Data: angledelta=360

Steps: 2037
Sending Data ....
TCP Connection ID: 0 , DataLength: 33 ,
Data: Stepper Motor Moved: 360 Degrees
```

The ArduCAM Library and GitHub

For this book I used version 3.4.7 of the ArduCAM library that was released on 8/8/2015 for most of the hands-on examples involving the ArduCAM Mini OV2640 and the ArduCAM Mini OV2640 Plus cameras. For the following hands-on example that involves the ESP8266 processor on the NodeMCU I used the most recent version of the ArduCAM library that was available which was version 4.1.2 released on 10/15/2018. The reason was that this updated library contained the needed code to support the ESP8266 processor.

ArduCAM has a software repository on GitHub where you can download the current and previous versions of the ArduCAM library. If you are having trouble getting the camera examples to work with the most current version of the ArduCAM library then try installing the exact versions that I used for each of the hands-on examples.

- Main Repository: https://github.com/ArduCAM
- Arduino Repository: https://github.com/ArduCAM/Arduino

Hands-on Example: The Wireless NodeMCU and ArduCAM Mini OV2640 Plus Surveillance System

In this hands-on example I show you how to build and operate a wireless remote-controlled video surveillance system using a NodeMCU and ArduCAM Mini OV2640 Plus digital camera.

Parts List

- 1 Android cell phone
- 1 NodeMCU microcontroller
- 1– ArduCAM Mini OV2640 Plus digital camera
- 1 Breadboard
- 1 Package of wires to connect the components

***Note:* Installation of a version of the ArduCAM library which supports the ESP8266 processor such as ArduCAM library version 4.1.2 that was released on 10/15/2018 is required.**

Setting Up the Hardware

ArduCAM Mini OV2640 Plus

1. Connect the SDA pin on the camera to pin D2 on the NodeMCU.
2. Connect the SCL pin on the camera to pin D1 on the NodeMCU.
3. Connect the MOSI pin on the camera to pin D7 on the NodeMCU.
4. Connect the MISO pin on the camera to pin D6 on the NodeMCU.

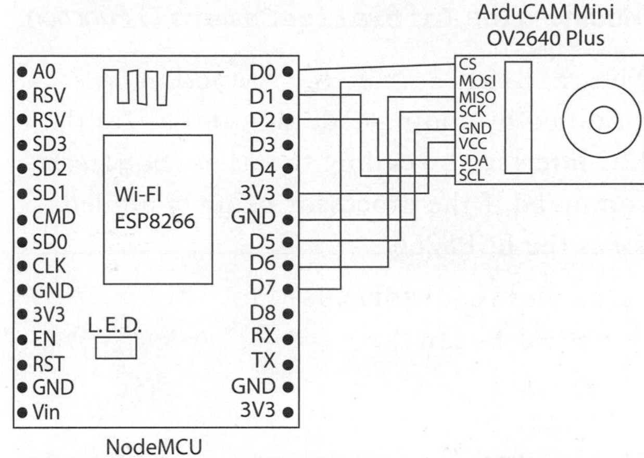

Figure 9-22 The wireless remote-controlled NodeMCU and ArduCAM Mini OV2640 Plus surveillance system.

5. Connect the SCLK pin on the camera to pin D5 on the NodeMCU.
6. Connect the CS pin on the camera to pin D0 on the NodeMCU.
7. Connect the VCC pin on the camera to a 3.3-V power pin on the NodeMCU.
8. Connect the GND pin on the camera to a GND pin on the NodeMCU.

See Figure 9-22.

Setting Up the Software

The Arduino program for this hands-on example is based on the NodeMCU basic framework version 2.0 of the server presented previously in Chapter 8. In this section I will cover only the new code that is specific to this hands-on example using the ArduCAM Mini with the NodeMCU.

The `SPI_CS` variable which holds the pin on the NodeMCU that serves as the chip select pin and is used to select if the ArduCAM is processing data coming from the SPI bus. The default value is set to pin D0 on the NodeMCU.

```
const int SPI_CS = D0;
```

Modifying the `InitializeCamera()` Function

The `InitializeCamera()` function is modified by adding code that initializes the I2C interface by calling the `Wire.begin()` command if the processor being compiled for is the ESP8266.

```
#if defined(ESP8266)
  Wire.begin();
#endif
```

The `SendBinaryDataToClient()` Function

The `SendBinaryDataToClient()` function sends binary data to a client by doing the following:

1. Searching through the `Clients` array and attempting to find the name of the target client.
2. If the client is found then send the binary data to the client by calling the `write()` function with a pointer to the beginning of the array of data and the length of the data as parameters.
3. If the target client is found then return a 1 value otherwise return a 0 value.

See Listing 9-12.

Listing 9-12 The `SendBinaryDataToClient()` Function

```
int SendBinaryDataToClient(String
ClientID, byte* Data, int DataLength)
{
 boolean found = false;
 int result = 0;

 // send Data to client with ClientID
 for (int i = 0; (i < MaxClients) &&
!found; i++)
 {
 if (Clients[i].ID == ClientID)
 {
 Clients[i].ClientObject.write(Data,
DataLength);
 found = true;
 }
 }
 if (found)
 {
 result = 1;
 }

 Serial.print(F("Sending Binary Data
to Client ID: "));
 Serial.println(ClientID);

 return result;
}
```

Modifying the `ReadFifoBurst()` Function

The `ReadFifoBurst()` function was modified from a version used with the Arduino Mega 2560 and ESP-01 Wi-Fi module. The change is that the `SendTCPBinaryData()` function is replaced by the `SendBinaryDataToClient()` function to send binary data. See Listing 9-13. The changes are shown in bold print.

Modifying the `TransmitImageSize()` Function

The `TransmitImageSize()` function is modified by replacing the `SendTCPMessage()` function by the `SendDataToClient()` function in order to send data to clients. The modified code is in bold print. See Listing 9-14.

Modifying the `setup()` Function

The `setup()` function is modified by calling the `InitializeCamera()` function in order to initialize the ArduCAM Mini digital camera.

Modifying the ProcessIncomingClientData() Function

The camera-related commands are added into the `ProcessIncomingClientData()` function. The commands are functionally the same as those in the camera surveillance program for the Arduino Mega and ESP-01 Wi-Fi module. See Listing 9-15.

Listing 9-13 The `ReadFifoBurst()` Function

```
uint8_t ReadFifoBurst(String ClientIndex)
{
 uint32_t length = 0;
 uint8_t temp,temp_last;

 long bytecount = 0;
 int total_time = 0;

 boolean is_header = false;


 length = myCAM.read_fifo_length();
 if(length >= 393216)//384 kb
 {
  Serial.println(F("Not found the end."));
  return 0;
 }

 Serial.print(F("READ_FIFO_LENGTH() = "));
 Serial.println(length);

 total_time = millis();

 myCAM.CS_LOW();
 myCAM.set_fifo_burst();

 temp = SPI.transfer(0x00); // Added in to original source code, read in dummy byte
 length--;

 // Reset TCP Buffer
 TCPBufferIndex = 0;
 TCPBufferLength = 0;
 while( length-- )
 {
  temp_last = temp;
  temp = SPI.transfer(0x00);

  // Total Bytes read in from ArduCAM fifo
  bytecount++;

  // New Code for Arducam Mini Plus
  if (is_header == true)
  {
   TCPBuffer[TCPBufferIndex] = temp;
   TCPBufferLength++;
  }
  else if ((temp == 0xD8) & (temp_last == 0xFF))
  {
```

```
    is_header = true;
    Serial.println(F("ACK IMG"));

    TCPBuffer[TCPBufferIndex] = temp_last;
    TCPBufferLength++;
    if (TCPBufferLength == MaxPacketSize)
    {
     // Buffer Full so Send Data
     //SendTCPBinaryData(ClientIndex, TCPBuffer, TCPBufferLength);
     SendBinaryDataToClient(ClientIndex, TCPBuffer, TCPBufferLength);

     // Reset Buffer
     TCPBufferIndex = 0;
     TCPBufferLength = 0;
    }
    else
    {
     // Buffer Not full so continue reading in Data from
     // ArduCAM FIFO Memory
     TCPBufferIndex++;
    }

    TCPBuffer[TCPBufferIndex] = temp;
    TCPBufferLength++;
    }

    // Process TCP Buffer
    if (TCPBufferLength == MaxPacketSize)
    {
     // Buffer Full so Send Data
     //SendTCPBinaryData(ClientIndex, TCPBuffer, TCPBufferLength);
     SendBinaryDataToClient(ClientIndex, TCPBuffer, TCPBufferLength);

     // Reset Buffer
     TCPBufferIndex = 0;
     TCPBufferLength = 0;
    }
    else
    {
     // Buffer Not full so continue reading in Data from
     // ArduCAM FIFO Memory
     TCPBufferIndex++;
    }

    // Commented out because it causes loss of data in transfer
    // so that data does not match amount of data read from read fifo length command
    //if( (temp == 0xD9) && (temp_last == 0xFF) )
    // break;
    delayMicroseconds(10);
   }
```

```
  // Check to see if data is still in TCP Buffer
  if (TCPBufferLength != 0)
  {
   // Send remaining data in buffer
   //SendTCPBinaryData(ClientIndex, TCPBuffer, TCPBufferLength);
   SendBinaryDataToClient(ClientIndex, TCPBuffer, TCPBufferLength);
  }

  myCAM.CS_HIGH();

  Serial.print(F("ByteCount = "));
  Serial.println(bytecount);

  total_time = millis() - total_time;
  Serial.print("Total time used:");
  Serial.print(total_time/1000, DEC);
  Serial.println(" seconds ...");
}
```

Listing 9-14 The `TransmitImageSize()` Function

```
void TransmitImageSize(String
ClientIndex)
{
 // Get Image Size
 uint32_t FIFOImageLength =  myCAM.
read_fifo_length();
 //long Size = FIFOImageLength - 1;
// Account for 1 dummy byte at
beginning of image
 long Size = FIFOImageLength; // 1
dummy byte removed from Android Plus
Version of Camera

 Serial.print(F("FIFO RAW IMAGE SIZE = "));
 Serial.println(FIFOImageLength);

 Serial.print(F("Transmitting Image
Length = "));
 Serial.println(Size);

 // Transmit Image Size to Android
 String returnstring = String(Size) +
"\n";
 //SendTCPMessage(ClientIndex,
returnstring);
 SendDataToClient(ClientIndex,
returnstring);
}
```

Listing 9-15 Adding in the Camera Commands

```
if (Data == "vga")
{
 if (Resolution != VGA)
 {
  Resolution = VGA;
  SetCameraResolution();
 }
 CaptureImage();
 TransmitImageSize(ClientID);
}
else
if (Data == "qvga")
{
 if (Resolution != QVGA)
 {
  Resolution = QVGA;
  SetCameraResolution();
 }
 CaptureImage();
 TransmitImageSize(ClientID);
}
else
if (Data == "qqvga")
{
 if (Resolution != QQVGA)
 {
  Resolution = QQVGA;
  SetCameraResolution();
 }
 CaptureImage();
 TransmitImageSize(ClientID);
}
```

```
else
if (Data == "GetImageData")
{
  ReadFifoBurst(ClientID);
}
```

Operating the Surveillance System

Download and install the surveillance program for the NodeMCU and ArduCAM Mini Plus for this chapter on your NodeMCU. Start up the Serial Monitor and you should see something like the following if you are using a Windows 10 system. If you are using an XP operating system you might not see any start-up text.

```
**************************************
****************** NodeMCU ESP-12E
Access Point Server Program ***********
************** Basic Framework v2.0
*************************
* For Use With Android Basic Wireless
Framework Version 2.0 *
**********  ArduCAM Mini Plus
Surveillance Program  *********
ArduCAM Starting .........
OV2640 detected ..........
TCP Server started, Server IP Address:
192.168.4.1
WIFI Mode: 2
************* System is Now Active
*****************
```

Connect your Android device to the Wi-Fi access point on the NodeMCU. Start up the basic wireless framework application and connect to the TCP server running on the NodeMCU. The name of the android device should be changed and a "Client ID Received" text string should be sent back to the Android controller.

```
**************************************
* NodeMCU ESP-12E Access Point Server
Program **********
************** Basic Framework v2.0
*************************
* For Use With Android Basic Wireless
Framework Version 2.0 *
**********  ArduCAM Mini Plus
Surveillance Program  *********

TCP Server started, Server IP Address:
192.168.4.1
WIFI Mode: 2

[Client connected]
Waiting to Recieve New CLient's ID
Incoming Client's ClientID: android
Number of Clients Killed: 0
Sending Text to Client, ClientID:
android, Text: Client ID Recieved ...
```

Next, take a photo using the camera at qvga resolution by sending a "qvga" command to the NodeMCU. You should see something like the following on the Serial Monitor. The photo should be transferred to and displayed on your Android device.

```
ClientID: android Incoming Data: qvga
............ Starting Image Capture
..............
Start Capture
Capture Done!
FIFO RAW IMAGE SIZE = 10248
Transmitting Image Length = 10248
Sending Data to Client ID: android ,
Data: 10248

ClientID: android Incoming Data:
GetImageData
READ_FIFO_LENGTH() = 10248
ACK IMG
Sending Binary Data to Client ID: android
Sending Binary Data to Client ID: android
Sending Binary Data to Client ID: android
Sending Binary Data to Client ID: android
Sending Binary Data to Client ID: android
Sending Binary Data to Client ID: android
ByteCount = 10247
Total time used:0 seconds ...
```

Next, take a photo using the qqvga resolution by sending a "qqvga" command to the NodeMCU. You should see something like the following on the Serial Monitor:

```
ClientID: android Incoming Data: qqvga
............ Starting Image Capture
..............
```

```
Start Capture
Capture Done!
FIFO RAW IMAGE SIZE = 4104
Transmitting Image Length = 4104
Sending Data to Client ID: android , Data: 4104

ClientID: android Incoming Data:
GetImageData
READ_FIFO_LENGTH() = 4104
ACK IMG
Sending Binary Data to Client ID: android
Sending Binary Data to Client ID: android
Sending Binary Data to Client ID: android
ByteCount = 4103
Total time used:0 seconds ...
```

Next, take a picture in vga resolution by sending a "vga" command to the NodeMCU. The picture may be very dark at first so issue several vga commands to get a normal lightened picture.

```
ClientID: android Incoming Data: vga
............. Starting Image Capture
...............
Start Capture
Capture Done!
FIFO RAW IMAGE SIZE = 9224
Transmitting Image Length = 9224
Sending Data to Client ID: android ,
Data: 9224

ClientID: android Incoming Data:
GetImageData
READ_FIFO_LENGTH() = 9224
ACK IMG
Sending Binary Data to Client ID: android
Sending Binary Data to Client ID: android
Sending Binary Data to Client ID: android
Sending Binary Data to Client ID: android
Sending Binary Data to Client ID: android
ByteCount = 9223
Total time used:0 seconds ...
```

An example of a screen capture from the Android controller displaying a photo inside the basic wireless framework version 2.0 application is shown in Figure 9-23.

If you select the store photo option from the Android basic wireless framework menu then incoming images will be saved to the Android's storage space under the Pictures/CAMPics directory. An example of such a picture is shown in Figure 9-24.

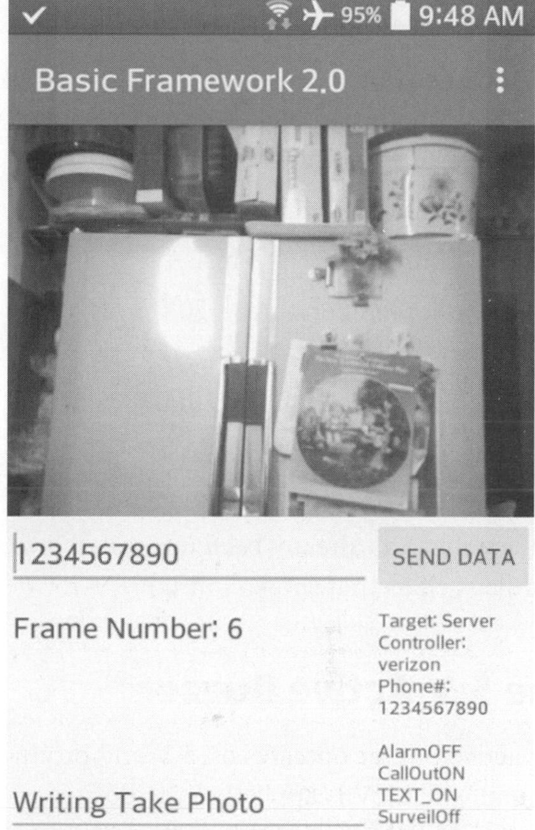

Figure 9-23 Android screen capture of a VGA photo from the NodeMCU and ArduCAM Mini Plus camera.

Figure 9-24 A saved VGA photo in 640 by 480 pixel resolution.

The MQ-2 Smoke Detector

The MQ-2 sensor is capable of detecting smoke as well as other gases such as methane and propane. The specifications for the sensor are:

- Operating voltage: 5 V
- Operating current: 160 mA
- Operating temperature: –20 through 50 degrees Centigrade
- Power consumption: Less than 800 mW
- Preheat time: Over 24 hours

The easiest way to use the MQ-2 sensor is to buy one that has already been integrated into a module such as that shown in Figure 9-25.

The 5-V Active Buzzer

The active buzzer operates on 5 V and produces a sound when 5 V is applied to the positive terminal of the buzzer. See Figure 9-26.

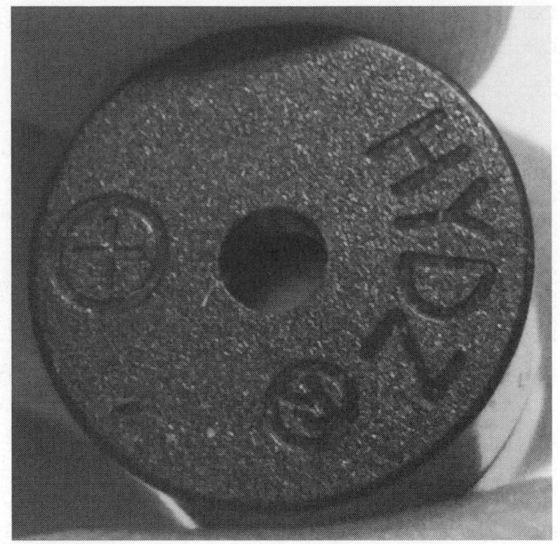

Figure 9-26 The 5-V active buzzer.

Hands-on Example: The Wireless Remote Controlled Smoke Detector Alarm System Using the Arduino Mega 2560 and the ESP-01 Server

In this hands-on example I show you how to build and operate a wireless remote controlled smoke detector alarm system using an Arduino Mega 2560 and an ESP-01 Wi-Fi module configured as a server.

Figure 9-25 The MQ-2 smoke detector.

Parts List

- 1 Android cell phone
- 1 Arduino Mega 2560 microcontroller
- 1 5- to 3.3-V step-down voltage regulator
- 1 Logic level converter from 5 to 3.3 V
- 1 ESP-01 Wi-Fi module
- 1 Arduino development station such as a desktop or a notebook
- 1 Breadboard
- 1 MQ-2 smoke detector
- 1 5-V active buzzer
- 1 Power supply module for a breadboard (such as the YwRobot 545043 power supply module)
- 1 Package of wires to connect the components

Setting Up the Hardware

Step-Down Voltage Regulator Connections

1. Connect the Vin pin on the voltage regulator to the 5-V power pin on the Arduino Mega 2560.

2. Connect the Vout pin on the voltage regulator to a node on your breadboard represented by a horizontal row of empty slots. This is the 3.3-V power node that will supply the 3.3-V power to the ESP-01 module as well as provide the reference voltage for the logic level shifter.
3. Connect the GND pin on the voltage regulator to the GND pin on the Arduino or form a GND node similar to the 3.3-V power node in the previous step.

ESP-01 Wi-Fi Module Connections

1. Connect the Tx or TxD pin to the Rx3 pin on the Arduino Mega 2560. This is the receive pin for the serial hardware port 3.
2. Connect the EN or CH_PD pin to the 3.3-V power node from the step-down voltage regulator.
3. Connect the IO16 or RST pin to the 3.3-V power node from the step-down voltage regulator.
4. Connect the 3V3 or VCC pin to the 3.3-V power node from the step-down voltage regulator.
5. Connect the GND pin to the ground pin or node.
6. Connect the Rx or RxD pin to pin B0 on the 3.3-V side of the logic level shifter. The voltage from this pin will be shifted from 5 V from the Arduino to 3.3 V for the ESP-01 module.

Logic Level Shifter

1. Connect the 3.3-V pin to the 3.3-V power node from the step-down voltage regulator.
2. Connect the GND pin on the 3.3-V side to the ground node for the Arduino.
3. Connect the B0 pin on the 3.3-V side to Rx or RxD pin on the ESP-01. This corresponds to the A0 pin on the 5-V side.
4. Connect the 5-V pin to the 5-V power pin on the Arduino.
5. Connect the GND pin on the 5-V side to the ground node for the Arduino.
6. Connect the A0 pin on the 5-V side to Tx3 pin on the Arduino. This pin corresponds to the B0 pin on the 3.3-V side.

MQ-2 Smoke Sensor

1. Connect the GND pin to the GND node for the Arduino.
2. Connect the VCC pin to the 5-V output on the power supply module on the breadboard.
3. Connect the OUT pin to digital pin 7 on the Arduino.

The 5V Active Buzzer

1. Connect the GND pin to the GND node for the Arduino.
2. Connect the positive terminal to digital pin 8 on the Arduino.

YwRobot 545043 Power Supply Module for Breadboard

1. Put the power supply module on the breadboard so that the power and ground pins line up with the power and ground columns on the breadboard.
2. Make sure that the jumper selection for the output power pins you will be using are set to a 5-V output.
3. Connect the 5-V output of the power supply module to the Vcc pin on the smoke detector.
4. Connect the GND on the power supply to the ground node for the Arduino.

See Figure 9-27.

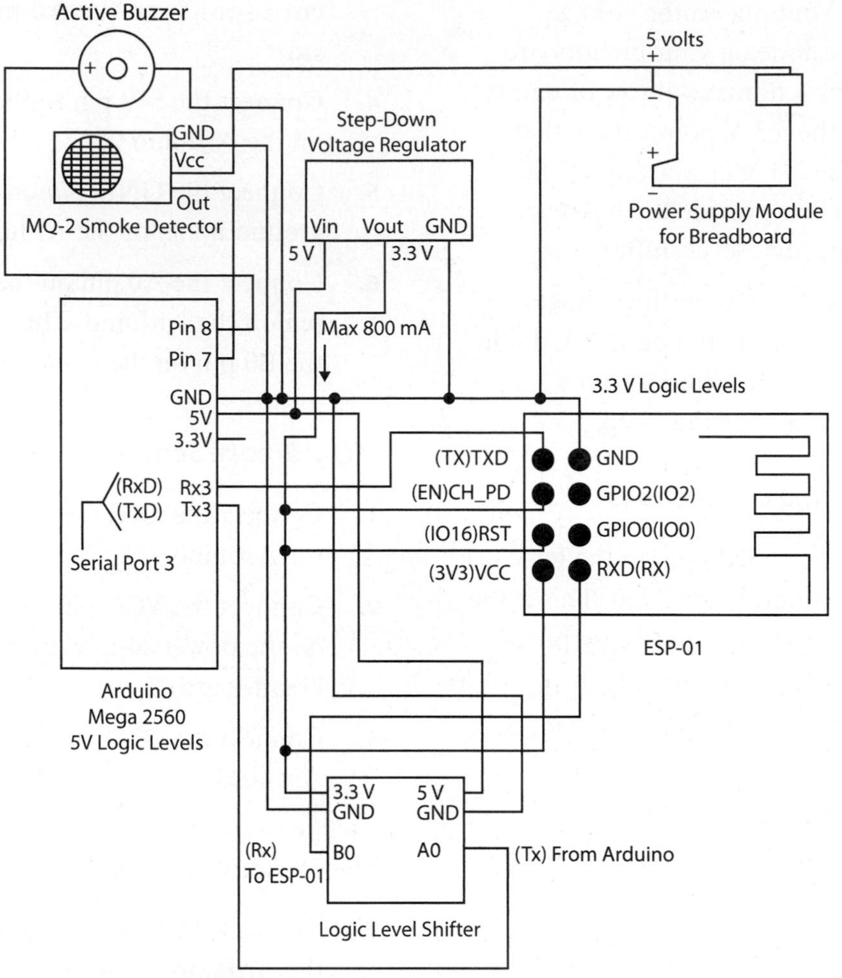

Figure 9-27 The wireless remote controlled smoke detector alarm system.

Setting Up the Software

The section discusses the new code that is specific to the smoke detector alarm system example.

The `BuzzerPin` holds the Arduino digital pin number that is connected to the positive terminal of the active buzzer and is set by default to pin 8.

```
int BuzzerPin = 8;
```

The `SensorPin` variable holds the Arduino digital pin that is connected to the smoke detector sensor's output pin and is set by default to pin 7.

```
int SensorPin = 7;
```

Modifying the `ProcessSensor()` Function

The `ProcessSensor()` function is modified from previous versions by setting the Arduino pin that is connected to the active buzzer to a HIGH voltage value when the smoke detector detects smoke. This activates the buzzer and it produces a loud sound to indicate that the alarm has been tripped. The code changes are highlighted in bold print. See Listing 9-16.

Modifying the `InitializeSensor()` Function

The `InitializeSensor()` function is modified by adding code to set the Arduino pin connected to the buzzer to be an output pin. The modified

Listing 9-16 The `ProcessSensor()` Function

```
void ProcessSensor()
{
 // Put Code to process your sensors or other hardware in this function.
 if ((AlarmState == Alarm_ON) || (AlarmState == Alarm_TRIPPED))
 {
  // Process Sensor Input
  SensorValue = digitalRead(SensorPin);
  if ((PrevSensorValue == EventNotDetected) && (SensorValue == EventDetected))
  {
   unsigned long ElapsedTime = millis() - MinDelayBetweenHitsStartTime;
   if (ElapsedTime >= MinDelayBetweenHits)
   {
    // Event Just Detected
    NumberHits++;
    Serial.print(F("SensorValue: "));
    Serial.print(SensorValue);
    Serial.print(F(" , Event Detected ... NumberHits: "));
    Serial.println(NumberHits);
    PrevSensorValue = EventDetected;

    // Time this hit occured
    MinDelayBetweenHitsStartTime = millis();

    // Turn on Active Buzzer
    digitalWrite(BuzzerPin, HIGH);
   }
  }
  else
  if ((PrevSensorValue == EventDetected) && (SensorValue == EventNotDetected))
  {
   Serial.print(F("SensorValue: "));
   Serial.println(SensorValue);
   PrevSensorValue = EventNotDetected;
  }
 }
}
```

code is highlighted in bold print. See Listing 9-17.

Listing 9-17 The `InitializeSensor()` Function

```
void InitializeSensor()
{
 // Initialize Sensor Pin
 pinMode(SensorPin, INPUT);

 // Initialize Buzzer Pin
 pinMode(BuzzerPin, OUTPUT);
}
```

Modifying the `CheckAndProcessIPD()` Function

The `testbuzzer` command is added in to the `CheckAndProcessIPD()` function. This command activates the buzzer for 3 seconds and then sends a text string back to the Android controller

Listing 9-18 Adding the `testbuzzer` Command

```
if (Data == "testbuzzer\n")
{
 // Turn on Active Buzzer
 digitalWrite(BuzzerPin, HIGH);

 // Play tone for 3 seconds
 delay(3000);

 // Turn off Active Buzzer
 digitalWrite(BuzzerPin, LOW);

 String returnstring = "Buzzer
Tested.\n";
 SendTCPMessage(ClientIndex,
returnstring);
}
```

indicating that the buzzer has been tested. You should hear a loud tone generated from the buzzer if the test was successful. See Listing 9-18.

Operating the Smoke Detector Alarm System

Download and install the smoke detector alarm example for this chapter on your Arduino. Activate the Serial Monitor and you should see something similar to the following:

```
****** ESP-01 Arduino Mega 2560 Server
Program ********
****************************************
***************
** Basic Framework 2.0 Smoke Detection
Alarm System ***

AT Command Sent: AT+CIPMUX=1
ESP-01 Response: AT+CIPMUX=1

OK

ESP-01 mode set to allow multiple
connections.

AT Command Sent: AT+CWMODE_CUR=2
ESP-01 Response: AT+CWMODE_CUR=2

OK

ESP-01 mode changed to Access Point
Mode.

AT Command Sent: AT+CIPSERVER=1,80
ESP-01 Response: AT+CIPSERVER=1,80

OK
Server Started on ESP-01.

AT Command Sent: AT+CWSAP_CUR="ESP-
01","esp8266esp01",1,2,4,0
ESP-01 Response: AT+CWSAP_CUR="ESP-
01","esp8266esp01",1,2,4,0

OK

AP Configuration Set Successfully.

AT Command Sent: AT+CWSAP_CUR?
ESP-01 Response: AT+CWSAP_CUR?

+CWSAP_CUR:"ESP-
01","esp8266esp01",1,2,4,0

OK

This is the curent configuration of the
AP on the ESP-01.

AT Command Sent: AT+CIPSTO?
ESP-01 Response: AT+CIPSTO?

+CIPSTO:180

OK

This is the curent TCP timeout in
seconds on the ESP-01.

************** System Has Started
*******************
```

Connect your Android cell phone to the Wi-Fi access point on the ESP-01 module. Start up the Android basic framework application version 2.0 and connect to the TCP server running on the ESP-01. Once connected, the name of Android device on the network is changed. In my case the name is changed to `verizon`.

```
Notification: '0,CONNECT'
Processing New Client Connection ...
NewConnection Name: 0

Notification: '
'Notification: '
'Notification: '+IPD,0,13:name=verizon
'ClientIndex: 0
DataLength: 13
Data: name=verizon
Client Name Change Requested,
ClientIndex: 0, NewClientName: verizon
Client Name Change Succeeded, OK...
Sending Data ....
TCP Connection ID: 0 , DataLength: 3 ,
Data: OK
```

Next, test the active buzzer by issuing a "`testbuzzer`" command from the Android application. You should hear the buzzer generate a sound and then receive a text string back to your Android indicating that the buzzer has been tested.

```
Notification: '
'Notification: '
'Notification: '+IPD,0,11:testbuzzer
'ClientIndex: 0
DataLength: 11
Data: testbuzzer

Sending Data ....
TCP Connection ID: 0 , DataLength: 15 ,
Data: Buzzer Tested.
```

Next, turn off the call out and the text message features on the Android application. Then, activate the alarm system on the Arduino from the Android application. After the 10-second wait period the smoke alarm is fully activated.

```
Notification: '
'Notification: '
'Notification: '+IPD,0,15:ACTIVATE_ALARM
'ClientIndex: 0
DataLength: 15
Data: ACTIVATE_ALARM

Sending Data ....
TCP Connection ID: 0 , DataLength: 40 ,
Data: Alarm Activated...WAIT TIME:10 Seconds.
Notification: '
'Sending Data ....
TCP Connection ID: 0 , DataLength: 20 ,
Data: ALARM NOW ACTIVE...

ALARM WAIT STATE FINISHED ... Alarm Now
ACTIVE ....
```

Next, the server is by default the target device so select "Monitor Target Device" from the basic wireless framework's main menu to set the server as one of the devices that will be monitored. Then activate the network monitoring to begin reading in the number of hits that the alarm system has generated. The number of hits should be 0 now if there is no smoke in the room. Note: The smoke detector might need a few minutes to warm up so you may get some false positives for a few minutes or other erratic behavior.

```
Notification: '
'Notification: '
'Notification: '+IPD,0,5:hits
'ClientIndex: 0
DataLength: 5
Data: hits

Sending Data ....
TCP Connection ID: 0 , DataLength: 16 ,
Data: HITS DETECTED:0
```

Next, use something that will generate smoke such as lighted incense stick to trigger the smoke detector. I used an incense stick and brought the smoke just under the smoke detector. This should trigger the sensor. When the Android sends the next `"hits"` command the Arduino will send an updated value of 1. The Android will then reset the number of hits back to zero on the Arduino by sending a `"reset"` command to the Arduino.

```
Notification: '
'SensorValue: 1 , Event Detected ...
NumberHits: 1
SensorValue: 0
Notification: '
'Notification: '+IPD,0,5:hits
'ClientIndex: 0
DataLength: 5
Data: hits
Sending Data ....
```

```
TCP Connection ID: 0 , DataLength: 16 ,
Data: HITS DETECTED:1

Notification: '
'Notification: '
'Notification: '+IPD,0,6:reset
'ClientIndex: 0
DataLength: 6
Data: reset

Reset NumberHits to 0.
```

Next, pull the smoke source away from the sensor and put it out. Deactivate the alarm using the Android application.

```
Notification: '
'Notification: '+IPD,0,17:DEACTIVATE_
ALARM
'ClientIndex: 0
DataLength: 17
Data: DEACTIVATE_ALARM

Sending Data ....
TCP Connection ID: 0 , DataLength: 32 ,
Data: Alarm Has Been DE-Activated ...
```

The MQ-2 Smoke Detector (Analog Version)

The MQ-2 smoke detector can produce analog output as well as digital output. The MQ-2 module shown in Figure 9-28 has both digital and analog outputs.

Figure 9-28 The MQ-2 smoke detector (with analog output).

Figure 9-29 The back of the MQ-2 smoke detector (with analog output).

For this model of MQ-2 sensor there is a power LED to indicate that the sensor is on and a DOUT LED to indicate if the digital output senses smoke. The analog pin on the module is labeled A0. See Figure 9-29.

Hands-on Example: Wireless Remote Controlled Smoke Detector (Analog) Alarm System Using Arduino and ESP-01

In this hands-on example I show you how to build and operate a wireless remote controlled smoke detector (analog output) alarm system using an Arduino Mega 2560 and an ESP-01 Wi-Fi module configured as a server. The key difference between this example and the last one is that the MQ-2 sensor has an analog output that gives you more control over the amount of smoke that triggers the alarm.

Parts List

- 1 Android cell phone
- 1 Arduino Mega 2560 microcontroller
- 1 5- to 3.3-V step-down voltage regulator
- 1 Logic level converter from 5 to 3.3 V
- 1 ESP-01 Wi-Fi module

- 1 Arduino development station such as a desktop or a notebook
- 1 Breadboard
- 1 MQ-2 smoke detector module (with analog output)
- 1 5-V active buzzer
- 1 Power supply module for a breadboard (such as the YwRobot 545043 power supply module)
- 1 Package of wires to connect the components

Setting Up the Hardware

Step-Down Voltage Regulator Connections

1. Connect the Vin pin on the voltage regulator to the 5-V power pin on the Arduino Mega 2560.
2. Connect the Vout pin on the voltage regulator to a node on your breadboard represented by a horizontal row of empty slots. This is the 3.3-V power node that will supply the 3.3-V power to the ESP-01 module as well as provide the reference voltage for the logic level shifter.
3. Connect the GND pin on the voltage regulator to the GND pin on the Arduino or form a GND node similar to the 3.3-V power node in the previous step.

ESP-01 Wi-Fi Module Connections

1. Connect the Tx or TxD pin to the Rx3 pin on the Arduino Mega 2560. This is the receive pin for the serial hardware port 3.
2. Connect the EN or CH_PD pin to the 3.3-V power node from the step-down voltage regulator.
3. Connect the IO16 or RST pin to the 3.3-V power node from the step-down voltage regulator.
4. Connect the 3V3 or VCC pin to the 3.3-V power node from the step-down voltage regulator.
5. Connect the GND pin to the ground pin or node.
6. Connect the Rx or RxD pin to pin B0 on the 3.3-V side of the logic level shifter. The voltage from this pin will be shifted from 5 V from the Arduino to 3.3 V for the ESP-01 module.

Logic Level Shifter

1. Connect the 3.3-V pin to the 3.3-V power node from the step-down voltage regulator.
2. Connect the GND pin on the 3.3-V side to the ground node for the Arduino.
3. Connect the B0 pin on the 3.3-V side to Rx or RxD pin on the ESP-01. This corresponds to the A0 pin on the 5-V side.
4. Connect the 5-V pin to the 5-V power pin on the Arduino.
5. Connect the GND pin on the 5-V side to the ground node for the Arduino.
6. Connect the A0 pin on the 5-V side to Tx3 pin on the Arduino. This pin corresponds to the B0 pin on the 3.3-V side.

MQ-2 Smoke Sensor

1. Connect the GND pin to the GND node for the Arduino.
2. Connect the VCC pin to the 5-V output on the power supply module on the breadboard.
3. Connect the AO pin to analog pin A12 on the Arduino.

The 5-V Active Buzzer

1. Connect the GND pin to the GND node for the Arduino.

2. Connect the positive terminal to digital pin 8 on the Arduino.

YwRobot 545043 Power Supply Module for Breadboard

1. Put the power supply module on the breadboard so that the power and ground pins line up with the power and ground columns on the breadboard.
2. Make sure that the jumper selection for the output power pins you will be using are set to a 5-V output.
3. Connect the 5-V output of the power supply module to the Vcc pin on the smoke detector.
4. Connect the GND on the power supply to the ground node for the Arduino.

See Figure 9-30.

Setting Up the Software

This section will cover only the changes in code from the previous hands-on example of the digital smoke sensor.

The `SensorPin` variable holds the Arduino pin that is connected to the smoke detector's analog output pin and is set by default to analog pin A12.

```
int SensorPin = A12;
```

The `SensorTripLevel` variable holds the level at which a reading from the analog

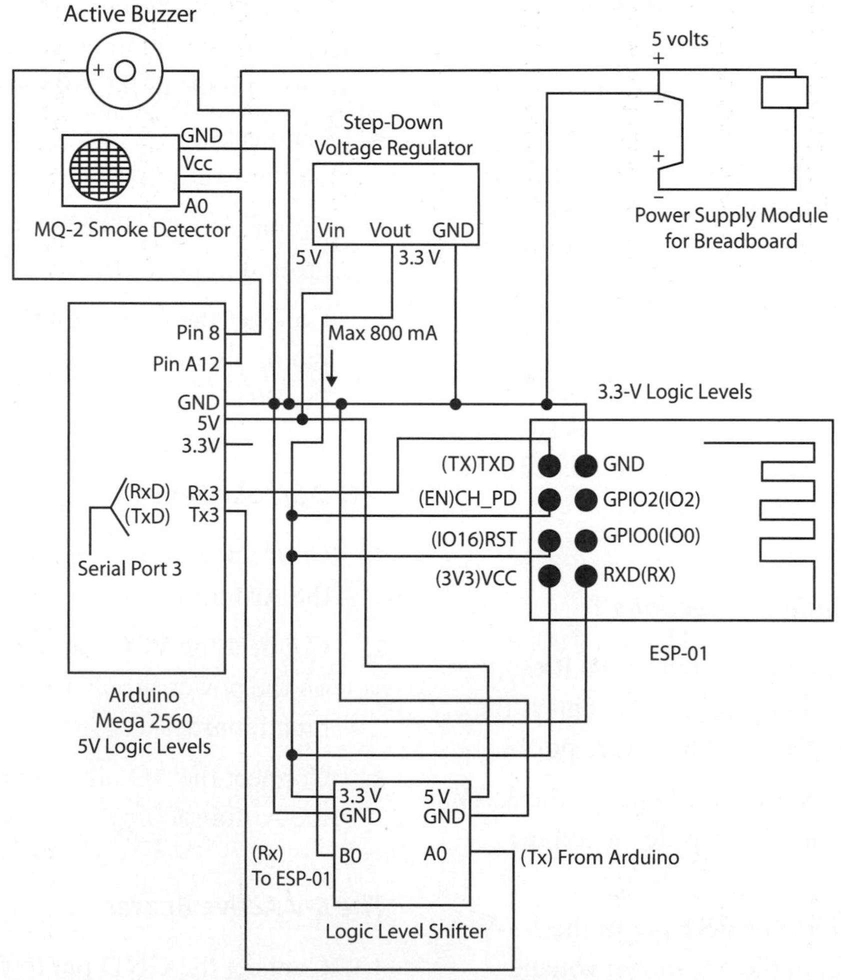

Figure 9-30 The wireless remote controlled analog smoke alarm system.

smoke sensor will trip the alarm. The default value is set to 300. A lower value increases the sensitivity of the smoke detector.

```
int SensorTripLevel = 300;
```

The `TestSensor` variable is true if the smoke sensor is in test mode which will continuously read in the value from the smoke sensor and output it to the Serial Monitor. The default value is false or off.

```
boolean TestSensor = false;
```

The `MinDelayBetweenTests` variable holds the minimum time between reads that the smoke sensor will be read in test mode. The default value is 1 second or 1000 milliseconds.

```
unsigned long MinDelayBetweenTests =
1000UL;
```

The `MinDelayBetweenTestsStartTime` variable holds the time that the last read from the smoke sensor occurred while in test mode.

```
unsigned long
MinDelayBetweenTestsStartTime = 0;
```

The `ProcessSensorTest()` Function

The `ProcessSensorTest()` function processes the sensor test function by doing the following:

1. If the sensor is in test mode then continue the execution of the function otherwise exit the function.
2. Calculate the elapsed time since the last test reading from the sensor.
3. If the elapsed time is greater than or equal to the minimum time between tests then read the smoke sensor, print out the result to the Serial Monitor and then set the `MinDelayBetweenTestsStartTime` variable to the current time.

See Listing 9-19.

Listing 9-19 The `ProcessSensorTest()` Function

```
void ProcessSensorTest()
{
 int TestValue = 0;

 if (!TestSensor)
 {
  return;
 }

 unsigned long ElapsedTime = millis()
- MinDelayBetweenTestsStartTime;
 if (ElapsedTime >=
MinDelayBetweenTests)
 {
  // Read Analog sensor data and
 print out to Serial Monitor
  TestValue = analogRead(SensorPin);
  Serial.print(F("Analog Sensor
 Value: "));
  Serial.println(TestValue);

  MinDelayBetweenTestsStartTime =
 millis();
 }
}
```

The `ProcessSensor()` Function

The `ProcessSensor()` function is very similar to the `ProcessSensor()` function discussed in previous examples throughout the book. However, there are a few changes that are made which are:

1. The `analogRead()` function is used to read the value from the sensor's analog output pin.
2. If the value read in from the sensor is greater than or equal to the `SensorTripLevel` value then the amount of smoke in the room is enough to trip the alarm.
3. If the alarm is tripped then the buzzer is activated by writing a HIGH voltage value to positive terminal of the buzzer.

The code modifications are highlighted in bold print. See Listing 9-20.

The `InitializeSensor()` Function

The `InitializeSensor()` function is modified from the previous version by printing out the sensor trip level to the Serial Monitor. Code modifications are highlighted in bold print. See Listing 9-21.

Listing 9-20 The `ProcessSensor()` Function

```
void ProcessSensor()
{
 // Put Code to process your sensors
or other hardware in this function.
 if ((AlarmState == Alarm_ON) ||
(AlarmState == Alarm_TRIPPED))
 {
  // Process Sensor Input
  SensorValue = analogRead
 (SensorPin);
  if (SensorValue >= SensorTripLevel)
  {
   unsigned long ElapsedTime =
  millis() - MinDelayBetween
  HitsStartTime;
   if (ElapsedTime >=
  MinDelayBetweenHits)
   {
    // Event Just Detected
    NumberHits++;
    Serial.print(F("SensorValue: "));
    Serial.print(SensorValue);
    Serial.print(F(" , Event Detected
   ... NumberHits: "));
    Serial.println(NumberHits);

    // Time this hit occured
    MinDelayBetweenHitsStartTime =
   millis();

    // Turn on Active Buzzer
    digitalWrite(BuzzerPin, HIGH);
   }
  }
 }
}
```

Modifying the `CheckAndProcessIPD()` Function

The `triplevel` command sets the level of smoke that will trigger the alarm and returns a text string to the Android controller indicating the new trip level. See Listing 9-22.

The `testsensoron` command activates the test mode for the sensor. See Listing 9-23.

Listing 9-21 The `InitializeSensor()` Function

```
void InitializeSensor()
{
 // Print out Sensor Initialization
Information
 Serial.println(F("Initializing
Sensors ..."));
 Serial.print(F("SensorTripLevel: "));
 Serial.println(SensorTripLevel);

 // Initialize Sensor Pin
 pinMode(SensorPin, INPUT);

 // Initialize Buzzer Pin
 pinMode(BuzzerPin, OUTPUT);
}
```

Listing 9-22 The `triplevel` Command

```
if (Data.indexOf("triplevel=") >= 0)
{
 int indextriplevel = Data.
indexOf("=");
 String triplevel = Data.
substring(indextriplevel+1);
 SensorTripLevel = triplevel.toInt();

 String returnstring = "";
 returnstring = returnstring +
F("Sensor Trip Level set to: ") +
SensorTripLevel + F("\n");
 SendTCPMessage(ClientIndex,
returnstring);
}
```

Listing 9-23 The `testsensoron` Command

```
if (Data == "testsensoron\n")
{
 TestSensor = true;
 String returnstring = "Activating
Sensor Test.\n";
 SendTCPMessage(ClientIndex,
returnstring);
}
```

Listing 9-24 The `testsensoroff` Command

```
if (Data == "testsensoroff\n")
{
 TestSensor = false;
 String returnstring = "Deactivating
Sensor Test.\n";
 SendTCPMessage(ClientIndex,
returnstring);
}
```

Listing 9-25 The `loop()` Function

```
void loop()
{
 // Process Wifi Notifications
 ProcessESP01Notifications();

 // Update Alarm Status
 UpdateAlarm();

 // Process Sensor Input
 ProcessSensor();

 // Process Sensor Test if Needed
 ProcessSensorTest();
}
```

The `testsensoroff` command deactivates the test mode for the sensor. See Listing 9-24.

The loop() Function

The `loop()` function is modified to support the testing mode of the analog smoke alarm by calling the `ProcessSensorTest()` function. The modified code is shown in bold print. See Listing 9-25.

Operating the Smoke Alarm

Download and install the analog smoke alarm project on your Arduino. Start up the Serial Monitor and you should see something like the following initialization sequence:

```
****** ESP-01 Arduino Mega 2560 Server
Program ********
****************************************
***************
******** Android Basic Wireless
Framework 2.0 *********
********* Analog Smoke Detection Alarm
System *********


Initializing Sensors ...
SensorTripLevel: 300

AT Command Sent: AT+CIPMUX=1
ESP-01 Response: AT+CIPMUX=1

OK

ESP-01 mode set to allow multiple
connections.

AT Command Sent: AT+CWMODE_CUR=2
ESP-01 Response: AT+CWMODE_CUR=2

OK

ESP-01 mode changed to Access Point
Mode.

AT Command Sent: AT+CIPSERVER=1,80
ESP-01 Response: AT+CIPSERVER=1,80

OK

Server Started on ESP-01.

AT Command Sent: AT+CWSAP_CUR="ESP-
01","esp8266esp01",1,2,4,0
ESP-01 Response: AT+CWSAP_CUR="ESP-
01","esp8266esp01",1,2,4,0


OK

AP Configuration Set Successfullly.
```

```
AT Command Sent: AT+CWSAP_CUR?
ESP-01 Response: AT+CWSAP_CUR?

+CWSAP_CUR:"ESP-
01","esp8266esp01",1,2,4,0

OK

This is the curent configuration of the
AP on the ESP-01.

AT Command Sent: AT+CIPSTO?
ESP-01 Response: AT+CIPSTO?

+CIPSTO:180

OK

This is the curent TCP timeout in
seconds on the ESP-01.

************** System Has Started
*******************
```

Connect your Android cell phone to the Wi-Fi access point on the ESP-01 module. Start up the Android basic framework application version 2.0 and connect to the TCP server running on the ESP-01. Once connected the name of Android device on the network is changed. In my case the name is changed to "`lg`".

```
Notification: '0,CONNECT'
Processing New Client Connection ...
NewConnection Name: 0

Notification: '
'Notification: '
'Notification: '+IPD,0,8:name=lg
'ClientIndex: 0
DataLength: 8
Data: name=lg

Client Name Change Requested,
ClientIndex: 0, NewClientName: lg
Client Name Change Succeeded, OK...
Sending Data ....
TCP Connection ID: 0 , DataLength: 3 ,
Data: OK
```

Next, test the active buzzer by issuing a "`testbuzzer`" command from the Android application. You should hear the buzzer produce a tone for a few seconds and then stop.

```
Notification: '
'Notification: '
'Notification: '+IPD,0,11:testbuzzer
'ClientIndex: 0
DataLength: 11
Data: testbuzzer

Sending Data ....
TCP Connection ID: 0 , DataLength: 15 ,
Data: Buzzer Tested.
```

Next, let's read and display the values coming from the smoke sensor by issuing a "`testsensoron`" command using the Android application. The specification for this sensor recommends a 24-hour burn-in period before the best results can be achieved but the sensor should be able to be useable after a few minutes of being turned on. If you are using the sensor for the first time then you should see the results of this test as rapidly decreasing until it becomes somewhat stable. For example, after a few minutes of being turned on for the first time my results eventually settled around 238 after decreasing from about 400.

```
Notification: '
'Notification: '
'Notification: '+IPD,0,13:testsensoron
'ClientIndex: 0
DataLength: 13
Data: testsensoron

Sending Data ....
TCP Connection ID: 0 , DataLength: 24 ,
Data: Activating Sensor Test.

Analog Sensor Value: 238
Notification: '
'Analog Sensor Value: 239
Analog Sensor Value: 238
Analog Sensor Value: 238
Analog Sensor Value: 238
```

```
Analog Sensor Value: 238
Analog Sensor Value: 238
Analog Sensor Value: 238
Analog Sensor Value: 238
Analog Sensor Value: 238
Analog Sensor Value: 237
Analog Sensor Value: 238
Analog Sensor Value: 238
Analog Sensor Value: 238
Analog Sensor Value: 238
```

Next, stop the sensor test by issuing a "`testsensoroff`" command from the Android application.

```
Notification: '
'Notification: '+IPD,0,14:testsensoroff
'ClientIndex: 0
DataLength: 14
Data: testsensoroff

Sending Data ....
TCP Connection ID: 0 , DataLength: 26 ,
Data: Deactivating Sensor Test.
```

Next, turn off the call out and the text message features on the Android application. Then, activate the alarm system on the Arduino from the Android application. After the 10-second wait period the smoke alarm is fully activated.

```
Notification: '
'Notification: '
'Notification: '+IPD,0,15:ACTIVATE_ALARM
'ClientIndex: 0
DataLength: 15
Data: ACTIVATE_ALARM

Sending Data ....
TCP Connection ID: 0 , DataLength: 40 ,
Data: Alarm Activated...WAIT TIME:10
Seconds.

Notification: '
'Sending Data ....
TCP Connection ID: 0 , DataLength: 20 ,
Data: ALARM NOW ACTIVE...

ALARM WAIT STATE FINISHED ... Alarm Now
ACTIVE ....
```

On the Android application the server is by default the target device so select "`Monitor Target Device`" from the basic wireless framework's main menu to set the server as one of the devices that will be monitored. Then activate the network monitoring to begin reading in the number of hits that the alarm system has generated. The number of hits should be 0 now if there is no smoke in the room. Note: The smoke detector might need a few minutes to warm up so you may get some false positives for a few minutes or other erratic behavior.

```
Notification: '
'Notification: '
'Notification: '+IPD,0,5:hits
'ClientIndex: 0
DataLength: 5
Data: hits

Sending Data ....
TCP Connection ID: 0 , DataLength:
16 , Data: HITS DETECTED:0
```

Next, create a smoke source such as a lighted incense stick and bring it near the smoke detector. The alarm should eventually trip and the next time the Android polls for the number of hits the value should be greater than 0. In my case the smoke produced a sensor value of 300 which tripped the alarm. A sound effect should then play on the Android device indicating that smoke has been detected. A "`reset`" command is then issued by the Android application to reset the number of hits back to 0.

```
Notification: '
'SensorValue: 300 , Event Detected ...
NumberHits: 1
Notification: '
'Notification: '+IPD,0,5:hits
'ClientIndex: 0
DataLength: 5
Data: hits
```

```
Sending Data ....
TCP Connection ID: 0 , DataLength:
16 , Data: HITS DETECTED:1

Notification: '
'Notification: '
'Notification: '+IPD,0,6:reset
'ClientIndex: 0
DataLength: 6
Data: reset

Reset NumberHits to 0.
```

Next, stop the network monitoring using the Android application. Then, deactivate the alarm.

```
Notification: '
'Notification: '+IPD,0,17:DEACTIVATE_
ALARM
'ClientIndex: 0
DataLength: 17
Data: DEACTIVATE_ALARM

Sending Data ....
TCP Connection ID: 0 , DataLength: 32 ,
Data: Alarm Has Been DE-Activated ...
```

Summary

The main focus of this chapter was on Arduino projects that required a separate power source for the hardware being controlled. These projects involved a servo, a DC motor with a fan attachment, a stepper motor, a digital smoke sensor, and an analog smoke sensor. An emergency backup battery power system was also presented. In addition, a video surveillance project involving an ArduCAM mini plus camera and a NodeMCU was given. For the downloadable content for this chapter I included the APK files for all the Android basic wireless framework projects. I also included the Android Studio project for the final version of the Android basic wireless framework using a more recent version of Android Studio. I also presented some troubleshooting tips for those having problems getting the examples to work correctly.

Index

Symbols and Numbers

A

G

H

I

J

K

L

M

N

X

Y